Practical Pharmaceutical Chemistry

Practical Pharmaceutical Chemistry

Fourth edition,
in two parts

Edited by
A. H. BECKETT
O.B.E., Ph.D., D.Sc., F.P.S.,
C.Chem., F.R.S.C.

Emeritus Professor of Pharmaceutical Chemistry
King's College London, University of London

and

J. B. STENLAKE
C.B.E., Ph.D., D.Sc., F.P.S.,
C.Chem., F.R.S.C., F.R.S.E.

Honorary Professor of Pharmacy
University of Strathclyde, Glasgow

Part Two

The Athlone Press
London

First published in Great Britain 1988
by The Athlone Press, 44 Bedford Row
London WC1R 4LY
Copyright © 1988 A. H. Beckett and J. B. Stenlake

British Library Cataloguing in Publication Data

Beckett, A. H.
  Practical pharmaceutical chemistry.—
  4th ed.
  Pt. 2
  1. Drugs—Analysis
  I. Title   II. Stenlake, J.B.
  615'.19015   RS189

  ISBN 0-485-11323-8

Library of Congress Cataloging in Publication Data

Beckett, A. H. (Arnold Heyworth)
  Includes bibliographies and indexes.
  1. Chemistry, Pharmaceutical. I. Stenlake, J.B.
(John Bedford) II. Title [DNLM: 1. Chemistry,
Pharmaceutical. QV 744 B396p]
RS403.B43 1988   615'.1901   87-14470
ISBN 0-485-11323-8 (pt. 2)

Typeset by KEYTEC, Bridport, Dorset
Printed in England at the University Press, Cambridge

Contents

Preface to the Fourth Edition

A spate of official publications, including the *British Pharmacopoeia 1980*, and its Addenda (1981, 1982, 1983 and 1986), the *British Pharmacopoeia (Veterinary)* 1985, the combined *United States Pharmacopoeia* XXI and *National Formulary* XVI and the second edition of the *European Pharmacopoeia*, call for yet another revision of *Practical Pharmaceutical Chemistry*. In this, the fourth edition, we have endeavoured to reflect the growing international convergence of policy and practice both in the change of subtitle of Part 1 to Pharmaceutical Analysis and Quality Control, and in the breadth of its content.

The objectives of this revision have been achieved in part by reducing the heavy dependence of earlier editions on the methods of the *British Pharmacopoeia*. Wherever possible, examples are based on drugs and dosage forms that are in widespread and common use in Britain, continental Europe and North America. Additionally, some reference to veterinary pharmaceuticals is made where they provide appropriate examples. As in previous editions, substances that are the subject of monographs in the British Pharmacopoeia are denoted by their British Approved Names (BAN), and are distinguished from United Adopted Names (Usan) where these are given by setting BANs in italics.

The discussion of drug registration has also been broadened to include reference to FDA and EEC procedures, the control of veterinary as well as human medicines, and the need in the United Kingdom for biologically based products to be manufactured in conformity with the requirements of the *Biological Compendium* (1977).

The detailed chapter-by-chapter revision in Part 1 encompasses the changeover in European analytical practice from NORMALITY to MOLARITY, which despite the reservations of some analysts, particularly in relation to oxidation-reduction titrations, is now virtually complete. The expansion of rapid complexometric titration methods, with the consequent almost complete demise of the older much slower gravimetric methods, is reflected in the regrouping of complexometric, argentometric and gravimetric methods into a single chapter. A brief treatment of variables in quantitative analysis appropriate to the application of chemical methods has been included for the first time, and the chapter on the analysis of dosage forms has been updated to better reflect both the range of products and the methods used in their control. In this respect the need for control analysts to embrace an appreciation of biological methods is reflected in the inclusion of short sections on sterility testing, on microbiological contamination and challenge tests for antimicrobial preservatives, on microbiological assays and on enzymes in pharmaceutical analysis.

There have been few, if any, major innovations in physical methods applicable to pharmaceutical analysis since the publication of the third edition in 1976. However, there have been a number of improvements and changes of emphasis. These are reflected in Part 2 of the present edition, which has been extensively revised in an endeavour to give a broader coverage of the most widely used techniques and a better balance of material that fairly reflects modern practice. In particular, the steady growth in importance of quantitative chromatographic techniques is recognised in the broader coverage and depth accorded to gas chromatography and high performance liquid chromatography, the inclusion of a short section on capillary column gas chromatography, and the transfer from Part 1 of important sections on ion exchange and size exclusion chromatography. The treatment of NMR spectroscopy has also been extended to include a brief introduction to ^{13}C NMR, and the coverage of radiopharmaceuticals increased to include radionuclide generators and quality control of radiopharmaceuticals – a subject of special interest to those engaged in hospital pharmacy. To balance these expansions, coverage of electrochemistry and polarography has been compressed into a single chapter – a degree of emphasis that is more in keeping with the rather modest role that these techniques continue to play in the practice of pharmaceutical analysis.

Two new chapters have been added to improve cohesion in Part 2. The first, by way of introduction, sets out the contribution and role of physical methods of analysis in the various phases of drug development, in quality control within the factory and in independent control laboratories, and in the clinic. The second, by way of conclusion, consists of a series of 'workshop style' exercises, with separate solutions, to illustrate and give practice in the application of spectroscopic techniques in structural elucidation and verification of identity.

Many analytical methods serve a common purpose in their suitability both for the quality control of pharmaceutical products and the study of their absorption, distribution, metabolism and excretion whether in laboratory animals or human subjects. The importance of developing sensitive, cold methods, as opposed to those based on radiolabelled compounds, is widely recognised in clinical pharmacology. Hospital pharmaceutical departments well equipped for essential quality control work are in a unique position with staff, equipment and laboratory facilities to undertake pharmacokinetic and metabolic studies, and even to provide a routine pharmacokinetic monitoring service if this is required. With such developments in mind, we have continued to feature applications of the various separation and spectroscopic methods to drug metabolism and pharmacokinetics together with relevant practical exercises.

J.B.S.

Part 2
Physical Methods of Analysis

Revised by
J.B.STENLAKE
C.B.E., Ph.D., D.Sc., F.P.S., C.Chem., F.R.S.C., F.R.S.E.

with contributions by

A.G.DAVIDSON
B.Sc., Ph.D., M.P.S.
University of Srathclyde, Glasgow

J.R.JOHNSON
B.Pharm., Ph.D., M.P.S.
University of Strathclyde, Glasgow

R.T.PARFITT
B.Pharm., Ph.D., M.P.S.
University of Western Australia, Nedlands, Western Australia

E.G.SALORE
B.Sc., Ph.D., M.P.S.
University of Strathclyde, Glasgow

G.A.SMAIL,
B.Sc., Ph.D., A.R.C.S.T, M.P.S.
University of Strathclyde, Glasgow

T.L.WHATELEY
B.A., M.A., M.SC., Ph.D., C.Chem., M.R.S.C.
University of Strathclyde, Glasgow

G.C.WOOD
B.Sc., Ph.D.
University of Strathclyde, Glasgow

Acknowledgements

Permission to reproduce the following Tables and Figures to gratefully acknowledged:

Tables
3.6, Stanton-Redcroft Ltd; 14.2, *Analytical Chemistry*.

Figures
3.5, HIAC/Royco Instruments Corp,; 3.6, Malvern Instruments; 3.17, Stanton-Redcroft Ltd; 3.12, 3.13, 3.14, 3.15, 3.16, 3.19, 10.3, 11.15 and 11.25, Perkin-Elmer Ltd; 6.14 and 6.15, Pye Unicam; ***, 8.1, and 8.3, Corning-EEL; 9.1, American Instrument Co Inc.; 10.4, N.B.Colthup, American Cyanamide Co; 11.3 and 12.7, Varian Associates; 11.4, *Angew. Chemie*, International Edition and Professor J.D.Roberts; 11.5 and 11.6 *Chemical and Engineering News*; 11.18b and c from *Introduction to Practical High Resolution Nuclear Magnetic Resonance* by D.Chapman and P.D.Magnus, Academic Press; 11.22, 11.28 and 11.31, *Journal of Chemical Education*; 11.26, *Chemical and Engineering News* and Dr.F.A.Bovey; 11.27, adapted from Chan and Hill, *Tetrahedron*, Pergamon Press; 11.29 and 11.30, *Jurnal of Pharmaceutical Sciences*; 11.33, 12.10 and 12.19, Wiley-Interscience; 12.1, adapted from K.Biemann, Mass Spectrometry, McGraw-Hill Publishing Co Ltd; 12.2 John Wiley and Sons and Drs.R.M.Silverstein and G.C.Bassler; 12.3, A.E.I. Ltd and Heyden and Son Ltd; 12.4 and 12.6, Finnegan Corporation; 12.8, *Journal of the American Chemical Society*; 12.9 adapted from *Industrial Research*; 12.11, *Biomedical Mass Spectrometry*; 12.12, Heydon and Son Ltd; 12.13, The Chemical Society and Dr.R.I.Reed; 12.14, McGraw-Hill Publishing Co Ltd and Professor K. Biemann; 12.18, *Arzneim. Forsch.*; 12.20, *Analytical Chemistry*; 13.9, J.R.French and the Radiochemical Centre, Amersham.

1

Instrumental methods in the development and use of medicines

J. B. STENLAKE

Introduction

Quality assurance plays a central role in determining the safety and efficacy of medicines. Highly specific and sensitive analytical techniques hold the key to the design, development, standardisation and quality control of medicinal products. They are equally important in pharmacokinetics and in drug metabolism studies, both of which are fundamental to the assessment of bioavailability and the duration of clinical response.

Modern physical methods of analysis are extremely sensitive, providing precise and detailed information from small samples of material. They are for the most part rapidly applied, and in general are readily amenable to automation. For these reasons they are now in widespread use in product development, in the control of manufacture and formulation, as a check on stability during storage, and in monitoring the use of drugs and medicines.

Product characterisation for drug development

Once a new drug candidate has been identified for development and clinical trial, physico-chemical methods are called into play at every stage of the development program. In the first instance, they are essential for product characterisation. This is a basic prerequisite to ensure that the animal toxicological studies required to establish safety prior to clinical trial are soundly based. Such studies are time consuming and grossly expensive, and have proved to be one of the prime factors in raising drug development costs to their currently near-prohibitive levels. The product under test must therefore be precisely characterised, and its quality clearly defined to ensure that the maximum amount of useful information is obtained from each test. In particular, product characterisation is necessary to eliminate variations in pharmacological and therapeutic responses and to determine the origin of any toxic effects that may be uncovered. It is also an essential means of ensuring that successive batches of material used in toxicological studies are of like quality, and, equally important, that a reasonable correlation of analaytical profiles is established between material used in animal toxicology and that later produced for clinical trial and marketing.

Such demanding requirements can only be met by employing the full range and sensitivity of modern physico-chemical separative and analytical techniques. Similarly, sensitive and specific methods for identifying decomposition rates are essential for monitoring stability during toxicological and clinical testing, and in determining the ultimate shelf life of the product. They are equally important in establishing metabolic profiles, for studying release rates from oral solid dosage forms, and in acquiring pharmacokinetic data.

The first step in product characterisation is to establish the precise chemical identity of the product. It is important to determine whether the material is a compound, i.e. a single chemical entity, a mixture of closely related compounds, a mixture of isomers, or merely a loose molecular complex of readily dissociable components. Such information is fundamental to a proper evaluation of the biological properties of the material. Thus *Gentamicin*, for example, consists of a mixture of variable composition dependiong on the source, of the closely related components Gentamicin C_1, C_{1a}, C_2, and C_{2a}, which differ in antibiotic activity. Likewise, only the (E)-isomer of *Triprolidine Hydrochloride* is an effective antihistamine, and *Crotamiton*, which is used in the treatment of scabies, consists solely of the (E)-isomer. Similarly, the sympathomimetic effect of adrenaline resides almost exlusively in the (R)-isomer. Such important physical characteristics could only be precisely, accurately, and rapidly determined by instrumental methods.

For compounds of synthetic origin, identity is usually clearly defined in the great majority of cases by the synthetic route employed. However, it is essential not only that identity be confirmed by alternative means but that the means employed should be capable of providing rapid verification whenever this may be required at any stage of the development program. Modern spectroscopic techniques, such as as[1]H and ^{13}C NMR (Chapter 11) and mass spectrometry (Chapter 12), together with ultraviolet (Chapter 7) and infrared spectroscopy (Chapter 10) are sensitive tools for such purposes. This is shown in the many examples described in the relevant chapters, and in the exercises in their application (Chapter 14). They are invaluable in the resolution of ambiguities where two or more alternative products may result from synthesis, and in the precise characterisation of complex natural products. Thus, the small structural differences that characterise the various components of *Gentamicin* are readily detectable by ^{1}H NMR, so much so that methods based on these differences have been applied for controlling the composition of the mixture, though superior methods based on high performance liquid chromatography are now available for routine use.

The interpretation of spectroscopic data obtained from compounds, however, is wholly dependent on the knowledge that the material under study is homogeneous. Powerful separative techniques, particularly chromatography in all its many forms (Chapter 4), provide sensitive methods for both purification and analysis of product homogeneity. Gas and high performance liquid chromatography, which make use of electronic recorders, are also eminently suitable and widely used for quantitative determination of the composition of mixtures of related compounds such as the gentamicins and even mixtures of isomers. The speed and high separative power of capillary gas chromatography (Chapter 4) make it a particularly useful, if highly specialised, technique for the separation of complex mixtures during the research phase of drug development.

Where the product consists of more than one isomer, the isomers must be capable of separate identification and measurement to establish means of ensuring batch to batch consistency of isomer composition. For most optically active compounds, a simple polarimetric measurement of the specific optical rotation at the wavelength of the sodium D-line will suffice (Chapter 2). Occasionally, however, where the measured rotation is small, measurements on a more sensitive instrument at other wavelengths either directly or after derivatisation may be necessary to secure adequate control of the product. Thus, contamination of *Ethambutol Hydrochloride*, which is the *S,S*-isomer, by its inactive *R,S*-isomer is limited in a test that requires the formation of a complex and measurement of its optical rotation using a spectropolarimeter.

In exceptional cases, the identification of optical isomers differing in only one of several chiral centres may call for the use of optical rotatory dispersion or circular dichroism (Chapter 2) to provide a degree of sensitivity that cannot be obtained from simple measurements of optical rotation. For example, the antibiotic *Phenethicillin Potassium*, which is prepared synthetically from 6-aminopenicillanic acid and racemic phenoxypropionyl chloride, is a mixture of the two expected diasterioisomers, but varies in composition depending on the solvent and crystallisation procedures used in its manufacture. The two isomers show differences in their antibacterial spectrum and the product may vary widely in composition. The two isomers give distinct peaks in their circular dichroism spectra, but of opposite sign, from which the composition of the isomer mixtures may be readily determined, and these correlate well with the results of microbiological assay (Stenlake *et al.*, 1972). NMR using chiral lanthanide shift reagents (Chapter 11) also provides a useful alternative method for the examination of diastereoisomeric mixtures.

NMR is particularly valuable in distinguishing geometrical isomers, as for example the *cis-* and *trans-*2,6-dimethyl-1-benzylpiperidines (Chapter 11); also *Crotamiton* (Chapter 14, Compound 16). Similarly, the correct assignment of the C-8 proton signals to the *cis-* and *trans-*isomers of the muscle relaxant *Atracurium Besylate* required the use of a combination of NMR nuclear Overhauser experiments (Chapter 11) and synthesis of a model compound, laudanosine methiodide, with ^{13}C-enriched methyl iodide (Lindon and Ferrige, 1980). Likewise, high resolution ^{1}H NMR and high performance liquid chromatographic characteristics of the *cis-cis, cis-trans-* and *trans-trans-*isomers of atracurium besylate provide a basis for the development of sensitive, accurate and rapid methods capable of establishing the isomer composition of the material (Stenlake *et al.*, 1984).

It is important not only to establish the chemical identity of any new drug substance but also to identify and quantify potential impurities at an early stage in development. These impurities relate to the source materials (i.e. the substance itself if it is a natural product, or starting materials for synthesis), the manufacturing process and the stability of the product.

Relatively unsophisticated techniques such as those of thin-layer chromatography are now in widespread use for the detection and quantitation of organic impurities, but in the development phase formulation of such tests necessarily rests on the formal identification of the chemical structure of each impurity, and this calls for the same heavy dependence on spectroscopic techniques as is required for the characteristation of the investigational drug itself. Inorganic impurities that might interfere with the assessment of toxicological profiles are confined to residues from toxic elements arising from catalysts and reagents used in synthesis. Trace amounts of toxic metal catalysts, such as nickel and platinum, are readily determined directly at parts per million levels by atomic absorption spectrophotometry (Chapter 8). Traces of toxic non-metals such as boron, derived from the use of borohydride reagents, are not amenable to direct measurement in this way, and require preliminary treatment to destroy the organic matter before determination by some suitable spectroscopic means.

Product development

The design of appropriate dosage forms for any new therapeutic entity is just as important and just as challenging as the other aspects of development. In this, the solid state characters of the active ingredients are especially relevant to the efficacy of oral dosage forms, such as tablets, capsules, oral suspensions, preparations for topical use and aerosols.

Compounds that are water-soluble are usually released readily to solution and hence are available for systemic absorption. Poorly soluble substances, however, may be only marginally effective or may produce erratic clinical responses, unless presented in an appropriate physical form. The definition and control of polymorphism (Chapter 3) can be critical for some compounds, e.g. *Chloramphenicol Palmitate*, which can exist in two polymorphic forms, one of which is biologically inactive. The preparation of dosage forms such as *Chloramphenicol Palmitate Mixture* (Chloramphenicol Palmitate Oral Suspension USP) requires careful control to minimise the formation of the inactive form. Infrared absorption (Chapter 10) and thermal analysis (Chapter 3) provide sensitive methods for the detection and control of such polymorphism. The former is used to control the limit of the inactive polymorph A in *Chloramphenicol Palmitate Mixture*.

The particle size of relatively insoluble compounds is also a critical factor in ensuring effective and consistent response in products for oral administration (e.g. *Digoxin* and *Griseofulvin* in tablets), injectable suspensions (e.g. *Insulin Zinc Suspension*) and topical preparations (e.g. *Hydrocortisone Acetate Ointment*). Similar control of particle or droplet size for all compounds, irrespective of solubility, is essential for the production of aerosols. The preparation of lipid suspensions for total parenteral nutrition is also dependent on adequate means of controlling droplet size. Particle size, and particle size distribution, can be determined (Chapter 3) by simple manual methods using microscopy and the measurement of sedimentation volumes. Instrumentation methods based on the measurement of electrical conductivity as in the Coulter counter, or requiring the measurement of specific surface area from the permeability to gas flow, offer considerable advantages. Such instrumental methods are particularly useful where large numbers of samples need to be examined, as for example in in-process controls of injectable suspensions and in routine checks for unwanted particulate contamination of injection solutions.

Viscosity (Chapter 2) measurements are used as a means of determining the molecular size of high molecular weight materials to control their properties, as in the various grades of *Methylcellulose* and *Dextran Intravenous Infusions*. When applied by methods suitable for measurements on non-Newtonian systems, viscosity also provides an important way of controlling the flow properties of semi-solid preparations such as ointments and creams.

Production and pharmacopoeial controls

Once established as marketable products, all drug substances and their formulated dosage forms are subject to strict quality assurance procedures. These encompass manufacturing methods, in-process controls, release specifications, and finally check specifications that are applicable throughout the shelf life of the product. In this respect it is important to emphasise the distinction between analytical procedures suitable for use by the manufacturer within the factory in connection with in-process controls and quality control at the time of manufacture (release specifications) on the one hand, and on the other hand procedures suitable for incorporation into a published pharmacopoeial monograph (check specifications) that are capable of independent assessment. Clearly, the latter must necessarily take into account any deterioration in the product during storage, so that the extent to which check specifications differ from release specifications depends on the stability and the intended shelf life of the product. Pharmacopoeial standards must also take account of sample size, which may well be smaller than desirable for analytical accuracy. Tolerances are therefore often wider than in release specifications.

There may also be differences so far as methodology is concerned. The manufacturer, with his complete knowledge of the drug and the additives used in its formulation, is in the best possible position to choose the most specific and sensitive procedures. Sophisticated instrumental methods capable of automation to deal with large numbers of samples linked to computerised data recording systems for automatic calculation and recording of results are rapidly becoming the norm. Pharmacopoeial methods, on the other hand, must be such that they are widely available and in common use in control laboratories, whether these be in hospitals, in government departments or independent organisations. Such methods are therefore often simpler.

As a consequence of these latter requirements, the wet chemical methods described in Part 1, whether of classical origin or more recent innovations, have generally been favoured in pharmacopoeial specifications. Even so, such methods have always been supplemented by the measurement of physical constants wherever appropriate, since the facilities for the measurement of, for example, melting and boiling range, refractive index and optical rotation with relatively unsophisticated equipment are readily available in most laboratories. The widespread availability of relatively cheap ultraviolet and infrared spectrometers has brought a degree of precision and simplification into the characterisation and standardisation of drugs, and such techniques are now widely used in both the factory and independent control laboratories. Thin-layer chromatography, likewise, is commonplace, sensitive, quick and easy to apply, and hence is also in widespread use both for verification of identity, and in control tests to limit the degree of contamination by related substances and decomposition products, in both factory and pharmacopoeial specifications.

The much higher instrumental costs of advanced analytical methods, such as nuclear magnetic resonance spectroscopy, mass spectrometry, and gas and high performance liquid chromatography, tend to limit the use of these techniques in pharmacopoeial monographs to the control of complex molecules that cannot be sufficiently closely identified or standardised by simpler methods. Thus NMR is used to identify the antibiotics *Framycetin Sulphate* and *Gentamicin Sulphate*. High performance liquid chromatography is in widespread use because of its versatility in the analysis of complex antibiotics and peptide hormones in addition to its separative power and sensitivity for the analysis of synthetic isomer mixtures; it is increasingly used in the control of related substances in both medicinal agents and their formulated products.

The move towards instrumental methods, particularly in the drive to reduce dependence on biological assays and limit tests requiring animals, has led to the acceptance of a number of less widely used methods. These include, for example, electrophoresis on polyacrylamide gel to control related proteins (arginylinsulin, insulin ethyl ester and proinsulin) in *Insulin*, size exclusion chromatography (Chapter 4) to control higher molecular weight proteins in *Insulin*, and spectrofluorimetry (Chapter 9) to control related substances in the veterinary anthelmintic *Haloxon*. The sensitivity of spectrofluorimetry is also the reason for it being chosen as the most suitable technique for the control of Al levels in *Haemodialysis Solutions*. These are used in large volume and regularly over long periods of time. Hence the need for stringent limits on the permitted levels of this toxic metal. Classical methods for the analysis of aluminium are not suitable, but fluorescence measurement of its 8-hydroxyquinoline derivative provides a sensitive method for control at levels below 15 ppm of aluminium.

One major disadvantage of many of these methods is the need to compensate for instrument and operator variables, both within and between laboratories. For this reason, many of these techniques require the use of reference compounds or internal standards of known quality and composition. The preparation, storing and distribution on an international basis of large numbers of official reference compounds (British, United States and European Pharmacopeia Reference Substances) is expensive, not just in the materials themselves, but also in distribution costs and in staff time devoted to verification of quality and stability during storage. The distribution of reference samples of controlled drugs (drugs of abuse) is also undesirable, and is subject in some countries to special conditions.

Some progress in overcoming such difficulties has been made in recent years, for example in thin-layer chromatography by use of the parent compound under test at multiple plate loadings (e.g. 1%, 0.01% and 0.05%) to provide a basis for assessing the density of chemically related impurity spots as in *Dextromoramide Tartrate*. However, care is necessary in setting standards when this technique is used, since the sensitivities of the parent and the related substances to the visualising reagent may differ, thereby complicating the problem of setting precise limits for the impurities concerned.

The issue of reference spectra by the British Pharmacopeia Commission for use in the verification of identity in place of reference compounds is a major step forwards in overcoming some of the problems of standardising methods to achieve comparable results on an inter-laboratory basis. In particular it overcomes the need for actual sample distribution of controlled drugs.

Notwithstanding the obvious disadvantages attending the use of complex analytical instrumentation in small control laboratories, their selectivity, sensitivity, and speed of operation ensure their continued use on an expanding scale. As already mentioned, they offer, increasingly, meaningful alternatives to distasteful biological methods. The following chapters describe the more important physico-chemical methods of analysis currently in use in connection with the development, control and use of pharmaceutical products.

Drug metabolism and pharmacokinetics

Investigation of the metabolism and pharmacokinetics form an integral part of all drug development programs. Such studies are conducted initially in laboratory animals, but as development proceeds they must be capable of extension to the clinical level in human volunteers and finally in patients undergoing treatment. The scale of difficulty that attaches to such studies is related to both dosage and the complexity of the metabolic pathways, but highly sensitive methods are essential. For this reason, radiolabels, primarily ^{14}C and ^{3}H, are widely used in primary animal studies (Chapter 13) of tissue distribution, blood levels, and excretion via the lungs, liver and kidney. Recovery studies are necessarily dependent on the use of satisfactory extraction methods (Part 1, Chapter 9) and the powerful separative techniques of thin-layer, gas and high performance liquid chromatography (Chapter 4) to separate labelled metabolites from unchanged drug for identification and quantitative determination.

Human metabolic studies, on the other hand, are constrained by the unacceptability of exposure to radioactivity. Pharmacokinetic studies, too, often require measurements in the microgram, nanogram or picogram/ml range. Much depends on the capability of the separative and ultimate detection system. Gas chromatography, capillary gas chromatography and high performance liquid chromatography all have their place, but much use is now made of such powerful coupled separative–detector systems as gas chromatography–mass spectrometry (GC–MS) and liquid chromatography–mass spectrometry (LC–MS) (Chapters 4 and 11). These systems are extremely versatile but very expensive, and much valuable work is still achieved with the less sophisticated and less versatile though sensitive detection capabilities of ultraviolet spectrophotometry, spectrofluorimetry (Chapter 9) and polarography (Chapter 5). In a few special cases where there is sufficient clinical demand for the development of a sensitive routine method for pharmacokinetic monitoring of the patient's clinical status, radioimmunoassay (Chapter 13) is a valuable tool as in the case of *Digoxin*.

References

Stenlake, J.B., Wood, G.C., Mital, H.C. and Stewart, S. (1972) *Analyst* **97**, 639.
Stenlake, J. B., Waigh, R.D., Dewar, G.H., Dhar, N.C., Hughes, R., Chapple, D.J., Lindon, J.C., Ferrige, H.G. and Cobb, P.H. (1984) *Eur.J.Med.Chem.* **19**, 441.

2
General physical methods

G. C. WOOD

Density

Density

Density, ρ, is the mass of a unit volume of a material. The millilitre is usually chosen to express volume, this being the volume of 1 g water at 3.98°C, the temperature of maximum density of water. Thus the density of water at 3.98°C is 1.0000 g ml^{-1} (for most purposes the difference between densities expressed in g ml^{-1} and in g cm^{-3} can be neglected). Density depends on temperature which is therefore specified by a subscript (ρ_t), t being in degrees centigrade. The density of water at various temperatures is given in Table 2.1.

Table 2.1 Densities of water (g ml^{-1}) at various temperatures

t°C	0	3.98	10	15	20	25
ρ_t	0.99987	1.0000	0.99973	0.99913	0.99823	0.99707

t°C	30	40	50	60	70	80
ρ_t	0.99569	0.99224	0.98807	0.98324	0.97781	0.97183

Specific gravity

Specific gravity or relative density, $d_{t_1}^{t_2}$, is the ratio of the mass of a certain volume of the material at a particular temperature (t_2) to the mass of an equal volume of water at the same or some other specified temperature (t_1).

Thus, $d_{t_1}^{t_2} = \rho_{t_2}/\rho_{t_1 H_2O}$. It follows that d_4^t is numerically equal to ρ_t (the difference between the densities of water at 4°C and 3.98°C being neglected) though, unlike density, it is dimensionless.

Molar volume

The molar volume of a compound, a quantity often used in calculations, is expressed as:

$$\text{molar volume} = \text{molecular weight/density}$$

Physically the molar volume is a measure of molecular volume plus any free space between the molecules. Attempts have been made to use molar volumes as an additive property of the number and types of atoms and groupings in a molecule, but the additivity is only approximate.

Partial molar volume

When solutions are formed from pure components, unless the solution is ideal (obeying Raoult's Law, and giving no heat or volume changes on mixing the components), the final volume is not simply the sum of the constituent volumes. Volume changes nearly always occur on mixing. It is rare to find that:

$$V = n_1V_1 + n_2V_2 \qquad (1)$$

where V is the total volume of solution, and V_1 and V_2 are the molar volumes of components 1 and 2 respectively. For non-ideal solutions, which are those normally encountered, eq. (1) becomes:

$$V = n_1\bar{V}_1 + n_2\bar{V}_2 \qquad (2)$$

where $\bar{V}_1$ and $\bar{V}_2$ are the partial molar volumes of components 1 and 2 and n_1 and n_2 are the numbers of moles of components 1 and 2. A partial molar volume is the change in volume when one mole of a particular component is added to an infinitely large volume of solution at constant temperature and pressure.

For component 1:

$$\bar{V}_1 = \left(\frac{\delta V}{\delta n_1}\right)_{T,P,n_2} \qquad (3)$$

A simple means of evaluating the partial molar volume is to plot V, the volume of solution containing containing 1000 g solvent, against the molality, m, of the solute. The slope of (or tangent to) the line at a particular value of m, gives $\bar{V}_2$ at this concentration.

Interconversion for concentrations

Solution concentrations are expressed in a number of different ways, and it is often useful to be able to convert from one system to another. Consider a solution of total volume, V, and density, ρ, in which:

W_1 = weight of solvent of molecular weight M_1
W_2 = weight of solute of molecular weight M_2
Total weight of solution, $W = W_1 + W_2$
number of moles of solvent, $n_1 = W_1/M_1$
number of moles of solute, $n_2 = W_2/M_2$

Molarity (C):

C = number of moles of solute per litre of solution

$$C = 1000n_2/V = \frac{1000n_2\rho}{W_1 + W_2} = \frac{1000\rho W_2}{M_2(W_1 + W_2)} \qquad (4)$$

Molality (m):

m = number of moles solute per 1000 g solvent

$$m = \frac{1000n_2}{W_1} = \frac{1000W_2}{M_2W_1} \tag{5}$$

Interconversion of molarity and molality:

$$C = \frac{1000m\rho}{1000 + M_2m} \tag{6}$$

Interconversion of percentages:

$$(\%w/w) = (\%w/v)/\rho \tag{7}$$

Mole fraction:

$$X = \frac{\text{number of moles of component}}{\text{total number of moles present}}$$

$$\text{Mole fraction of solvent} = X_1 = n_1/(n_1 + n_2) \tag{8a}$$
$$\text{Mole fraction of solute} = X_2 = n_2/(n_2 + n_2) \tag{8b}$$

Interconversion of molarity and mole fraction:

$$C = \frac{1000X_2\rho}{(X_1M_1 + X_2M_2)} \tag{9}$$

Practical experiments

Experiment 1 *Determination of the specific gravity and the density of a liquid*

A pycnometer is used for the accurate determination of liquid densities. A convenient form is shown in Fig. 2.1, although many different types are available. For example, the vessels for volatile liquids are

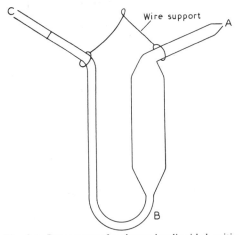

Fig. 2.1. Pycnometer for determining liquid densities

fitted with ground glass caps to prevent evaporation. For rapid, low precision determination of liquid densities, hydrometers are convenient. As the temperature coefficient of density is about 0.1% per degree, the temperature must be carefully controlled during measurements. To obtain an accuracy of about 0.0001 g ml^{-1} in ρ, temperature should be controlled to $\pm 0.1°$.

Method Clean the pycnometer with chromic acid, and wash it with *water*. Rinse it with ethanol and then acetone, and dry by drawing air through it by means of a filter pump. Determine the weight (W_1) of the pycnometer plus the support by which it is attached to the balance arm. Fill the pycnometer by slowly drawing *water* in through A; hold it inverted until the liquid level reaches B, then turn it upright and allow the *water* to reach mark C. Suspend it in a thermostat bath at 25°, so that the ends A and C are clear of the water in the thermostat. After temperature equilibrium has been attained (20 min), adjust the volume of *water* in the pycnometer so that it is completely full from the mark near C to the tip A, using either filter paper to remove water from A, or a fine bore pipette to add *water* through C.

Remove the pycnometer from the bath, carefully dry the outside, and determine the weight of vessel, support, and water (W_2). Clean and dry the vessel as before, fill with a 25% w/w glycerol/water mixture, and repeat the above procedure to obtain the weight of pycnometer, support, and liquid (W_3).

Treatment of results

1. Calculate the specific gravity of the mixture.

$$d^{25}_{25} = \frac{W_3 - W_1}{W_2 - W_1}$$

2. Calculate the density of the mixture.

The approximate volume of the pycnometer is given by the expression:

$$V_{\text{approx}} = (W_2 - W_1)/\rho_{t\text{H}_2\text{O}}$$

and approximate density is:

$$\rho' = (W_3 - W_1)/V_{\text{approx}}$$

For a liquid the British Pharmacopeia refers to 'weight per millilitre'. This is equal to the density ρ', uncorrected for buoyancy. More accurate values for the density can be obtained by correcting the weighings for the difference between the weight of air displaced by the balance weights, and that displaced by the pycnometer and liquid. Displacement of air by the glass of the pycnometer is the same for the dry pycnometer and for the vessel full of liquid, and can be neglected.

$$\text{Volume of weights used to weigh water} = (W_2 - W_1)/\rho_{\text{brass}}$$
$$= (W_2 - W_1)/8.4$$
$$\text{Volume of water in the pycnometer} = (W_2 - W_1)/\rho_{t\text{H}_2\text{O}}$$

Hence, the buoyancy correction, b, is given by:

$$b = (W_2 - W_1)\left\{\frac{1}{\rho_{t\text{H}_2\text{O}}} - \frac{1}{8.4}\right\}(\text{density of air}) \tag{10}$$

The density of air can be taken as 0.0012 g ml^{-1}, though it varies with temperature and barometric pressure.

$$\therefore \quad \text{True weight of pycnometer plus liquid} = (W_3 - W_1) + b,$$
$$\text{and, true volume of pycnometer} = [(W_2 - W_1) + b]/\rho_{t\text{H}_2\text{O}}$$
$$\rho = \text{corrected density} = \frac{(W_3 - W_1) + b}{[(W_2 - W_1) + b]/\rho_{t\text{H}_2\text{O}}}$$

The application of the buoyancy correction may be simplified by using the above equations in equivalent form:

$$\rho = \rho' + C' \qquad (10a)$$

where

$$C' = (\text{density of air})\left\{1 - \frac{\rho'}{\rho_{tH_2O}}\right\}$$

the correction, C' (which may be negative) being added to the appropriate density, ρ'.

The determination of solid densities is not usually so precise as that of liquids, and, unless special methods are used, buoyancy corrections are not significant compared with other errors.

Experiment 2 *Determination of the partial molar volume of glycerol in aqueous solution*

Prepare accurately by weight solutions containing about 5, 10, 15, and 20% w/w glycerol in *water*, using 100 ml volumetric flasks. Determine the density of each solution at 25°, using the method of Experiment 1.

Treatment of results

1. Calculate the density of each solution.

2. Construct a graph of V against molality of glycerol (see p.10), and determine the slope at 1, 2, and 3 molal. Use the result of Experiment 1 to provide an additional point on the graph. Using 92.09 for the molecular weight of glycerol, and 1.2583 g ml^{-1} for its density at 25°, calculate the molar volume, and compare this with the partial molar volumes at the three molalities.

Experiment 3 *Determination of the density of sodium chloride*

Densities of solids may be determined by flotation techniques; in these a few particles of the solid are placed in a liquid mixture, in which they are quite insoluble, and the density of the mixture is altered by changing the proportions of the liquids, until the solid particles neither rise nor fall. Under these conditions, the density of the solid is equal to that of the liquid, and the latter can be measured in a pycnometer.

A second method, used in this experiment, is to measure the volume of liquid displaced by a known weight of solid. For success, both methods depend on removal of air from around the solid particles; this can be achieved by covering the solid with a layer of liquid, and evacuating the system.

Method Clean, dry, and weigh a 50 ml density bottle fitted with a ground-in stopper pierced by a fine hole (W_1). Fill the bottle with *water*, insert the stopper, and place in a thermostat bath at 25° until water ceases to emerge through the stopper. Dry the top of the bottle with filter paper. Remove the bottle from the bath, dry the outside, and weigh (W_2). Dry the bottle inside and out, and repeat the procedure using dry benzene to obtain the weight of bottle plus benzene (W_3). Benzene is used as the displacement liquid.

Place about 30 g sodium chloride in the dried bottle and weigh (W_4); just cover the solid with benzene and swirl gently. Fit up a filter pump with a piece of clean smooth pressure tubing whose end had been cut off square. Press the end of the tube to the bottle top, and outgas the liquid gently for five minutes. Allow the solid to settle, fill up the bottle with benzene, insert the stopper, place in the thermostat bath for 20 min, and weigh (W_5). It may be necessary to allow the benzene to warm up at intervals during the outgassing procedure.

Treatment of results

1. Determine the volume of the bottle using the formula

$$\frac{(W_2 - W_1)}{0.9971}$$

and calculate the density of the benzene used (ρ).

2. Find the volume of benzene displaced by sodium chloride ($W_4 - W_1$) which is given by

$$\frac{(W_3 - W_1)}{\rho} - \frac{(W_5 - W_4)}{\rho} \tag{11}$$

Hence calculate the density of the solid.

Solubility

Solubilities of pharmacopoeial substances are provided primarily for information, and are increasingly being expressed in descriptive terms (Part 1, Chapter 1). Specific pharmacopoeial statements of solubility as weight of solute in a specified volume of solvent at a particular temperature only apply at this temperature (see 'Effect of Temperature' below). If the temperature is not stated, the solubilities stated apply at room temperature. Solubilities are also expressed as 'one part of solute in x parts solvent', meaning either 1 g solid solute or 1 ml liquid solute in x ml solvent.

The usual expression is in terms of weight of solute per volume or per weight of solvent or solution; usually as g solute/100 g solvent, although the normal means of expressing solution concentration such as molar, molal, are also used. Only solutions of liquids in liquids, and solids in liquids will be dealt with here.

Effect of temperature

Whether solubility increases or decreases with increasing temperature will depend on the nature of the solute and solvent, but in general solubility increases with temperature rise. The rate of solution is also increased by heating. Studies of the variation of solubility with temperature can be used to evaluate the heat of solution, ΔH from *either*

$$\frac{d \ln S}{dT} = \frac{\Delta H}{RT^2} \tag{12a}$$

or from the integral form (12b), derived on the assumption that ΔH is independent of temperature

$$2.303[\log S_2 - \log S_1] = \frac{\Delta H(T_2 - T_1)}{RT_2 T_1} \qquad (12b)$$

where S_1 and S_2 are the solubilities at two temperatures T_1 and T_2. The value of ΔH obtained can give information on the mechanism of solution.

Effect of particle size

While small particles of a solid solute may often dissolve more rapidly than large ones (Chapter 3), particle size has little effect on solubility as such in the size range normally encountered. Only in the colloidal particle size range does solubility increase significantly with decrease of particle size.

Solubility and purification procedures

Recrystallisation is used for purifying a vast range of organic compounds; the basis of the technique is that, for a mixture of solutes, the required one will crystallise from a particular solvent, while the others remain (largely!) in solution. Recrystallisation is only really effective when most of the major contaminants have already been removed from the solute, as large amounts of impurities may either crystallise out with the desired material, or may inhibit its crystallisation.

The decreasing solubility of polymers with increasing molecular weight is used as a means of fractionation. The polymer is dissolved in a solvent, some precipitating liquid added, and the fraction of polymer precipitated is centrifuged off. This fraction has the highest molecular weight. Addition of a little more precipitating liquid gives a fraction of lower molecular weight, and a range of fractions can be obtained by repeating this process. In protein chemistry much use is made of precipitation for purification, generally using salts as precipitants, and arranging experimental conditions which will leave the impurities in solution as far as possible.

Solubility of liquids in liquids

Information on the mutual solubility of immiscible liquids can be obtained by shaking them together at a controlled temperature, allowing separation to occur, and analysing samples from each layer for each component. For liquids which become miscible at accessible temperatures, it is often convenient to determine their mutual solubility curve. Mixtures of the two liquids of known composition are prepared, and heated until a homogeneous phase appears (a saturated solution of one liquid in the other). The highest temperature at which two phases disappear, measured as a function of composition, is the critical solution temperature; above it the two liquids are miscible in all proportions, no matter what the composition.

Practical experiments

Experiment 4   *Determination of the mutual solubility curve of phenol and water*

Method Use test tubes of 20 ml capacity, fitted with ground glass stoppers. Weigh accurately about 1, 2, 3, 4, 5, 6, 7, 8, and 9 g phenol into the tubes. Add sufficient *water* to each tube to make the total contents weigh about 10 g, and weigh accurately. Fit up a 2 l beaker with a hand stirrer, and a thermometer (0–100°, graduated to 0.1°). Attach a piece of copper wire to the tube containing the most dilute solution of phenol in water, and hang it on the rim of the beaker. Place sufficient water in the beaker to bring the level to the bottom of the ground glass stopper of the tube. Heat the water in the beaker, shaking the tube frequently, and raising the temperature slowly when the turbidity of the phenol–water mixture shows signs of disappearing. Determine the temperature at which the turbidity disappears on shaking. Return the tube to the beaker, allow the temperature to fall, and note the temperature at which the turbidity just reappears. Repeat the heating and cooling, and take the mean of all four temperature readings. Detertmine the temperature at which miscibility occurs for the other solutions (cooling may be required for the most concentrated solutions).

Treatment of results Plot the temperatures at which a single phase occurs as ordinates, against percentage (w/w) phenol in water as abscissae. Determine the critical solution temperature from this graph.

Experiment 5   *Determination of solubility of adipic acid in water*

A simple quick method for determining the solubility of a solid in a liquid is to prepare a set of test tubes each containing a weighed amount of solid, add varying quantities of solvent, shake thoroughly, and note the concentration at which all the solid dissolves.

For accurate solubility determinations it is necessary to ensure that the solvent is completely saturated with solute at a particular tempera- ture, withdraw a sample of saturated solution without disturbing the equilibrium, and analyse it to determine concentration. The solid and solvent are mixed by stirring or shaking in a thermostat. Samples are withdrawn at intervals and assayed. Equilibrium is reached when there is no further uptake of solute by solvent. This method can also be applied to the determination of liquid–liquid solubilities.

Method Place adipic acid (about 4 g, roughly weighed) in each of two flasks (100 ml) fitted with B24 size ground glass stoppers, and add *water* (about 60 ml) to each flask. Immerse one flask at a level just above the bottom of its stopper in a thermostat bath set at 20°, warm the second flask at 50° for 10 min, shaking it occasionally, and then place it similarly in the thermostat bath. This procedure is adopted so that equilibrium is approached from both over- and under-saturation. Shake the flasks every 20 min for 2 h. Allow the flasks to remain undisturbed for 10 min. Fit a one inch length of polythene tubing to the tip of a 10 ml pipette, and pack the tubing with cotton wool to act as a filter. Cautiously withdraw a sample of solution from the flask, examining the contents of the pipette against a bright light to make sure the filtration was efficient. Remove the filter, adjust the volume of the solution in the pipette to 10 ml, and run the solution out into a tared flask. Weigh the sample.

Titrate the sample with standard 0.2M sodium hydroxide using phenolphthalein as indicator. Continue to shake the flasks during a further hour, withdraw a sample from each, and re-assay. Equilibrium should have been reached after 2 h, as shown by agreement between the four sets of results.

If the heat of solution is to be determined, repeat the experiment at 30°. The pipette should be preheated to the temperature of the experiment.

Treatment of results

1. Calculate the solubility of adipic acid in g solute/100 g water, and $mol\,l^{-1}$ solution (molecular weight of adipic acid = 146). Present the results to show the equilibrium has been reached.

2. Calculate ΔH from eq. (12b) using the solubility in $mol\,l^{-1}$.

Molecular weight

Excluding very large molecules, the determination of the molecular weight of compounds is based on the use of colligative properties, generally freezing point depression or boiling point elevation. Osmotic pressure can be used in the 5000–300 000 molecular weight range.

The presence of a non-volatile solute in a solution lowers the vapour pressure from p_1^0 (that of the pure solvent) to p_1 (that of the solution). For ideal solutions, Raoult's law gives

$$p_1 = X_1 p_1^0 = (1 - X_2)p_1^0 \tag{13}$$

where X_1 is the mole fraction of solvent, and X_2 that of the solute. In dilute solution the term W_2/M_2 in the denominator of X_2 can be neglected:

$$X_2 = \frac{W_2/M_2}{W_1/M_1 + W_2/M_2} \simeq \frac{W_2 M_1}{M_2 W_1} \tag{14}$$

giving for the relative lowering of the vapour pressure

$$\frac{p_1^0 - p_1}{p_1^0} = \frac{W_2 M_1}{M_2 W_1} \tag{15}$$

Equation (15) can be used to determine M_2 from measurements of vapour pressure, assuming that no dissociation or association of solute occurs.

Using Fig. 2.2 consider the boiling point elevation and freezing point depression. AB represents the sublimation curve of pure solid solvent, and BC the vapour pressure curve of liquid solvent, reaching the boiling point at C. When solute is present, the vapour pressure of the solution is lower than that of the solvent (DE), giving a freezing point at T_f, a depression of the freezing point $\Delta T_f = T_0 - T_f$; the boiling point of the solution is raised to T_b, giving elevation of boiling point $\Delta T_b = T_b - T_{bo}$.

At the freezing point of a solution, there is an equilibrium between solid solvent and solution, such that the chemical potential of the separated (frozen) solvent and of solvent in the solution are equal. By considering the variation of chemical potential with temperature the following relationship is obtained

$$\Delta T_f = \frac{RT_0^2 X_2}{\Delta H_{fus}} \tag{16}$$

where ΔH_{fus} is the latent heat of fusion per mole of solvent. In dilute solution eq. (14) is used for the mole fraction of solute, and taking molality from eq. (5):

$$X_2 = M_1/m1000$$

hence:

$$\Delta T_f = \frac{RT_0^2 M_1 m}{\Delta H_{fus}1000}$$

and

$$\Delta T_f = K_f m \tag{17}$$

in which all the constants for a particular solvent are combined in a new constant K_f.

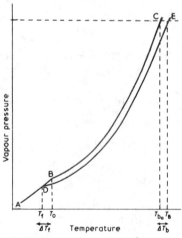

Fig. 2.2. Vapour pressure-temperature curves for pure solvent and solution

By an analogous treatment, considering the equilibrium between solvent vapour and liquid solvent in solution at the boiling point, the chemical potential of solvent in the vapour phase must be the same as that in the liquid phase, and

$$\Delta T_b = \frac{RT_b^2 X_2}{\Delta H_{vap}} \tag{18}$$

where ΔH_{vap} is the molar latent heat of vaporisation. Simplifying as before for dilute solution:

$$\Delta T_b = \frac{RT_b^2 M_1 m}{\Delta H_{vap}1000}$$

and

$$\Delta T_b = K_b m \qquad (19)$$

Examples of molal cryoscopic and ebullioscopic constants are given for commonly used solvents in Table 2.2. The table shows that for a certain solute concentration, the depression of the freezing point will be greater than the elevation of the boiling point; freezing points are also easier to determine than boiling points.

Table 2.2 Cryoscopic and ebullioscopic constants

Solvent	T_f	T_b	K_f	K_b
Acetic acid	16.7	118	3.9	2.93
Benzene	5.4	80	5.12	2.53
Chloroform		61		3.63
Ethanol		79		1.22
Water	0	100	1.86	0.51

The osmotic pressure can be considered in relation to the other colligative properties using Fig. 2.3. Solvent and solution are placed in the two arms of the vessel shown. As $p_1^0 > p_1$ solvent distils through the vapour space, which in this case is acting like a semi-permeable membrane. To prevent distillation, the pressure on the solution must be increased; this raises the chemical potential of the solvent in solution, so that at equilibrium there is no further distillation.

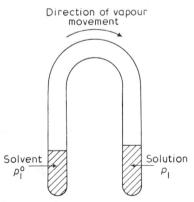

Fig. 2.3. Apparatus for equilibrium of solvent and solution

From a thermodynamic consideration of the effect of pressure

$$\Pi \overline{V}_1 = RT \ln (p_1^0/p_1) \qquad (20)$$

where Π = osmotic pressure and $\overline{V}_1$ = partial specific volume of solvent. If it is assumed that Raoult's law holds, and that solutions are dilute, this equation can be simplified:

$$\Pi = W_2RT/M_2 \tag{21}$$

This enables the molecular weight of the solute to be calculated from measurements of osmotic pressure, W_2 being the concentration of solute in $g\,l^{-1}$.

Solute mixtures

If the solute under test is impure an average molecular weight will be obtained. The average is due to materials of different molecular weight depressing the freezing point (or elevating the boiling point) by different amounts. Colligative methods depend on the number of molecules present in the solution, and the methods give a number average molecular weight, M_N,

$$M_N = \frac{\Sigma n_i M_i}{\Sigma n_i}$$

where n_i = number of molecules of molecular weight M_i, and so on. Hence contamination of the material under test with low molecular weight impurities can give erroneous values for the molecular weight.

Practical experiments

Experiment 6   *Determination of molecular weight by freezing point depression*

The conventional apparatus of Beckmann (Fig. 2.4) consists of an outer bath, A, holding the freezing mixture, which cools the solution in the tube, B, which is surrounded by an air jacket, C. A Beckmann thermometer, D, and stirrer, E, dip into the solution. Samples may be introduced through the side arm, F. The freezing bath is covered by a lid, and kept mixed by a stirrer, G.

When using this apparatus, some supercooling of the sample occurs, as separation of solid solvent does not take place until the temperature is a little below the freezing point. The latent heat of crystallisation then raises the temperature. This effect may be minimised if the temperature of the freezing mixture is too low, and a temperature lower than the true freezing point may be recorded. The freezing mixture should be kept 3–4° below the freezing point of the solution. The degree of supercooling should not be allowed to exceed 0.3–0.5°, otherwise a great deal of solvent may crystallise out, and seriously affect the concentration of the solution.

The Beckmann thermometer   This is a sensitive thermometer, graduated in 0.01°, and capable of being read to 0.002°. The scale length of the thermometer is 5–6°, and provision is made for adjusting the amount of mercury in the bulb, so that the thermometer can be used over different temperature ranges. Place the bulb in a beaker of water whose temperature has been adjusted to the freezing point of the solvent to be used. The mercury should stand near the top of the scale. If it stands above, place the thermometer in water warm enough to cause the mercury to rise up and form a drop at the end of the capillary (A) (Fig. 2.5). Shake off the drop and test again in the cool water. If necessary repeat this procedure until the mercury stands at the correct place on the scale. The initial test may show that there is too little mercury in the thermometer bulb (mercury below scale); in this case warm up the thermometer until drops form at the capillary, invert it, and tap gently to make mercury from the reservoir join with mercury in the capillary. Re-invert, and place the thermometer in a water bath held 2° above the freezing point of the solvent; the cooling draws mercury from the reservoir into the bulb. When this process is complete, tap the upper part of the thermometer gently to make excess mercury separate from the end of the capillary. Test the thermometer in water at the freezing point of the solvent to make sure that the setting is correct, i.e. the mercury is on the scale. If it is a little above, remove the excess dropwise to the reservoir as described above. Beckmann thermometers require careful handling as they are expensive, and easily broken.

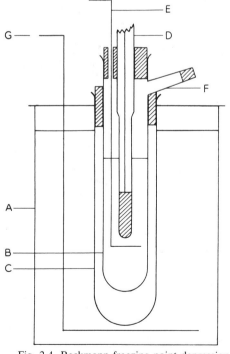

Fig. 2.4. Beckmann freezing point depression apparatus

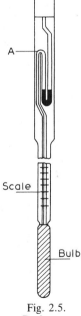

Fig. 2.5. Beckmann thermometer

Determination of freezing point Remove the tube B from the apparatus. Clean, dry, and weigh the tube. Introduce dry, crystallisable benzene (15–20 g), re-weigh and insert the Beckmann thermometer and the stirrer. Pack the bath with a freezing mixture of ice and water controlled at about 2°. Place the tube in the freezing mixture, and when solid separates, remove it, dry the outside and place it in the air jacket. Stir the solution about once every second (faster stirring can generate undue heat), and note the constant (highest) temperature on the Beckmann thermometer. The first reading is usually approximate. Remove the tube from the apparatus, and warm gently with stirring, until the temperature is about 1° above the freezing point. Replace it in the apparatus, and stir; the supercooling should not be more than 0.5°, crystallisation of solvent being induced by a bout of vigorous stirring, or by adding a crystal of benzene. Follow the temperature rise accompanying crystallisation, noting the highest temperature, and tapping the thermometer with the fingernail to prevent the mercury from sticking in the stem.

Repeat the procedure until three consistent readings for the freezing point of pure solvent have been obtained. Introduce a weighed tablet of naphthalene (about 0.12 g) through the side arm, dissolve it in the benzene with gentle warming, and determine the freezing point of the solution. When consistent results have been obtained, a second weighed tablet should be introduced to obtain a second reading.

Treatment of results Calculate the molality of the solution in both parts of the experiment using eq. (17) and the observed ΔT_f values. Calculate M_2 from eq. (5).

Experiment 7   *Detection of association on solution*

Many carboxylic acids associate in non-polar solvents, generally to form dimers. Acetic, benzoic and phenylacetic acids are among those showing this phenomenon.

Method Weigh 1, 2 and 3 g of dry phenylacetic acid into each of three graduated flasks (50 ml), make up approximately to volume with dry crystallisable benzene, stopper, and re-weigh (weight concentrations are required; the graduated flasks are merely convenient containers). Determine the freezing point of each solution, and of a sample of the solvent, by the technique of the preceding experiment.

Treatment of results
1. Calculate the apparent molecular weight, M, for phenylacetic acid in each solution. Compare this with the monomer molecular weight of 136.
2. Calculate the degree of association, α, of the phenylacetic acid in each solution, assuming that dimers are formed, e.g. $2C_6H_5CH_2COOH \leftrightharpoons (C_6H_5CH_2COOH)_2$.

For one mole of solute $(1 - \alpha)$ moles are unassociated, and $\frac{1}{2}\alpha$ are associated. The total number of moles is therefore $\frac{1}{2}\alpha + (1 - \alpha)$. Using the usual freezing point equation, calculate the freezing point depression $(\Delta T_f)_0$ assuming that no association took place, i.e., for the monomer of molecular weight, M_0.

Now,

$$(\Delta T_f)_{exp} = \left[(1 - \alpha) + \frac{\alpha}{2}\right](\Delta T_f)_0$$

$$\therefore \quad (1 - \alpha/2) = \frac{(\Delta T_f)_{exp}}{(\Delta T_f)_0}$$

$$= \frac{M_0}{M}$$

Hence,

$$\alpha = \frac{2(M - M_0)}{M} \tag{22}$$

Experiment 8 *Micro method for molecular weight determination (Rast)*

A number of terpenoid substances which have very large K_f values, e.g. camphor 40, camphene 35, are useful for determing molecular weights by freezing point depression, using small quantities of material. Generally about 10% solutions in camphor are prepared; the method assumes that the freezing point depression equation is obeyed, and that pure camphor separates on freezing. K_f for camphor varies from sample to sample, so either a sample of known K_f value must be used, or it must be measured using naphthalene as solute (10% − 15% by weight) in camphor for the method given below.

Method Seal one end of a clean, thin-walled glass tube (1 × 10 cm), allow to cool, and weigh to 0.1 mg. Introduce about 15 mg of acetanilide, ensuring that the sample is all placed at the bottom of the tube, weigh accurately, add about 150 mg camphor and re-weigh. Carefully seal the end of the tube, place it in an oil bath, and heat until the contents melt. Rotate the tube to mix the materials thoroughly. For a second determination prepare another tube with acetanilide (20 mg) and camphor (150 mg). Allow the tubes to cool in the oil bath.

   Determine the melting points of the two acetanilide camphor mixtures and of camphor itself in capillary tubes using an electrical melting point apparatus; note the temperature at which the last crystal of each sample disappears. Slow raising of the temperature is essential in the region of the melting point.

Treatment of results
   1. Calculate K_f from the naphthalene experiment, if this is necessary.
   2. From the melting point difference in the acetanilide experiment, calculate the molecular weight using both results.

Mass spectrometry

Precise molecular weights may now be determined by mass spectrometry, as described in Chapter 12.

Refractometry

Index of refraction

When a ray of light passes from one medium (1), Fig. 2.6, into another medium (2) it undergoes refraction. The ray travels at a lower velocity in the optically more dense medium (2) than in medium (1) which is less optically dense. The angle of incidence (i) and angle of refraction (r) are related by Snell's Law where n is the refractive index of medium (2) relative to medium (1), and the subscripts indicate the direction of the ray. The refractive indices of liquids are nearly always referred to that of air.

$$\frac{\sin i}{\sin r} = {}_1n_2 \tag{23}$$

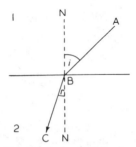

Fig. 2.6. Index of refraction

The critical angle is used extensively in refractometry. For a narrow cone of rays, a−b, placed close to the boundary between media (1) and (2) (Fig. 2.7), and observed at C, a band of light is seen. This band has a sharp edge at b, where the actual ray which is incident along the surface of medium (2) (b−b) is observed. No rays are present in the b−b′ region. Now

$$_1n_2 = \frac{\sin i}{\sin r} = \frac{\sin 90°}{\sin \theta} = \frac{1}{\sin \theta} \tag{24}$$

A measurement of the critical angle θ can therefore give the refractive index of medium (2).

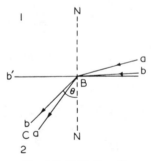

Fig. 2.7. Critical angle

Effect of wavelength

As refractive index generally decreases with increase of the wavelength of light used, a monochromatic source must be used for precision work. Alternatively the refractometer must be capable of compensating for the use of white light. The Cauchy formula

$$n = A + B/\lambda^2 + C/\lambda^4$$

where *A*, *B* and *C* are constants, expresses the variation of refractive index with wavelength when the medium does not absorb light. When the medium and the wavelength are such that absorption occurs, the variation of refractive index with wavelength is 'anomalous' (Fig. 2.8).

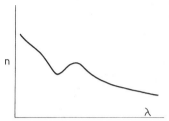

Fig. 2.8. Refractive index dispersion in the region of an absorption band

Effect of temperature

The refractive index of a liquid decreases as temperature increases, the percentage decrease ranging from 0.01 to 0.1% per degree, depending on the nature of the liquid. Since refractive indices are easily measurable to within $\pm 2 \times 10^{-4}$, good temperature control is essential. The temperature of measurement is specified as a superscript numeral on n, and the wavelength of light used as a subscript capital, e.g. n_D^{20}, where D refers to the sodium D line ($\lambda = 589.3$ nm).

Specific refractive index increment

The refractive index of a solvent generally changes if a solute is dissolved in it. The specific refractive index increment (dn/dc) is the change in refractive index of a solvent with change of solute concentration

$$\frac{dn}{dc} = \frac{n_{solution} - n_{solvent}}{\text{concentration of solute}}$$

the concentration being expressed in $g\,ml^{-1}$, giving dn/dc in $ml\,g^{-1}$.

A knowledge of dn/dc permits the concentration of a solution to be determined by measuring its refractive index.

Molar refraction

Lorentz and Lorenz found that the specific refraction

$$[n] = \frac{n^2 - 1}{n^2 + 2} \cdot \frac{1}{\rho}$$

where ρ is the density of a liquid of refractive index n, was almost independent of temperature. The molar refraction:

$$R = \frac{n^2 - 1}{n^2 + 2} \cdot \frac{M}{\rho} \tag{25}$$

has the dimensions of molar volume, and is a property of the number and type of atoms and groupings in the molecule. This additivity arises from the property of polarisability of molecules when subjected to electromagnetic radiation. The incident radiation sets the electrons vibrating and induces dipoles in the molecule. The total polarisability of the molecule is related to the sum of all the induced dipoles, and hence to the number and type of constituent atoms. Polarisability is in turn related to refractive index.

Some atomic refractivities evaluated at the sodium D line are given in Table 2.3. When the empirical formula of a compound is known and its molar refraction measured, the additivity of values for atoms in the molecule can be a useful guide in structure determinations.

Table 2.3 Atomic refractivities

C	2.418	O (− OH)	1.525	Br	8.748
H	1.100	O (− OR)	1.643	N (− NH_2)	2.322
C (C = C)	1.733	O (= O)	2.211	N (− NHR)	2.502
C (C ≡ C)	2.398	Cl	5.967	N (− NR_2)	2.840

Refractive index of water

Some values for the refractive index of water relative to dry air at 760 mm mercury pressure are given in Table 2.4.

Table 2.4 Refractive indices of water at various temperatures

15°	20°	25°	30°
1.33339	1.33299	1.33250	1.33194

Practical experiments

Experiment 9 *Determination of refractive index using the Abbé refractometer*

The optical system of the Abbé refractometer, illustrated in Fig. 2.9, employs the critical angle principle. The liquid under examination is placed between the two prisms. The top face of the lower prism is ground, to diffuse the light rays in all directions. Rays passing from the liquid to the upper prism (which must have a higher refractive index than the liquid) may be refracted in the normal way, giving a bright field in the eyepiece. Rays which strike the liquid/glass interface at grazing incidence give rise to the critical ray. These effects result in the field of view containing a dark and a light area, with a sharp dividing line.

The optical diagram for the upper prism is drawn out in Fig. 2.10. Let

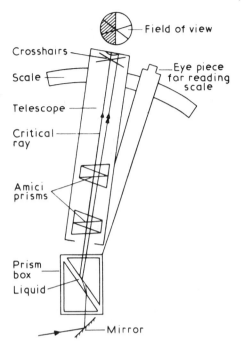

Fig. 2.9. Optical system of the Abbé refractometer

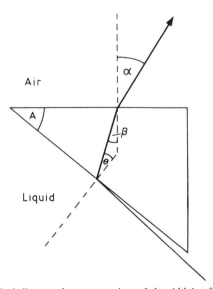

Fig. 2.10. Optical diagram for upper prism of the Abbé refractometer

the liquid have a refractive index n (air to liquid) and let N be the refractive index of the prism compared to air,

then,

$$N = \sin \alpha / \sin \beta$$

$$\sin \theta = {}_{glass}n_{liquid} = n/N$$

and

$$A = \beta + \theta$$

Now

$$n = N \sin (A - \beta)$$
$$= N \sin A \cos \beta - N \cos A \sin \beta$$

As

$$\sin \alpha = N \sin \beta$$

$$\cos \beta = \sqrt{\left\{1 - \frac{\sin^2\alpha}{N^2}\right\}}$$

Substituting:

$$n = N \sin A \sqrt{\left\{1 - \frac{\sin^2 \alpha}{N^2}\right\}} - \cos A \sin \alpha$$

$$n = \{\sin A \sqrt{(N^2 - \sin^2 \alpha)}\} - \sin \alpha \cos A \qquad (26)$$

Equation 26 gives the refractive index of the liquid relative to air in terms of two constants for a particular prism (A and N), and a measurable angle, α. The Abbé refractometer measures the angle α between the critical ray emergent from the upper side of the prism and the normal. Using the constants A and N this angle has been converted directly into refractive index, which is printed on the calibrated scale of the instrument.

The telescope of the instrument is fixed, and the prism box is attached to the scale. The prism box is rotated until the critical ray falls on the cross hairs of the telescope. The value of n at this setting is read from the scale.

Calibration of the instrument may be checked by using standard liquids of known refractive index as given in the European Pharmacopoeia (Table 2.5).

Table 2.5 Refractive indices of European Pharmacopoeia Reference Liquids

Standard liquid	n_D^{20}
Trimethylpentane	1.3915
Carbon tetrachloride	1.4603
Toluene	1.4969
α-Methylnaphthylamine	1.6176

Method For accurate measurements attach the prism box to a thermostat bath regulated at 25°. Open the prism box, and place a few drops of liquid on the lower prism; close the box. Adjust the mirror to give a bright illumination of the field. Turn the knurled knob until the field has a light and a dark section. If there is a coloured fringe between the two areas, adjust the Amici prisms until the boundary is sharp and black, set it on the cross hairs, and note the reading of refractive index. The Amici prisms deviate all wavelengths except those of the sodium D lines, and may also be used to determine the dispersion of the liquid.

Open the prism box, and wipe off the liquid with cotton wool moistened with acetone. Use first *n*-octane, and then *n*-octene as the liquid, and determine the refractive index of each.

Treatment of results

1. Octane has $M = 114.2$, $\rho_{25} = 0.699\,\mathrm{g\,ml}^{-1}$, and octene has $M = 112.2$, $\rho_{25} = 0.715\,\mathrm{g\,ml}^{-1}$. Calculate the molar refraction of each liquid, using the observed refractive indices.

2. Compare these values with those calculated from the atomic refractivities on p.00, and note the contribution of the double bond.

Note A simple immersion model of the Abbé refractometer is available consisting of the upper prism fixed to a telescope, and a scale. The prism is dipped in a beaker of liquid, the scale reading observed, and converted to refractive index using tables.

Experiment 10   *Determination of critical micelle concentration by refractometry*

Substances which form micelles in water generally have two distinct regions in their molecules, the hydrophobic portion (hydrocarbon chain) and the hydrophilic portion (polar group). The formation of micelles from a number of monomers places all the hydrocarbon chains together in the centre of the micelle, thus lowering the free energy of the system. The concentration at which micelles are first detectable is known as the critical micelle concentration (CMC). In general, the physical properties of solutions of soaps (or of other substances forming micelles) undergo sharp changes at the CMC. A plot of refractive index against concentration should show a change in slope at the CMC.

Method Prepare 200 ml of a 25% w/v solution of butyric acid in *water*. Accurately prepare 2.5, 5, 7.5, 10, 15 and 20% solutions of butyric acid in water by measuring appropriate volumes from a burette into 50 ml volumetric flasks, and making up to volume with *water*. Measure the refractive index of each solution, and of the original 25% solution at 25°, using the Abbé refractometer. Both dilutions and measurements must be very carefully made. Measure also the refractive index of water.

Treatment of results Plot refractive index as ordinate and concentration as abscissa. Two straight lines should be obtained, intersecting at the CMC.

Note Butyric acid forms micelles in aqueous solution in a concentration region suitable for measurements with the Abbé refractomerer. The majority of soaps form micelles at low concentrations, and the precision of the Abbé is insufficient to detect the CMC. In these cases an interference refractometer, such as the Rayleigh, must be used.

Optical activity

The electric fields associated with a beam of monochromatic light vibrate in all directions perpendicular to the direction of propagation of the light. Certain crystalline materials such as Iceland spar have different refractive indices for light whose field vibrates parallel or perpendicular to the principal plane of the crystal. As a result a Nicol prism constructed of this material transmits only light whose electric field oscillates in one plane (Fig. 2.11). Optical activity concerns the interaction of such plane-polarised light with certain materials, particularly solutions of some organic compounds.

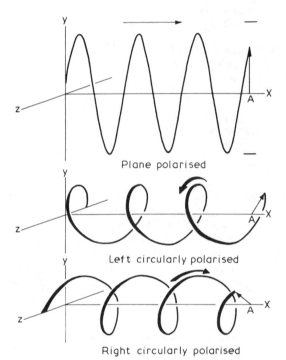

Plane polarised

Left circularly polarised

Right circularly polarised

Fig. 2.11. The electric field of plane-polarised light and its two circularly polarised components. The electric field of the plane-polarised beam at point A is the vector sum of the electric fields of the circularly polarised beams (indicated by arrows)

The electric field of plane-polarised light can be considered to be composed of two components of fixed magnitude rotating in opposite sense to one another (left circularly polarised and right circularly polarised) and the plane-polarised beam is the vector sum of these components (Fig. 2.11).

When plane-polarised light passes through a medium it is retarded to an extent which is indicated by the refractive index of the medium. When the latter is optically inactive both circularly polarised components are retarded to the same extent and the beam emerges from the medium polarised in the same plane as the incident beam. If the medium is optically active the components are retarded to different extents because the refractive indices of the medium for left circularly polarised light (n_L) and right circularly polarised light (n_R) differ. As a result of this circular birefringence $(\Delta n = n_L - n_R)$ the beam emerges from the medium still plane-polarised but with the plane of polarisation inclined at an angle α degrees to the plane of polarisation of the incident beam, given by:

$$\alpha = \frac{1800}{\lambda} l \, \Delta n \qquad (27)$$

where l is the light-path in dm and λ the wavelength in cm. α is the optical rotation of the medium and is positive when the plane of polarisation is rotated clockwise relative to that of the incident beam when viewed looking towards the light source (dextrorotation) and negative when rotation is anticlockwise (laevorotation).

If the wavelength of the light is such that the medium absorbs a fraction of the radiation, an optically active medium may show a second physical effect arising from unequal absorption of left and right circularly polarised light. As a result of this circular dichroism the beam emerges from the medium elliptically polarised. The ellipticity, ψ degrees, of the emergent beam is given by:

$$\psi = \frac{1800}{\lambda} l \, \Delta \kappa \qquad (28)$$

where l and λ have the same meaning as in eq. 27, for optical rotation, and $\Delta \kappa = \kappa_L - \kappa_R$, the difference between the absorption indices* for left and right circularly polarised light. Using the more familiar absorbance, A (Chapter 7):

$$\psi = \frac{2.303 \times 1800}{4\pi} \Delta A \qquad (29)$$

Any medium that shows circular dichroism must at the same time show circular birefringence and hence optical rotation. Both result from unequal interaction of the medium with left and right circularly polarised light and hence are closely related phenomena.

A more detailed account of the relationship between optical rotation and circular dichroism is given by Foss (1963).

---

*κ is defined by the equation:

$$I = I_0 e^{-4\pi\kappa/\lambda}$$

where I and I_0 are the intensities of the transmitted and incident light

$$\therefore \quad \kappa = \frac{2.303\lambda}{4\pi l} A$$

Effect of concentration, solvent and temperature on optical activity

The dependence of optical rotation and circular dichroism of a solution of an optically active substance on concentration may be taken into account by calculating the specific optical rotation and ellipticity or the molecular rotation and ellipticity as shown in Table 2.6.

Table 2.6 Dependence of optical rotation and circular dichroism on concentration

Specific rotation	Specific ellipticity
$[\alpha]_{\lambda}^{t} = \dfrac{\alpha_{\lambda}^{t}}{lc}$	$[\psi]_{\lambda}^{t} = \dfrac{\psi_{\lambda}^{t}}{lc}$
units $-$ deg cm^2 decagram^{-1}	

Molecular rotation	Molecular ellipticity
$[\Phi]_{\lambda}^{t} = \dfrac{M[\alpha]_{\lambda}^{t}}{100}$	$[\Theta]_{\lambda}^{t} = \dfrac{M[\psi]_{\lambda}^{t}}{100}$
units $-$ deg cm^2 dmol^{-1}	

α_{λ}^{t} = optical rotation $\left.\right\}$ in degrees at temperature $t°$C
ψ_{λ}^{t} = ellipticity $\quad$ and wavelength λ nm
$\quad c$ = concentration, g cm^{-3}, of solute
$\quad l$ = light-path, dm
$\quad M$ = molecular weight of solute

It may also be shown that

$$[\Theta] = 3300 \, \Delta\epsilon$$

where $\Delta\epsilon = \epsilon_{L} - \epsilon_{R}$, the difference between the molar-absorptivities of left and right circularly polarised light.

Measurements of optical rotation are frequently made with sodium D light and usually, though not necessarily, at 20°, and specific rotations based on such measurements are reported as $[\alpha]_{D}^{20}$. Specific rotation is generally concentration-dependent and in dilute solution an equation of the type

$$[\alpha] = A + Bc + Cc^2 \qquad (30)$$

can be used to describe results, A, B and C being constants. It is necessary always to state the concentration at which specific rotation was measured unless a procedure for extrapolating $[\alpha]$ to zero concentration and quoting $[\alpha]_{c = 0}$ is followed.

Temperature may have a pronounced effect on specific rotation, and suitable temperature control is required for precise work. Effects may arise due to changes of intermolecular interactions with temperature, or to changes in equilibria between configurations. Tartaric acid provides an example of the latter effect, the variation in $[\alpha]$ being about 10%/°C, owing to the equilibrium between two forms with different optical rotatory power varying with temperature.

The particular solvent used can have a large effect on the results, e.g.

a 20% w/w solution of nicotine in chloroform has $[\alpha]_D^{20} \simeq + 4°$, while the same concentration in water gives $[\alpha]_D^{20} \simeq + 10°$. Chloramphenicol gives a change in sign, for example $[\alpha]_{25}^D = + 19°$ in ethanol changes to $- 25°$ in ethyl acetate. These effects, which can be very large, indicate that a statement of the solvent used must always accompany any report of $[\alpha]$.

For pure liquids $[\alpha]_D^{20} = \alpha/ld^{20}$ where $d^{20} =$ relative density of the substance.

Optical Rotary Dispersion (ORD) and Circular Dichroism (CD) spectra

Information about the structure of organic compounds and their optical purity can be gained from measurements of optical rotation at a single wavelength. More detailed and precise information can be gained by recording the variation of α with wavelength (ORD spectra) or of ψ with wavelength (CD spectra).

The ORD, CD and absorption spectra of D-camphor-10-sulphonic acid (Fig. 2.12) illustrate many of the general features of optical activity. At wavelengths well clear of the absorption bands of the carbonyl group ($\lambda > 450$ nm) $[\Phi]$ increases in magnitude as λ decreases and, as in many other examples, the spectrum can be fitted by an equation of the form

$$[\Phi] = \frac{K}{\lambda^2 - \lambda_0^2} \quad \text{(one-term Drude equation)} \quad (31)$$

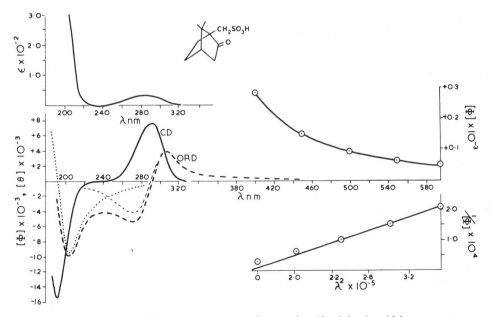

Fig. 2.12. Absorption, ORD and CD spectra of D-camphor-10-sulphonic acid in water at 27°

where K and λ_0 are constants, as shown by the linearity of the $1/[\Phi]$ *vs* λ^2 plot. For some substances more complicated Drude equations are required. The ORD spectra of many optically active compounds are of this 'plain' type which may be positive, as in this case, or negative. One obvious advantage in analytical work of measuring α at wavelengths lower than the sodium D line (589.6 nm) is the gain in sensitivity; the molecular rotation of D-camphor-10-sulphonic acid at 370 nm is about ten times its value at 589.6 nm.

In the region of the absorption band (λ_{max} = 287 nm) D-camphor-10-sulphonic acid shows circular dichroism, the shape of the CD spectrum between 240 nm and 320 nm being similar to that of the absorption spectrum. The ORD spectrum in this region takes a sigmoid course (*cf* the variation of refractive index with wavelength, Fig. 2.8) with a distinct peak and trough (extrema) at 306 nm and 270 nm and change of sign at 290 nm. This is known as a Cotton effect and by convention, when the sign of the CD is positive and the higher wavelength extremum of the ORD spectrum is positive, the Cotton effect itself is said to be positive.

Below 240 nm D-camphor-10-sulphonic acid has a second optically active absorption band ($\lambda_{max} \simeq 190$ nm) but this time the Cotton effect is negative. The CD and the higher wavelength extremum of this part of the ORD spectrum (204 nm) are both negative (the lower wavelength extremum is inaccessible).

In this relatively simple example the two absorption bands associated with the single carbonyl chromophore are well resolved, as are the corresponding CD peaks. Circular dichroism and optical rotation both result from the unequal interaction of left and right circularly polarised light with chromophores in chiral molecules and are closely related phenomena. The ORD spectrum corresponding to a particular CD band can be calculated from the CD spectrum by means of a Kronig–Kramers transform:

$$[\Phi]_\lambda = \frac{2}{\pi} \int_0^\infty [\Theta]_{\lambda'} \frac{\lambda'}{\lambda^2 - \lambda'^2}\, d\lambda' \qquad (32)$$

where $[\Phi]_\lambda$ is the molecular rotation at wavelength λ, $[\Theta]_{\lambda'}$ is the molecular ellipticity at wavelength λ', and λ and λ' are the main variable and parameter of integration respectively. The dotted lines in Fig. 2.12 are the approximate contributions to the ORD spectrum calculated from the two CD bands of D-camphor-10-sulphonic acid. The ORD Cotton effects are incompletely resolved since they spread on either side of the corresponding CD bands. In more complex molecules, possibly with several optically active chromophores, the higher resolving power of CD is advantageous in determining the contribution of each chromophore to optical activity. On the other hand, many optically active compounds have absorption bands at such low wavelengths that their solutions show no CD in the accessible spectral region; the ORD Cotton effects associated with these absorption bands extend into the accessible spectral region so that the optical activity of such substances can be studied by means of optical rotation. CD and ORD are thus complementary techniques.

More detailed descriptions of the relationship between ORD and CD spectra and the stereochemistry of organic compounds are given in a number of reviews (e.g. Snatzke, 1967; Crabbé, 1971).

Instrumentation

(a) Visual polarimetry.

The optical system for a simple visual polarimeter is shown in Fig. 2.13.

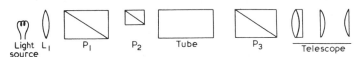

Fig. 2.13. Polarimeter optical system

The light source is usually a sodium vapour lamp whose light is made parallel by lens L_1 and polarised by the Nicol prism, P_1. In the absence of optically active material in the sample tube a second Nicol prism, P_3, can be rotated about the axis of the instrument so that the principal planes of P_1 and P_3 are mutually at right angles; as a result no light is transmitted by P_3 and the field of view in the telescope is dark — P_1 and P_3 are then said to be crossed. If now an optically active sample is placed in the tube, light will once more emerge from P_3. Extinction can be restored by rotating P_3 through an angle, measured on a calibrated circular scale, which will be equal in magnitude and sign to the rotation of the sample. It is difficult to determine precisely at which angular position of P_3 the illumination of the field is at a minimum because the transmitted intensity change per unit angular rotation is very small near the extinction point. A third prism, P_2 (or a half wave plate) is therefore inserted to cover half the field. This causes two plane-polarised rays to pass through the tube, differing in phase by half a wavelength.

If the analysing prism is crossed with the unobstructed part of the field, the part is darkened, as in A, Fig. 2.14, but some light is present in the obstructed half. Rotation of the analyser in the correct direction

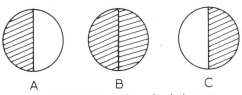

A   B   C

Fig. 2.14. Fields of view of polarimeter

gives a field as in B, in which both beams are almost completely extinguished, and both areas of the field are of equal intensity. Further rotation gives C, in which light from the obstructed half of the field is eliminated, but some rays from the unobstructed half pass through. This device of obstructing half the field gives a means of finding the balance point (B), by judging when the illumination of the two areas is of equal intensity.

(b) Photoelectric spectropolarimeters.

The light source is usually a xenon arc lamp which emits continuously between 190 nm and 700 nm. The light passes through a monochromator and the monochromatic beam then passes through the polariser, sample cell and analyser (see Fig. 2.15) as in the visual polarimeter.

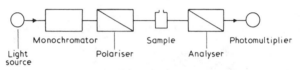

Fig. 2.15. Diagram of the optical components of a spectropolarimeter

Two methods have been used to determine the position of the analyser which gives extinction, both employing a photomultiplier to detect the transmitted light. In the simpler of these the polariser is caused mechanically to oscillate through a small angle (say 2°). The plane of polarisation of the beam passing through the sample is thus modulated through the same angle. To position the analyser at its extinction angle it is rotated until the intensity of the light reaching the photomultiplier (and hence the photocurrent) is the same for both extreme positions of the polariser.

In the second class of instruments the modulation of the plane of polarisation is achieved not mechanically but electrically by interposing in the light path between the polariser and the analyser a Faraday cell. This consists of a silica cylinder surrounded by a coil through which is passed an alternating current (typically 60 Hz). The magnetic field of the coil induces in the silica optical rotation whose sign depends on the direction of the current through the coil. The plane of polarisation of the beam which leaves the Faraday cell is thus modulated. When the analyser is not in its extinction position this causes an asymmetric a.c. current in the photomultiplier which is detected and used to drive the analyser to its extinction position. ORD spectra may be recorded automatically.

(c) Circular dichroism.

One of several methods for measuring circular dichroism is shown diagrammatically in Fig. 2.16. A beam of plane-polarised light is passed

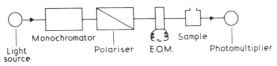

Fig. 2.16. Diagram of the optical components of a circular dichrometer

through an electro-optic modulator (EOM) or Pockels cell. This is a 0°-Z-cut plate of potassium dideuterium phosphate oriented so that the incident light is parallel to either its X or Y crystallographic axes. When an alternating electric field (typically 325 Hz) is applied to transparent conducting films on the opposing Z-surfaces of the plate it transmits alternately the left and the right circularly polarised components of the plane-polarised beam. When these are passed through a circularly dichroic sample, the photocurrent from the photomultiplier contains a steady component proportional to the average transmittance of the sample and a component, alternating with the frequency of EOM voltage, proportional to the difference in sample transmittance for left and right circularly polarised light. The electrical system processes the alternating component so as to record the circular dichroism of the sample.

More detailed descriptions of ORD and CD instrumentation are given in several reviews (e.g. Chignell and Chignell, 1972).

Practical experiments

Experiment 11 *Determination of concentration dependence of specific rotation*

Method Prepare accurately 25 ml each of solutions of 1, 2, 3, 4 and 5% w/v quinidine sulphate in 0.05 M sulphuric acid. Check the polarimeter to see that the field is evenly illuminated (otherwise adjust the light source), and that the telescope is focussed on the dividing line of the field of view. Determine the zero of the instrument as follows.

Clean the polarimeter tube, A (Fig. 2.17) and screw on one end plate; set the thread gently finger tight otherwise some polarisation may arise due to strain in the glass.

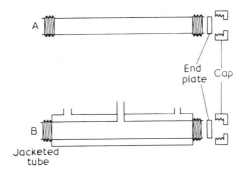

Fig. 2.17. Two types of polarimeter tube

Holding the capped end downwards, fill the tube with 0.05 M sulphuric acid and screw up the second cap. Dry the outside of the tube, and clean the end plates with lens tissue. Set the tube in the instrument, so that any air bubbles lie in the enlargement at the side of the tube. [Some tubes have no accommodation for bubbles. Fill these with one cap in place until a meniscus is built up at the open end of the tube. Slide the glass end plate artfully through the meniscus and screw on the cap.]

Rotate the analyser until the two areas of the field are illuminated with equal intensity, and read both verniers. Take three readings of the balance point approaching from one side, and three approaching from the other. Rest the eyes now and again, otherwise some sensitivity is lost. Calculate the mean readings, i.e. one mean of approximately 0°, and one of approximately 180°.

Repeat this procedure with each of the five solutions of quinidine sulphate in 0.05 M H_2SO_4 in turn. Note also the sign of the rotation.

Treatment of results

1. Correct the observed rotations of the solutions for the rotation of the solvent, and calculate $[\alpha]_D$ for each solution.
2. Plot $[\alpha]_D$ against concentration, and note the shape of the graph. Use the general equation (26) to obtain a straight line graph, and find the value of $[\alpha]_D$ at zero concentration.

Experiment 12   *The optical activity of Levodopa*

The specific rotation of Levodopa in the visible region is low ($[\alpha]_D^{20} = - 12°$ in M hydrochloric acid). Assay of optical purity is facilitated by formation of a complex with hexamine which has enhanced optical rotation; this is the basis of the BP method for control of optical rotation in Levodopa.

Method Prepare a 2.5% (w/v) solution of Levodopa in M hydrochloric acid. Also prepare a 0.8% (w/v) solution of Levodopa–hexamine complex by dissolving 0.2 g Levodopa and 5 g hexamine in 10 ml M hydrochloric acid, making up to 25 ml with M hydrochloric acid and allowing to stand for 3 h protected from light. Using a photoelectric polarimeter, determine the ORD spectra of both solutions between 600 nm and 350 nm.

Comment on the relative magnitudes of the spectra in relation to assay of optical purity. Does your material meet the BP specification?

Dilute the 2.5% solution of Levodopa to 0.01% with M hydrochloric acid and the 0.8% solution of the complex to 0.008% with M hydrochloric acid. Record the absorption, ORD and CD spectra of the two dilutions between 350 nm and 230 nm.

Comment on the relationship between the three spectra for each solution. What advantages might be gained by utilising observations in the lower spectral region for assay of optical purity of Levodopa?

Experiment 13   *Polarimetric study of the transformation of the gelatin helix in solution*

In many macromolecules optical activity stems, not only from asymmetric carbon atoms but also from coiling the polypeptide chain into a helix, which may be right or left handed and optically asymmetric. The helix is generally stabilised by hydrogen bonding. On heating, increased thermal agitations can break down the hydrogen bonds, giving a change in optical rotation as the helical form is destroyed.

Method Prepare a 5% w/v solution of gelatin in *water*. Cool the solution, and pour it into a jacketed polarimeter tube, B (Fig. 2.17). Leave the tube in a refrigerator overnight.

Set up a thermostat bath with an external circulating device, e.g. 'Circotherm', adjacent to the polarimeter. Fill the thermostat bath with cold tap water and also sufficient ice to make the temperature 10°. Place the jacketed tube in the polarimeter, and circulate the cold water through it. When temperature equilibrium has been reached (after about 20 min) take readings of the rotation as described in Experiment 11. Raise the temperature of the bath by five degree intervals up to 30°, and measure the rotation at each temperature. Take a final reading at 40°. The zero of the tube should strictly be measured at each temperature, but it is sufficiently accurate to determine a single zero reading at 25°, and use this for correcting the other measurements.

Treatment of results Calculate the specific rotation at each temperature, and plot a graph of $[\alpha]_D$ against temperature. The curve should show a decrease in specific rotation as temperature rises. Comment on the changes in slope.

With a spectropolarimeter this experiment can be done with more dilute gelatin solutions since the specific rotation of gelatin increases markedly as the wavelength is reduced. Thus $[\alpha]_{313}$ and $[\alpha]_{208}$ are respectively about $7 \times$ and $100 \times [\alpha]_D$. Whereas a 5% solution is required to give reasonable results, using a 0.5 dm tube in a visual polarimeter, as described above, a 0.1% solution is adequate with a 1 cm cell in a spectropolarimeter at 313 nm.

ORD and CD are used extensively in studying the conformation of protein molecules in solution.

Further applications

The recent commercial availability of precise circular dichrometers has made it possible to exploit the higher resolving power of circular dichroism in the assay of optical purity of some compounds whose structure is such that there is little difference between the ORD spectra of their enantiomers. One example is the determination of the proportion of the two diastereoisomers in *Phenethicillin Potassium* (Stenlake *et al.*, 1972). The strong association of some small optically inactive organic molecules with chiral macromolecules has been shown to induce optical activity in the small molecule. Such effects have proved useful in studying interactions of drugs with plasma proteins (Chignell and Chignell, 1972; El-Gamal *et al.*, 1983; Fehske and Muller, 1981; Wood and Stewart, 1971).

Viscosity

The internal friction of liquids, due to intermolecular attractions, is known as viscosity. In a flowing liquid each layer of molecules exerts a drag on the next and, to cause the liquid to flow, work must be done to push the layers past one another. The lines in Fig. 2.18 represent the moving layers, in contact with one another over an area A; the velocity gradient is du/dx, x being chosen at right angles to the direction of flow. Newton showed that the applied force, F, was proportional to A and to

du/dx, the proportionality constant being the coefficient of viscosity, η; hence,

$$F = \eta \frac{du}{dx} A \qquad (33)$$

Liquids which obey this equation when flowing are said to give streamlined or Newtonian flow. Very rapid flow can cause Newtonian flow to become turbulent flow, in which the energy causing the liquid to flow is no longer used exclusively to slide the planes of molecules over one another, but part is dissipated as eddies and turbulence.

x Direction

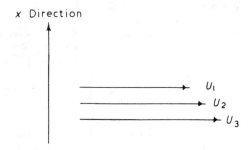

Fig. 2.18. Velocity gradient in fluid flow. (Horizontal arrows represent velocities of layers of liquid.)

Poiseuille's Law.   The capillary viscometer

When a liquid flows in a tube (Fig. 2.19) the layer adjacent to the wall is stationary, whilst that in the centre flows fastest. The flow can be considered to consist of concentric rings of molecules moving over one another. The velocity of flow, v, at a distance, r, from the centre of the tube of radius, a, is given by

$$v = \frac{P}{4\eta l}(a^2 - r^2)$$

in which P is the pressure difference maintained between the ends of the tube, of length, l. Integration of this equation to find the total volume, V, flowing in unit time through the tube gives Poiseuille's law

$$V = \frac{Pa^4}{8\eta l} \qquad (34)$$

Fig. 2.19. Flow of liquid in a tube. (length of horizontal lines represent velocities of flow)

For a liquid flowing under its own pressure head, the pressure may be expressed in terms of density, ρ, acceleration due to gravity, g, and height; remembering V was defined per unit time, we can rewrite eq. (34) as:

$$\eta = k\rho t \tag{35}$$

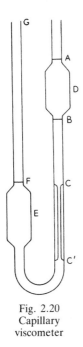

This equation applies to the capillary viscometer, which is shown in Fig. 2.20. The simple Ostwald viscometer has the form of a U-tube, with bulbs at D and E, a capillary tube C–C' and marks at A, B and F. Liquid is forced up bulb D to above mark A, and the time taken for the meniscus to fall from A to B noted. If a liquid (1) of known viscosity is compared with another (2) of unknown viscosity, in the same apparatus under identical experimental conditions, then:

Known liquid: $\eta_1 = k\rho_1 t_1$
Unknown liquid: $\eta_2 = k\rho_2 t_2$

$$\therefore \quad \eta_2 = \frac{\eta_1 \rho_2 t_2}{\rho_1 t_1} \tag{36}$$

Fig. 2.20
Capillary
viscometer

The constant (k) for a particular viscometer need not be determined, as it cancels (eq. (36)). To obtain readily measured flow times, viscometers with narrow capillary tubes are used for liquids of low viscosity, while wider capillary tubes are used in viscometers for high viscosity liquids.

Units of viscosity

The SI unit of viscosity (dynamic viscosity) is $1\,N\,m^{-2}\,s = 1\,Pa\,s$ (pascal second), but dynamic viscosity is often expressed in cgs units, i.e. 1 poise $= 1\,g\,cm^{-1}\,s^{-1} = 10^{-1}\,N\,m^{-2}\,s$. For water at $20°$, $\eta = 0.01002$ poise $= 1.002$ centipoise. Kinematic viscosity, v, is dynamic viscosity divided by density. The SI unit is $m^2\,s^{-1}$ but again it is frequently expressed in cgs units, i.e., 1 stoke $= 1\,cm^2\,s^{-1} = 10^{-4}\,m^2\,s^{-1}$. For water at $20°$, $v = 0.01004$ stokes $= 1.004$ centistokes. Another term sometimes used is the fluidity, which is the reciprocal of dynamic viscosity.

Other viscosity functions which are used, particularly in studying macromolecules in solution, are:

$$\text{relative viscosity} = \eta_{\text{rel}} = \frac{\eta}{\eta_0} = \frac{t\rho}{t_0\rho_0}$$
$$\text{(viscosity ratio)}$$

$$\simeq \frac{t}{t_0} \text{ (dilute solutions)}$$

$$\text{specific viscosity} = \eta_{\text{sp}} = \eta_{\text{rel}} - 1$$

$$\text{reduced specific viscosity} = \frac{\eta_{\text{sp}}}{c}$$
$$\text{(viscosity number)}$$

$$\text{intrinsic viscosity} = [\eta] = \lim_{c \to 0}\left(\frac{\eta_{\text{sp}}}{c}\right) \tag{37}$$

where η and ρ are the viscosity and density of a solution of concentration $c\,\text{g cm}^{-3}$ and η_0 and ρ_0 are the viscosity and density of the solvent. $[\eta]$ is a function of the size and shape of macromolecules in solution.

Viscosity and temperature

Viscosity decreases by about 2% per degree rise in temperature. Strict control of temperature during measurements is essential. The relationship between viscosity and temperature is expressed by an equation of the form:

$$\eta = A\,e^{B/RT} \tag{38}$$

A and B being constants for a particular liquid. A plot of log η against $1/T$ should therefore give a straight line.

The Couette viscometer

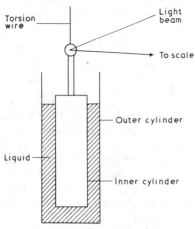

With non-Newtonian fluids, whose viscosity depends on shear rate, capillary viscometers can give misleading results. In the Couette viscometer the liquid is contained in the annular space between two coaxial cylinders (Fig. 2.21) and the rotational force on the inner cylinder is measured when the outer one is rotated at known angular velocity, ω. By making measurements at different values of ω the viscosity of the liquid may be determined at different shear rates.

Fig. 2.21. Scheme of Couette viscometer

The falling ball viscometer

By allowing a solid sphere to fall through a liquid, and measuring its limiting velocity, v, the gravitational force pulling the sphere down is equal to $mg = \frac{4}{3}\pi r^3(\rho_2 - \rho_1)g$, where r is the radius of the sphere and $(\rho_2 - \rho_1)$ the density difference between sphere and liquid. The gravitational force is balanced by the viscous drag, which, from Stokes' Law, is $6\pi\eta rv$. Equating and solving for η, gives the expression:

$$\eta = \frac{2(\rho_2 - \rho_1)r^2 g}{9v} \qquad (39)$$

Thus a measurement of the limiting velocity can be used to determine the viscosity of the liquid, η.

Provided that the velocity of fall is not too rapid, the main theoretical difficulty is to account for wall effects. In narrow tubes additional resistance effects are present, and v is smaller than expected. This makes the apparent viscosity calculated from eq. (39) too high. A number of equations are available for correction for wall effects: they generally involve terms in $2r/D$ where D is the diameter of the tube. Such a correction is given in the Pharmacopoeial formula for the falling ball viscometer which is derived from eq. (39) using the diameter of the sphere in place of the radius, and expressing the result as the kinematic viscosity (see p.41) in centistokes.

$$v = \frac{d^2(\rho_2 - \rho_1)g \times 0.867}{0.18v\rho_1}$$

Practical experiments

Experiment 14 *Calibration and use of capillary viscometers*

(a) *Determination of the intrinsic viscosity of dextran in Dextran Intravenous Infusions*

Method Clean and dry a size A Ostwald viscometer (capillary bore $\simeq 0.5$ mm). Prepare dilutions containing 2.0, 1.0, 0.5 and 0.25% w/v of dextran from the Dextran Infusion provided. Using a plumb line, set the viscometer upright in a thermostat set at 37°. Fill the viscometer through tube G with saline solution to mark F (Fig. 2.20). When the apparatus has come to temperature, adjust the liquid level exactly to this mark. Slip a short length of clean rubber tubing onto A and apply suction until the liquid rises above mark at A. Place the finger on the end of this tube to prevent the liquid level falling while removing the rubber tubing. Find the time taken for the liquid level to fall between marks A and B. Repeat until results agreeing to 0.1 s are obtained. Remove the viscometer from the bath, wash it out, dry it and repeat the procedure using each of the dextran solutions in turn. Calculate the specific viscosity of each solution and by plotting η_{sp}/c *vs* c (eq. (37)) determine the intrinsic viscosity of the dextran. Compare it with the BP requirement for the product examined.

(b) *Determination of viscosity of Methylcellulose*

Method Prepare a 2% solution of Methylcellulose in *water* by weighing 2.0 g into a 250 ml conical flask (containing a magnetic follower) fitted with a ground glass stopper, adding

100 ml *water* heated to 85–90°, inserting the stopper, and stirring for 10 min. Place in ice and continue stirring until the solution is of uniform consistency. Allow the solution to attain room temperature.

The suspended level viscometer (Fig. 2.22) is used in this experiment. Introduce sufficient liquid through tube A to ensure that the bulb D will be full. Set the viscometer in the thermostat at 20°. After temperature equilibrium has been attained, fit a clean piece of rubber tubing onto B. Close the tube C with the finger and suck liquid up above mark E. Close B by 'nipping' the rubber tubing, open C, quickly detach the tubing, and time the fall of the meniscus between marks E and F. This procedure leaves a hanging column of liquid in BG. This type of capillary viscometer has the advantages that precise filling is not required, errors in setting the viscometer exactly upright are minimised, and dilutions may be made within the viscometer.

Determine reproducible flow times for the methylcellulose solution and for a solution of glycerol of known concentration (between 90–95% w/w). If necessary determine the density at 20° of the glycerol solution used for calibration.

Treatment of results Find the viscosity of the glycerol solution used from the data in Table 2.7.

Table 2.7 Viscosities and densities of some glycerol solutions

% glycerol w/w	90	91	92	93	94	95
Viscosity (cp, 20°)	234.6	278.4	328.4	387.7	457.7	545
ρ_{20}, g/ml	1.2347	1.2374	1.2401	1.2428	1.2455	1.2482

(A convenient means of interpolation is from a graph of $\log \eta$ against ρ_{20}.) Calculate both the absolute and kinematic viscosities of the methylcellulose solution, using $\rho_{20} = 1.005$ g ml^{-1} for methyl cellulose solution.

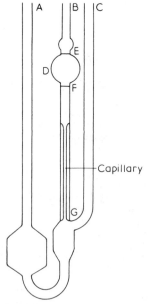

Fig. 2.22. Suspended
level viscometer

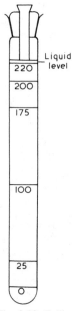

Fig. 2.23. Falling
ball viscometer.

Experiment 15 *The falling ball viscometer—wall effects*

A convenient form of the falling ball viscometer is shown in Fig. 2.23, having a tube graduated at lengths of 0, 25, 100, 175, 200 and 220 mm.

Method Fill the tube with glycerin (>99.5% w/w). Place the tube in a thermostat at 20°, and allow 20–30 min for temperature equilibrium. Use $\frac{1}{16}$ inch (1.59 mm) diameter steel spheres, pre-wetted with the liquid whose viscosity is to be determined, also at 20°. Introduce a sphere into the funnel at the top of the apparatus, and time its fall between the 200 mm and 0 mm marks. Repeat until consistent results are obtained. As the limiting velocity, v, is used in the equation for determining viscosity by this method, it should be ascertained that this velocity has been obtained, by timing spheres between the 200 mm and 100 mm marks, then between the 100 mm and 0 mm marks. Repeat the experiment with $\frac{1}{8}$ inch diameter (3.176 mm) spheres. Measure the internal diameter of the fall tube; for the BP apparatus this is 2.5 cm.

Repeat the whole experiment using a large measuring cylinder (500 ml, internal diameter 4–5 cm) fitted with a large rubber bung and a funnel. Calibrate the cylinder to allow a 20 cm fall to be used in the first part of the experiment.

The falling ball method is used to determine the viscosity of a pyroxylin solution, prepared as BP using 1.59 mm diameter balls in the apparatus used in the first part of the experiment, and timing the fall between 175 mm and 25 mm marks. The viscosity in centistokes is calculated from the basic formula (39) with the substitution of diameter for radius, and the inclusion of a correction factor for wall effects.

Treatment of results
1. Calculate the apparent viscosities of the glycerin from eq. (39) for the four experiments (two sizes of sphere in two cylinders). Note that the largest ball in the smallest cylinder gives the highest viscosity (take $\rho_2 - \rho_1 = 6.49$ g ml^{-1}).
2. Plot the apparent viscosity against the ratio $2r/D$ and extrapolate to $2r/D = 0$. This is a hypothetical case where the diameter of the cylinder is infinite and wall effects are negligible. Hence the correct viscosity should be obtained from the intercept on the viscosity axis. The apparent viscosity may be multiplied by the correction factor F,

$$F = 1 - 2.104d/D + 2.09d^3/D^3$$

in which d is the diameter of the sphere, and D that of the cylinder. Calculate F for each of the four cases mentioned above, and calculate the viscosity in each case.

Surface tension

In the bulk of a liquid, a molecule is, on average, subjected to attractions in all directions. At a liquid surface, a molecule is not completely surrounded by others, and is subject to a resultant attraction acting inwards towards the bulk of the liquid. Because of this inward force, the surface has a tendency to contract. To increase the area of the surface, work must be done against the inwardly directed attractive forces.

Surface energy is the work required for unit increase in surface area.

Surface tension (γ) is the force acting at right angles to a line of unit length, present in the surface. As work is force × distance, extension of surface area by 1 m^2 can be considered as moving a line of length 1 m

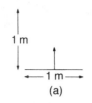

Fig. 2.24. Diagrammatic representation of extension of surface area by $1\,m^2$

through a distance of $1\,m$ (Fig. 2.24). The surface energy therefore has the same numerical value as the surface tension.

Following similar arguments, the work necessary to extend the interface between two immiscible liquids by unit amount is the interfacial energy, and the interfacial tension is the force acting at this interface. Surface and interfacial tension are important in pharmacy in studies of surfactants, the properties of emulsions, and to provide information on liquid surfaces. The SI unit of surface tension is $N\,m^{-1}$, but values are often given in $mN\,m^{-1}$.

Effect of temperature

In general both surface and interfacial tension decrease with a rise in temperature. The surface tension of water at various temperatures is given in Table 2.8.

Table 2.8 Surface tension of water at various temperatures

$t^\circ C$	0.20	4.99	10.02	20.00	40.20	60.00
γ $(mN\,m^{-1})$	75.66	74.96	74.26	72.79	69.57	66.23

The surface tension decreases by approximately $0.15\,mN\,m^{-1}\,deg^{-1}$.

Effect of solute

While it is difficult to generalise on the effect of solutes on surface or interfacial tension, two classes of substances give specific behaviour. Inorganic salts, like sodium chloride or calcium chloride, in fairly high concentrations, raise the surface tension of water. This is probably due to the surface layer being of pure water, and there being a lack of solute in it. Gibbs' equation relates the change of surface tension, $d\gamma$, to the surface excess concentration of solute, Γ_i, and change in chemical potential, $d\mu_i$, so for a solute species i

$$-\,d\gamma = \Gamma_i d\mu_i$$

If the surface tension increases with concentration, Γ_i is negative, i.e. the solute has a smaller concentration in the surface than in the bulk of the solution.

The second important class of substances is the surfactants. These are strongly adsorbed at surfaces and decrease the surface tension. For a dilute solution of a non-electrolyte surfactant, giving a fairly high adsorption at the surface, Gibbs' equation reduces to

$$-\,d\gamma = \Gamma_2 2.303RT\,d\log C_2 \tag{40}$$

where Γ_2 and C_2 are the surface excess and bulk concentrations of solute respectively, concentrations having been substituted for activities. This form of equation is also applicable to ionised surfactants in salt solutions, but for ionised surfactants in pure water the additional factor 2 is required on the right hand side.

The progressive addition of a surfactant to water lowers the surface tension (A to B, Fig. 2.25), the solute distributing itself between the bulk and surface of the solution. At B, the solute molecules in the bulk begin to aggregate to form micelles, and the addition of further solute (B to C) has little effect on the surface tension. The surface excess, Γ_2, may be determined from the slope of the γ *vs* $\log C_2$ curve in the A–B region; the concentration at which micelles first form (critical micelle concentration = CMC) can be found from the break in the curve at point B.

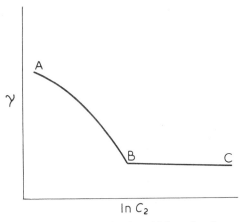

Fig. 2.25. Change in surface tension by addition of surfactant to water

Application of Gibbs' equation is one of the most important methods of studying adsorption of soluble solutes at liquid surfaces.

Dynamic and static surface tension

Often when measurements are made of the surface tension of a surfactant solution, the values obtained change with time, owing to slow adsorption at the surface or the presence of impurities. These values are dynamic surface tensions, while that obtained after full equilibrium of surface and bulk phases, which no longer changes with time, is the static value.

In pharmacy, there is more interest in the surface tension of solutions than that of pure liquids. Hence a method like capillary rise, which is excellent for pure liquids but not for solutions, is not considered. The Wilhelmy plate method, and to a lesser extent the drop weight method, can be used for dynamic as well as static measurements.

Practical experiments

Experiment 16   *Surface tension of water by the Wilhelmy Plate Method*

When a thin platinum plate dips vertically into the surface of a liquid, the surface tension pulls the plate downwards, acting all round its perimeter. If l is the perimeter of the plate, the downward pull due to surface tension = $l\gamma$. To counteract this downward pull an upward force of mg must be applied

$$l\gamma = mg$$

or

$$\gamma = mg\, l^{-1} \qquad (41)$$

To avoid having to apply buoyancy corrections for the volume of plate submerged in the liquid, a technique is used which ensures that the bottom of the plate just touches the surface. The apparatus is shown in Fig. 2.26.

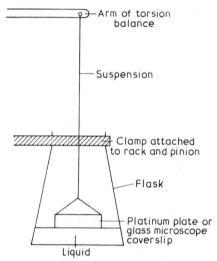

Fig. 2.26. Wilhelmy plate apparatus for determination of surface tension

Method Clean the depolished platinum plate by dipping it in chromic acid for 2–3 min, and washing it in six changes of water (*Note* 1) in a beaker. Dry the plate with filter paper. Do not touch it with the fingers; it can be lifted by means of the nylon suspension. Attach the suspension to the arm of the (0–500 mg) torsion balance, unclamp the beam, set the pointer to a reading of zero mg, and adjust the zero control until the indicating pointer comes to the zero mark. Clamp the beam, ensuring that it remains at the zero position.

Clean a 250 ml Erlenmayer flask with chromic acid, wash well with water, and place about 50 ml water in the flask. Clamp the flask to the rack and pinion device, and lift it until the plate hangs about 0.5 cm above the surface of the water. Rack up the flask very slowly. This is a convenient time to ensure that the bottom of the plate is parallel to the surface, by observing the degree of parallelism of the bottom of the plate and its image in the surface. If it is not parallel, remove it from the flask, gently bend the wire part of the suspension. Continue racking up the flask until the plate just touches the surface. Unclamp the balance arm.

Rotate the handle of the balance, until the indicating pointer is exactly on the zero mark. Note the reading in mg. Turn the handle back to zero mg, so that the plate sinks in the water, and obtain two agreeing readings for the pull. By this method the plate is always raised in the surface to take a reading, which ensures that the contact angle between water and plate is correct (0°). At the end of the experiment measure the temperature of the water.

Treatment of results Calculate (*Note* 2) the surface tension of water from eq. (41).

Note 1 Water must be freshly redistilled from permanganate solution, and stored in a well seasoned glass container.
Note 2 If the perimeter of the plate is not given, it must be measured with a cathetometer.

Experiment 17   *Determination of area/molecule and CMC from surface tension measurements*

Fill a burette with a 5% solution of the nonionic detergent polysorbate 80 ('Tween' 80) in 0.1 M potassium chloride. Prepare 50 ml of the following dilutions: 2%, 0.1%, 0.05%, 0.005%, 0.0005% and 0.0001%, making the flasks up to volume with 0.1 M potassium chloride solution.

Clean the plate and flask as in Experiment 16. Rinse out the flask with two 5 ml portions of the most dilute solution, place the remainder in the flask, and measure its surface tension. With some samples of polysorbate 80, changes of surface tension with time may be noted. Continue to take readings until a constant value is obtained within 2 mg. Clean the flask and plate as before, and measure the surface tension of each solution.

Treatment of results
1. Calculate the equlibrium surface tension of each solution from eq. (41), and plot γ against log (% concentration).
2. Determine the CMC.
3. Determine $d\gamma/d \log C$ from the slope of the graph in the concentration region just below the CMC. Taking $R = 8.31 \text{ J mol}^{-1} \text{ deg}^{-1}$, calculate Γ_2 from eq. (40) (mol cm^{-2}). Calculate the area (nm^2) per molecule.

Experiment 18   *Determination of the surface tension of water by the drop weight method*

A hanging drop of the liquid is slowly formed on a circular glass tip. The force causing the drop to adhere to the tip is:

$$\text{circumference} \times \text{surface tension} = 2\pi r \gamma$$

where r is the radius of the tip. The force causing the drop to fall is *mg*, where m is the weight of the drop. At the instant when the drop falls:

$$2\pi r \gamma = mg$$

$$\gamma = \frac{mg}{2\pi r}$$

This simple theory is complicated by the fact that, during detachment, some of the drop remains on the tip, so the simple formula is modified by the introduction of a correction factor, F.

$$\gamma = \frac{mgF}{r} \tag{42}$$

(2π has been taken into the correction factor.) F was originally determined by measuring the drop weight of liquids whose surface tensions were known, falling from tips of known radius. F is found by calculating the volume of the drop, V, from its weight, working out V/r^3, and reading off F from Table 2.9.

Table 2.9 Correction factors for the drop weight method (Harkins and Brown)

V/r^3	4.65	3.98	3.43	3.00	2.64	2.24
F	0.254	0.257	0.259	0.261	0.262	0.264
V/r^3	2.09	1.88	1.55	1.31	1.21	1.12
F	0.265	0.265	0.266	0.265	0.264	0.263

Method The apparatus is shown in Fig. 2.27. The polythene tubing connecting the reservoir and tip must be well extracted with water. The end of the tip must be ground flat, and its radius determined with a cathetometer.

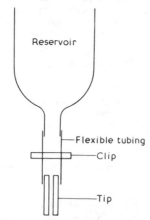

Fig. 2.27. Drop weight apparatus for determination of surface tension

Immerse the tip in chromic acid for 2–3 min. Wash down the outside of the tip, and run about 50 ml *water* through the apparatus. Close the clip, and put about 20 ml *water* into the reservoir. Adjust the clip until drops are falling at about 2 min intervals. Collect three drops in a tared weighing bottle, and re-weigh. Repeat for a further three drops. Avoid vibrations of the apparatus while drops are being collected. At the end of the experiment note the temperature of the water in the reservoir.

Treatment of results Using the density data in Table 2.1, and the mean drop weight of each set collected, calculate V/r^3. Calculate the surface tension from eq. (42) and Table 2.9.

References

Chignell, C.F. and Chignell, D.A. (1972) *Methods in Pharmacology*, ed. C.F. Chignell, Appleton–Century-Crofts, New York, Vol. 2, p.111.

Crabbe, P. (1971) *Determination of Organic Structures by Physical Methods*, ed. F.C. Nachod and J.J. Zuckerman, Academic Press, New York, p.134.

El-Gamal, S., Wollert, U. and Muller, W.E. (1983) *J.Pharm.Sci.* **72**, 202.

Fehske, K.J. and Muller, W.E. (1981) *J.Pharm.Sci.* **70**, 549

Foss, J.G. (1963) *J.Chem.Ed.* **40**, 592

Snatzke, G. (1967) *ORD and CD in Organic Chemistry*, Heyden.

Stenlake, J.B., Wood, G.C., Mital, H.C. and Stewart, S. (1972) *Analyst* **97**, 639

Wood, G.C. and Stewart, S. (1971) *J.Pharm.Pharmac.* **23**, 248S.

3
Analysis of drugs and excipients in the solid state

E. G. SALOLE

Introduction

Modern concepts of quality differ considerably from those of a decade or two ago. We are now concerned not only with chemical purity but also with those other characteristics of pharmaceutical materials which may influence safety, efficacy, formulation and processing of medicines. Many physico-chemical properties of drugs in their solid state have thus assumed importance.

The gastrointestinal absorption of a drug from a solid dosage form involves the following steps:

$$\text{drug in dosage form} \xrightarrow{\text{rate of dissolution}} \text{drug in solution in gastrointestinal lumen} \xrightarrow{\text{rate of absorption}} \text{drug in blood}$$

If the rate of dissolution is slow then this may become the rate-limiting step in the absorption process. Since the rate of dissolution of a solid is directly proportional to exposed surface area, the particle size of a poorly soluble drug may influence the rate and extent of absorption and this may be reflected in differences in clinical response or toxicity. The rate of dissolution may also be determined by the crystal form of the drug.

A complete profile of the solid state properties of a drug would include determination of particle size distribution, melting point, rate of dissolution, equilibrium solubility and crystal properties, the range of parameters determined for each drug depending on its subsequent formulation or use. This chapter describes some of the techniques used to assess the particle size and crystal properties of drugs and excipients.

Particle size analysis

Particle size is an important property, affecting characteristics as diverse as the flavour and 'mouth feel' of chocolate to the setting qualities of cement; in pharmaceutical systems it influences the flow and packing properties of powder mixtures, the colour and covering capacity of pigments and drug bioavailability. For example, inhalation aerosols fail to penetrate into the bronchioles if larger than about 5 μm and poorly soluble drugs like griseofulvin need to be finely divided. Some examples of official requirements and descriptions of drug particle size are listed in Table 3.1.

Table 3.1 Some official requirements for particle size

Substance or preparation	Remarks
Dispersible Aspirin Tablets	Aspirin in *fine powder* form.
Cascara Liquid Extract	The dried bark for percolation is in *coarse powder* form.
Cortisone Tablets	*Fine powder* to be used for preparation of tablets.
Cortisone Injection	Crystalline particles rarely exceeding 30 μm in length.
Griseofulvin Tablets	Particle size determined from disintegrated tablet generally up to 5 μm in maximum dimension with larger particles occasionally exceeding 30 μm.
Hydrocortisone Ointment	Hydrocortisone incorporated as *very fine powder*.
Biphasic Insulin Injection	Majority of particles rhombohedral crystals with maximum dimension greater than 10 μm but rarely exceeding 40 μm.
Isophane Insulin Injection	Rod-shaped crystals, the majority not less than 5 μm and rarely exceeding 60 μm, free from large aggregates.
Methisazone Mixture	Weight median diameter of particles in suspension not greater than 15 μm as measured by electrical zone sensing.
Liquid Paraffin and Phenolphthalein Emulsion	Microcrystalline phenolphthalein has not more than an occasional particle with a diameter greater than 15 μm.
Sodium Cromoglycate Insufflation	The capsule contents immediately after dispersion in pentan-1-ol by exposure for 20 s to low intensity ultrasonic waves exhibit two types of particles: small rounded particles not greater than 10 μm, often present as loose agglomerates up to 30 μm; the contents of capsules also containing lactose also exhibit larger angular particles, usually axehead in shape, of length 10–150 μm but mostly within the range 20–80 μm.

The following terms are used *inter alia* in the description of powders in the British Pharmacopoeia:
Coarse Powder: a powder of which all the particles pass through a 170 μm sieve and not more than 40% pass through a 355 μm sieve.
Fine Powder: a powder of which all the particles pass through a 180 μm sieve.
Very Fine Powder: a powder of which all the particles pass through a 125 μm sieve.

Concepts of particle size: the equivalent sphere

The notion of 'particle size' is not as straightforward as might at first appear. Consider the geometrically regular parallelepiped in Fig. 3.1a:

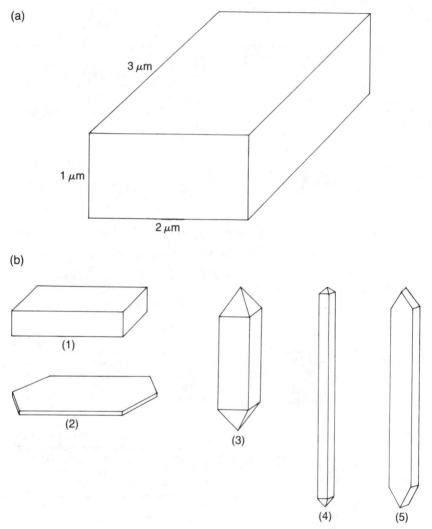

Fig. 3.1. (a) An idealised particle. (b) Different *habits* of crystalline particles: (1) tabular, (2) platy, (3) prismatic, (4) acicular, (5) bladed

clearly no one linear dimension characterises its size. Real particles are even more difficult to define because they adopt a variety of shapes (Fig. 3.1b), which are rarely geometrically regular or bound by smooth surfaces. To circumvent this difficulty the concept of an **equivalent**

sphere is utilised because, uniquely, size parameters such as surface area and volume are defined by the diameter; essentially the 'size' allocated to a particle is the diameter of a hypothetical sphere which exhibits the same measurable property as the particle. Since different techniques of particle size analysis measure different parameters, a number of **equivalent sphere diameters** are possible:

Sieve diameter (d_s):	the nominal sieve aperture width through which the particle just passes.
Projected area diameter (d_p):	diameter of a sphere of projected area equal to the projected area of the particle resting in its most stable position.
Stokes' diameter (d_{St}):	diameter of a sphere of equal density having the same settling velocity as the particle in a fluid medium within the range of Stokes' Law.
Volume diameter (d_v):	the diameter of a sphere of equivalent volume.
Volume-surface diameter (d_{vs}):	the diameter of a sphere with the same ratio of volume to surface area.

These of course differ numerically from each other, the discrepancy increasing the more anisodiametric the particle; e.g. for the particle in Fig. 3.1a:

$$d_s = 2\,\mu m$$
$$d_p = 2.76\,\mu m$$
$$d_{St} = 2.08\,\mu m \ (\rho = 1\,\mathrm{g\,cm^{-3}}),$$
$$d_v = 2.25\,\mu m$$

It is important to remember therefore that the validity of particle size data is linked to the particular technique used, which if possible should be chosen in view of the end-use of the data; e.g. d_{St} may be more appropriate for a powder destined for formulation as a suspension. Semi-empirical mathematical procedures are available for the interconversion of equivalent sphere diameters.

Size distribution

Consider a sample of powder examined under a microscope from which 500 particles are sized by comparing them with some suitable scale so that the particles can be grouped into classes of different sizes. The number of particles lying in each interval of $2\,\mu m$ is shown in Table 3.2. The particles in this particular example all lie between 2 and $22\,\mu m$. The data can be expressed in the form of a histogram (Fig. 3.2) where the abscissa represents the particle size interval and the ordinate the frequency per interval. However, the histogram does not provide a unique pattern for a given particle size distribution, since its shape varies if the scale of particle size intervals is changed.

Table 3.2 Size distribution in a sample of 500 particles

Size interval in μm	2–4	4–6	6–8	8–10	10–12	12–14	14–16	16–18	18–20	20–22
Number of particles	25	88	107	110	55	45	30	18	12	10
% Frequency in each interval	5	17.6	21.4	22	11	9	6	3.6	2.4	2
% Cumulative undersize	5	22.6	44	66	77	86	92	95.6	98	100

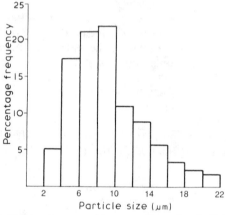

Fig. 3.2. Number-frequency histogram (from data in Table 3.2)

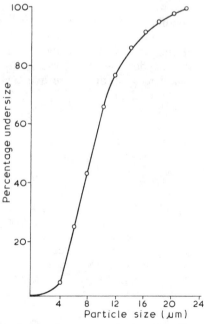

Fig. 3.3. Cumulative number-frequency undersize curve (from data in Table 3.2)

An alternative method is to express the data in the form of a cumulative distribution curve where the cumulative percentage larger (oversize) or smaller (undersize) is plotted against the particle size as shown in Fig. 3.3.

The example just discussed was a size distribution based on numbers of particles. In sieving and sedimentation methods the distribution is calculated on a weight basis. If it is necessary to convert from a size distribution by number to one by weight, the conversion is usually made by assuming that all the particles have the same shape and density.

Mean size of a particulate system

The concept of *mean size* requires some explanation since it may be calculated on the basis of numbers of particles, weight of particles or surface area of particles. This acquires importance in correlation problems.

The simplest average diameter is the arithmetic mean diameter $\bar{d}_n$ which we define for two diameters d_1 and d_2 as

$$\bar{d}_n = (d_1 + d_2)/2$$

In general terms we define the **length-number mean diameter** by

$$\bar{d}_{ln} = \frac{\Sigma \, n \cdot d}{\Sigma \, n} \tag{1}$$

where n is the number of particles with diameter d. In many pharmaceutical applications the surface area of the particles is important and the **volume-surface mean diameter** can be employed. This is defined as

$$\bar{d}_{vs} = \frac{\Sigma \, nd^3}{\Sigma \, nd^2} \tag{2}$$

because the surface area is proportional to nd^2. The significance of volume-surface mean diameter $\bar{d}_{vs}$ is that the **specific surface area** (surface area per unit weight) can be calculated from it by the equation

$$S = \frac{\Sigma \, n \, \pi d^2}{\Sigma \, n \, (\pi/6)d^3\sigma} = \frac{6}{\bar{d}_{vs}\sigma} \tag{3}$$

where σ is the density of the solid. This equation is strictly correct only for spherical or cubical particles but may be used without correction for shape if the particles are not too asymmetrical.

If the weight of each fraction, rather than number or surface area, is noted then the **weight-moment mean diameter** is obtained from

$$\bar{d}_{wm} = \frac{\Sigma \, nd^4}{\Sigma \, nd^3} \tag{4}$$

This quantity emphasises the larger particles in a sample.

It is important that the terminology of mean diameters used above is strictly adhered to. For example, the weight-moment mean diameter is not the same as the diameter of a particle of mean weight. The latter quantity is termed the **mean weight diameter** and is equal to

$$\sqrt[3]{\left(\frac{\Sigma\ nd^3}{\Sigma\ n}\right)} \tag{5}$$

a quantity used when one is concerned with the number of particles per unit weight of material.

From the data in Table 3.2 the following mean diameters are obtained:

$$\overline{d}_{\text{ln}} = 9.28\ \mu\text{m}$$

$$\overline{d}_{\text{vs}} = 12.9\ \mu\text{m}$$

$$\overline{d}_{\text{wm}} = 14.3\ \mu\text{m}$$

Specific surface area = $186.0\ \text{m}^2\,\text{kg}^{-1}$ ($\sigma = 2.5 \times 10^3\ \text{kg}\,\text{m}^{-3}$).

Methods of particle size analysis

1. Sieving

Sieves are generally used for grading coarser powders although they are capable of separating fractions as fine as $45\ \mu\text{m}$. Such fine powders, however, may clog the sieve apertures and *wet sieving* methods may be necessary. The lower limit of $45\ \mu\text{m}$ for dry sieving may be reduced to about $20\ \mu\text{m}$ if special sieves prepared by the electro-deposition of metals are used in conjunction with an air flow through the sieves (as in the Alpine Air Jet Sieve).

Specifications for test sieves appear in British Standards, to which reference should be made for further details. To obtain reproducible particle size analyses the detailed procedure laid down in British Standard 410 (1962) should be carefully adhered to. The powder is passed through a number of sieves of successively smaller mesh size and the weight remaining on each sieve is determined. The method of shaking the sieves is important to obtain rapid sieving and it is usual to use a mechanical shaker that imparts gyratory and vibratory movement to spread the material over the whole mesh. Errors can arise from overloading the sieves and not allowing sufficient time for the passage of particles.

2. Microscope method

Direct observations of the particles would appear to be very reliable, but unless the technique of measurement is carefully standardised considerable errors may be introduced. The sources of error include sampling, technique of slide preparation and choice of diameter. Under the microscope a particle is seen as a projected area whose dimensions depend on the orientation of the particle on the slide. The equivalent diameter that is frequently used is the diameter of a circle whose area is equal to the projected area of the particle. The diameter is estimated by using a graticule placed in the microscope eyepiece (Fig. 3.8). British Standard 3406: Part 4 (1963) describes in detail a graticule technique. Less tedious methods involve commercially available instruments fitted with image-shearing eyepieces and fully automatic quantitative image analysers.

3. Sedimentation methods

These methods, which have been critically reviewed (Analytical Methods Committee, 1968), are based on the free settling of particles individually dispersed in a suitable fluid. A sphere of diameter d_{St}, falling slowly in a fluid, reaches a terminal velocity given by Stokes' Law:

$$v = \frac{d_{St}^2 g(\sigma - \rho)}{18\eta} \tag{6}$$

where v = terminal velocity, g = acceleration due to gravity, σ = density of the solid, ρ = density of the dispersion medium, and η = dynamic viscosity of the dispersion medium.

The equation is only applicable to streamline motion of particles, i.e. Reynolds Number (Re), calculated from the equation $d\rho v/\eta$ = Re, must not exceed 0.2. For large dense particles it may be necessary to increase either the density or the viscosity of the suspending fluid to maintain streamline flow. This limits the maximum size of particle to about 60 μm for a solid of density 3×10^3 kg m^{-3} settling in water.

To determine the size distribution, in 'homogeneous' techniques the powder is dispersed uniformly through the fluid and at suitable time intervals the concentration of powder at a given depth in the suspension is determined. Sampling can be done with a suitable pipette, e.g. the Andreasen pipette, and from the weight of solid in the sample the percentage of undersize particles can be calculated. Pipette sampling is an *incremental* method and gives a direct measure of the particle concentration. In *cumulative* methods, e.g. with a sedimentation balance, the proportion of material falling on a balance pan as a function of time must be subjected to a differentiation process in order to calculate the percentage of oversize material. To avoid particle-particle interaction the concentration of solid in the suspension should not be more than about 1% w/v.

4. Electrical sensing zone method   (The Coulter counter)

The instrument is shown diagrammatically in Fig. 3.4. As a suspension of particles in electrolyte flows through a small aperture having an immersed electrode on either side, the passage of each particle displaces electrolyte within the sensing zone, momentarily changing the resistance between the electrodes and producing a voltage pulse of magnitude proportional to particle volume. When the controlled external vacuum is applied, the mercury column in the manometer is displaced (Fig. 3.4) and electrolyte is drawn through the orifice. Closure of the stopcock permits the mercury to rebalance, thus drawing the sample suspension through the aperture and permitting a count to be taken on a known volume of suspension by actuation of the start/stop-count electrodes.

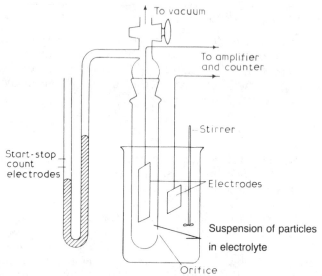

Fig. 3.4. Schematic diagram of the Coulter counter

The series of pulses so obtained are amplified and fed to a threshold circuit which discriminates against particles below, and permits counting of those which exceed, the preset threshold value. Successive counts at different threshold levels give the data for determining the cumulative frequency oversize curve.

Orifice tubes are available with diameters ranging from $30 \mu m$ to $2000 \mu m$. Each aperture is able to measure particles of equivalent diameters between 2% and 60% of its stipulated diameter. The instrument has to be calibrated by using a reference powder of known particle size. It is advisable to pre-saturate the electrolyte for accurate analysis of even 'insoluble' drugs; soluble powders may be examined in suspension in non-aqueous electrolytes, e.g. 4% w/v ammonium thiocyanate in propan-2-ol.

5. Optical sensing zone and light diffraction methods

Optical sensing zone instruments operate by detecting the attenuation of a collimated beam (light blockage) produced by particles in suspension as they tumble, in single-file, through an illuminated sensing zone (Fig. 3.5). Electronic pulses equivalent to the maximum projected areas of particles are recorded and analysed to provide size distribution data.

The method can be considered the optical analogue of the resistive pulse technique, with the advantages of not requiring special dispersing media (transparent particles can be suspended in liquid with a different refractive index), being unintrusive, capable of handling high volume throughput and therefore being readily adaptable to on-line analysis. Instruments operating on this principle are also available for assessment of atmospheric contamination in **clean room** areas.

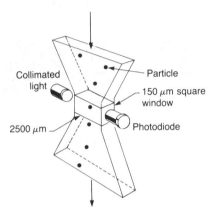

Fig. 3.5. Schematic diagram of the HIAC CMB-150 optical sensing zone particle sizer (courtesy of HIAC/Royco Instruments Corporation)

When a monochromatic beam impinges on an opaque particle some of the light is scattered (reflected, refracted and diffracted in all directions) to a degree partly dependent on the effective cross-sectional area of the particle. If the particle size is much larger than the incident wavelength (roughly, $d > 4\lambda$), then the fundamental principles of **Fraunhofer diffraction theory** can be applied to analysis of the pattern resulting from the forward-direction, low-angle scatter. The operational characteristics of a commercial particle sizer are illustrated in Fig. 3.6. As particles (solid or liquid droplets) cross the expanded and collimated beam from a low power (2 mW) laser, light is diffracted (at angles inversely proportional to particle size, and energy directly proportional to the number of particles and their cross-sectional area) and focussed on a multi-element detector placed at the focal plane of the lens. Since the detected light energy distribution has maxima at radii specific for particular particle sizes, rapid scanning and computer analysis of the output from the radially distributed detectors on the multi-element detector provide size distribution data within seconds (for a $0.6328\,\mu m$ source on commercial instruments the size limit is reduced to $1\,\mu m$ by including correction factors). The 'Fraunhofer Diffraction equivalent sphere diameter' is probably similar to an orientation-averaged d_p. Diffraction instruments possess all the advantages of light blockage methods and are more versatile. As they are based on fundamental optical principles, calibration is unnecessary, and aerosol sprays can be analysed because (since the detector lies in the focal plane of the lens) the focussed diffraction from equivalent particles is superimposed irrespective of their relative motion or position in space.

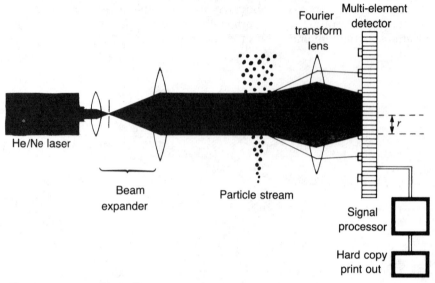

Fig. 3.6. Schematic diagram of the Malvern particle and droplet sizer (courtesy of Malvern Instruments)

Sampling procedures

As with all methods of analysis it is essential that the sample taken is truly representative of the bulk material under examination. In the case of aerosol formulations, for example, the sample taken when the can is full may not be identical to that taken when the can is nearly exhausted. Sampling procedures to obtain unbiased samples for sizing are described in detail in B.S.3406: Part 1.

If very small samples are required, as for example in microscopical examination, the method of sampling is critical and the reader is referred to the precise procedure laid down in B.S.3406: Part 4 (1963). Some pharmaceutical preparations (e.g. aerosol inhalations, ointments and creams) require special methods of sampling. It is also essential that the sampling technique, or indeed measuring technique, does not alter particle size (as may occur with emulsions on dilution into the electrolyte required for the Coulter counter).

Scope of the methods

Most methods of particle size analysis have limitations in the range of sizes which they can cover. While sieving is capable of assessing powders over a wide range of sizes, other methods such as visible light microscopy are more limited. In Table 3.3 the approximate limits are quoted as a guide.

Table 3.3 Scope of some methods of particle size analysis

Method of analysis	Useful range (μm)
Sieving:	
B.S. sieves	75–1000
Electroformed sieves	20–100
Sedimentation:	
Gravitational (liquid medium)	3–200
Microscope (optical)	2–150
Coulter counter	0.6–800

Practical experiments

Experiment 1 *Size analysis of calcium carbonate by sedimentation using the Andreasen pipette*

Selection of suitable dispersion medium. The following tests can be applied to detect whether the suspension is deflocculated.

(*a*) *Rheological behaviour* Gradually add the dispersion medium to the powder on an ointment tile, and work the mass with a spatula to a pasty consistency.

A good dispersion flows off the spatula in long syrupy threads and has dilatant properties, i.e. stiffens up on being worked. A flocculated paste does not flow off the spatula even on tapping and the mass appears dull and pasty in contrast to the dilatant paste. This test is useful in making a preliminary selection of dispersing agent.

(*b*) *Microscopical examination* Place a small drop of dilute suspension on a microscope slide; carefully cover with a cover slip and examine at a suitable magnification.

A flocculated suspension will show clumping of the particles while a deflocculated suspension will show the particles individually dispersed.

(*c*) *Sedimentation volume* This method can be used to determine quantitatively the optimum amount of dispersing agent. Suspensions (V_o) are prepared using different amounts of dispersing agent and the sediment volume (V_s) is measured after the suspensions have been allowed to stand for a suitable time. The maximum sedimentation volume (V_s/V_o) is obtained with the best dispersing agent. Prepare 100 ml quantities of the following solutions in distilled water:

> Sodium pyrophosphate 10^{-2}, 10^{-3}, 10^{-4}, 10^{-5} M
> Cetrimide 2, 0.1, 0.01, 0.001% w/v

Use these solutions to prepare 2% w/v suspensions of calcium carbonate. Transfer the suspensions to 100 ml stoppered graduated cylinders. Shake the suspensions and then allow them to sediment for several hours or overnight. From the appearance of the suspensions determine the optimum concentration of dispersing agent. Confirm your results by applying tests (*a*) and (*b*).

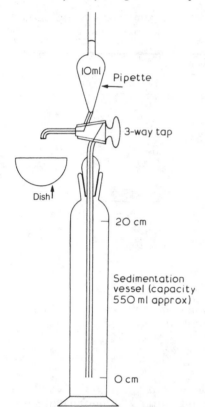

Andreasen pipette method The pipette is shown in Fig. 3.7. The tip of the pipette is 20 cm below the zero mark. Prepare 600 ml of sodium pyrophosphate solution of concentration found to be most effective in (c) above. Mix calcium carbonate (5 g) with a little of the solution to form a smooth paste in a mortar using a brush to break up agglomerated particles. Make the suspension up to 600 ml with more solution and transfer it to the Andreasen sedimentation vessel up to the 20 cm mark. Reserve the excess suspension for determination of dissolved carbonate. The temperature of the suspension should be kept constant for the whole period of the experiment, and, if the room temperature is likely to vary by more than a few degrees, the sedimentation vessel should be placed in a thermostated bath. Note the temperature of the suspension. Agitate the suspension with a plunger type stirrer, replace the pipette head and start a stopclock. Immediately take the first sample by sucking the suspension into the pipette and then discharge it into a conical flask. Rinse the pipette with a little water and add the rinsings to the flask. Take subsequent samples at 2, 4, 8, 16, 32, 64, 128 and 256 min. Take further samples the following day if necessary. Determine the amount of calcium carbonate in each sample by adding 0.2 M hydrochloric acid (10 ml) to each flask, boil to remove carbon dioxide, and back titrate with 0.1 M sodium hydroxide (methyl orange as indicator). Owing to the slight water solubility of the carbonate, a blank should be carried out on the aqueous phase after removal of the solid by centrifuging. The excess suspension previously reserved can be used for this. The titration figure for the first sample is proportional to the initial powder weight whilst the subsequent samples give the cumulative weight undersize corresponding to the values of the Stokes' diameters.

Fig. 3.7. The Andreasen pipette

Calculation of Stokes' diameters These diameters are calculated from the Stokes' equation most conveniently expressed as follows from eq. (6):

$$d_{St} = 17490 \left\{ \frac{h\eta}{(\sigma - \rho)t} \right\}^{\frac{1}{2}} \tag{9}$$

where d_{St} = Stokes' diameter (μm), η = viscosity of the dispersion medium (water = 0.001 kg m^{-1}s^{-1} at 20°C), σ = density of the solid (kg m^{-3}), ρ = density of the dispersion medium (water = 998 kg m^{-3} at 20°C), h = mean depth at which sample is taken (cm), and t = time of sedimentation for each sample (min).

Since the level of the suspension falls as samples are taken, a mean depth is used in the above calculation. For instance if the level falls 0.6 cm on removing the first sample, the mean depth of sampling is 19.7 cm. The mean depths for the subsequent samples are 19.1, 18.5 cm and so on.

Determination of density of solid Introduce about 10 g of calcium carbonate into a 25 ml density bottle. Weigh the bottle and contents together with stopper. Add sufficient dispersing fluid so that the bottle is about half filled. Place the bottle without stopper in a vacuum desiccator and gradually evacuate it. When no further air is released from the powder, add sufficient dispersing fluid to fill the bottle. Replace the stopper, dry and weigh. If a liquid other than water is used its density must also be determined. The density of the powder is calculated from the formula:

$$\sigma = \frac{\rho(W_2 - W_1)}{V\rho - (W_3 - W_2)} \tag{10}$$

where W_1 = weight of dry density bottle, W_2 = weight of density bottle + powder, W_3 = weight of density bottle + suspension, ρ = density of dispersion medium, and V = volume of density bottle.

Results Plot the percentage of solid in each sample against the corresponding Stokes' diameter to obtain a cumulative curve (undersize), and construct a histogram to show the size frequency distribution. An approximate value for the specific surface area of the powder can be obtained from the histogram assuming that the particles are cubical. The surface area contributed by the particles in each block of the histogram is given by $6.10^4 f/\sigma d$ where f is the percentage by weight of the particles in each block, d is their mean size and σ is the density. The specific surface $(m^2 kg^{-1})$ is obtained by the summation of the surface areas so obtained.

Experiment 2 *Determination of size distribution by microscopy*

As an introductory exercise it is suggested that a fractionated sample of powder is examined, e.g. < 200 > 300 sieve mesh size (i.e., 53-75 μm), so that the sizing can be done at one magnification. Where the size range is large it is necessary to count fields of different sizes using appropriate magnifications to ensure that sufficient large particles have been counted in relation to the smaller ones. The 'diameters' of particles can be determined from their mean projected areas by the use of graticules mounted in the eyepiece of the microscope, so that particles can be matched up with a circle of known diameter on the calibrated graticule. If a projection microscope is available, the image of the graticule and the images of the particles can be viewed together. About 600 particles should be counted.

In the following exercise, a bench microscope fitted with an eyepiece graticule is used to measure the size distribution of a fractionated sample of lactose or magnesium carbonate powder.

Apparatus
 Bench microscope This should be provided with (i) coarse and fine focussing; (ii) focussing and centering substage condenser; (iii) an adjustable substage condenser diaphragm. A mechanical stage giving two graduated movements at right angles, each capable of being read to 0.1 mm by scale or micrometer screw, is desirable but not essential. It is convenient to have a graduated draw tube so that the magnification can be varied in order to match up with an exact distance on the stage micrometer (see B.S.3406: Part 4) but this is not essential. Eye piece ($\times$ 20) fitted with standard graticule; objective ($\times$ 4 or $\times$ 5).
 Standard graticule The graticule (Fig. 3.8) conforms to B.S.3625. The relative dimensions of the graticule are such that the reference circles

increase in geometrical progression with the constant ratio of $\sqrt{2}$. Two calibration marks are inscribed on the longer axis of the grid to enable easy calibration of the reference circles.

Stage micrometer The micrometer, 0.1 mm scale length, is subdivided into 100 μm and 10 μm divisions.

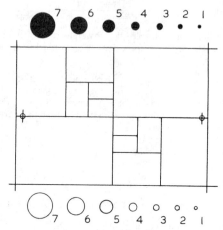

Fig. 3.8. Microscope eyepiece graticule (B.S.3625)
Distance between calibration marks = 60.4 units
Diameter of circles 1–7: 1.00, 1.41, 2.00, 2.83, 4.00, 5.66, 8.00 units

Method

(*a*) *Microscope* Adjust the condenser and light source to obtain even illumination of suitable intensity.

Place the stage micrometer on the microscope stand and align it with the longer axis of the grid on the graticule. Adjust the magnification by altering the tube length until there is an exact correspondence between the distance separating the calibration marks on the graticule and the length on the stage micrometer. For instance, if the distance between the calibration marks is 800 μm then the sizes of the reference circles are:

Circle	7	6	5	4	3	2	1
μm	106	75	53	37	26	19	13

(*b*) *Preparation of sample* Mix a little of the lactose sample with liquid paraffin on a watch glass using a glass rod. When satisfactory mixing has been achieved, a drop of the suspension should be removed with a dropping rod and transferred to a clean microscope slide. Gently lower a cover slip on the sample. The drop size should be such that no surplus liquid spreads outside the edges of the cover slip.

(*c*) *Counting procedure* The graticule is subdivided into four large rectangles. Count the particles within these areas. If the sample on the slide is examined over an area of 20 mm × 20 mm it is possible to examine say 25 fields by moving the slide 4 mm at a time by reference to the scales of the micrometer screws on the mechanical stage. The number of particles within each rectangle should on average not be more than 6.

A particle is recorded as belonging to the field area if it lies wholly within the boundary lines of the field area and also if it lies on either of two adjacent sides or the corner formed by these sides. Particles touching the other two sides or the other three corners do not belong to the field area and are not recorded.

Examine every particle present in each field area individually and mentally compare its area with the areas of the graticule circles. A particle whose area is estimated to be smaller than that of say circle 6 but larger than that of circle 5 is assigned to the class interval defined by the diameters of circles 5 and 6, and so on for other size classes.

(*d*) *Calculation of size distribution* The counts for each size class of particle have been observed over the same area of slide so that the percentage of number in each class can be calculated directly (B.S.3406: Part 4 gives the method of calculating if different magnifications have been used for different size classes).

(i) Draw a graph showing the percentage undersize by number as a function of size, as shown in Fig. 3.3.

(ii) Calculate the weight size distribution by converting the numbers of particles to weight, in each size range, by cubing the diameter of the particle *d* and multiplying by the number of particles *n*. Σnd^3 will be the relative total weight of the sample, and hence the percentage weight of each fraction will be obtained by $nd^3/\Sigma nd^3 \times 100$. This calculation assumes that the density and shape of particles are independent of size.

Experiment 3 *Measurement of the particle size of hydrocortisone in Hydrocortisone Suppositories*

Method Melt one suppository with the aid of gentle heat, dilute the melt in a suitable volume of liquid paraffin, and determine the particle size distribution as described in Experiment 2.

Experiment 4 *Measurement of the particle size of methisazone in Methisazone Mixture using the Coulter counter*

The following method is suitable for water-insoluble powders that disperse readily in aqueous solutions of sodium chloride. A small amount of surface active agent is generally added to ensure adequate dispersion. The approximate particle size range of the sample should be known since the size of the largest particle should not exceed 40% of the orifice diameter.

Method Prepare a 0.9% w/v sodium chloride solution containing 0.005% w/v polysorbate 80 as dispersant, and pre-saturate with methisazone. Transfer about 200 ml of the saturated solution, filtered through a 0.2 μm membrane immediately before use, to the sample beaker of a Coulter counter fitted with a 280 μm orifice tube. Add an amount of Methisazone Mixture B.P. equivalent to about 1 mg of solid drug and ensure thorough dispersion using the stirrer attached to the instrument. Adjust the instrument controls and take particle counts at threshold settings of 8, 10, 15, 25, 35, 45, 55, 70, 85 and 100 μm. Replace the tube with a 50 μm orifice, pass the sample suspension through a coarse filter, and count particles at 1, 1.5, 2.5, 4.5 and 8 μm thresholds. Plot the size distribution of methisazone particles and determine the median diameter.

Thermal methods of analysis

Thermal methods of analysis comprise a group of techniques in which a physical property of a substance (and/or its reaction products) is measured as a function of temperature while the substance is subjected to a controlled temperature programme. Several dozen techniques have so far been developed, some of which are listed in Table 3.4.

Table 3.4 Classification of thermoanalytical techniques

Physical property	Technique (abbreviation)	Instrument
Acoustic characteristics	Thermoacoustimetry	
	Thermosonimetry	
Dimensions	Thermodilatometry	Dilatometer
Enthalpy	Differential scanning calorimetry (DSC)	Differential calorimeter
Mass	Evolved gas analysis (EGA)	Evolved gas detector
	Thermogravimetry (TG)	Thermobalance
Mechanical characteristics	Thermomechanical measurement	
Optical characteristics	Thermoptometry	
Temperature	Differential thermal analysis (DTA)	DTA apparatus

The techniques most commonly used in pharmacy are thermomicroscopy (a thermoptometric technique), DTA, DSC and TG, which collectively can provide a comprehensive and wide-ranging analysis of, for example, the identity and purity of drugs, the melting and crystallisation characteristics of waxes and polymers, phase transformations and the physicochemical compatibility of tentative formulations. Meisel (1982) has reviewed some of the other techniques and their applications.

Thermomicroscopy

This technique (also known as **hot-stage microscopy**) essentially involves the observation of a sample through a microscope fitted with a stage that can be heated (or cooled) at a controlled rate. Apart from the obvious advantages of direct visual examination of a sample as its temperature changes, thermomicroscopy supports and can elucidate results from other thermoanalytical methods. For instance the desolvation of a crystalline solvate, which ambiguously appears as a pre-melting endothermic peak on DTA and DSC curves and a minor weight loss on a TG curve, is often distinctly observed as *pseudomorphosis* (the sudden apparent darkening of the crystal due to light scatter caused by micro-fractures produced on the egress of solvent); sublimation and crystal–melt–crystal transformations, both of which are often difficult to distinguish on DTA, DSC or TG curves, are also readily noted.

Modern hot-stages, which can cover the range − 180° to + 600°, with facilities for controlled atmospheres, special illumination, photographic recording and the continuous measurement of changes in transmitted, reflected or emitted light, make thermomicroscopy a versatile technique, and particularly useful for the examination of liquid crystals and phase transformations. The monograph by Kuhnert-Brandstatter (1971) is recommended as an introduction to pharmaceutical applications of the technique.

Differential thermal analysis

In differential thermal analysis a record (the DTA or differential thermal curve) is made of the **temperature difference** (ΔT) between the sample and a reference material, against time or temperature, as the two specimens are subjected to an identical controlled temperature regime. The reference material, e.g. alumina, is a substance which does not undergo any physical or chemical change in the temperature range of interest, and which ideally should have heat transfer properties similar to those of the sample.

Apparatus for DTA consists of sample and reference holders, a furnace for programmed heating, a detector of thermal differences created in the differential thermocouple, an amplifier and a recorder. Commercial instruments capable of working from $- 150°$ to $+ 1600°$ are available. A typical arrangement of the head assembly is shown in Fig. 3.9. Heat is supplied at a constant rate to sample S and reference R metal cups and the thermocouple junction detects the difference in temperature ΔT (i.e. $T_S - T_R$) as the two samples are heated. A direct heating curve and a differential heating curve are shown in Fig. 3.10: the direct trace shows an endothermic change in the sample temperature, resulting in the typical differential trace shown. Until the sample undergoes an enthalpic change, $\Delta T = 0$; melting transitions are always endothermic. An idealised DTA curve is shown in Fig. 3.11.

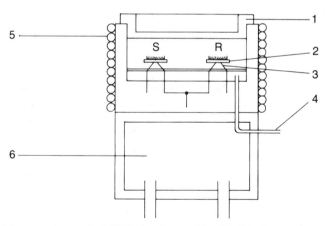

Fig. 3.9. Diagram of a typical DTA head assembly: (1) lid with Pyrex window, (2) specimen platform, (3) chromel–alumel thermocouple welded to platform, (4) purge-gas inlet, (5) insulated heating coils around metal chamber, (6) cooling chamber with liquid nitrogen vents

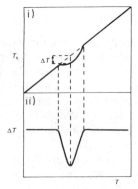

Fig. 3.10. (i) A plot of sample temperature (T_S) versus temperature of chamber, showing an endothermic change in sample; (ii) the corresponding differential curve ($\triangle T$) of this sample compared with a reference substance treated in the same way

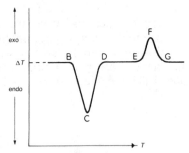

Fig. 3.11. An idealized DTA curve showing an endotherm (BCD) and exotherm (EFG) with increasing temperature. The important features of the DTA curve are the points of deviation from base line B and E and the peak areas, which correspond to the temperatures and heats of reaction respectively

Samples may be studied in the form of powders, fibres, single crystals, polymer films, semi-solids or as liquids. It is essential for meaningful results that regard is paid to a number of factors which can affect the results; these include sample weight, particle size, heating rate, atmospheric conditions surrounding the sample, and conditions of sample packing into the dishes. In some cases it is desirable to control the atmosphere surrounding the sample and reference material and most commercial instruments have facilities for running samples under reduced pressure or in inert atmospheres. When decomposition of the sample occurs it is advisable to crimp the specimen dishes, although in some cases a pin-hole should be left to avoid build-up of pressure; it is best to first investigate the effect of crimping on the results. When the sample is fibrous or fluffy, white kieselguhr may be a suitable reference substance as an alternative to alumina. Instrumental differences and operating conditions are such that the reproducibility of temperature measurement in DTA is ± 5°; individual workers on a single instrument should, however, achieve a precision better than ± 2.5°. In order to achieve accuracy in recording temperature and heats of reaction, calibration materials have to be used, preferably over a range of temperatures; Table 3.5 lists some useful compounds.

Table 3.5 Materials for temperature and heat of reaction calibration of DTA and DSC instruments

Reference compound	T_m (°C)	H_f (kJ mol^{-1})
Stearic acid	72.0	56.66
m-Dinitrobenzene	90.0	17.39
Benzoic acid	122.4	18.07
Urea	132.7	14.53
Indium	156.4	3.27
Salicylic acid	159.5	—
Anthracene	216.2	28.86
Tin	231.9	7.20

In the analysis of pharmaceuticals several thermal effects may be noted; the most important of these and their influence on the DTA curve are listed in Table 3.6.

Table 3.6 Qualitative interpretation of DTA and DSC curves

	Thermal effect		
Phenomenon	Endothermic	Exothermic	Peak type
Dehydration	+		Broad, large
Desorption	+		Broad with no pronounced peak
Crystalline transition	+	+	Sharp, small
Melting	+		Sharp, medium to large
Sublimation	+		No typical peak
Decomposition	+	+	Generally large
Oxidative degradation	+	+	Generally broad, large

In addition to indicating the temperatures at which phase transitions and reactions occur, DTA can also provide quantitative data. One of the simplest expressions for the area under the ΔT-time (t) DTA curve is the equation

$$\frac{m\,\Delta H}{g\lambda_s} = \int_{t_1}^{t_2} \Delta T \cdot \mathrm{d}t \qquad (14)$$

where m is the mass of sample in g, ΔH is the enthalpy change per g of substance, g is a constant dealing with the effect of the geometry of the sample arrangement and construction on heat transfer and other factors, and λ_s is the coefficient of thermal conductivity of the sample. If A is the peak area, $\Delta H = A/K$, where K is $m/g\lambda_s$, and (being temperature-dependent) should ideally be determined over a range of temperature using standards as similar to the sample as possible. Areas under the ΔT − t curve can be obtained by cutting and weighing copies of the trace. However, the problems of heat transfer in the system sometimes render application of this equation difficult. The effect of heating rate is obviously of importance since the rate of reaction depends on the temperature of the sample; if the temperature rise from the external heater is considerable during the period of the reaction then the DTA curve will differ from that obtained at a slower rate of heating. Cooling cycles are often of use in studying phase transitions. Rates of cooling can be critical if supercooling effects are to be avoided; where super-cooling occurs a shift in the position of the exotherm may occur. ΔH_f values determined by DTA or DSC for some drugs are listed in Table 3.7.

Table 3.7 ΔH_f values for selected drugs

Drug	ΔH_f (kJ mol^{-1})	Source
Chlordiazepoxide	30.5	MacDonald, Michaelis and Senkowski (1972)
Diazepam	24.7	MacDonald, Michaelis and Senkowski (1972)
Levallorphan Tartrate	43.5	Rudy and Senkowski (1973)
Paracetamol	28.5	Fairbrother (1974)
Phenacetin	32.4±2.5	Marti (1972)
Sulphamethoxazole	31.4	Rudy and Senkowski (1973)
Testosterone	27.6	Perkin-Elmer Corp. literature (1966)

Differential scanning calorimetry

In contrast to DTA, in this technique the difference in **energy inputs** into a substance and reference material is measured as a function of temperature as the specimens are subjected to a controlled temperature programme. Depending on the method of energy measurement, **heat-flux** and **power-compensation DSC** can be distinguished; the latter mode is illustrated in Fig. 3.12.

DSC can be used for the same analyses as DTA, with the advantage that, since power is measured directly, the curves are quantitative for heat of reaction, specific heat etc. It should be noted that, unlike DTA, conventions for representing heat changes have not as yet been agreed and endothermic deviations may appear on the positive or negative side of the ordinate. Some applications of DSC are illustrated in Figs. 3.13–3.15; the technique is of particular use in the determination of purity.

The presence of minute amounts of impurity in a substance broadens its melting range and lowers its melting point by an amount ΔT, which may be related to the mole fraction (x_2) of the impurity by the van't Hoff relation

$$\Delta T = \frac{RT_1^2 x_2}{\Delta H_{f,1}} \cdot \frac{1}{F} \tag{15}$$

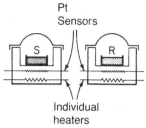

Fig. 3.12. Schematic representation of a power-compensation DSC system (courtesy of Perkin-Elmer Ltd). Nominally isothermal conditions ($T_S = T_R$) are maintained by the platinum resistance thermometers (in close contact with sample (S) and reference (R) holders) operating a servo system supplying different amounts of heat to each specimen. The difference in power input, $d(\triangle Q)/dt$, is recorded against T (or t)

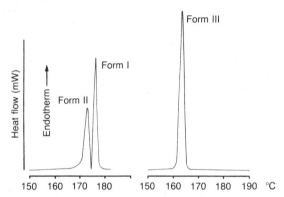

Fig. 3.13. Polymorphism of phenobarbitone studied by DSC at a heating rate of $5°\,\mathrm{min}^{-1}$. The low temperature peak is the melting of a metastable crystal form which reverts to the most stable form, which melts at 176° (redrawn, with permission, from Perkin-Elmer Corp. literature.)

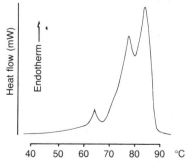

Fig. 3.14. DSC trace of carnauba wax run at $10°\,\mathrm{min}^{-1}$, illustrating the complex, but characteristic, melting profiles of waxes (redrawn, with permission, from Perkin-Elmer Corp. literature.)

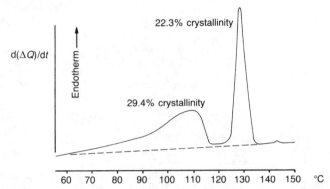

Fig. 3.15. Illustration of the use of DSC in analysis of polymer blends, which are often difficult to distinguish from copolymers by usual analytical techniques. The figure shows the melting region of a blend of 25% linear and 75% branched polyethylene. The high melting crystallites produce a peak well-resolved from that of the crystallites consisting of shorter linear and branched molecules. Total crystallinity 51.7%, from which pharmaceutically relevant parameters such as relative water vapour permeability may be calculated (redrawn, with permission, from Perkin-Elmer Corp. literature.)

where $\Delta H_{f,1}$ is the heat of fusion of the sample in $J\,mol^{-1}$, R the gas constant, T_1 (K) is the melting point of a pure sample, T_S is the temperature of the sample, F is the fraction of sample melted at T_S, and $\Delta T = T_1 - T_S$. On rearrangement and substitution we obtain

$$T_S = T_1 - \frac{RT_1^2 x_2}{\Delta H_{f,1}} \cdot \frac{1}{F}$$

For compounds which are 99.5 mol % pure ($x_2 = 0.005$) or more, the melting point depression is very small but even at this level of purity the melting range will have increased considerably. It is thus the range combined with the melting depression that is used in the assessment of thermally stable compounds of high purity by DSC. In the *dynamic method* the small (1–3 mg) sample is heated slowly ($\sim 1°C\,min^{-1}$) through the melting range in order to minimise thermal lag and allow equilibrium. The fraction melted (F) at any temperature T_S is obtained from the ratio of the area under the curve (AUC) at T_S to the total AUC of the endotherm. A plot of sample temperature versus $1/F$ should be a straight line with a slope equal to $-RT_1^2 x_2/\Delta H_{f,1}$ and an intercept of T_1. The procedure is illustrated in Fig. 3.16; in this example the slope is 0.227, and from the DSC curve AUC $\Delta H_{f,1} = 27.6\,kJ\,mol^{-1}$. Substituting the values in eq. (15)

$$0.227 = \frac{8.314 \times (428.28)^2}{27.6 \times 1000} x_2$$

$$\therefore x_2 = 0.004$$

$$\therefore \quad \text{sample purity} = 99.6\,mol\%$$

The method has been critically reviewed by van Dooren and Muller (1984).

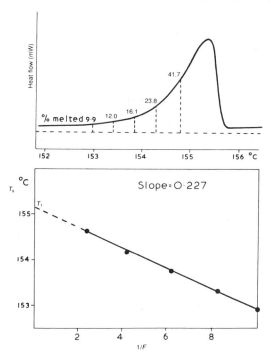

Fig. 3.16. Purity determination of testosterone by DSC. The lower diagram is a plot of the reciprocal of the fraction melted ($1/F$) as a function of the temperature of a sample of testosterone heated at $1.25°\,min^{-1}$

Thermogravimetry

In this technique the mass of a sample is monitored while it is being subjected to a controlled temperature programme. Although only events associated with changes in mass under *dynamic* or *isothermal* conditions are recorded, TG may be considered the modern equivalent of gravimetry and is one of the most widely used thermoanalytical methods.

The essential parts of a thermobalance are the programmable furnace, microbalance and recorder (Fig. 3.17). Commercially available instruments can heat samples to $1500°$ at rates of $2.5°\,min^{-1}$, under special atmospheric conditions (e.g. inert gas, vacuum), and detect mass changes as small as $0.1\,\mu g$. The rate of mass loss can also be monitored, and the exhaust gas from the furnace purged into a gas chromatograph or mass spectrometer for analysis.

The usefulness of thermogravimetry is illustrated by the TG curve obtained on heating hydrated calcium oxalate (Fig. 3.18a): from ambient temperature to $100°$ the monohydrate is stable, losing its water of crystallisation between 100 and $226°$ to give the anhydrous form, which in turn remains stable to $420°$, when it decomposes to calcium carbonate and carbon monoxide, the residual solid finally decomposing to calcium oxide with the evolution of carbon dioxide between 660 and $840°$. The

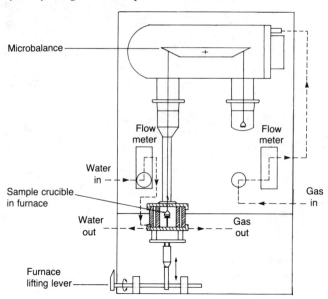

Fig. 3.17. Schematic diagram of a thermobalance (courtesy of Stanton-Redcroft Ltd)

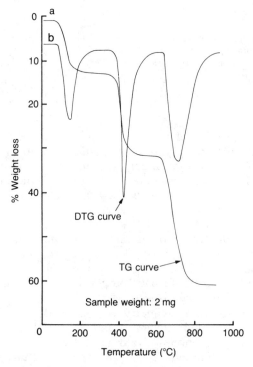

Fig. 3.18. Thermogravimetric analysis of calcium oxalate monohydrate: (a) TG curve; (b) derivative TG curve (courtesy of Stanton-Redcroft Ltd)

relatively complicated behaviour of hydrated calcium oxalate may also be represented by plotting the first derivative of the TG curve (i.e. rate of mass loss against temperature, Fig. 3.18b), a facility available with modern instruments. The advantages of DTG curves are that they highlight the rate and extent of mass loss and often closely resemble the DTA curves for specimens.

Practical experiments

Experiment 5   *Demonstration of a pre-melting crystalline transition by DSC/DTA*

Method Run a sample of potassium nitrate (unground) in static air at $10°\,min^{-1}$ over the temperature range ambient to 350°. After the temperature has reached about 350°, allow the sample chamber to cool to about 100° by passing liquid nitrogen through the cooling assembly and repeat the run up to 400°.

Note the peak melting and crystalline transition temperatures and comment on the effects of the cooling and reheating cycles.

Experiments 6 and 7   *Detection of polymorphism and pseudo-polymorphism in pharmaceuticals by DSC or DTA*

Triamcinolone Forms A and B. A large number of pharmaceuticals exhibit polymorphism and most can be studied by DSC or DTA techniques. Triamcinolone exists in two forms, A and B, obtained by recrystallisation from 60% aqueous propan-2-ol and from dimethylacetamide-water mixtures respectively.

Method Run samples of triamcinolone Form A and Form B from ambient to 350°. With a new sample of Form A, programme the instrument to stop the heating cycle at 270° and cool to ambient. Start the heating cycle to 350° and compare the curves with those of Form A and B. The favoured form of the compound is the form to which the melt reverts on heating and cooling.

7   *Ampicillin trihydrate*

Method Run a sample of ampicillin trihydrate from ambient to 200°. Endotherms due to the desolvation of water and to the melting transition should be observed.

Experiment 8   *Determination of the purity of a sample of testosterone by DSC*

Method Run the drug sample (about 1.5 mg) from 150° to 156° at a slow heating rate $(1.25°\,min^{-1})$. Make a Xerox copy of the chart paper. The complete curve can be cut out to the base line (obtained in the absence of sample) and the cut-out weighed (x mg). Cut out the area under the curve from the start of the trace up to 153° and weigh this (y mg). The fraction (F) melted is obtained from y/x. Cut out the area between 153° and 153.4°, and weigh this, adding the weight to that of the first portion ($y + y^1$ mg). Repeat this procedure at 153.8, 154.4 and 155°, and calculate successive F values. Plot T_S versus $1/F$. Determine T_1 by extrapolation as shown in Fig. 3.16 and calculate the slope of the line and carry out the calculation of mole fraction of the impurity (x_2) as explained in the text.

X-ray powder diffraction

From a morphological viewpoint, a crystal may be defined as the regular polyhedral form, bounded by smooth surfaces, which is assumed by a chemical compound under the influence of its interatomic forces when passing (under suitable conditions) from the gaseous or liquid state to the solid state. The overtly geometric shapes of crystals reflect an internal symmetry of atoms and molecules arranged in a regular and repeated pattern in space (Fig. 3.19); it is this long-range order which distinguishes **crystalline** solids from **amorphous** ones (e.g. glass). In practice the physical characteristics of crystals depend very much on the conditions of crystallisation and subsequent processing; therefore surfaces are usually rough, the crystals are structurally flawed internally and they can adopt a variety of **habits** depending on features of processing such as the type of solvent and degree of agitation (Fig. 3.1b). An important aspect of the solid state is the ability of compounds to crystallise in a variety of symmetrical arrangements of their molecules in space, i.e. **polymorphic forms**, which are often quite different from each other in physical characteristics (e.g. habit, melting point and solubility) although chemically identical, as the diamond and graphite forms of carbon illustrate. It has been estimated that some 20% of all organic compounds exhibit polymorphism; for instance, 70% of the barbiturates group are polymorphic, several exhibiting more than two forms (11 polymorphs of phenobarbitone, with melting points ranging from 176° to 112°, have been reported). The pharmaceutical relevance of polymorphism lies in the fact that the differences in physico-chemical

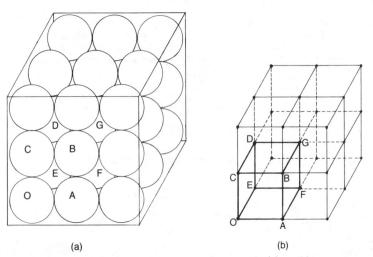

(a)   (b)

Fig. 3.19. Illustration of the internal structure of a crystal: (a) packing arrangement of molecules; (b) *space lattice* made up of a plot of the centres of molecules—the bold parallelepiped marks the basic repeating pattern, the *unit cell* (redrawn from Davidson, 1980.)

properties between forms may influence manufacturing processes and, importantly, dissolution rate and therefore bioavailability, e.g. the A (or β) and B (or α) forms of chloramphenicol palmitate (Chapter 10). Although infrared spectroscopy and thermal methods of analysis can provide useful information, X-ray diffraction methods are necessary to distinguish between different crystallographic arrangements. Of the various techniques available, X-ray powder diffraction has the advantages of speed, availability of equipment and relatively straightforward assimilation of data.

Production of X-rays

X-rays are electromagnetic waves similar in character to light, but of much shorter wavelength, generated when a beam of high velocity electrons from a heated filament impinge on a small (1 × 12 mm) target in a sealed diffraction tube (Fig. 3.20). Two kinds of X-rays are distinguished, depending on the mechanism of production (Fig. 3.21): the shaded region under curve (a) comprises the **white** or **general** radiation resulting from interaction of the electrons with target nuclei, and beyond a certain critical voltage the sharp peaks of **characteristic** X-rays, associated with the target element, are superimposed. The critical voltage needed to generate these rays is that which gives a bombarding particle (i.e. an electron) sufficient energy to eject one of the shell electrons from the target atom; when this happens, an electron

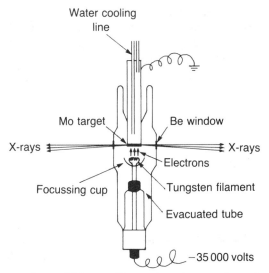

Fig. 3.20. Diagram of a molybdenum X-ray tube: electrons from the heated tungsten filament bombard the target, generating X-rays which emerge through the low-absorption beryllium windows and which are focussed onto the specimen (redrawn from Davidson, 1980.)

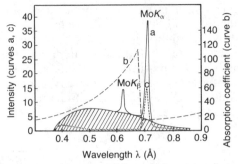

Fig. 3.21. X-ray spectrum from a molybdenum tube: (a) variation of intensity of radiation with wavelength—the shaded region represents *white radiation*; (b) variation of the mass absorption coefficient of zirconium with wavelength; (c) filtering effect of a thin film of zirconium on white and K_β radiation (redrawn from Davidson, 1980.)

from one of the outer shells immediately negotiates a 'jump', replacing the dislodged electron and emitting a quantum of radiation. If the initial electron is ejected from the K shell and the replacement electron is from the L shell, the radiation is termed K_α; if the replacement electron arises from the M shell, the radiation is termed K_β, etc. As there are two L electrons in slightly different energy states which may fill a vacancy in the K shell, K_α radiation is really a close doublet, K_{α_1} and K_{α_2}, which for most powder diffraction work is not resolved and a weighted-average wavelength of the two is used. Table 3.8 records the K_{α_1,α_2} and K_α (unresolved) wavelengths and other data for some of the target elements used in diffraction work. For most powder work a monochromatic beam is needed; the simplest way of achieving this is to introduce a thin ($\sim2.0\,\mu$m) selective filter into the beam where it emerges from the tube (Fig. 3.21).

Table 3.8 Useful X-ray wavelengths of elements in diffraction studies

Element	K_{α_1} (Å)	K_{α_2} (Å)	Unresolved K_α (Å)*	K_{β_1} (Å)	Excitation potential (kV)
Ag	0.55941	0.56380	0.56084	0.49707	25.52
Mo	0.70930	0.71359	0.71073	0.63229	20.00
Cu	1.54056	1.54439	1.54184	1.39222	8.98
Ni	1.65791	1.66175	1.65919	1.50014	8.33
Co	1.78897	1.79285	1.79026	1.62079	7.71
Fe	1.93604	1.93998	1.93735	1.75661	7.11
Cr	2.28970	2.29361	2.29100	2.08487	5.99

* Weighted means of K_{α_1} and K_{α_2}, $K_{\alpha1}$ being given twice the weight of K_{α_2}.

Copper K_α radiation is much used in diffraction studies for two major reasons. First, the relatively long wavelength spreads the diffraction pattern over a wider angular range than does molybdenum radiation for example, and yet copper rays are not appreciably absorbed in air; the greater 2θ interval means that, with an instrument of given geometry, copper radiation will permit greater accuracy in measurement of *d* values. Secondly, owing to its outstanding thermal conductivity, a copper tube can safely be operated at higher voltages than others and thereby provides more intense X-ray beams.

X-ray powder diffraction

X-rays are diffracted because crystalline solids are constructed from a relatively simple assemblage of components (atoms and atomic group-ings) repeated at regular intervals in three dimensions (Fig. 3.19), and X-ray wavelengths are the same order of magnitude as the spacing of atom centres (a necessary condition for interference between electro-magnetic waves).

Consider a two-dimensional array of atom centres, in planes parallel to a crystal face, with X-rays incident at angle θ to the plane *pp* (all points on the wavefront AA′ in phase) and reflected along CD (Fig. 3.22). In order for the reflected wavelet from B′ to reinforce the one reflected at C it must arrive at C in phase with the wave ABC, i.e. the path difference must be a whole number of wavelengths

$$B'C - BC = n\lambda$$

By simple trigonometry,

$$B'C = d/\sin \theta$$

where *d* is the interplanar space, and

$$BC = B'C \cos 2\theta = d(\cos 2\theta)/\sin \theta$$

Therefore, by substitution, the condition for constructive interference is

$$n\lambda = 2d \sin\theta$$

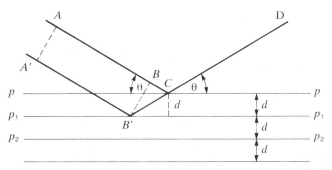

Fig. 3.22. Geometry of the Bragg 'reflection' analogy: the monochromatic X-ray beam AA′ impinges at an angle θ onto a series of lattice planes (*pp*, p_1p_1, . . .) parallel to a crystal face and separated by distance *d*

This is the **Bragg equation** and is the fundamental relation underlying all X-ray diffraction measurements. It is important to note that only for angles of incidence such that $\sin \theta = n\lambda/2d$ will X-rays be reflected; at all other angles destructive interference occurs.

In powder work, irradiated samples consist of small ($<50\,\mu$m) crystallites in *random orientation* so that the Bragg condition is fulfilled for every set of lattice planes, the reflected rays from each lying along the surface of a cone having its apex at the specimen, its axis in the direction of the X-ray beam and its semi-vertical angle equal to 2θ (Fig. 3.23). In the **powder camera** (e.g. Debye-Scherrer, Guinier) technique, segments of the cones of diffracted rays fall on a strip of photographic film to give a pattern of lines of varying intensity, the **powder photograph**; by measuring the distances of lines from the direct-beam position, and knowing the camera radius and irradiating wavelength, d values can be calculated (and corresponding line intensities measured microdensitometrically or visually). On the other hand, with a **powder diffractometer** the intensities of diffracted rays are directly measured by a radiation detector (e.g. a scintillation counter) which moves in a defined arc around the specimen, the diffraction for a range of Bragg angles being recorded as peaks on a chart calibrated in values of 2θ (Fig. 3.24). The advantages of this technique over the powder camera include accurate measurement of both the angle and intensity of diffracted radiation, and speed (typically 0.5 h per scan, cf. 5–9 h film exposures for organic specimens).

The most widespread use of powder diffraction is for identification of crystalline materials, using their diffraction pattern as a 'fingerprint'. The simplest way of doing this is by direct comparison of the specimen powder photograph or diffractometer trace with a standard; for acceptable identification all the lines and peaks must match in position and relative intensity. If the identity of the specimen is a more open question it can usually be identified by reference to libraries such as the **Powder Data File** of the Joint Committee on Powder Diffraction Standards (JCPDS) which contains the diffraction patterns of some tens of thousands of substances in numerical form, i.e. lists of *d*-spacings and corresponding intensities of reflections. The presence of additional lines on the photograph of the specimen, not present on an otherwise matching standard, will suggest the presence of an impurity which can then be further identified. With the enhanced 'search–match' capabilities of modern automated and computerised diffractometers, this aspect of powder diffraction is useful for the routine quality control of raw materials. For instance it is more practicable than other physical techniques for analysis of high volume (e.g. 10–20 ton per week) throughputs of cosmetic talc, in which the contaminant tremolite (a potentially carcinogenic amphibole) can be detected at levels of 0.3% w/w in specimens of only a few milligrams.

Another important use of powder diffraction is in the characterisation of the crystallinity of known substances. The different lattice arrangements of polymorphic forms of drugs are reflected in different patterns, i.e., *d*-values and relative line or peak intensities. Whereas the absence

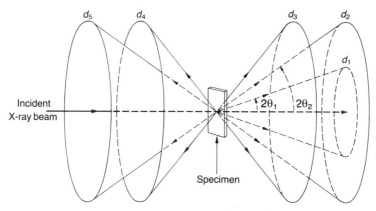

Fig. 3.23. Diffraction of X-rays by a powder sample: in the camera technique the cones, which have a common apex and correspond to particular lattice planes, intercept a photographic film strip to leave a pattern of lines of characteristic spacing and intensity (redrawn from Klug and Alexander, 1974.)

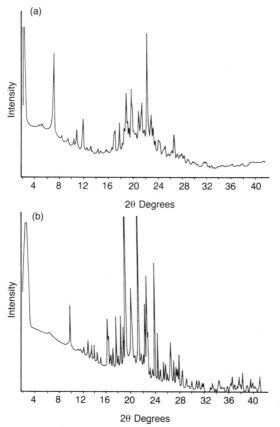

Fig. 3.24. X-ray powder diffraction patterns for chloramphenicol palmitate polymorphs: (a) Form B (or α); (b) Form A (or β) (redrawn from Szulzewsky *et al.*, 1982.)

of a diffraction pattern indicates lack of sufficiently long-range order, i.e. that the solid is essentially amorphous, broad, diffuse reflections suggest specimen particle size $\ll 0.1\ \mu$m or a strained lattice imposed by, say, rapid precipitation or severe comminution.

References

Particle Size Analysis Sub-Committee of the Analytical Methods Committee (1968) *Determination of Particle Size. Part 1. A Critical Review of Sedimentation Methods,* Society for Analytical Chemistry, London.

British Standards Institution (1963) *B.S.3406. Methods for the Determination of Particle Size of Powders, Parts 1–4.*

Davidson, W.L. (1980) *Physical Methods in Modern Chemical Analysis,* Vol. 2, Academic Press, New York.

Fairbrother, J.E. (1974) in *Analytical Profiles of Drug Substances,* Vol. 3, ed. K. Florey, p 1, Academic Press, New York.

Klug, H.P. and Alexander, L.E. (1974) *X-Ray Diffraction Procedures for Polycrystalline and Amorphous Materials,* 2nd edn, Wiley, New York.

Kuhnert-Brandstatter, M. (1971) *Thermomicroscopy in the Analysis of Pharmaceuticals,* Pergamon Press, Oxford.

MacDonald, A., Michaelis, A.F. and Senkowski, B.Z. (1972) in *Analytical Profiles of Drug Substances,* Vol. 1, ed. K. Florey, pp. 15 and 79, Academic Press, New York.

Marti, E.E. (1972) *Thermochim. Acta* **5**, 173.

Meisel, T. (1982) *Fresenius' Z. Anal. Chem.* **312**, 83.

Rudy, B.C. and Seenkowski, B.Z. (1973) in *Analytical Profiles of Drug Substances,* Vol. 2, ed. K. Florey, pp. 339 and 467, Academic Press, New York.

Szulzewsky, K., Kulpe, S., Schulz, B. and Fichtner-Schmittler, H. (1982) *Acta Pharm. Suec.* **19**, 457.

van Dooren, A.A. and Muller, B.W. (1984) *Int. J. Pharm.* **20**, 217.

General reading

Allen, T. (1981) *Particle Size Measurement,* 3rd edn, Chapman and Hall, London.

Barth, H.G. (ed.) (1984) *Modern Methods of Particle Size Analysis,* Wiley, New York.

Blazek, A. (1972) *Thermal Analysis,* Van Nostrand Reinhold, London.

Duval, C. (1976) in *Wilson and Wilson's Comprehensive Analytical Chemistry,* Vol. 7, ed. G. Svehla, p 1, Elsevier, Amsterdam.

Edmundsen, I.C. (1967) *Advances in Pharmaceutical Sciences,* **2**, 95, Academic Press, London.

Garn, P.D. (1965) *Thermoanalytical Methods of Investigation,* Academic Press, New York.

Herdan, G. (1960) *Small Particle Statistics,* 2nd edn, Butterworths, London.

Klug, H.P. and Alexander, L.E. (1974) *X-Ray Diffraction Procedures for Polycrystalline and Amorphous Materials,* 2nd edn, Wiley, New York.

Kuhnert-Brandstatter, M. (1971) *Thermomicroscopy in the Analysis of Pharmaceuticals,* Pergamon Press, Oxford.

Mackenzie, R.C. (ed.) (1970, 1972) *Differential Thermal Analysis,* Vol. 1 and 2, Academic Press, London.

4
Chromatography

A.G. DAVIDSON

Introduction

Chromatography is essentially a group of techniques for the separation of the compounds of mixtures by their continuous distribution between two phases, one of which is moving past the other. The systems associated with this definition are:

(a) a solid stationary phase and a liquid or gaseous mobile phase (adsorption chromatography)
(b) a liquid stationary phase and a liquid or gaseous mobile phase (partition chromatography)
(c) a solid polymeric stationary phase containing replaceable ions, and an ionic liquid mobile phase (ion exchange chromatography)
(d) an inert gel which acts as a molecular sieve, and a liquid mobile phase (gel chromatography).

The basis of the separation of the components of a mixture may be defined in terms of one of these four modes of separation, or by a combination.

Advances in technology since the first simple applications of chromatography were recorded have resulted in a wide range of techniques varying in complexity, separating ability, sensitivity and cost. The modern instrumental techniques of gas–liquid chromatography and high performance liquid chromatography provide excellent separation and allow the accurate assay of very low concentrations of a wide variety of substances in complex mixtures. The older inexpensive chromatographic techniques, such as column chromatography are used in analytical and preparative separations which do not require the resolution and sensitivity or justify the expense of the instrumental techniques.

In this chapter the treatment of the principal chromatographic techniques in pharmaceutical chemistry is based primarily on the equipment used. Where appropriate, the basis of the separation is discussed with reference to the different chromatographic materials available for use in each of the techniques. To illustrate the improvements in sensitivity and resolution which have resulted from technological progress, the techniques are discussed in approximately the order of their historical development.

Column chromatography

Adsorption chromatography

The technique was originally developed by the Russian botanist Tswett in 1906 during the course of an investigation into the nature of leaf pigments. He found that leaf pigments extracted with light petroleum were adsorbed on the top of a column of calcium carbonate supported in a glass tube. As more solvent was allowed to percolate through the column the region of pigmentation became broader and finally separated into distinct and differently coloured bands. Prolonged washing with solvent caused complete separation of the bands, which could be eluted separately. It is one of the simplest laboratory exercises to illustrate the use of column chromatography. Tswett's work attracted little attention and it was not until 1931, when polyene pigments were investigated by Kuhn and Lederer, that interest in chromatography was renewed.

The principle underlying the separation of the compounds is adsorption at the solid-liquid interface. For successful separation, the compounds of a mixture must show different degrees of affinity for the solid support (or adsorbent) and the interaction between adsorbent and component must be reversible. As the adsorbent is washed with fresh solvent the various components will therefore move down the column until, ultimately, they are arranged in order of their affinity for the adsorbent. Those with least affinity move down the column at a faster rate than, and are eluted from the end of the column before, those with the greatest affinity for the adsorbent. The technique in which the individual components of a mixture are separated by eluting the column with fresh solvent is **elution analysis** (Fig 4.1).

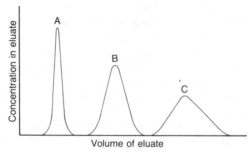

Fig 4.1. Elution analysis: the separate elution of three substances A, B and C

An adsorbent which is already saturated with respect to one substance may take up a small quantity of a second. The latter displaces the former and consequently, if a solution of a mixture is percolated continuously throught the column and the eluate is examined for the presence of substances, a plot of amount of substance (per ml of eluate) against volume of eluate will appear as in Fig. 4.2. This technique was redeveloped by Tiselius in 1940 and is known as **frontal analysis**. It is convenient for the determination of the number of components in a mixture as each is represented by a step in the chromatogram; the height of each step is proportional to the concentration of that component in the original mixture. In this method some of the least strongly adsorbed component only (A in Fig. 4.2) is obtained pure. Closely allied to this technique is **displacement analysis** in which a small volume of the mixture is added to the column which is developed by a solution of a substance which is capable of displacing all the components of the mixture. When the result of the experiment is plotted as described for Fig. 4.2 a diagram similar to that in Fig. 4.3 is obtained. Although the division between components is sharp there are no intervening fractions of solvent only, and some mixing is bound to occur between pairs unless more displacing agents are used to separate A from B, B from C, and C from D. The technique is useful for preparative work but the investigation of conditions for carrying out this method requires considerable time.

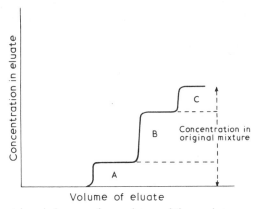

Fig. 4.2. Frontal analysis: curve for a mixture of three substances, A, B and C

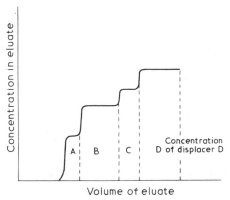

Fig. 4.3 Displacement analysis: curve for a mixture of three substances, A, B and C displaced by a fourth, D

Adsorbents and solvents

Many adsorbents of varying degrees of activity are available and they should preferably conform to the following requirements. They should be insoluble in solvents, chemically inert, active but not so active that no movement of components occurs, colourless to facilitate observation of zones, allow suitable flow of mobile phases, and have reproducible properties from batch to batch. It is not always possible for adsorbents to comply with all these requirements as the mobile phases and particle size also play a part.

The amount of a substance adsorbed from solution by an adsorbent can be determined by shaking a known weight of adsorbent with a known volume of solution at a fixed temperature until equilibrium is attained. The adsorbent is filtered off and the concentration of the substance in the filtrate is determined by any suitable means. If this procedure is carried out with solutions of different concentrations the results can be expressed graphically by plotting the amount of substance adsorbed per gram of adsorbent against the concentration of substance remaining in solution. The curves so obtained represent adsorption isotherms, which are important in explaining the appearance of chromatograms.

The curves may take any one of the three forms (Fig. 4.4) each of which explains a characteristic appearance of peaks in a chromatogram.

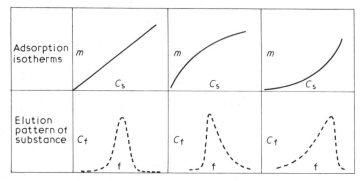

Fig. 4.4. Adsorption isotherms and related elution patterns (diagrammatic) of substances from a column of adsorbent; m, weight of substance adsorbed per g of adsorbent; C_s, concentration of solution; C_f, concentration of each fraction; f, number of each fraction

(a) **Linear adsorption isotherms.** These are obtained when the amount of substance adsorbed per gram of adsorbent is proportional to the concentration of solution. The adsorption coefficient (m/C_s) is a constant value independent of the initial amount of the substance shaken with the adsorbent and, in this respect, is analagous to the partition coefficient of a substance which partitions between two liquids (p.92).

When a substance moves as a band through a column of adsorbent there is no tendency for any portion of the band to be adsorbed more strongly than another. Therefore, a symmetrical peak is obtained as the eluate from the column is examined. The adsorbent and solvent for adsorption chromatography should be chosen to maximise the differences in adsorption coefficients of the components of the mixture.

(b) **Convex adsorption isotherms**. These are obtained when adsorption from weak solutions is greater than from strong solutions. Therefore, even if the pattern of the band of substance is initially symmetrical, the substance, in low concentration, at the front of the band is held more strongly by adsorption than is the centre of the band as the band of substance moves down the column. Therefore the centre of the band 'catches up' with the front and a sharp leading edge to the band is obtained. By a similar argument, the tail of the band becomes longer and longer. This appearance (Fig. 4.4) is one which is met frequently in practice.

(c) **Concave adsorption isotherms**. These, on the other hand, are obtained when adsorption from strong solutions is greater than from weak solutions so that the appearance of an initially symmetrical band of substance after passage through the column is once more characteristic (Fig. 4.4). It is not often that this type occurs.

The most commonly used chromatographic adsorbent is silica (silicic acid (SiO_2), silica gel) which adsorbs polar and unsaturated substances by the formation of hydrogen bonds with the hydroxyl groups on the silicon atom. Other adsorbents used in column chromatography are listed in Table 4.1.

Table 4.1 Adsorbents and solvents

	Adsorbent		Solvent
Weak	Sucrose Starch Inulin Talc Sodium carbonate	↑ increasing eluting power	Petroleum ether Carbon tetrachloride Cyclohexane Carbon disulphide Ether (ethanol-free)
Medium	Calcium carbonate Calcium phosphate Magnesium carbonate Magnesium oxide Calcium hydroxide		Acetone Benzene Toluene Esters Chloroform
Strong	Activated magnesium silicate Activated alumina Activated charcoal Activated magnesia Activated silica		Acetonitrile Alcohols Water Pyridine Organic acids Mixtures of acids or bases with ethanol or pyridine

Although the adsorption forces involved in chromatography are weak, cognisance must be taken of undesirable chemical changes that might occur because of the properties of the column material itself. Thus, an alkaline grade of alumina may cause hydrolysis of esters or lactones. Other changes associated with a poor choice of column are isomerisation, neutralisation of acids or bases and decomposition of compounds. The last may be put to good use in certain preparations, e.g. cadalene from oil of cade forms a crystalline picrate which is decomposed on a short column of alumina. The pure cadalene is eluted, leaving the picric acid fixed on the column.

The strongest adsorbents are silica and alumina activated by heating to about 200° to remove water. Alumina may be rendered acidic or basic prior to activation. Careful addition of water to the treated alumina allows different degrees of activity to be obtained and reproduced from batch to batch. Preliminary treatment of the material in this way often overcomes the property which leads to the undesirable effects noted above. It may, indeed, introduce increased specificity of the column for certain compounds, e.g. when silica gel is freshly prepared in the presence of propyl orange and the dye is finally removed by elution, the column has a greater affinity for propyl orange than it has for the methyl, ethyl and butyl analogues. A similar situation obtains when the silica gel is prepared in the presence of one enantiomorph of an optically active compound, e.g. laevorotatory quinine.

Adsorption is most powerful from non-polar solvents such as petroleum ether or benzene and a single solvent may often be effective in developing the chromatogram. The rate of movement of the compounds down the column can be increased by the addition of a second solvent to the mobile phase; the second solvent is usually more polar than the first. Strain (1942) has arranged both adsorbents and solvents in order of adsorptive and eluting power respectively and Table 4.1 lists the series. It is usual to redistil all solvents before use, so that traces of non-volatile matter, e.g. grease, are completely absent. The change from one solvent to another should be gradual, e.g. the change-over from petroleum ether to toluene should be done in proportions such as the following (petroleum ether first) 100:0, 95:5, 90:10, 80:20, 60:40, 40:60, 10:90, 0:100. Such a procedure is time-consuming and a more rapid elution may be achieved by the addition of about 0.5 to 1.0% of ethanol to the first non-polar solvent used.

Alteration in the composition of the eluting solvent may also be achieved by adding the second solvent gradually to a reservoir of the first with efficient mixing; the solvent entering the column therefore becomes gradually and continuously richer in the second solvent. This technique is known as **gradient elution** which, with proper choice of adsorbent and gradient, often reduces tailing of the compounds on the column.

Preparation of the column

A typical arrangement for column chromatography is shown in Fig.4.5.

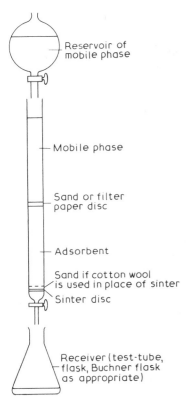

Reservoir of mobile phase

Mobile phase

Sand or filter paper disc

Adsorbent

Sand if cotton wool is used in place of sinter

Sinter disc

Receiver (test-tube, flask, Buchner flask as appropriate)

Fig. 4.5. Apparatus for column chromatography

Method Prepare the chromatographic column by mixing the adsorbent into a slurry with the solvent and pouring the mixture into the glass tube which contains solvent. The sand serves to give a flat base to the column of adsorbent when cotton wool is used instead of a sinter disc. After the adsorbent has settled, add a filter paper disc and sand then run off the supernatant liquid until the level falls to about 1 cm above the top layer of sand. The filter paper disc and sand is one means of avoiding disturbance of the adsorbent as fresh mobile phase is added to the column in the initial stages of development. The level of solvent must never be allowed to fall below the level of adsorbent, otherwise the latter develops cracks and becomes useless for chromatography because the solvent runs through the cracks rather than between the particles of adsorbent.

The preparation of the slurry may prove difficult with dense adsorbents and it is convenient to pour the powder directly into the solvent in the tube. Frequent tapping of the tube and stirring of the mixture assists in even packing and removal of air bubbles or pockets. Alternatively, the tube may be packed with the dry powder and the solvent allowed to percolate through with the stopcock open until the level falls to about 1 cm above the adsorbent.

The dimensions of the column and quantity of adsorbent depend upon the nature and amount of the substance to be chromatographed but a rough guide is given in Table 4.2.

Use of column Wash the column with about 50 ml of the mobile phase used to prepare it, which should be the least polar solvent in which the mixture will dissolve. Add the mixture dissolved in a small volume of solvent and carefully allow it to run into the sandy layer by opening the stopcock. Add a small volume of solvent and wash in the mixture. Repeat with gradually increasing quantities of solvent and develop the chromatogram, collecting the eluate in appropriate receivers if the components are to be eluted from the column.

Table 4.2 Column characteristics	
Adsorbent/adsorbate weight ratio*	30:1
Length/diameter ratio†	10–15:1
Column length	
(a) multi-component system	long column
(b) components with similar affinities for adsorbent	long column
(c) components with different affinities for adsorbent	short column

*The 30:1 weight ratio is suitable for preparative separations. For analytical purposes the ratio (30:1) is much too small, and often mg quantities of substance are chromatographed on 20 g or more of adsorbent (see Experiment 3).

†In general, narrow columns give better separations than wide columns.

Detection and recovery of components

For those mixtures which are coloured, visual examination of the column is usually sufficient to locate the coloured components. Colourless components may also be detected visually if they fluoresce, e.g. quinine and ergometrine. Recovery of the components after detection on the column requires extrusion of the column of adsorbent and isolation of each zone for extraction with solvents. If plastic tubing is used instead of glass tubes the zones are conveniently isolated by cutting the tubing into sections.

It is however, more convenient to complete the chromatogram by eluting the various components with solvents. For colourless compounds the eluate is collected as a large number of fractions, each of small volume.

Automatic fraction collectors enable large numbers of fractions to be obtained without the tedium associated with manual collection. The large number of fractions also assists in obtaining better separation of components, providing attention is directed to correct choice of flow rate. Each fraction is examined appropriately for the presence of a compound. The examination may be by evaporation of the solvent from each fraction and weighing the residue, by simple spot tests, by examination of the fraction by paper or thin layer chromatography or by spectrophotometry, either directly or after addition of reagents.

Partition chromatography

All partition chromatographic separations are based upon the differences in partition characteristics (partition coefficients) of the individual components of a mixture between a liquid stationary phase and a gaseous or liquid mobile phase. In column partition chromatography, the mobile phase is a liquid.

The theoretical principles of partition chromatography may be readily understood by considering the partitioning behaviour of substances between two immiscible liquids. Few substances, when shaken with two immiscible liquids, partition completely into one or other of the liquids. Instead, most distribute themselves between the liquids such that the partition coefficient (the ratio of **concentrations** of the substance in each phase) is a constant value independent of the total amount, provided neither phase is saturated with the substance.

Substances with large differences in their partition coefficients may be completely separated by simple solvent extraction techniques involving few (one to three) extractions (Part 1, Chapter 9). As the differences in partition coefficients of a mixture of substances decrease, the number of solvent extractions necessary to achieve complete separation increases. In theory, it is possible to exploit even small differences in partition coefficients to separate chemically similar substances by carrying out a sufficiently large number of extractions. The discontinuous counter-current distribution (CCD) technique invented by Craig in 1944 is based on this principle.

Essentially, the CCD apparatus comprises a large number (up to 1000) extraction tubes in which the sample is distributed between two immiscible solvents. The mixture of components is added to tube 1 (Fig. 4.6) and shaken until equilibrium is reached between the concentrations of each component in the two liquids. The upper phase is then drained into tube 2 which contains fresh lower phase, upper phase is added to tube 1 and then both tubes are shaken as before. The process is repeated many times; each time the upper phase containing its dissolved components drains into the next higher tube and fresh upper phase is added to tube 1. Thus, the individual fractions of upper phase move progressively along the row of tubes while the fractions of lower phase remain in their own tubes. The process is shown diagrammatically in Fig. 4.6.

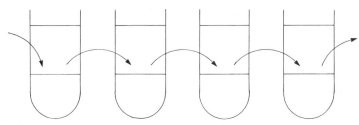

Fig. 4.6. Counter-current distribution apparatus showing the movement of the upper layer of two immiscible solvents along the row of tubes

The components of the mixture travel along the row of tubes at a rate depending on their partition coefficients. Those components whose partition coefficients favour the upper phase travel further along the row of tubes than those whose coefficients favour the stationary lower phase. The technique gives excellent resolution of similar substances, in particular of very polar compounds which are not well separated by column adsorption chromatography. For example, the lower fatty acids—ethanoic, propanoic, butanoic and pentanoic acids—have been separated using 25 extraction tubes and the immiscible solvent combination isopropyl ether–phosphate buffer pH 5.2. However, the complexity of the apparatus and the cost of the solvents have proved to be major disadvantages of the technique.

Similar work by Martin and Synge on a 'cascade extraction train' extraction procedure resulted in their first applications of column partition chromatography in 1941. In the latter technique, one of the immiscible liquids is distributed as a thin layer on particles of a solid support in a glass column while the other liquid percolates down through the column under the action of gravity. During the chromatographic process the two liquid phases are in intimate contact and this allows the partitioning of the components of the mixture to occur rapidly. The components whose partition coefficients favour the moving liquid travel down the column faster than those whose coefficients favour the sorbed stationary phase. The components thus emerge from the column in the order of their partition coefficients.

Tailing of the bands, which is commonly encountered in adsorption chromatography (Fig. 4.4), is rarely seen in partition chromatography (because the partition coefficients of substances do not vary with concentration) unless adsorption effects are also present. Consequently, the narrower bands obtained in partition chromatography permit the separation of closely related chemical substances, whereas this is often not possible in adsorption chromatography, which in general is used to separate compounds in different chemical classes.

Plate theory of chromatography

Martin and Synge in 1941 developed the concept of the 'theoretical plate' in order to establish a satisfactory theory for partition chromatography. The column is considered as being made up of a large number of parallel layers or 'theoretical plates', and when the mobile phase passes down the column the components of a mixture on the column distribute themselves between the stationary and mobile phases in accordance with their partition coefficients. The rate of movement of the mobile phase is assumed to be such that equilibrium is established within each plate. The equilibrium, however, is dynamic and the components move down the column at a definite rate depending on the rate of movement of the mobile phase. The R value of a component is

$$R = \frac{\text{rate of movement of component}}{\text{rate of movement of mobile phase}}$$

$$= \frac{\text{distance moved by component}}{\text{distance moved by front of mobile phase}} \tag{1}$$

The interrelationship between R and the partition coefficient K can be shown to be

$$R = \frac{A_m}{A_m + KA_s} \tag{2}$$

where

A_m = average area of cross-section of mobile phase

A_s = average area of cross-section of stationary phase

$K = \dfrac{\text{concentration of component in stationary phase}}{\text{concentration of component in mobile phase}}$

Although the formula is of potential value in devising optimum conditions for chromatography, these are more frequently determined empirically.

The thickness of the layer of the column in which one 'partition' is considered to occur (and which is analogous to an extraction tube in the CCD apparatus described above) is called the **height equivalent to a theoretical plate** (HETP). The smaller that the HETP is, the greater is the number of theoretical plates in the column and the greater is the separating ability of the column. For example, if the HETP is 0.5 mm and the length of the packed column is 10 cm, there are 200 theoretical plates in the column which will give the same separation as a CCD apparatus comprising 200 extraction tubes. The **efficiency** of a chromatographic column is measured by its number of theoretical plates.

Supports and liquid phases

In partition chromatography the solid material of the column functions solely as a support for a thin layer of liquid phase which is ordinarily polar in character, e.g. water or aqueous buffer solutions, the mobile phase being one of those normally used in adsorption chromatography. The two liquid phases must be in equilibrium and this is accomplished by shaking them together in a separator and allowing them to separate. The aqueous layer is used to coat the support which is afterwards transferred to the chromatographic tube and packed firmly. The other layer is used as developing solvent. A change from one solvent to another during development is inappropriate here (compare adsorption chromatography) as the equilibrium would be disturbed. Of the solid supports, silica gel is probably most convenient for aqueous systems as it is capable of holding a considerable volume of water whilst retaining its powder form. Cellulose powder is also frequently used.

The technique has been extended to enable organic liquids to be used as the stationary phase with kieselguhr commonly used as a support. It is made water-repellent by treating the dry powder with an ethereal solution of dichlorodimethylsilane (CARE), allowing the ether to evaporate in a fume cupboard and washing the residue with methanol until free from acid. Dry the powder and store in a well-stoppered jar. The column is prepared in the way described above but the equilibrated organic phase is used as the stationary phase. This method is known as **reversed phase chromatography**.

The procedure for column partition chromatography differs little from that of adsorption chromatography and a typical elution pattern from a mixture of diphenylamine (0.1 mg) and phenothiazine (0.1 mg) is shown in Fig. 4.7. Each fraction (10 ml) in this experiment (Experiment 3) was examined spectrophotometrically at 253 nm.

Fractions 10–14 contained diphenylamine.

Fractions 20–29 contained phenothiazine.

On a larger scale the appropriate fractions may be evaporated for recovery of the components.

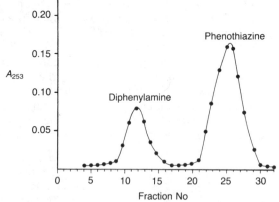

Fig. 4.7. Chromatogram for a mixture of diphenylamine and phenothiazine.

Ion exchange chromatography

Separation of ionic substances may be carried out in glass columns similar to those described for adsorption and partition chromatography. The chromatography medium (the stationary phase) is an ion exchange resin which is a polymer containing fixed charged groups and replaceable **counter ions** of the opposite charge. When a sample containing organic or inorganic ions is passed down the column the ions of the same charge as the counter ions displace the counter ions into the mobile phase and are retained on the column. Cationic and anionic exchange resins have positively and negatively charged counter ions respectively, and retard the migration of the sample cations and anions respectively. The mobile phase in ion exchange chromatography is usually an aqueous solution containing one or more electrolytes.

Ion exchange resins

Modern resins are based on cross-linked polystyrene prepared in bead form by the copolymerisation of styrene and divinylbenzene (DVB). The cross-linking on the polymer provides structural rigidity and the degree of cross-linking is controlled by the proportion of DVB which is added in concentrations of 4–12%. Most commonly used resins are prepared with approximately 8% DVB.

Strong cation exchange resins are prepared by sulphonating the free benzene rings. In water the beads of the resin swell and the sulphonic acid ionises, the $-SO_3^-$ forming the fixed charged group and the H^+ forming the counter ion (Fig. 4.8). Alternatively, the resin may be converted to the Na^+ form by treatment with sodium chloride solution which replaces the H^+ with Na^+ and releases hydrochloric acid into the eluate. Strong anion exchange resins containing quaternary ammonium residues are prepared by chloromethylation of the free benzene rings followed by treatment with a tertiary amine salt, e.g. trimethylamine hydrochloride. Anion exchange resins are available in either the Cl^- (Fig. 4.9) or OH^- form, the latter being obtained by treatment of the resin with sodium hydroxide.

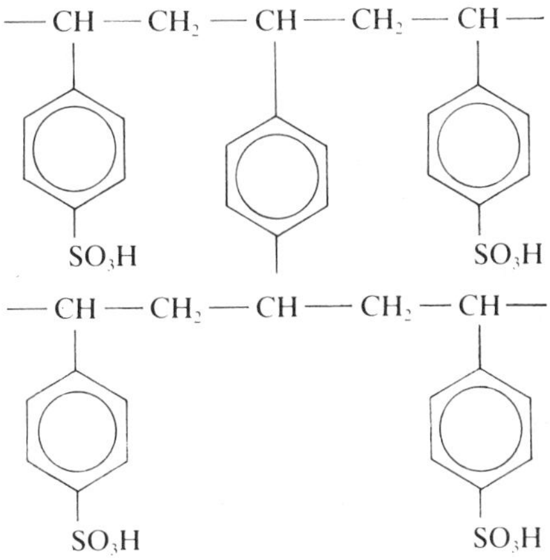

Fig. 4.8. Strong cationic exchange resin

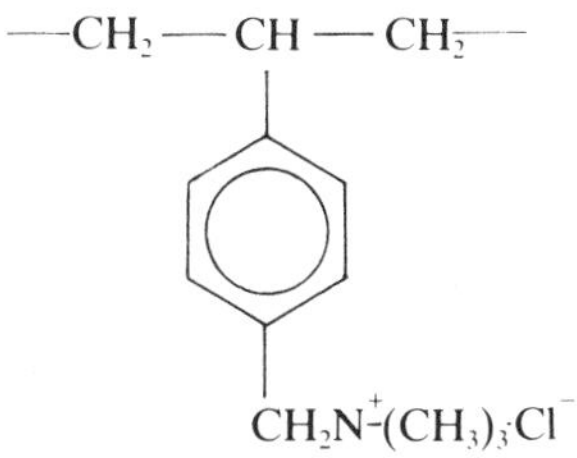

Fig. 4.9. Strong anionic exchange resin

The strength and exchange capacities of ion exchange resins depend on the acidic or basic strength of the fixed charged group. Thus, the strongly acidic sulphonic acid and strongly basic quaternary ammonium groups give strong ion exchange resins with a high exchange capacity. Weaker exchange resins containing the weakly acidic carboxylic acid ($-COOH$) or weakly basic derivatives of ammonia (e.g. $-NHR_2{}^+Cl^-$) generally have a lower exchange capacity.

The strong ion exchange resins may be used over the pH range 1–14 because the sulphonic acid and quaternary ammonium groups are fully ionised throughout this range. In contrast, the weak ion exchange resins are limited to the pH region where the functional groups are ionised, i.e. above pH 5 for weak cation exchangers and below pH 9 for weak anion exchangers.

Applications

The versatility of ion exchange methods derives from the number of experimental parameters which affect the retention of the ionic components of the sample, i.e. type of resin (whether it is a strong or weak cation or anion exchanger), mesh size, degree of cross-linking, exchange capacity, counter ion, pH and ionic composition of the mobile phase and temperature.

The strong ion exchange resins are used more frequently than the weak exchangers because of their wider ranges of maximum capacity and pH and because they give better resolution of substances with similar charge. The weak exchange resins are used in separations which do not require the high resolution of the strong exchangers or where strongly acidic or basic solutes are retained too firmly on a strong exchanger.

The separation of organic and/or inorganic cations on a cation exchange resin in a glass column involves passing a small volume (typically 1–10 ml) of the sample onto the top of the resin in which the counter ion has been converted to H^+ or Na^+. The movement of the cations in the sample as it passes down the column is retarded by displacing the counter ions into the mobile phase. The ion exchange is an equilibrium reaction which may be illustrated for a sample containing a sodium salt chromatographed on a strong cation exchange resin in the H^+ form.

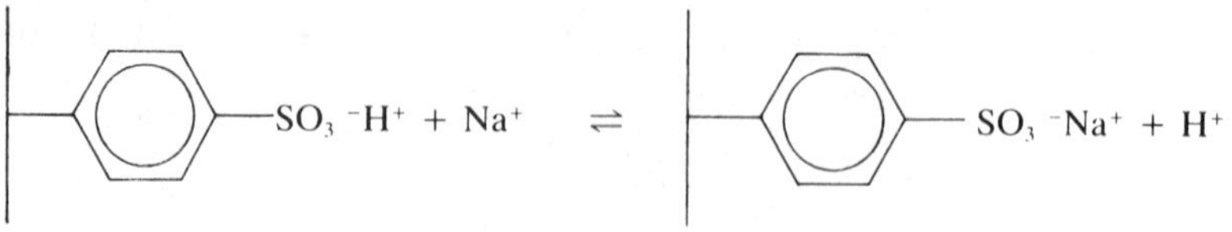

A simple analytical application of this technique is the assay of a sodium salt by titration of the acid, which appears in the eluate, against standard sodium hydroxide solution. Later, the sodium can be displaced from the resin and the column reconstituted in the H^+ form by passing through the column 4M hydrochloric acid. The high concentration of H^+ drives the equilibrium position in the reaction shown above towards the left hand side. Mixtures of metals can be separated on a cation exchange column if there are reasonable differences in the charge-to-radius ratios of their ions or if, in the presence of the eluting solvent, they form complexes with different charge. An example of the latter is the separation of Cd^{2+} from Cu^{2+} using 0.5M hydrobromic acid as the eluting solvent. The reduction in charge on Cd as a result of the formation of $CdBr_2$, $CdBr_3^-$ and $CdBr_4^{2-}$ complexes allows Cd to be eluted before Cu, which is retained more firmly on the column. The copper is later recovered by displacement with 3M nitric acid.

Alkaloids and other nitrogenous substances can be isolated from complex samples (e.g. plant material) on cation exchange resins. The pH of the sample and of the mobile phase has to be adjusted to a value at least 1 pH unit lower than the pK_a values of the substances to maintain them substantially in the protonated (ionised) form which will be retained on the column while non-ionic components of the sample are eluted.

Applications of anion exchange chromatography include the assay of total halogenic salts using a resin in the OH^- form. The anions are retained on the resin, displacing an equivalent amount of OH^- into the eluate. Titration of the hydroxyl ions in the eluate against standard mineral acid yields the concentration of anions in the sample

$$-CH_2NR_3^+OH^- \; + \quad X^- \quad \rightleftharpoons \quad -CH_2NR_3^+X^- + OH^-$$

| strong anion exchange resin | halogen (or other anion) |

An interesting application of anion exchange chromatography is the separation of morphine from its 3-methyl ether (codeine). In alkaline solution the codeine is unionised and so is not retained on the column whereas the phenolic group of morphine ($pK_a = 9.9$) is ionised as the phenoxide ion and is retained on the column. The morphine is later displaced by passing a concentrated solution of sodium chloride through the column. The separated alkaloids can be assayed spectrophotometrically at their wavelengths of maximum absorption in the ultraviolet region.

Anion exchange is also used to separate neamine, and neomycin C from neomycin B to test for neomycin C in *Framycetin Sulphate* and *Neomycin Sulphate*. The test provides a means of distinguishing between the two antibiotics, the level of neomycin C being not greater than 3% in the former and between 3% and 13% in the latter.

One of the most useful applications of ion exchange resins is the removal of unwanted ions from water. Softening of water on the industrial or domestic scale is carried out using potassium or sodium aluminosilicate cation exchange material (e.g. Permutit or Delcalso) which removes the Ca^{2+} and Mg^{2+} ions, the cause of the hardness, in exchange for K^+ or Na^+. Mixed resins containing a strong cation exchanger in the H^+ form and a strong anion exchanger in the OH^- form provide a rapid and inexpensive means of de-ionising water for use in situations where the presence of ions is undesirable.

Gel chromatography (size exclusion chromatography)

The introduction in 1959 of cross-linked dextran (Sephadex) gels as a packing material for column chromatography established a new mode of chromatographic separation in which substances are separated according to their molecular size.

The stationary phases in gel chromatography are inert gels of dextran (a polyglucose, Fig 4.10) or other polymers such as agarose and polyacrylamides, in which the macromolecules are cross-linked to give a porous three-dimensional structure. The degree of cross-linking and consequently the sizes of the pores within the gels are controlled during the manufacture. The gels are hydrophilic and swell in the presence of water. The gels with a low degree of cross-linking and a large pore size require a larger quantity of water (water regain) to fill the pores within the gel structure than tightly linked gels (Table 4.3). The mobile phases in gel chromatography are usually buffered aqueous solutions, although modified gel materials (e.g. Sephadex LH) are available for use with organic solvents.

The ability of gels to retard the movement of substances down a column packed with the gel depends on the molecular size of the substance relative to the pore sizes within the gel matrix. If a substance has a molecular size smaller than the largest pores of the gel it will penetrate the pores and move more slowly down the column than a substance of high molecular weight which, because it is unable to diffuse into the pores, passes down the column in the channels between the gel grains. The exclusion of substances of molecular size larger than the pores results in their elution from the column in the void volume ahead of the small molecular weight components. Substances are thus eluted in the order of the decreasing molecular size.

The liquid absorbed by the polymer granules is available in varying degree as solvent for solute molecules in contact with the gel. The distribution of solute between the inside and outside of the gel granules is a function of the space available and the distribution coefficient between granular and interstitial aqueous phases is independent of pH, ionic strength and concentration of the solvent.

The partition ratio between the granular and interstitial aqueous phases in a Sephadex column is defined by the distribution coefficient K_D, which can be calculated from the relationship $K_D = (V_e - V_o)/V_i$, where V_e is the elution volume of the substance undergoing gel filtration; V_o is the void or interstitial volume; and V_i is the inner volume, i.e., the volume of liquid taken up by the gel granules. Low molecular weight substances which can diffuse freely into the gel grains (i.e. $V_e = V_o + V_i$) have K_D values between 0.8 and 1.0. Values of K_D greater than 1 (i.e. $V_e > V_o$) indicate that there is absorption to the gel, and where there is no absorption (i.e. $V_e = V_o$), $K_D = 0$.

Table 4.4 gives a selection of K_D values obtained on gel filtration through different grades of Sephadex. The highly cross-linked gels (G-25 and G-50) provide effective separation of proteins from amino acids, and for the desalination of proteins and carbohydrate materials. Fractionation of amino acids by molecular weight is not possible, but preferential adsorption of aromatic and heterocyclic amino acids within the gel grains permits their separation from aliphatic amino acids which are eluted first. Peptide fractionation, and the separation of monosaccharides and oligosaccharides (e.g. in blood or urine) from low molecular weight polysaccharides can be accomplished with these polymer

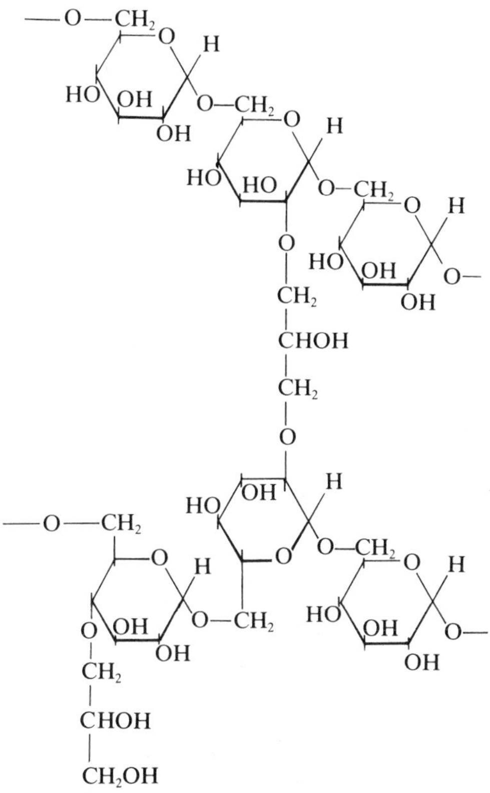

Fig. 4.10. Sephadex

Table 4.3 Properties of dextran polymer gels

Grade	Water regain (g/g)	Molecular weight excluded	Bed volume ml g^{-1} dry polymer	Swelling times (h)
Sephadex G-10	1.0	700	2–3	3
Sephadex G-15	1.5	1 500	2.5–3.5	3
Sephadex G-25	2.5	5 000	5	12
Sephadex G-50	5.0	10 000	10	12
Sephadex G-75	7.5	50 000	13	24
Sephadex G-100	10	100 000	17	48
Sephadex G-200	20	200 000	30	72

grades, while fractionation of proteins, and high molecular weight polysaccharides, can be achieved with gel columns of Sephadex G-75, G-100 or G-200.

Size exclusion chromatography of *Insulin* in 1M acetic acid on Sephadex G-50 is used to limit the amount of higher molecular weight proteins (1%) that may be present.

Table 4.4 K_D values obtained on gel filtration with Sephadex

Substance	K_D for various grades of Sephadex				
	G-25	G-50	G-75	G-100	G-200
Ammonium sulphate	0.9	–	–	–	–
Potassium chloride	1.0	–	–	–	–
Glycine	0.9	–	1.0	–	–
Phenylalanine	1.2	1.0	–	–	–
Tyrosine	1.4	1.1	–	–	–
Tryptophan	2.2	1.6	1.2	–	–
Pepsin	0	0	0.3	–	–
Chymotrypsin	0	0	0.3	0.5	0.7
Trypsin	0	0	0.3	0.5	0.7
Serum albumin	0	0	0	0.2	0.4
γ-globulin (19S type)	0	0	0	0	0
γ-globulin (7S type)	0	0	0	0	0.2

Apparatus and techniques

Chromatographic columns are generally of glass with a diameter to height ratio of between 1:10 and 1:20. Separations involving substances with small differences in K_D may, however, require much longer columns with diameter to height ratios as high as 1:100. A plug of cotton or glass wool at the bottom of the tube supports the gel bed; the tube should be partly filled with water or buffer before insertion of the plug in order to avoid entrapping air bubbles.

The dry polymer must be allowed to swell completely in water or electrolyte solution before filling into the tube. The use of electrolyte solution prevents the beads of polymer from sticking together as may occur in water. Most grades of gel swell rapidly, but appreciable time is required for the gel to reach equilibrium. Fine particles, which if present in appreciable quantity can decrease the flow rate, should be removed by decantation, and the column filled by pouring in the gel suspension consisting of about one part of gel to two of fluid. Even packing is essential to avoid channelling of the column, and to assist this, the flow of eluate should be commenced slowly and carefully as soon as a few centimetres of gel have collected at the bottom of the tube. Continue the addition of gel to the column at a rate to provide a steady rise in the column bed, the surface of which should remain both even and horizontal throughout, until packing is complete. The upper surface of the column should be protected from disturbances by a filter paper and/or a plastic net. When the column is completely filled, connect it to a suitable reservoir of eluant, and allow the column to run freely overnight before use.

Remove most of the excess eluant from the top of the column by suction. Allow the remaining eluant to drain down to the surface of the gel, and just as the level of eluant reaches the surface of the gel, stop the flow of effluent and carefully run in the sample solution from a pipette. The sample should be applied in small volume and in as narrow a zone as possible if the maximum column efficiency is to be achieved. Open the stopcock to re-commence the effluent flow, and as the last of the sample solution reaches the surface of the gel, wash the last traces from the top surface with a small volume of eluant; add sufficient eluant to give roughly a 5 cm head of fluid, and maintain column flow and elution by re-connecting to the reservoir.

The effluent is best collected in fixed volumes by an automatic fraction collector, and may be monitored by chemical or preferably physico-chemical techniques, such as ultraviolet absorption spectrophotometry.

Practical experiments

Experiment 1   *Determination of the percentage w/w of strychnine in syrup of iron (II) phosphate with quinine and strychnine*

The assay depends upon the observation that strychnine hydrochloride is soluble in chloroform.

Method Introduce into a chromatograph tube a layer of Celite 545 (1 g) previously mixed with *water* (1.0 ml). Pack well with the aid of a flat-ended glass rod and add a second layer of Celite 545 (3 g) well mixed with 2M hydrochloric acid (3.0 ml). Pack firmly. Mix the sample (about 1 g, accurately weighed) with Celite 545 (5 g), 2M hydrochloric acid (2.0 ml), and *water* (1.0 ml) and transfer quantitatively to the column using more Celite 545 (1 g) to clean the beaker. Transfer the Celite 545 to the column, clean the beaker with a small wad of cotton wool and use the latter to push down any traces of Celite 545 adhering to the sides of the chromatography column.

Wash the column with anaesthetic ether saturated with *water* (125 ml in portions) and elute the strychnine with chloroform saturated with *water* (150 ml) in portions. Combine the chloroform eluates and remove the solvent with the aid of gentle heat and a stream of air. Dissolve the residue in 0.1M hydrochloric acid with the aid of heat, cool, transfer to a 25 ml volumetric flask and dilute to the mark with acid.

Measure the absorbance at the peak absorption (254 nm) and calculate the percentage w/w of strychnine, using $A(1\%, 1\,\text{cm})$ (p.277) for strychnine at 254 nm, as 375. For this calculation the weight per ml of the syrup must also be determined.

Garratt (1964) refers to the use of a three-point correction procedure (p.288) to allow for irrelevant absorption and gives the equation

$$A_{\text{corrected}} = 6.9(A_{254} - 0.533A_{247} - 0.467A_{262})$$

Experiment 2   *Determination of the primary and secondary glycosides in digitalis leaf, both calculated as digitoxin*

Method Weigh the powdered leaf (4 g), add ethanol (70%, 40 ml), shake gently for 1 hour and filter. To the filtrate (5 ml), add *water* (35 ml), solution of lead acetate (15%, 5 ml) shake well and filter (no. 54 paper). Transfer the filtrate (40 ml) to a separator, add ethanol (95%, 20 ml) and extract with chloroform (3 × 25 ml). Bulk the extracts, dry with sodium sulphate, filter and adjust the filtrate to 100 ml with chloroform. Evaporate 25 ml of the solution to dryness at a temperature not exceeding 65°. Dissolve the residue as completely as possible in ethanol-free chloroform (2 ml) and transfer to a column of cellulose powder (3 g) previously prepared with the aid of ethanol-free chloroform. Elute with ethanol-free chloroform (30 ml), using the first 10 ml to wash out the flask used for evaporating the chloroform solution. The eluate is fraction A. Continue the elution with chloroform: ethanol (90:10, 30 ml), collecting the eluate in the flask used for evaporating the 25 ml of chloroform solution. This is fraction B. Evaporate each fraction to dryness.

Dissolve each residue in separate portions of methanol (each of 5 ml) and add to each, sodium picrate reagent (1% picric acid in 0.5% sodium hydroxide freshly prepared, 5 ml). Measure the absorbance at 485 nm after 25 min using a blank of reagents and calculate the amount of secondary glycosides (fraction A) and primary glycosides (fraction B) both calculated as digitoxin. $A(1\%, 1\,cm)$ for digitoxin is 190 in this reaction.

Note The rather large quantity of leaf and ethanol is to ensure that a representative sample is obtained and to allow for adequate filtrate. Similarly, 40 ml of filtrate from the reaction with lead acetate (a decolorising agent) is used to avoid the difficult washing procedure which would be required if the whole of the filtrate were to be used. The chloroform used in preparing the column and for eluting fraction A must be free from ethanol, otherwise some of the primary glycosides will appear in that fraction. To confirm that separation is complete, examine fractions A and B obtained in a separate experiment, by paper or thin-layer chromatography.

Experiment 3   *Determination of phenothiazine in the presence of diphenylamine and carbazole*

Crude phenothiazine may contain diphenylamine, carbazole and pheno-thiazone as likely impurities and Bailey *et al.*, (1963) developed an elegant chromatographic method to determine phenothiazine in the presence of these impurities. Although in practice this assay would now be carried out using HPLC with spectrophotometric detection, the experiment is given here to illustrate good column chromatographic technique.

Solvent system Shake acetonitrile (1 volume) with hexane (10 volumes); use the lower layer as stationary phase and the upper layer to develop the chromatogram.

Column Mix Celite 545 (25 g) with stationary phase (12.5 ml) and transfer to the chromatographic tube (70 × 2 cm) in small portions. Pack each portion firmly with a flat-ended glass rod. The prepared Celite 545 is bulky and a large tube is required.

Method Dissolve the sample (about 100 mg accurately weighed) in methanol and dilute accurately to 100 ml with methanol. Transfer the solution (10 ml) to a 100 ml volumetric flask and dilute to volume with methanol. Transfer 1.0 ml of the dilution (equivalent to 0.1 mg of phenothiazine) to a small beaker and evaporate to dryness. Dissolve the residue in stationary phase (1.0 ml), add Celite 545 (2 g), mix well and transfer to the column. Clean the beaker with more Celite (1 g) and add to the column. Add the developing solvent carefully to the column and adjust the flow rate to about 2 drops per second. Collect about 36 fractions each of 10 ml and measure the absorbance of each at 253 nm against developing solvent as reference. Repeat the procedure using a carefully purified phenothiazine as standard.

Calculate the percentage of phenothiazine in the sample by means of the formula:

$$\% = \frac{A_T \times W_s \times 100}{A_s \times W_T}$$

where A_T = sum of absorbances obtained from chromatogram band for sample;

A_s = sum of absorbances obtained from chromatogram band for standard; W_T = weight of sample (mg); and W_s = weight of standard (mg).

Plot the readings against fraction number and identify any impurities by the position of the peaks:

e.g. diphenylamine is eluted first (fraction numbers 8–12 approximately) followed by phenothiazine (13–17) and carbazole (20–30).

Note The authors of this method used acid-washed Celite 545 for the separation. In the absence of this material, silane-treated Celite (40–60 mesh) proved very satisfactory, the separation of the components being slightly better than that quoted in the original paper.

Experiment 4 *Determination of phytomenadione in phytomenadione injection*

Chromatography is carried out on alumina containing water (4%) to control adsorption, with precautions against exposure to bright light.

Method Prepare a column (17 mm internal diameter) in the normal way with alumina (10 g) and 2,2,4-trimethylpentane. Dilute a measured volume of injection with *water* to produce a solution containing 2 mg ml^{-1}. Extract 3 ml of the solution with 2,2,4-trimethylpentane (25 ml) in the presence of saturated sodium sulphate solution (5 ml) and anhydrous sodium sulphate (5 g), shaking for 3 min. Centrifuge to separate the two phases, transfer the trimethylpentane solution to the column, and elute until about 25 ml of trimethylpentane has passed through the column. Repeat the extraction and chromatography with 4×25 ml of trimethylpentane, and continue the elution until 200 ml has been collected. Dilute the eluate (25 ml) with trimethylpentane (50 ml), and measure the absorbance at about 248 nm in a 1 cm cell using 420 as the A (1%, 1 cm) at 248 nm.

Experiment 5 *Determination of quinine in ethanolic solution by ion exchange chromatography*

A weak cation exchanger (H-form) is used so that the displacement can be carried out with as small a volume of liquid as possible. As absorption is rather slow, a fairly long resin column is needed if the solution is to be put through the column at a reasonable rate. To avoid the need of a correspondingly greater volume of eluting solution (ethanol saturated with ammonia) the column is spilt in two, the bulk of the quinine being retained in the upper column.

Column preparation Use a burette plugged at its lower end with glass wool. Add solvent until the tube is one third filled, and then add a slurry or resin, which has previously been soaked in solvent, until the required length of column is obtained. Wash the resin thoroughly with solvent, maintaining a head of 1 cm of liquid above the resin to avoid drying out.

Method Prepare two 10 cm columns using a weak cation exchange resin (H-form) in 96% ethanol and mount one column directly above the other. Pipette 10 ml of an approximately 10 mg ml^{-1} solution of quinine in 96% ethanol on to the upper column and allow the effluent from this to flow on to the lower column at a rate of one drop per second. Adjust the flow from the lower column to the same speed. When most of the solution has passed through the resin, add 10 ml ethanol (96%) to the upper tube and allow this to pass through the columns in series at the same rate. Repeat twice and reject the effluents from the lower column. Separate the two columns and place them side by side so that their further effluents can be collected in a single vessel. Elute each at the same flow rate with

6 × 10 ml of ethanol (96%) saturated with ammonia. Evaporate the combined effluents to dryness for determination by weighing, or to small volume (2–3 ml) for polarimetric determination. In the latter case make the residual solution with ethanol (96%) washings of the container, up to 25 ml and determine the rotation in a 2 dm polarimeter tube.

Paper chromatography

Paper partition chromatography was developed by Consden *et al.* (1944) as a technique for the separation of amino acids. Paper is used as the support or adsorbent but partition probably plays a greater part than adsorption in the separation of the components of mixtures, as the cellulose fibres have a film of moisture round them even in the air-dry state. The technique is therefore closely allied to column partition chromatography, but whereas the latter is capable of dealing with a gram or more of material, the former requires micrograms. It is therefore an extremely sensitive technique of enormous value in chemical and biological fields.

As in column chromatography, aqueous buffer solutions may be incorporated in ordinary papers to assist in the separation process, or they may be further modified by incorporating organic liquids such as liquid paraffin and hydrocarbon grease to give supports for reversed phase chromatography.

The movement of components on the paper depends on the amount and nature of stationary phase compared with the amount of mobile phase in the same part of the paper, and also on the partition coefficient. R values are frequently called R_F values in paper chromatography and eq. (3) is commonly employed for R_F. Strictly, however, because the rate of movement of the mobile phase at the solvent front tends to be faster than at the position of the component on the paper, it is better to define R_F as

$$R_F = \frac{\text{distance travelled by centre of component}}{\text{distance travelled by solvent front}} \tag{3}$$

R_F values are of considerable importance in paper chromatography, as much information is available concerning the movement of compounds on paper with various solvent systems. They offer a means of tentative identification of components and simplify experimental work in reducing the number of reference substances used in any one experiment, e.g. if examination of a trial chromatogram of a protein hydrolysate reveals no spot corresponding to an R_F value of about 0.2 with liquefied phenol as mobile phase, it is unlikely that aspartic acid is present. Reliance should not be placed upon one solvent only but the behaviour of the sample in several such systems should be correlated with that of reference compounds in the same systems. Some differences in R_F values between those found and those quoted in the literature may well occur, as the conditions used are probably slightly different, e.g. in temperature, stationary phase, size of spots, quality of paper and equilibration times. In order to eliminate or minimise these variables the R_F values are sometimes related to that for a standard substance, e.g. glucose in the investigation of carbohydrates.

The ratio:

$$R_X = \frac{\text{distance substance moves from origin}}{\text{distance reference substance moves from origin}} \qquad (4)$$

is known as the R_X value where X refers to the reference substance, e.g. when glucose is used as a reference substance the term would be R_g value. The ratio is particularly useful where the chromatogram must be allowed to continue for such a time that solvent reaches the end of the paper and drips off the end. R_F values are clearly not possible under these conditions.

R_F values are also used in the determination of structure of certain groups of compounds; this is done indirectly via R_M values. Martin (1949) has stated that the differences between the logarithms of the partition coefficients for adjacent members of a homologous series is constant. Therefore by applying the relationship

$$R_F = \frac{A_m}{A_m + KA_s} \text{ (substituting in equation 2)}$$

$$K = \frac{A_m}{A_s}\left(\frac{1}{R_F} - 1\right)$$

$$\therefore \quad \log K = \log \frac{A_m}{A_s} + \log\left(\frac{1}{R_F} - 1\right)$$

The term $\log(1/R_F - 1)$, is known as R_M. The change in this value (ΔR_M), when a substituent or functional group is added to compounds, should be the same for all the compounds in the same solvent system. Mikes (1966) and Lederer and Lederer (1957) give a table of ΔR_M values for various solvent systems.

Apparatus

The paper functions in the same way as a column but, since evaporation of the mobile phase would occur as development progressed, it must be enclosed in a chamber to prevent such loss. Chromatography tanks are therefore required and the assembly for the two types of paper chromatography are shown in Fig. 4.11 (descending technique) and Fig. 4.12 (ascending technique). Both these methods use vertical papers, but horizontal chromatography is also used for rapid separation of mixtures into their components. In one form with specially cut papers, up to five mixtures may be examined at one time (Fig. 4.13). The solvent rises to the centre of the paper and spreads radially by capillary action to develop the chromatogram.

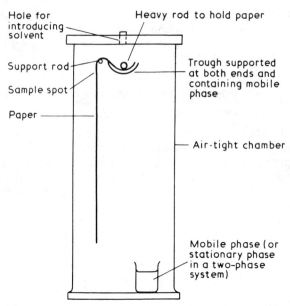

Fig. 4.11. Apparatus for descending paper chromatography

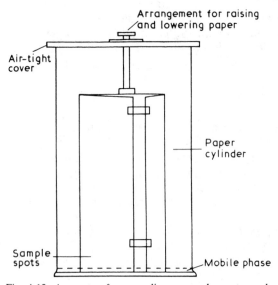

Fig. 4.12. Apparatus for ascending paper chromatography

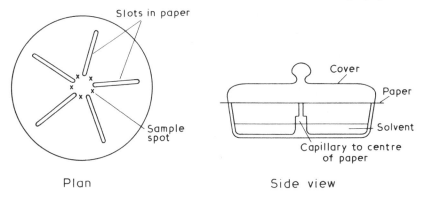

Fig. 4.13. Apparatus for circular paper chromatography

Methods

Descending chromatography Draw a pencil line about 12 cm from one end of the selected paper and apply 2–10 μl volumes of solutions of the mixtures and reference samples to this line. The spots should not be larger than 5 mm and spaced at about 5 cm intervals on the line or, if a larger number of samples is examined, not closer than 2 cm. Prepare the tank by putting some mobile phase or equilibrated stationary phase (as appropriate) in the bottom. Place the paper in the tank so that the end is held firmly in the *empty* trough and the line on which the spots are located is about 3 cm below the support rod (Fig. 4.11). Seal the tank and allow to stand for several hours, or overnight, for equilibrium between paper and volatile phases to be attained. Add the remainder of the mobile phase to the trough as quickly as possible via the stoppered hole immediately above the trough (Fig. 4.11) and allow development to proceed until the solvent front is about 5–10 cm from the end of the paper. If the R_F value is not required, and separation of slow moving components is the main interest, the solvent may be allowed to run to the end of the paper and to drip from the end if necessary.

Remove the paper taking care not to splash mobile phase from the trough onto the paper and allow the paper to dry in a current of air or in an oven if the solvent boils at a high temperature. Locate the compounds by a suitable means (see below).

Ascending chromatography Draw a pencil line about 3 cm from one end of the selected paper and apply 2–5 μl volumes of solutions of sample and reference compounds at about 2 cm intervals on the line. Fold the paper, at right angles to the line, into cylindrical form and hold the edges together with clips.

Prepare the tank (Fig. 4.12) by placing the mobile phase in the bottom to a depth of about 1 cm, and line the tank with filter paper. The filter paper allows a rapid saturation of the atmosphere of the tank with solvent vapour. Unlined tanks may also be used; the solvent tends to run slightly slower under these conditions. Place the cylinder in the tank so that the lower end (with spots) is above the surface of the liquid and allow to equilibrate for 2 or 3 h. At the end of this time lower the paper into the mobile phase and allow development to take place until the solvent front has reached a suitable height (15–20 cm). Remove the paper, dry as described above and locate the components (see below).

Elaborate apparatus is not essential for this method and it can be conveniently carried out in readily available laboratory apparatus, e.g. strips of paper can be attached to corks and hence can be suspended with the ends dipping into solvent contained in test tubes or measuring cylinders. On this small scale, development times are short. Although not more than about two spots can be placed on the paper, several assemblies can be prepared and used at the same time, so that a fairly rapid assessment of solvent systems for the best separation of components can be made.

Two-dimensional chromatography Draw a pencil line along one side of the chromatography paper and place the sample solution (2–5 μl) on the line about 3 cm from its end. Develop the chromatogram in one solvent system by the method of ascending chromatography. Dry the paper and refold it into a cylinder at right angles to the first. Develop with a second solvent system, dry and locate the components of the mixture. Identify by treating reference compounds in the same way. The separation of a five component mixture might appear as shown in Fig. 4.14 and Experiment 6 (Polymyxin B Sulphate) is a typical example.

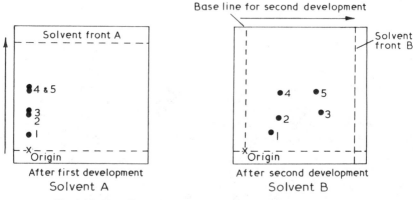

Fig. 4.14. Two-dimensional paper chromatography (diagrammatic)

Detection of components

Components of mixtures are located more readily in paper chromatography than they are in column chromatography, as it is easier to apply both chemical and physical methods if no visible colour is present.

Irradiation with light of 254 nm often reveals those compounds which absorb in the ultraviolet region, as dark areas on a pale purple background. The irradiation itself may also induce fluorescence either immediately, e.g. quinine and ergot alkaloids, or after several minutes, e.g. *p*-chloroacetanilide. Fluorescence is usually better observed with light of wavelength 365 nm, either before or after irradiation with light of shorter wavelength.

Chemical reagents for the development of colour are numerous as they react either with a specific group in the molecule to be detected or they are non-specific. Typical common reagents are listed in Table 4.5.

The reagents are applied most conveniently as a fine spray by means of an atomiser, except where vapour is employed, as with iodine. They are unpleasant in the mist form and should be used in a fume cupboard with a good draught. If a portion only of the chromatogram is to be developed, e.g. to detect the position of the required component on the chromatogram of the mixture, the position of the reference spot can be determined by laying a strip of the filter paper impregnated with the reagent along the development path of the reference substance. Alternatively, this path only may be sprayed if the remainder of the chromatogram is protected by means of stiff paper or cardboard.

Table 4.5 Common developing reagents for paper chromatography

Reagent	Remarks
Iodine vapour	Organic bases give brown spots
Potassium permanganate (1%) in acetone/water	General reagent: yellow or pale spots on pink background
Bromocresol green (0.05%)	Acids give yellow spots on green background
Ninhydrin (0.1%) in water-saturated butanol	Amino acids give purple spots on heating the paper for 5–10 min at 100°
Sodium nitrite (0.5%) followed by 0.1% *N*-(1-naphthyl) ethylene diamine dihydrochloride in 1M HCl*	Sulphonamides give purple or red spots if a primary aromatic amine is present
3,5-Dinitrobenzoic acid (1%) in 1M sodium hydroxide solution	Cardiac glycosides give red spots
2,4-Dinitrophenylhydrazine (saturated) in 2M HCl*	Aldehydes and ketones give yellow spots
Dragendorff's reagent diluted with glacial acetic acid (1:5)	Alkaloids give brown or orange spots

* The chemical reactions involved are described on p.300 onwards

Specialised means of detection may be used for some compounds, e.g. radioactive materials, by autoradiography or Geiger–Müller counter; and antibiotics by laying the chromatogram on nutrient agar inoculated with an appropriate strain of micro-organism—the zones of inhibition of growth indicate regions of active material.

Quantitative measurements

It is not possible to standardise the conditions of an experiment to such an extent that the results from one chromatogram can be applied to another carried out later. Therefore, the sample to be examined, and the standards, must be applied to the same paper, taking the utmost care to maintain the size of spots as uniform as possible. After development, the concentration of the required component can be determined by making use of a suitable test, e.g. ergometrine can be determined in extracts of ergot by means of its fluorescence in ultraviolet light. Visual comparison of sample with standard spots enabled Foster *et al.* (1949) to determine ergometrine with sufficient accuracy

(20%) for it to serve as a useful screening test for its presence in samples of ergot.

Determination of the component can also be carried out on the paper when it forms a coloured spot with a suitable reagent. This method generally involves measuring the area of the coloured spot and comparing it with the areas of standard spots. A graph is constructed by plotting the logarithm of the weight of component against the area of the spot.

The measurement of the area of spots can be carried out directly by means of a planimeter, or indirectly by cutting out the areas of paper and weighing them. Alternatively, squared paper can be laid over the spots and the area determined by counting the squares.

A more sensitive method than a simple visual comparison of the intensity of colour of spots (cf. fluorescence), is the measurement of the transmission of light through the spot by means of a photodensitometer. In order to reduce the general scattering of light by the paper in this measurement, it is convenient to treat the paper with an oil, providing it has no untoward effect on the spot, such as causing the colour to run. Reflectance measurements also offer a means of measurement.

For substances having characteristic absorption, the paper containing the required component can be cut out and the component eluted with a polar solvent for spectrophotometric examination. A blank should be carried out on a similar area of paper adjacent to the spot to allow for non-specific absorbing materials that may be eluted by the solvent. The solvent should preferably be added drop by drop to the paper and allowed to run into a suitable receiver. This avoids introducing much fibrous material into the solvent. Alternatively, the paper containing the component is cut into small pieces and a known volume of solvent added. Vigorous shaking followed by centrifugation yields a clear supernatant layer for spectrophotometric examination. This method is adopted for Methotrexate and its preparations.

Practical experiments

Experiment 6   *To identify the amino acids obtained on hydrolysis of Polymyxin B Sulphate*

Method Dissolve the sample (5 mg) in *water* (0.5 ml) and hydrochloric acid (0.5 ml) in a 1 ml ampoule. Seal the ampoule, wrap in a thin layer of cotton wool or copper foil and heat at 135° for 5 h. Allow the ampoule to cool, open carefully and transfer the contents to a small crucible. Evaporate to dryness on a water-bath and continue to heat until the odour of hydrochloric acid is no longer detectable. Dissolve the residue in 0.5 ml of *water* and apply the solution (5 μl) as a spot to the chromatography paper. On the same paper place separate spots (5 μl) of an aqueous solution containing 10 μg of leucine, phenylalanine, threonine, serine and α, γ-diaminobutyric acid respectively, and one spot containing 50 μg of a mixture of all five amino acids in equal amounts. Proceed by the general method for paper chromatography (ascending or descending technique, p.109) using the upper layer of the mixed solvent system butan-1-ol, glacial acetic acid, water (4:1:5) as the mobile phase, with the lower layer of the system placed in the chromatography tank. Dry and spray the paper with 0.1% ninhydrin in butan-1-ol saturated with water and heat at about 90° for 5–10 min.

Identify the amino acids in the hydrolysed sample by comparison of the spots obtained from the sample with those from the known amino acids.

Note A certain amount of butyl acetate is formed slowly, and the mixture should be allowed to stand for at least 48 hours before use in order that the equilibrium quantity of ester be present.

Similar separations of amino acids, achieved by thin-layer chromatography, are used in official identity tests for *Polymyxin B Sulphate, Bacitracin Zinc* and *Colistin Sulphate.*

Experiment 7 *To separate a mixture of potassium iodide and potassium iodate*

Method Prepare a 1–2% m/v solution of the sample in sodium bicarbonate solution (1% m/v) and place a spot on paper. On the same paper, place at well-spaced intervals, spots of potassium iodide (1% m/v) and potassium iodate (2% m/v). Proceed by the general method for ascending chromatography using the mobile phase, methanol: *water* (3:1). Allow to dry and detect the potassium iodide spots with filter paper impregnated with acetic acid and potassium iodate and the potassium iodate spots with filter paper impregnated with acetic acid and potassium iodide.

The experiment is simple but it forms the basis of a test for the radiochemical purity of *Sodium Iodide (^{131}I) Solution.* The radioactivity of the chromatogram obtained by adding potassium iodide, potassium iodate and sodium bicarbonate to the solution and treating as described above, should be located in one spot only—that corresponding to potassium iodide.

The addition of sodium bicarbonate is necessary to neutralise any trace of acid in the paper, otherwise iodine would be liberated.

COGNATE EXPERIMENT

Sodium Phosphate (^{32}P) Injection. Phosphoric acid solution is added to the injection and the solvent system is *t*-butanol, water, formic acid (8:4:1). All the activity must be associated with the phosphate spot, the position of which may be determined chemically by using an acid solution of ammonium molybdate.

Experiment 8 *To separate a mixture of oxalic, succinic and glutaric acids*

Method Dissolve the sample in water to give an approximately 3% solution and apply about 10 μl to paper using solutions of the acids (3%) as controls. Proceed as for the general method for ascending chromatography using the solvent system liquefied phenol: formic acid (99:1). Detect the acids by spraying with bromocresol green indicator. If the spots are indistinct, carefully expose the paper to ammonia vapour to give bright yellow spots on a blue background.

Experiment 9 *To examine extracts or tinctures of Digitalis Leaf*

Method Impregnate chromatographic paper with formamide by passing the paper through a shallow tray containing a solution of formamide (30%) in acetone. Allow the acetone to evaporate and apply 0.1 ml aliquots of the tincture, extract or fraction A (Experiment 2, p.103) in the form of a streak (2 cm) along the starting line. Apply in the same way reference solutions of Purpurea glycosides A and B, digitoxin, gitoxin, strospeside and digitalanum verum (or as appropriate for the species of digitalis examined) and develop with chloroform saturated with formamide (descending technique) or xylene: butanone (1:1) (ascending technique). Detect the zones by spraying with a mixture of 25% solution of trichloroacetic acid in ethanol: 3% aqueous chloramine (14:1) followed by heating at

120° for 4 min. Alternatively, a spray of xanthydrol (0.125% in glacial acetic acid) to which is added 1% v/v hydrochloric acid immediately before use, can be used.

To separate the primary glycosides the developing solvent is chloroform: tetrahydrofuran: formamide (50:50:6.5).

The colour developed is not intense and a streak, rather than a spot, gives more readily observed zones. The method is, in part, that of Cowley and Rowson (1963) and it can be applied as a thin-layer technique using cellulose powder (compare barbiturates, TLC Experiment 13, p.126).

Pharmacopoeial applications

Further examples of experiments involving paper chromatography may be obtained by consulting the monographs of the substances and preparations of the British Pharmacopoeia listed in Table 4.6.

Table 4.6 Paper chromatography of pharmacopoeial products

Substance	Method*	Tested for	Mobile Phase	Visualisation
Capreomycin Sulphate	D	Capreomycin I ($\not<$90%)	Propan-1-ol: water: glacial acetic acid: triethylamine (75 : 33 : 8 : 8)	254 nm radiation followed by quantitative spectrophotometry
Ergometrine Maleate	A	Ergot alkaloids and related substances	Special conditions	
Hydroxocobalamin	D	Coloured impurities	Special conditions	
Liothyronine Sodium	C	Di-iodothyronine (2%)	Amyl alcohol: 2-methylbutan-2-ol 13.5M ammonia : water ($5 : 5 : 3 : 3$) Upper layer	Ninhydrin (0.25%) in acetone containing glacial acetic acid (1%)
		Thyroxine sodium (5%)		
Methotrexate Injection Tablets	D	Assay	Special conditions	
Phenformin Hydrochloride	D	Related biguanides	Ethyl acetate: ethanol (95%) water	Special conditions
L-Selenomethionine (^{75}Se) Injection	D	Radiochemical purity ($\not<$90%)	Butan-1-ol: glacial acetic acid: water (60 : 15 : 25)	Ninhydrin (0.5%) in butan-1-ol Radioactivity
Sodium (^{125}I) Iodide and Sodium (^{131}I) Iodide Preparations	A	Radiochemical purity	Methanol : water (3 : 1)	See Experiment 7 Radioactivity

Table 4.6 (*cont*)

Substance	Method*	Tested for	Mobile Phase	Visualisation
Sodium Iodohippurate (^{131}I) Injection	A	Radiochemical purity ($\not<$95%)	Butan-1-ol : water : glacial acetic acid (120 : 50 : 30)	365 nm radiation Radioactivity
Sodium Pertechnetate (^{99m}Tc) Injections	D	Radiochemical purity ($\not<$95%)	Methanol : water : (4 : 1)	Radioactivity
Sodium (^{32}P) Phosphate Injection	A	Radiochemical purity	Propan-2-ol: water: 10M ammonia (75 : 25 : 0.3) + 5 g trichloroacetic acid	Acid molybdate Radioactivity
Vancomycin Hydrochloride	D	Identity	2-methylbutan-2-ol: acetone: water (2 : 1 : 2)	Bacillus subtilis
Vitamin A Ester Concentrate	D	Retinol	Dioxan: methanol: water containing butylated hydroxyanisole (1%) (70 : 15 : 5) Reversed phase system	365 nm radiation
Brilliant Green and Crystal Violet Paint	A	Identity	Ethyl acetate: pyridine: water (55 : 5 : 40). Use upper layer	Self-indicating

*D = Descending; A = Ascending; C = Circular

Thin-layer chromatography

The techniques of paper chromatography and thin-layer chromatography (TLC) are similar in that they are both 'open bed' techniques in which substances are separated by the differential migration that occurs when a solvent flows along a thin layer of paper (paper chromatography) or fine powder spread on a glass or plastic plate (TLC). For most substances TLC offers a faster and more efficient separation than paper chromatography, and the majority of paper chromatographic separations have now been superseded by TLC procedures.

Although the use of thin layers of adsorbent on glass plates was described as early as 1938, many difficulties were encountered and the technique in its present form was developed much later when suitable materials and apparatus became available as a result of the work of Kirchner (1951–8) and of Stahl (1956–8). TLC has achieved phenomenal success not only in its application to analytical problems (µg scale) and preparative work (mg scale) for which 'thicker' layers can be used, but also in investigating conditions for gravity-feed column and high performance liquid chromatographic methods.

The substances most frequently used as coating materials are silica gel, alumina and cellulose and to give stable layers they often contain binders such as calcium sulphate (gypsum) or starch. A test for *Adhesive Power* in the European Pharmacopoeia specifies that when a vertical jet of air (1 mm diameter at 2 atmospheres pressure) is directed onto the prepared plate in a horizontal position no particles should be displaced until the jet is not greater than 5 cm from the surface (Silica gel H) or 3 cm from the surface (Silica gel G). The coating material may also contain an inorganic fluorescent indicator (e.g. zinc silicate) which fluoresces when irradiated at a suitable wavelength. The wavelength of irradiation is specified by a subscript (e.g. silica gel GF_{254} which requires irradiation with the short wavelength line at 254 nm from a mercury lamp). Silica gel and alumina are available with different specific surface areas and these grades are identified by a number, e.g. silica gel 60 (or 40 or 150) which indicates the mean pore size in Angstroms $(10^{-10}$ m). The particle size of silica gel for TLC is 10–40 µm (average 15 µm).

Although the term 'adsorbent' is frequently used it must be remembered that adsorption may not always be the principle on which the separation of components of a mixture may be achieved. In fact, any of the four fundamental mechanisms of separation described above for column chromatography, i.e. adsorption, partition (including reversed-phase partition), ion exchange and molecular exclusion, may be carried out on thin layers of a suitable material. Athough a combination of mechanisms may be involved in any one chromatographic system, the principal mechanism is generally evident from the nature of the thin layer and the solvent system. For example, if the coating material is silica or alumina activated by heating to 110° for 1 h and the solvent system is a mixture of organic solvents, the principal mechanism is adsorption chromatography. Other examples are given in Table 4.7.

Table 4.7 Examples of materials for thin-layer chromatography

Coating substance	Solvent system	Mechanism of separation
Silica or alumina (activated)	Chloroform : methanol (9 : 1) toluene:acetone (1 : 1)	Adsorption
Diethylaminoethyl cellulose	0.1M aqueous NaCl	Ion exchange
Cellulose or silica (unactivated)	Butan-1-ol : acetic acid: water (4 : 1 : 5)	Partition
Paraffin oil, or silicone oil coated on silica	Acetic acid : water (3 : 1)	Reversed-phase partition
Sephadex G-50	Aqueous buffer	Molecular exclusion

Preparation of the plate

Although high quality pre-coated plates of adsorbent on glass, aluminium or plastic are readily available from laboratory equipment and chemical suppliers, the preparation of plates may be easily carried out in the laboratory for a fraction of the cost of commercial plates.

General method

The sizes of glass plates for use with commercially available spreaders are usually 20 × 20, 20 × 10 or 20 × 5 cm.

Mix the adsorbent (30 g) in a mortar to a smooth consistency with the requisite amount of water or solvent specified in the manufacturer's instructions and transfer the slurry quickly to the spreader. Spread the mixture over 4–5 plates (20 × 20 cm) or a proportionately larger number of smaller plates and allow the thin layers to set (about 4 min when $CaSO_4$ is present). Transfer the plates carefully to a suitable holder and after a further 30 min, dry at 100–120° for 1 h to activate the adsorbent. Cool and store the plates in a desiccator over silica gel. The thickness of the moist thin layer should be about 0.25 mm.

Special methods

(a) The adsorbent may be mixed with aqueous solution of salts, acids, bases, buffer solutions and water-soluble organic phases, e.g. formamide, for special purposes. The plate is prepared as described under the general method above except that the activation process is omitted. Partition rather than adsorption is the means whereby separation of the components of mixtures is achieved on these plates.

(b) Reversed phase systems. After preparation as described under the general method, dip the plate into a solution of the appropriate phase, e.g. silicone oil or hydrocarbon grease in light petroleum, and allow the solvent to evaporate. Alternatively, develop the plate with a solution of the appropriate phase and allow the solvent to evaporate.

(c) Preparative thin layer. The layers are 0.5–2 mm thick, prepared as described under the general method, but using a smaller quantity of water and allowing a longer time for the initial drying of the plate.

(d) In the absence of a commercial spreader, usable plates can be prepared by using strips of adhesive tape as guides for thickness of layer and for holding several clean plates in a row. The slurry is poured onto the row of plates and spread quickly over them by means of a glass rod pushed along the adhesive tapes. The tapes are removed and the plates dried and activated in the normal manner.

(e) Microscope slides are conveniently coated by a dipping technique in the following way: prepare a slurry of the adsorbent by shaking with chloroform or chloroform–methanol (2:1) and insert two microscope slides (back to back) into the slurry. Withdraw the slides, allow to drain, separate the slides and dry.

(f) The same technique is adopted for coating the outside of test tubes except that the test tube is inverted after removing it from the slurry.

(g) The slurry, prepared in the normal way, is sprayed onto the surface of glass plates, using a laboratory spray gun.

(h) The adsorbent, mixed with an organic solvent, e.g. chloroform or ethyl acetate, is distributed evenly over a glass plate by careful tilting, and, after evaporation of solvent, is dried in the normal way.

(i) Layers of loose adsorbent can be spread on plates using tapes as guides for thickness and a glass rod as scraper to smooth the powder. Alternatively, the glass rod can be lifted clear of the glass plate to the required degree, by means of thin rubber tubing at each end of the rod. The final plates must be developed in the horizontal or slightly inclined position.

In all methods the plates should be tidied before use by cleaning the edges and backs (microscope slides).

Development of chromatogram

Remove about 2 mm of adsorbent from each edge of the plate to give a sharply defined edge.

Apply 2–5 μl volumes of a 10 mg ml^{-1} solution of the mixture and of reference substances in an organic solvent to the plate (Fig. 4.15) with the aid of a template; the spot size should be about 0.3 cm. For weak solutions several applications may be necessary and each spot should be allowed to dry before applying another volume of solution to the same spot. This technique may also be adopted so as to obtain different concentrations of sample but it is not recommended; it is better to use single spots of solutions of different strength. Spot size is thus more uniform and results are found to be more consistent.

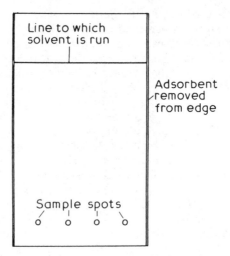

Fig. 4.15. Preparation of the plate for thin-layer chromatography

Allow the solvent to evaporate and transfer the plate to a previously prepared developing tank; this preparation is done about 30 min before insertion of the plate. The tank is similar to that used for paper chromatography (ascending technique) (Fig. 4.12) or may be one specially designed to allow a reproducible angle of inclination of the plate. It should be lined with filter paper dipping into the developing solvent so as to maintain an atmosphere saturated with the solvent vapour in the tank. This method helps to eliminate the 'edge effect' which sometimes occurs, particularly when mixed solvents are used. In this effect a substance moves faster when near the edge than when in the centre of the plate. Allow the solvent to rise a distance of about 10 to 15 cm, remove the plate and allow to dry using heat or a current of air as appropriate.

Detection of compounds

The two most important techniques used to locate colourless drug substances on a thin-layer chromatogram are:

(a) Spraying with a reagent that reacts with the substance to produce a coloured zone. A very large number of reagents has been proposed, which have varying degrees of selectivity for certain functional groups in the molecules. A useful compilation of over 300 spray reagents is the monograph *Dyeing Reagents for Thin Layer and Paper Chromatography* by Merck, Darmstadt, Germany. Some of the most commonly used reagents are listed in Table 4.5 (p.111). The use of highly corrosive non-selective spray reagents is possible with silica gel, alumina and kieselguhr (but not cellulose) thin layers, e.g. 70–80% sulphuric acid, which causes charring of organic compounds on heating to 100°.

(b) Examination under ultraviolet light of plates containing a fluorescent indicator. Any substance that absorbs at the wavelength of excitation (254 nm or 365 nm) quenches the fluorescence (p.365) of the indicator and is observed as a dark zone on a yellow–green background. The sensitivity of detection is thus related to the absorptivity (p.277) of the substance at the wavelength or irradiation.

Recovery of components

Recovery of the components of mixtures from the plates is conveniently done by means of a Craig tube (Fig. 4.16) to remove the adsorbent with component, followed by solvent extraction of the powder. It is particularly useful for preparative thin-layer chromatography in which the mixture is applied as a streak along the starting line and fairly large quantities of powder are therefore removed by suction.

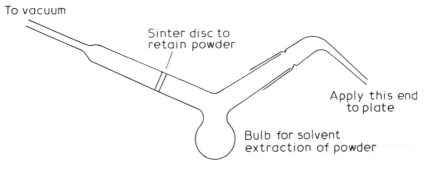

Fig. 4.16. Craig tube for removal of powder from thin-layer plates

Alternatively, the zones are removed by means of a spatula and extracted with a polar solvent such as ethanol. Glass sinters should be used to filter off the adsorbent to avoid contamination with traces of grease or fibres. The quantities of material involved are so small that it must be emphasised that all apparatus must be scrupulously clean.

Quantitative measurements

The techniques described above for the quantitative assay of substances separated on a paper chromatogram are also suitable for thin-layer chromatograms:

(a) Photodensitometric (in the transmission, reflectance or fluorescence modes) or visual comparison of the intensities of the sample spots with those of known loadings of the authentic substance.

(b) Elution of the sample spots into a suitable solvent and determination by an appropriate technique, e.g. visual spectrophotometry or spectrofluorimetry. The measurements should be made at as long a wavelength as possible because direct examination of the extracts in the ultraviolet region may give rise to errors due to irrelevant absorption by substances extracted from the adsorbents.

(c) Measurement of spot areas. The relationship between spot areas on the plate and the concentration of the component has been investigated and several methods of correlation have been proposed. Purdy and Truter (1962) found a linear relationship between the logarithm of the weight of the substance and the square root of the spot area and have provided formulae for general use.

Applications

Identification of substances

The quality control specifications for organic drug substances and their formulations frequently include a TLC test to confirm the identity of the substance or to confirm the presence of the correct drug(s) in the formulation. A confirmatory test of identity is necessary when the assay procedure (e.g. titrimetry or spectrophotometry) is not sufficiently specific for the substance. TLC is usually only one of several identity tests, including melting point determination, infrared spectrophotometry, ultraviolet spectrophotometry, optical rotation measurement and characteristic colour reactions, the results of which together can provide reasonable confirmatory evidence of identity. Confirmation of identity by TLC is based upon the coincidence of R_F or R_X values of the sample spots with those of authentic reference standards in one or more solvent systems (p.107). Many of the drugs and formulations which are the subject of a British Pharmacopoeia monograph include a TLC test among the identity tests. A representative selection of such tests is given in Table 4.8.

Table 4.8 Thin-layer chromatographic identity tests of pharmacopoeial products

Product	Stationary phase	Mobile phase	Detection
Stramonium Leaf	Silica gel G	Acetone: water: 13.5M ammonia (90:7:3)	Sodium iodobismuthate solution + 0.05M H_2SO_4
Tolnaftate	Silica gel GF_{254}	Toluene	254 nm
Gentamicin Cream	Silica gel 60	Chloroform: methanol: 13.5M ammonia (1:1:1) (lower layer)	0.25% Ninhydrin
Diphenhydramine Elixir	Silica gel G	Ethanol (96%): acetic acid: water (5:3:2)	0.25% Chloroplatinic acid 5% KI
Fixed oils	Kieselguhr G impregnated with liquid paraffin	Glacial acetic acid	Iodine vapour
Steroids and preparations	Kieselguhr G impregnated with formamide, liquid paraffin or propane-1,2-diol	Several	10% H_2SO_4 in ethanol, 120°/10 min viewed in daylight and under 365 nm
Phenothiazines and preparations	Kieselguhr G impregnated with 2-phenoxyethanol and propylene glycol 300	Diethylamine: petroleum spirit (2:100) saturated with 2-phenoxyethanol	365 nm also 10% H_2SO_4 in ethanol

A different approach is required for the identification of unknown components of the sample, e.g. manufacturing impurities or decomposition products in a drug substance or formulation, or drug metabolites in a biological specimen. Separation of the unknown substance in μg or mg quantities (the latter requiring preparative layer plates) is followed by elution into a suitable solvent and further treatment as appropriate to present the sample in a form suitable for structural analysis by infrared spectrophotometry, mass spectrometry and nuclear magnetic resonance spectrometry (Chapter 14).

Impurities
One of the most important aspects of the quality of a pharmaceutical material is its impurity pattern. The impurities can often be predicted from a knowledge of the synthetic route and of the stability of the material during the manufacturing process and subsequent storage. In this situation the levels of known impurities can be determined by any of the quantitative techniques discussed above (e.g. photodensitometry or elution/spectrophotometry) using suitable loadings of the authentic substance to calibrate the analytical measurement. Alternatively a limit test may be adopted in which the intensity of the impurity spot is required to be less than that of a specified loading of an authentic sample of the impurity.

If the synthetic route and manufacturing process or the stability profile are not available to the analyst, the identity of the impurities or decomposition products may also not be known and the assay of such impurities presents considerable difficulties. The usual approach to this problem is to establish a limit test for unknown impurities, in which the impurity spots in the chromatogram are required to be less intense than those of the principal spot in a specified loading of the sample after appropriate dilution. For example, in the test for related substances in *Pentazocine*, 10 μl of each of a 2.0% and 0.02% solution of the sample of pentazocine are applied separately to the plate. After development of the chromatogram and location of the spots by exposure to iodine vapour, any impurity spots in the chromatogram of the 200 μg loading are required to be less intense than the principal spot in the 2 μg loading of pentazocine. It is convenient to express the maximum permitted level of impurities in terms of the principal substance, e.g. the maximum permitted level of each impurity in the example above is 1% w/w of the concentration of pentazocine. However, it should be noted that there is an assumption that the impurities and parent substance have the same response to the visualizing reagent.

Table 4.9 records some examples of British Pharmacopoeia substances and preparations for which a TLC test for related substances and impurities is specified.

Table 4.9 Thin-layer chromatography of impurities in pharmacopoeial products

Substance	Tested for	Mobile phase	Detection
Amitriptyline Hydrochloride	Ketone (0.05%)	Carbon tetrachloride: toluene (3 : 7)	4% formaldehyde solution in sulphuric acid and examine with 365 nm radiation
Bisacodyl Suppositories Tablets	Foreign substances (1%)*	Xylene: butan-2-one (1 :)	254 nm radiation
Carbenoxolone Sodium	Related compounds (2%)*	Diethylamine: ethyl acetate: methanol (3 : 14 : 10) for plate prepared with phosphoric acid (0.25%)	254 nm
Cascara Tablets Dry Extract	Frangula	Ethyl acetate: methanol: water (100 : 17 : 30)	Nitrosodimethyl-aniline followed by potassium hydroxide in ethanol (50%) and heating at 105°
Chlorcyclizine Hydrochloride Tablets	N-Methylpiperazine	Chloroform: methanol: 13.5M ammonia (90 : 8 : 2)	Platinic chloride (0.16%): potassium iodide (2%)

Table 4.9 (*cont*)

Substance	Tested for	Mobile phase	Detection
Chlorpropamide Tablets	p-Chlorobenzene sulphonamide and NN'-dipropylurea (0.33%)	Chloroform: methanol: cyclohexane: 13.5M ammonia (100 : 50 : 30 : 11.5)	Sodium hypochlorite followed by potassium iodide in starch mucilage
Desipramine Hydrochloride	Iminodibenzyl (0.2%)	Toluene: ethyl acetate: ethanol: diethylamine (20 : 20 : 4 : 1)	Potassium dichromate (0.5%) in sulphuric acid: water (4 : 4)
Dichlorophen	4-Chlorophenol (0.1%)	Toluene	Ferric chloride: potassium ferricyanide (10% : 0.5%)
Diethylcarbamazine Citrate	N-Methylpiperazine (0.1%)	Ethanol (95%): glacial acetic acid: water (6 : 3 : 1)	Platinic chloride: potassium iodide
Doxycline Hydrochloride (on Kieselguhr G specially prepared)	Related compounds 1, 2 and 2% respectively	Ethyl acetate shaken with disodium edetate solution at pH 7	Ammonia fumes 365 nm
Emetine Hydrochloride	Other alkaloids (2.5%)	Chloroform: methoxyethanol: methanol: water: diethylamine (100 : 20 : 5 : 2 : 0.5)	Iodine in chloroform, heat and examine with 365 nm radiation
Ethambutol Hydrochloride Tablets	(+)-2-Amino-butanol (1%)	Ethyl acetate: glacial acetic acid: hydrochloric acid: water (11 : 7 : 1 : 1)	Cadmium and ninhydrin solution and heat at 90°
Nitrazepam Tablets	Decomposition and related substances 0.5%	Nitromethane: ethyl acetate (85 : 15)	254 nm radiation
Pentagastrin	Foreign substances (2%)*	Ether (anaesthetic): glacial acetic acid: water (10 : 2 : 1)	4-Dimethylamino-benzaldehyde reagent
Promethazine Hydrochloride Injection Tablets	Isopromethazine hydrochloride (1%) Other substances (0.5%)*	Hexane: acetone: diethylamine (85 : 10 : 5)	254 nm radiation

---

* Calculated on the basis that the impurities and parent substance are detected with equal sensitivity.

Special techniques

High performance thin-layer chromatography

Improved techniques for the manufacture of high performance liquid chromatographic grades of silica in which the particles are small (about

5 μm) and very uniform in size resulted in the introduction in the mid-1970s of similar grades of silica gel as adsorbents for thin-layer chromatography. These high performance silica gels give a more efficient and reproducible separation than conventional grades of silica. Consequently the plates are smaller, typically 10 cm in length (cf. 20 cm of conventional plates), and the development time is much shorter, typically only a few minutes. If small volumes (1–5 μl) of the samples are applied to the plates, the spots in the developed chromatogram are very compact and are quantified by means of a photodensitometer with greater sensitivity and precision than is possible with conventional plates. The term **high performance thin-layer chromatography** (HPTLC) is used for the technique in which substances are accurately and precisely assayed using high performance grades of silica gel.

Concentration zones

A number of suppliers of pre-coated plates produce plates with a concentration zone, as a modification of conventional chromatoplates. In these plates the lower 2–3 cm of silica is replaced by a layer of silica of very large pore diameter (about 50 000 Å), which has no chromatographic activity.

The sample is applied to the concentration zone in the normal way, and when the plate is placed in the solvent the sample is carried quickly to the boundary between the two layers of silica, where it is concentrated as a narrow band. When the separation commences in the upper layer of silica the substances migrate as very compact spots or bands which are able to be detected and quantified in the developed chromatogram with greater sensitivity and precision.

Practical experiments

Experiment 10   *To examine a petroleum ether extract of an umbelliferous fruit*

Plate Use kieselguhr G prepared with 0.05% aqueous fluorescein instead of water and chloroform:toluene, 1:1 by volume as the developing solvent.

Method Shake the coarsely powdered fruit (0.5 g) with petroleum ether (b.p. 60–80°, 5 ml) and allow the suspension to settle. Transfer approximately 30 μl of the supernatant extract to the plate by successive applications to produce a spot as small as possible. Develop the plate until the solvent has risen about 15 cm.

Allow the plate to dry, examine in ultraviolet light (365 nm) and note the presence of any dark fluorescence-quenching spots against the bright yellow fluorescent background. Expose the plate briefly to bromine vapour, re-examine under ultraviolet light and note any persistent fluorescein fluorescence against a dull background. Spray the plate with a saturated solution of 2,4-dinitrophenylhydrazine in 1M hydrochloric acid and note any orange spots. Finally, after air-drying the plate, spray with sulphuric acid containing vanillin (1%).

The results of the various methods of examination of the plate serve to identify many umbelliferous fruits, but do not distinguish caraway from dill nor anise from fennel. Table 4.10 correlates the results obtained with some fruits, and for a complete account the original paper should be consulted (Betts, 1964).

Table 4.10 Thin-layer characteristics of petroleum ether extracts of umbelliferous fruits

Fruit	Fluorescence	R_F	Compound	Br_2	2,4D	Vanillin
					Other tests	
Anise or Fennel	Quenched	0.70	Anethole	+		
		0.40	Anethole		+	
Parsley	Quenched	0.65	Apiole (?)	+	+	
Indian Dill	Quenched	0.60	Dillapiole (?)	+		
		0.40	Carvone	+	+	
Ajowan	Quenched	0.40	Thymol	+		red–mauve
Cumin		0.60	Cuminaldehyde	−	+	
		0.55				
	No quenching					
Caraway	but visible	0.4	Carvone		+	
Dill	after Br_2					mauve–
Coriander		0.25	Linalol			brown

The reasons underlying the adoption of the identification procedure are as follows. (a) Many substances are capable of quenching fluorescence and any such components of the oil show up as dark areas against the bright yellow fluorescence of fluorescein. (b) This test alone is not very informative, but additional information is obtained by exposing the plate *briefly* to bromine vapour. The sodium salt of fluorescein is converted to that of tetrabromofluorescein (eosin) which does not fluoresce, so that if any of the components absorb bromine preferentially, e.g. unsaturated systems, the fluorescein in the areas of those components remains unaffected and a persistent fluorescein fluorescence is observed. Over exposure of the plate to bromine will render this test useless; compare, in this respect, the use of ammonia vapour to render acid components visible as yellow spots against a blue background of indicator in Experiment 8 (paper chromatography, p.113). (c) Dinitrophenylhydrazine reacts with aldehydes and ketones to yield dinitrophenylhydrazones which vary in colour from yellow to orange–red or very dark red depending upon the carbonyl compound. Such components are therefore readily detected by the appearance of pale yellow to red spots on the plate. (d) 1% Vanillin in sulphuric acid is not a specific reagent for one class of compound but is useful in the present instance in giving a variety of colours, e.g. limonene (yellow to brown colour), linalol (mauve–brown), thymol (red–mauve) and khellin (bright yellow).

Experiment 11   *To detect* p-*chloroacetanilide in Paracetamol*

Plates Kieselgel G, 250 μm thick on 20 × 20 cm glass plates.
Solvent Cyclohexane, acetone, diisobutylketone, methanol and *water* (100:80:30:5:1).

Standard solution Dissolve 4-chloroacetanilide-free paracetamol (1.50 g) in methanol (8 ml) add 0.045% solution of 4-chloroacetanilide in methanol (1 ml) and dilute to 10 ml with methanol.
Sample solution Dissolve paracetamol (1.50 g) in methanol and dilute to 10 ml.

Method Apply as separate spots to the plate 2 μl of each of the standard and sample solutions.

Develop the plate until the solvent has risen 15 cm past the spots. Dry the plate in a stream of cold air and hold it within 2–3 cm of a source of ultraviolet light (254 nm) for 10 min and then examine under ultraviolet light (365 nm).

Compare the intensity of fluorescence of the spot due to *p*-chloroacetanilide in the standard with those (if any) in the samples.

The method gives an approximate estimate of the amount of *p*-chloroacetanilide in phenacetin and paracetamol as such or in tablets. Savidge and Wragg (1965) suggest, however, that it is better suited for use as a limit test.

Experiment 12   *To separate a mixture of sulphonamides*

Method Prepare a 0.5% solution of the sulphonamides in the solvent system acetone:diethylamine:methanol (20:4:3) or chloroform:methanol:*water* (32:8:5) and 0.1% solutions of reference sulphonamides in the same solvent. Apply 5–10 μl aliquots on a thin layer of silica gel (250 μm) previously activated by heating at 100–105° for 1 h and cooling in a desiccator. Develop in the solvent system above, and after air-drying, examine under ultraviolet light to locate and identify the spots as dark areas. Alternatively, detect the sulphonamides by spraying with 0.05% sodium nitrite followed by a solution of *N*-(1-naphthyl) ethylenediamine dihydrochloride (0.1 g) in water (30 ml) and hydrochloric acid (10 ml). The chemical reactions involved are discussed on p. 301.

Experiment 13   *To separate a mixture of barbiturates*

Method Prepare a 0.5% solution of the barbiturates in chloroform and 0.4% solutions of reference barbiturates. Apply 5–10 μl aliquots to a plate prepared with cellulose (without binder, 25 g) made into a slurry with 1M potassium nitrate (50 ml) and spread to a depth of 250 μm. Dry overnight. Develop with butan-1-ol:pentan-1-ol (1:1) which has been saturated with solution of ammonia.

Spray with a solution of mercury(I) nitrate in 0.1M nitric acid and identify the barbiturates as white spots.

Chatten and Morrison (1965) have separated barbiturates on a silica gel plate made up with 0.1M sodium hydroxide and have proceeded to determine the barbiturates quantitatively by isolation as the mercuric salts, treatment with dithizone and measurement of the absorbance of the mercuric-dithizone complex.

Experiment 14   *To show the presence of strychnine and brucine in Nux Vomica Mixture (acid or alkaline)*

Plate Silica Gel G.
Solvent Toluene, ethyl acetate, diethylamine (70:20:10).
Standard solutions 0.1% Strychnine hydrochloride and 0.1% brucine sulphate, both in ethanol.

Method Mix the preparation (10 ml) with dilute sulphuric acid (10 ml) and extract with two successive portions of chloroform each of 10 ml. Reject the chloroform extracts and make the aqueous solution alkaline wih dilute ammonia solution. Extract with four successive portions of chloroform each of 10 ml, evaporate the combined extracts to dryness, cool and dissolve the residue in ethanol (0.5 ml). Apply a portion (10 μl) to the prepared plate and, separately, 10 μl portions of the standard solutions. Develop the chromatogram, heat the plate at 105° for 30 min and spray with dilute potassium iodobismuthate solution. The spots in the test chromatogram should correspond in position and colour to the spots in the chromatograms of the standard solutions. Subsidiary spots should be ignored.

COGNATE EXPERIMENT

Ipecacuanha Mixture Paediatric Use the solvent system chloroform, diethylamine (90:10) along with cephaeline hydrochloride (0.1%) and emetine hydrochloride (0.1%) both in ethanol, as standards.

Experiment 15 *To examine Dexamethasone for the presence of related foreign steroids*

Method Prepare a silica gel G plate and lined tank as previously described using as the mobile phase dichloromethane, solvent ether, methanol and *water* (77:15:8:1.2). Apply separately to the plate, 1 μl of each of three solutions (1) 1.5% Dexamethasone, (2) 1.5% dexamethasone EPCRS and (3) 0.03% of each of prednisolone EPCRS, prednisone EPCRS and cortisone acetate EPCRS, all solutions being in chloroform: methanol (9:1). Allow the mobile phase to rise 15 cm from the line of application, remove the plate from the tank, allow to dry and detect the compounds with alkaline tetrazolium blue solution (Note 1).

The principal spot from solution (1) corresponds with that from solution (2) in position, colour and intensity (identity) and any other spot from solution (1) is not more intense than the proximate spot from solution (3) (limit on related foreign steroids) (Note 2).

Note 1 The reagent is prepared immediately before use by mixing tetrazolium blue (0.2% in methanol, 1 volume) with sodium hydroxide (12% in methanol, 3 volumes).

Note 2 The method described for Dexamethasone is Method A of the British Pharmacopoeia and this also applies to some others as indicated under cognate determinations. Method B, also indicated therein, requires a mobile phase of 1,2-dichloroethane, methanol and *water* (95:5:0.2) and for solution (3) a mixture (0.03%) of each of prednisone EPCRS, prednisolone acetate EPCRS, cortisone acetate EPCRS, and deoxycortone acetate.

COGNATE DETERMINATIONS

Beclomethasone Dipropionate (B)
Deoxycortone Acetate Implants (B)
Fludrocortisone Acetate (B)
Fluocinolone Acetonide (A)
Fluocortolone Hexanoate (B)
Fluocortolone Pivalate (B)
Hydrocortisone Hydrogen Succinate (A) For solution (3) use hydrocortisone EPCRS and hydrocortisone acetate (0.03% of each).

Hydrocortisone Sodium Succinate (A) Solution (3) is hydrocortisone acetate EPCRS (0.03%) and a fourth solution is also specified as hydrocortisone EPCRS (0.075%).

Hydroxyprogesterone Hexanoate Silica gel HF$_{254}$ is specified with mobile phase cyclohexane, ethyl acetate (1:1). This is an example of the substance itself being used at a lower concentration to serve as a limit on the steroids. The sample is applied at 1% and 0.01% levels and detected by quenching of fluorescence.

Megesterol Acetate The mobile phase is 1,2-dichloroethane, methanol and *water* (92:8:0.5) and detection is by development of a fluorescence (365 nm excitation) after spraying with ethanolic sulphuric acid (10%) and heating at 110° (10 min). Megestrol is used as a control at a limit corresponding to 0.5%.

Methylprednisolone   (A)
Prednisolone Pivalate   (B)

Experiment 16   *To determine the limit of detection of methimazole in carbimazole*

Method Prepare a 1% solution of Carbimazole in chloroform and 0.05, 0.01, 0.005 and 0.0025% solutions of methimazole in chloroform (Note). Use plates of silica gel GF$_{254}$ and silica gel G, a mobile phase of chloroform, acetone (9:1) and apply 10 μl of each solution to the plates. Develop in the normal manner, dry the plates in air and use the following methods of detection:

(a) by examination under radiation of 254 nm—both components show up as purple areas
(b) dilute potassium iodobismuthate solution—pink spots
(c) iodine vapour—brown spots
(d) dichlorobenzoquinonechlorimine—yellow spots.

Comment on the results and the specification limit of 0.5%.

Note Methimazole is readily prepared from Carbimazole by alkaline hydrolysis and extraction into chloroform, a reaction which serves as an identity test.

Gas chromatography

The suggestion that separation of the components of a mixture in the gaseous state could be achieved by partition column chromatography using a gaseous mobile phase was first made by Martin and Synge in 1941 although the first description of instrumentation and applications did not appear until 1952 (James and Martin). Since then the technique of gas chromatography has undergone several significant advances in instrumentation and now ranks among the most important techniques in pharmaceutical analysis.

The technique requires the vaporisation of the sample, which is carried through a prepared column, at a suitable temperature, by a stream of carrier gas (the mobile phase). During the passage of the vapour of the sample through the column, separation of the components of the sample occurs by adsorption effects (p.86) if the prepared column consists of particles of adsorbent only, or by partition effects (p.92) if the particles of adsorbent are coated with a liquid which forms a stationary phase. In the latter instance it is better to use the term **support** for the liquid phase rather than **adsorbent**, as adsorption effects are undesirable in partition columns, and supports are normally treated to eliminate, as far as possible, such effects. The two types of

gas chromatography are therefore gas–solid and gas–liquid chromatography respectively.

It is essential that the sample is stable when vaporised and during its
passage through the prepared, or **packed** column, in order to avoid
decomposition products and the production of a complex chromatogram
as the carrier gas elutes the products from the column. If unstable or
non-volatile compounds must be examined, a reasonable approach to
the problem is the preparation of derivatives of the compounds, e.g.
trimethylsilyl ethers of carbohydrates.

The basic apparatus for the technique is shown diagrammatically in
Fig. 4.17. The carrier gas, at a suitable pressure, passes through the
injection point which is heated to the temperature of the column or, if a
flash heater or specially heated injection block is used, to about 50°
above that of the column. The sample is instantaneously vaporised on
injection and passes down the column with the carrier gas. Ideally, the
various components of the sample are separated in so doing, and a
detector at the outlet of the column produces an electrical output
proportional to the amount of compound emerging from the column.

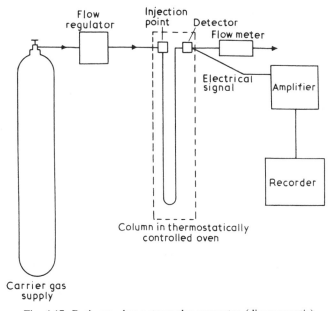

Fig. 4.17. Basic gas chromatography apparatus (diagrammatic)

The response of the detector is amplified and recorded to provide a
chromatogram. Some means of measuring the gas flow is desirable at
the outlet of the detector and the simplest form is a soap bubble flow
meter (Fig. 4.18). The soap solution provides soap films, the rate of
ascent of which through the calibrated tube is a measure of the gas flow.

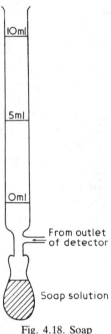

Fig. 4.18. Soap
bubble meter

Carrier gas

The gases used are argon, helium, nitrogen and occasionally hydrogen, but the selection of a carrier gas also depends on the type of detector, and it is convenient to discuss gases in association with detectors (p.134).

Columns

Packed columns are conveniently described as analytical columns when sample solutions of a few microlitres or less are used, and as preparative columns when larger volumes, i.e. 0.5 ml and more, are injected. In the latter instance, the components of a mixture can be collected at the outlet of the detector in a pure form in relatively large quantities, i.e. on a preparative scale. Analytical columns are generally 1–2 m long and about 3–6 mm outside diameter. The tubes from which the columns are made may be of glass or metal but the former is advisable if there is the possibility of interaction between the metal and sample, e.g. steroids are better examined in glass columns. Preparative columns are about 3–6 m long and about 6–9 mm outside diameter but, clearly, these values, and those for analytical columns, are representative only, as satisfactory results may be obtained with shorter or longer columns depending upon the types of compound to be separated and the packed columns available.

The selection of a packed column depends upon the type of analysis to be performed, and gas–solid chromatography is the method of choice for gases. The low solubility of gases such as oxygen, nitrogen, carbon dioxide, hydrogen and methane in liquids makes partition columns of little use. Reliance must be placed on adsorption effects and typical adsorbents are silica gel, alumina, charcoal and molecular sieves. They are affected by moisture and adsorbed gases, and must therefore be prepared for the column by heating and cooling in an appropriate carrier gas before use.

The appearance of the chromatograms often reflects the non-linearity of the adsorption isotherms (p.88, Fig. 4.4), but the disadvantage of tailing peaks can be overcome to a certain extent by 'poisoning' the adsorbent with a small amount of liquid stationary phase. **Active sites**, which are those local parts of a packed column where adsorption effects are more pronounced than elsewhere in the column, are reduced in activity. The columns prepared in this way may appear to be somewhat similar to partition columns (below), but the latter differ in that great care is taken to reduce the natural adsorption effect of the support by the method of manufacture and by preliminary treatment with a silicone compound. Further, with gases, as in this context, partition plays little or no part in the separation procedure.

Columns packed with highly porous organic polymer beads have proved very suitable for the analysis of water, alcohols and low molecular weight gas and liquid mixtures. Hollis (1966) suggests that adsorption plays little part in the separation of the components of the mixtures. The polymers in the Porapak series are based on ethylvinylbenzene–styrene–divinylbenzene (Porapak P), ethylvinyl-benzene–divinylbenzene (Porapak Q) and ethylvinylbenzene–divinyl-benzene modified with polar monomers (Porapak R, S and T). Janâk (1967) has pointed out the possibilities when these materials are applied as lipophilic stationary phases in thin-layer and column chromatography.

Gas–liquid chromatography offers considerably more scope for the analysis of mixtures than does gas–solid chromatography, because many stationary phases are now available. In fact, the number of liquid stationary phases in the catalogues of some suppliers of chromatographic materials exceeds 200 and this can make the task of selecting a stationary phase very difficult. The practice of many chromatographers now is to restrict the number of liquid phases used in their laboratories to approximately six. These preferred stationary phases provide the full range of polarity, from the non-polar Apiezon L grease to the polar polyethene glycol (Carbowax) 20M, and are considered to be adequate for the vast majority of separations that are required to be made by GLC (Table 4.11). The stationary phases that are used most extensively for drugs, particularly basic substances, are the low-polarity polysiloxane phases OV-1 (SE-30) and OV-17, whose structures are represented by

$$\left[\begin{array}{c} R \\ | \\ Si \\ | \\ R \end{array} - O - \begin{array}{c} R \\ | \\ Si \\ | \\ R \end{array}\right]$$

OV-1   R = CH$_3$

OV-17 R = 50% CH$_3$

50% C$_6$H$_5$

Moffat (1975) has recommended the use of 2–3% SE-30 as the preferred stationary phase for drugs and has published the retention indices (p.140) obtained with 480 drugs and commonly used chemicals using this packing material.

Table 4.11 Preferred stationary phases for gas–liquid chromatography

Stationary phase	Solvent for stationary phase	Temperature maximum (°C)	Uses
Apiezon L $\frac{1}{2}$%	Light petroleum	245	Hydrocarbons, steroids and esters
Carbowax 1000 20%	Chloroform	150	Alcohols, chloroform, camphor, essential oils
Carbowax 20M 10% with KOH 5%	Methanol	225	Volatile bases, e.g. amphetamines
Diethylene glycol succinate (DEGS) 10%	Chloroform	190	Methyl esters of fatty acids, nitriles, essential oils
SE-30 2.5%	Chloroform	350	Hydrocarbons, methyl derivatives of barbiturates, general purposes
OV-17 3%	Chloroform	375	General purposes

The support for the liquid phase in partition columns is generally based on silica, e.g. diatomaceous earth or glass beads of suitable mesh size. Diatomaceous earth is most useful because of its porosity and because liquid phases may be incorporated to the extent of 0.5% to 25%. It is a fine powder obtained by grinding the silicaceous skeletons of marine algae (diatoms) and calcining with a small amount of sodium carbonate. The best grades (denoted by AW-DMCS) are deactivated by acid washing and treatment with dimethyldichlorosilane to reduce adsorption on the active sites. Supports should be of a uniform particle size e.g. 80–100 mesh, 100–120 mesh.

The preparation of the packed column is conveniently described by reference to a column of 10% diethlyene glycol succinate (DEGS) on Gas-Chrom Q (80—100 mesh).

Method

Coating of support Dissolve DEGS (1 g) in chloroform (40 ml) in a 250 ml round-bottom flask and add Gas Chrom Q (80–100 mesh; 9 g) to the solution, slowly, and with gentle mixing. Place the flask on a warm water-bath and apply a moderate vacuum to assist evaporation of the solvent. Rotate the flask continually and gently during this stage to avoid depositing the DEGS as a film on the glass and to avoid production of fine particles by mechanical abrasion. After removal of the solvent, heat the residue under vacuum on a boiling water-bath for 1 h.

Packing of column For straight tubes, and for those with U-bends, attach a small funnel to the tube or limb of a U-tube, and pour in sufficient of the coated support to fill about 2.5 cm of tube. Settle the packing by tapping the tube carefully on a rubber bung, and also by means of a mechanical vibrator. Repeat this procedure until the tube or U-tube is filled. Plug the top of the tube or limbs with glass or asbestos yarn.

Preheating of column Insert the packed column in the oven or heating jacket of the chromatograph and pass a slow stream of argon or nitrogen through the column for 5 min; *do not let the issuing gas pass through the detector.* Raise the temperature to 20° above the operating temperature while continuing to pass inert gas, and allow to stand for at least 24 h under these conditions. Allow the column to cool, still with gas flowing, and connect the detector. The apparatus is now ready for use at an appropriate temperature.

The quantities given are sufficient for about 2 m of approximately 6 mm O.D. tubing, and the mesh size is suitable for fairly rapid rates of gas flow without excessive pressure at the injection head. For long columns, and high concentrations of stationary phase, larger mesh sizes, e.g. 60–80 or 40–60, are more convenient. For lower percentages of stationary phase, e.g. 0.5% Apiezon L, the evaporation of solvent requires the utmost care to avoid depositing a rim of the grease on the flask. For these small percentages of stationary phase another technique may be adopted, i.e. add the support to a strong solution of the stationary phase, filter rapidly on a Buchner funnel, allow to drain, and dry the residue. The exact amount of stationary phase on the support must be determined by a suitable assay procedure, e.g. solvent extraction.

Packing a straight tube or limbs of U-tubes offers no particular difficulty but coiled tubes are slightly more difficult to fill. However, by applying a vacuum to one end, the coated support is sucked into the coil which is ultimately filled. If a sufficient degree of vacuum is not obtained the coil must be filled by careful manipulation.

The pretreatment of the column is essential as volatile impurities are removed without contaminating the detector. The value of this treatment may not be particularly evident if non-polar samples are examined, but if polar materials, e.g. esters, are injected, tailing of the peaks may occur. This is often due to adsorption at active sites, but after continual use, the column shows a marked improvement in performance.

If no improvement in performance appears after several days use, several injections of ethereal solutions of dimethylchlorosilane or hexamethyldisilazane at the operating conditions may be tried, the column being disconnected from the detector whilst this treatment is carried out. Silylation should not be carried out on columns containing polyethylene glycol or other hydrogen-bonding stationary phases.

Injection systems

The injection system is an extremely important part of the apparatus and must be such that the sample enters the column as a small 'plug' of vapour. To inject liquid samples, microsyringes (1–10 μl volumes) with needles long enough to inject the sample into the injection port or directly onto the column without interruption of gas flow are ideal. For quantitative results, the silicone rubber septum through which the needle is inserted must be stout enough to remain leak-proof over several injections and while the sample is being injected.

Head-space analysis is a very effective technique for the analysis of volatile components in complex or viscous mixtures containing a high proportion of non-volatile material. The technique involves heating the sample in a closed container, e.g. a glass vial with an aluminium cap and PTFE-coated silicone rubber septum. The temperature (up to 150°) and time of heating (usually 5–20 min) are selected to produce an equilibrium of the volatile components between the sample and the vapour space above the sample. A sample of the head-space vapour (approximately 1 ml) is then injected onto the column either manually or automatically using an automatic analyser. Head-space sampling not only prevents contamination of the injection port and column with non-volatile components, but also increases the sensitivity of the analysis and allows the detection of much lower concentrations of the volatile components than would be possible by the injection of the liquid sample directly.

Detectors

The gas from the column is monitored constantly and any differences between it and the normal gas is recorded, i.e. detectors are, in most apparatus, differential. The chromatogram appears as a record of detector response against time, or, as the gas flow is known, against volume of carrier gas. Fig. 4.19 illustrates the trace obtained with a

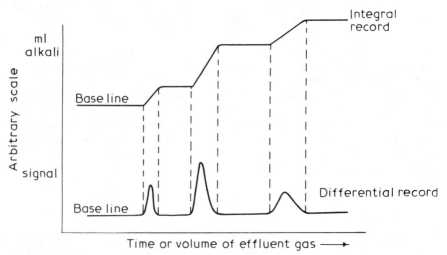

Fig. 4.19. Gas chromatogram of a mixture of three acids using differential and integral detectors

typical detector and includes the appearance of a corresponding curve for an integral detector such as was used by James and Martin in their first experiment with fatty acids using gas chromatography. The trace for the latter is a summation of some property related to the mass of the components passing through the column, e.g. volume of alkali required to neutralise the acid.

Few detectors are selective, i.e. when a peak is recorded no information is normally obtained on the type of compound which causes that peak (but see electron capture detector below). The detectors are therefore universal, which is an advantage. Among other desirable properties for a detector are:

(a) The sensitivity should be high and without instability at high sensitivities.
(b) The volume should be low so that the compound eluted from the column in a small 'plug' of carrier gas is not diluted further within the detector itself.
(c) The response should be rapid and linear with concentration of compound. In practice the detector is calibrated to determine the optimum range.
(d) The response should be unaffected by flow rate of carrier gas and temperature.

To accommodate all these requirements, many types of detector have been made and those in most frequent use are based upon thermal conductivity or thermal effects, and ionisation phenomena.

Katharometer

This is based upon the alteration of the thermal conductivity of the carrier gas in the presence of an organic compound. The principle of the detector is illustrated diagrammatically in Fig. 4.20. The platinum wires

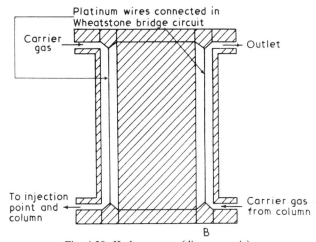

Fig. 4.20. Katharometer (diagrammatic)

are heated electrically and assume equilibrium conditions of temperature and resistance when carrier gas alone passes over them. They are mounted in a Wheatstone bridge arrangement and, when a compound emerges, the thermal conductivity of the gas surrounding wire B alters; hence the temperature and resistance of the wire changes with a concomitant out-of-balance signal, which is amplified and recorded.

The sensitivity of this detector is low by comparison with the sensitivities of other detectors and is affected by fluctuations in temperature and flow rate. Nevertheless it is of particular value for preparative work where a high concentration of a substance occurs in the gas phase. Hydrogen or helium are the best carrier gases for this detector, as nitrogen sometimes gives rise to peak reversal in the chromatogram. The sensitivity of the detector depends upon the difference between the thermal conductivity of the carrier gas and that of the compound eluted from the column. With nitrogen as carrier gas, compounds may have a greater or smaller thermal conductivity than that of nitrogen so that the response of the detector may be either positive or negative. Peak reversal is not likely when using hydrogen or helium. Helium is safer than hydrogen because there is no explosion hazard.

Flame ionisation detector

This detector is relatively simple in design (Fig. 4.21) and operates by change in conductivity of the flame as the compound is burnt.

The change in conductivity of the flame does not arise by simple ionisation of the compounds emerging from the detector. Partial or

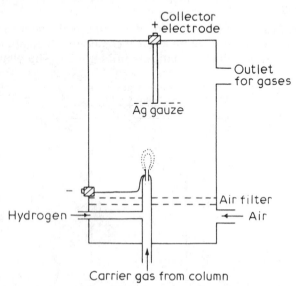

Fig. 4.21. Flame ionisation detector (diagrammatic)

complete stripping of the molecule appears to occur to give charged hydrogen-deficient polymers or aggregates of carbon of low ionisation potential.

The electrical resistance, and hence conductivity of the flame, therefore alters appreciably as a component emerges from the column. The carrier gas is usually nitrogen or argon and is mixed with hydrogen before passing to the burner tip, which is often made of a platinum capillary. This forms one electrode, the other being a silver gauze or brass collector electrode about 1 cm above the top of the flame. Designs vary, and some have the collector electrode as a cylinder surrounding the flame.

The detector is extremely sensitive, has a low background signal, is insensitive to small changes in carrier gas flow and water vapour, and responds to most organic compounds. The sample is, of course, lost in the flame but the advent of stream splitters at the column outlet combined with the high sensitivity of the detector enables a portion only of the effluent to be used for detection. The remainder passes on for recovery of samples if necessary.

The **alkali flame ionisation (thermionic) detector** is a flame ionisation detector with a source of alkali metal salt mounted between the flame jet and the collector electrode. The alkali source usually consists of a glass bead containing the non-volatile salt rubidium silicate fused on platinum wire and is maintained at a negative potential. The AFID can be made specific for nitrogen- or phosphorus-containing compounds or for phosphorus-containing compounds only by altering the polarity of the jet.

In the **phosphorus-only mode**, the jet is switched by means of an external circuit to a negative potential relative to the bead. The combustion of carbon–hydrogen-containing compounds in the detector does not give a signal because the electrons run to earth through the grounded jet. The combustion of phosphorus-containing compounds, however, produces thermionic electrons at the surface of the alkali bead and these electrons pass to the collector electrode and generate a signal. In this mode, an AFID is up to 50 times more sensitive to phosphorus compounds than an FID.

In the **nitrogen–phosphorus mode**, the jet is polarised to the same negative potential as the bead and the flow rate of the hydrogen is reduced to about 1–5 ml min^{-1}. This low flow rate cannot maintain a stable flame at the jet and the hydrogen burns on the surface of the electrically heated bead. The mechanism whereby nitrogen compounds produce a selective response is not fully understood but is thought to result from the migration to the collector electrode of negatively charged cyanide ions which are produced by the reaction of cyanide radical intermediates and the vaporised alkali. In this mode, the AFID is approximately 50 times more sensitive to nitrogen compounds and up to 500 times more sensitive to phosphorus compounds than a standard FID.

Electron capture detector

In this detector the carrier gas is ionised by means of a radioactive source. The potential across two electrodes is adjusted to collect all the ions and a steady saturation current is therefore recorded; the applied potential is normally about 20 V. If a compound containing an electronegative element such as oxygen or, in particular, a halogen, enters the detector, electrons are 'captured' by the compound and a negative ion is produced. Combination between positive and negative ions is very rapid as compared with recombination of positive ions and electrons so that a reduction in current occurs in the detector.

For compounds of high electron affinity argon may be used as carrier gas and, for those of lower affinity, nitrogen, hydrogen or carbon dioxide. The reason for this is that compounds of high electron affinity are able to capture electrons of high energy whereas those of low affinity, e.g. ethers and hydrocarbons, require electrons of low energy. The order of electron energy produced with various carrier gases is:

$$\text{argon} > \text{nitrogen} > \text{carbon dioxide}$$

Nomenclature

A number of terms relevant to gas chromatography (and also high performance liquid chromatography; pp.157–173) are used to describe the chromatographic behaviour of a compound or column.

Retention time (t_R)

This is the time of emergence of the peak maximum of a component after injection (Fig. 4.22). This is the sum of the times the component spends in the mobile phase (t_M) and in the stationary phase. The

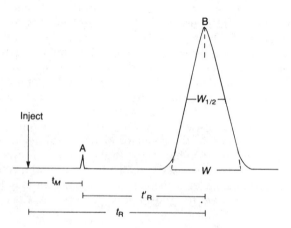

Fig. 4.22. Diagrammatic chromatogram of a pure substance (B) injected along with an unretained substance (A)

adjusted retention time (t'_R) is the time the component spends in the stationary phase and is given by

$$t'_R = t_R - t_M$$

The value t_M is obtained by measuring the time to elute an unretained substance, e.g. air or methane.

The *capacity factor (or ratio) (k)* is the ratio of the time the component spends in the stationary phase to the time in the mobile phase

$$k = \frac{t'_R}{t_M} = \frac{t_R - t_M}{t_M}$$

Retention volume (V_R)

This is the volume of carrier gas required to elute one half of the compound from the column as indicated by the peak maximum and is given by

$$V_R = t_R \times f$$

where f = flow rate of the carrier gas at the outlet pressure of the column and the temperature of the column.

Adjusted retention volume (V'_R)

This allows for the gas hold-up of the column which is due to the interstitial volume of the column and the volume of injector and detector systems. It is given by:

$$V'_R = t'_R \times f$$

Net retention volume (V_N)

The average flow rate of the carrier gas differs from the outlet flow rate because the gas is compressible and a pressure gradient exists down the column. A factor (j) must be used to correct the adjusted retention volume and therefore

$$V_N = V'_R \times j$$

$$V_N = V'_R \times \frac{3}{2}\left[\frac{(P_i/P_o)^2 - 1}{(P_i/P_o)^3 - 1}\right]$$

where P_i = presure of carrier gas at column inlet, and P_o = pressure of carrier gas at column outlet.

Specific retention volume (V_g)

This is the net retention volume per g of liquid phase at 0°

$$V_g = \frac{v_N 273}{TWt_L}$$

where Wt_L = weight of liquid phase, and T = temperature of column (degrees absolute).

Specific retention volumes are convenient data from which experimental retention volumes may be calculated for the particular column conditions in use. The other retention volumes described above apply to columns having particular gas hold-up volumes and amounts of liquid phase, and are therefore applicable only to those columns.

Relative retention volume

The determination of the various retention volumes requires a considerable amount of experimental work which is essential if the information is to be meaningful to other workers. A simpler treatment, which eliminates the need for calculating or measuring many of the parameters while retaining information of value, is to determine relative retention volumes. Retention volumes for compounds are expressed relative to the retention volume of a standard compound on the same column under the same conditions as the compound examined. Therefore, this ratio is given by:

$$\frac{V_N \text{ (sample)}}{V_N \text{ (standard)}} = \frac{V'_R \text{ (sample)} \times j}{V'_R \text{ (standard)} \times j} = \frac{t'_R \text{ sample}}{t'_R \text{ standard}}$$

Relative retention volumes can therefore be represented by ratios of the distances on the recorder chart and are the same as relative retention times.

Retention index

The retention index devised by Kováts (1958, 1965) makes use of the linear relationship between the logarithm of the adjusted retention volume V'_R and carbon number in straight-chain hydrocarbons. For any hydrocarbon C_zH_{2z+2} the retention index, I, is defined as $100z$. Thus for decane $I = 1000$.

The retention index for any drug is obtained from a chromatogram in which the drug is eluted between two straight-chain hydrocarbons C_zH_{2z+2} and $C_{z+1}H_{2z+4}$. Thus, if the compound appears between decane and undecane, the retention index lies between 1000 and 1100. The exact value is calculated from the general formula:

$$I_{\text{temp}}^{\text{St.phase}} = 100N\frac{\log t'_R \text{ (drug)} - \log t'_R \text{ } (A_1)}{\log t'_R \text{ } (A_2) - \log t'_R \text{ } (A_1)} + 100z$$

where t'_R is the adjusted retention time (Fig. 4.22); A_1 is the alkane with z carbon atoms eluting before drug; A_2 is the alkane with $(z + N)$ carbon atoms eluting after drug; and N is the difference in carbon number between the two alkanes.

Note that it is the adjusted retention time which is used in the calculation and this leads to some difficulty in that the gas hold-up time is required when examining the chromatogram. Conditions must therefore be identical for drug and hydrocarbon. For rapid calculation of retention indices, Caddy *et al.* (1973) used the plot of log t'_R *vs* carbon number as a nomogram. Multiplication of the carbon number found for log t'_R (drug) by 100 yielded a retention index of sufficient accuracy for tentative identification purposes in forensic science. The values were obtained on SCOT columns (support coated open tubular columns) of Apiezon L/KOH and Carbowax 20M/KOH. A plot of the retention indices obtained on the former column against those obtained on the latter gave a graph for use in the identification of certain amines.

ΔI values, e.g. ($I_{150°}^{\text{Carbowax}} - I_{150°}^{\text{Apiazon L}}$) have proved useful in the determination of the structure of compounds but the topic is beyond the scope of this chapter. Reference should be made to Kováts (1965) and references cited therein.

Height equivalent to a theoretical plate

A column may be considered as being made up of a large number of theoretical plates where distribution of sample between liquid and gas phase occurs (p.95). The number of theoretical plates (n) in a column is given by the relationship:

$$n = 16\left(\frac{t_R}{W}\right)^2 = 5.54\left(\frac{t_R}{W_{1/2}}\right)^2$$

where W is the peak width, i.e. the segment of the peak base formed by projecting the straight sides of the peak to the base line, and $W_{1/2}$ is the peak width at half height (Fig. 4.22). The latter is usually the easier to measure.

A similar equation which corrects for the hold-up volume (p.139) is used to calculate **effective efficiency** (NEFF)

$$\text{NEFF} = 16\left(\frac{t'_R}{W}\right)^2 = 5.54\left(\frac{t'_R}{W_{1/2}}\right)^2$$

The efficiency or effective efficiency of a column should be determined using a well retained peak, i.e. one having a capacity factor greater than 5. As t_R, t'_R, W and $W_{1/2}$ are proportional to the distances on the recorder chart, the correction for pressure variations in the column is not necessary for this calculation.

The height equivalent to a theoretical plate (HETP) is given by

$$\text{HETP} = \frac{\text{Length of column}}{n}$$

Temperature programming

It is possible for a mixture to contain compounds that boil over a very wide range of temperature, e.g. a mixture containing hydrocarbons from C_{20} to C_{35}. Consequently, the higher members of the mixture may require several hours to be eluted if a fixed temperature (isothermal operation) consistent with resolving the lower members is used. Under such circumstances the temperature of the column may be programmed so that a gradual rise in temperature occurs and elution of the compounds is complete within a reasonable time. Various types of programme may be used, e.g. temperature increasing linearly with time or a linear increase followed by a period of isothermal operation.

To obtain a steady base line, i.e. a base line free from drift, it may be necessary to operate temperature programming with two columns, one of which serves as a reference column so that the effects of such factors as the following are compensated for:

(a) Alteration in viscosity of mobile phase with increase in temperature.
(b) Alteration of gas flow with increase in temperature.
(c) Increase in 'bleed' rate of liquid stationary phase, i.e. the elution of traces of the liquid phase as the temperature increases. Ideally, this would be nil if the column is operated well below the limit for the stationary phase.

Quantitative analysis

The accuracy and precision of a gas chromatographic method depend on a number of factors, most of which are within the control of the analyst.

Column performance

The number of theoretical plates (p.141) in a column is a convenient criterion for assessing the performance or efficiency of a column. A value of 2500 (or more) for a 1.5 m column is indicative of good performance, but in the determination of this value an optimum time for the emergence of the compound occurs: generally, 10–15 min is a useful guide. Also, the value will vary according to the compound chosen for the experiment and the flow rate of the carrier gas. The experiment relates only to a single peak and therefore does not strictly refer to the ability of the column to separate mixtures of compounds; usually, however, the separating ability of the column increases with increase in n. In pharmacopoeial assays using gas chromatography, a minimum value of 600 theoretical plates per metre is required when a column other than the recommended one is used.

The performance of any column is also assessed by reference to the **symmetry factor** of a peak calculated from the expression

$$\frac{y_x}{2A}$$

where y_x is the width of the peak at one-twentieth of the peak height; and A is the distance between the perpendicular dropped from the peak maximum and the leading edge of the peak at one-twentieth of the peak height.

The factor is important in that, when its value lies between 0.95 and 1.05, the peak is reasonably symmetrical and peak heights, rather than areas, of sample and internal standard may be used in calculations (see

Experiment 5). Values outside this range indicate that tailing or fronting of peaks is occurring and that there may not be a proportional relationship between peak height ratio and concentration of the analyte. Peak areas are obtained by integrator, planimeter or by geometrical triangulation methods.

For satisfactory results in quantitative work the **resolution** between measured peaks on the chromatogram must be greater than 1.0. The British Pharmacopoeia states that the resolution should be calculated from the expression

$$\frac{2(t_{Rb} - t_{Ra})}{W_a + W_b}$$

where t_{Ra} and t_{Rb} are the distances along the base line between the points of injection and perpendiculars dropped from the maxima of two adjacent peaks; W_a and W_b are the respective peak widths.

Internal standard

An internal standard (or marker) is a compound added in constant amount to solutions in order to compensate for the small variations in volume of sample injected. These variations are almost inevitable when using volumes of 0.2 to 1 μl and could well exceed the tolerances allowed in pharmaceutical formulations. In calculations or preparation of calibration curves the ratio

$$\frac{\text{peak height of component}}{\text{peak height of internal standard}}$$

is used instead of peak height of component. Thus, as an extreme example, if a volume α μl were injected followed by a volume 2α μl, the peak heights for component and internal standard would be greater for the latter injection than for the former but the ratio should, ideally, be the same. If, however, a mixture consists of two components of similar physical characteristics (see Experiment 13) the composition may be determined by direct injection of the mixture and measurement of peak areas without resort to the use of an internal standard. In assays which use only a single mixture of analyte and internal standard as the standard solution, there is a requirement that the peak height ratio (or peak area ratio) be proportional to concentration. If a non-proportional relationship exists, a calibration series of four or five standard solutions is required.

Normalisation is a method for determining the proportion of one or more volatile components in a mixture, using the area(s) of the component(s) as a percentage of the total area of all the volatile components. The procedure requires the use of an electronic integrator, and a wide-range amplifier is also desirable. There is an assumption that the components have identical response factors in the detection system used and this assumption is approximately true with chemically similar components. The procedure is employed in the British Pharmacopoeia as a limit test for impurities in volatile liquids such as *Halothane*.

Detection

The most useful routine method in quantitative analysis is flame ionisation. Although the flame ionisation detector is normally regarded as being little affected by water, for quantitative results care must be taken to ensure that the water does not emerge at the same time as a component of whatever mixture is examined. For a discussion of the effect, see the reports by Foster and Murfin (1965) and Singer (1966). To determine the exact time at which the water emerges, place a small pellet of calcium carbide at the exit of the column and inject a small volume of water. Note the point of injection and the time required for a peak to emerge. The peak is due to ethyne produced by the action of water on the carbide. This interference occurs when Carbowax columns are used because water is eluted after the lower alcohols. In the official assay procedure on a Porapak column, water is eluted first.

Derivatives

Some classes of drugs such as barbiturates, basic nitrogenous compounds and phenols, even when examined on various columns under different conditions of temperature and gas flow, still exhibit non-linearity of response and tailing of peaks. This is attributable to adsorption of the compounds on the column. A derivative may solve the problem by reducing the polarity of the compound and by rendering it more volatile.

Trimethylsilyl derivatives are very useful for phenols (see Experiment 26) but their decomposition in the flame leads to deposits of silica in the detector; consequently, loss of sensitivity and lack of precision may be expected. These effects vary from instrument to instrument, depending upon the design of the detector. The only remedy to maintain precise and accurate results is to clean the collector electrode and burner tips frequently.

Trifluoroacetyl derivatives do not suffer from this disadvantage but note that silicone-based liquid stationary phases may bleed over a period of time and deposit silica when the flame is left on. The use of halogenated derivatives enables an electron-capture detector to be used for increased sensitivity but this is rarely necessary in pharmaceutical analysis, where relatively large amounts of active ingredient are available from formulations.

Barbiturates are generally methylated with trimethylanilinium hydroxide to give *N*-methyl derivatives of much better chromatographic characteristics than those of the parent compounds.

A detailed review of derivative formation procedures in the quantitative gas chromatographic assay of pharmaceuticals is given by Nicholson (1978).

The use of derivatives requires an initial period of investigation to determine the time required for complete reaction and, possibly, to check the stability of the product. It is wise to adopt a standard procedure at the chromatographic stage such that a check is obtained on incomplete reaction or instability. In any one assay, duplicate standard solutions (S_1 and S_2) and duplicate test solutions (T_1 and T_2) should be prepared and injected in a definite sequence as follows:

$$S_1, T_1, T_1, S_2, T_2, T_2, S_1$$

Use the mean ratio of compound/internal standard for S_1, S_2 and the mean ratio for T_1, T_1 to calculate the first result. Similarly, calculate a second result using the means for S_2, S_1 and T_2, T_2 respectively. Depending upon the time of elution of components, the period between the first injection of S_1 and the last could be up to several hours. If the ratios are in reasonable agreement, it can be concluded that nothing abnormal has occurred in the solutions during the assay.

Pharmaceutical applications

Specifications in the pharmaceutical industry are concerned not only with crude drugs and chemicals but also with intermediates used in the manufacture of the latter and formulations. Many of the intermediates are relatively simple compounds and amenable to analysis by GLC. The principle underlying the analysis is that *all* the sample injected into the GLC instrument is eluted and detected. Therefore, the total area of the main and subsidiary peaks represents 100% and a measure of the purity of the sample is readily determined by comparison of relative areas, i.e. by normalisation. Volatile substances such as *Ethanol*, *Halothane* and *Methoxyflurane* can be examined for related compounds in the same way (see Table 4.12).

Bulk preparation of crystalline chemicals generally involves crystallisation from a solvent, and the detection and determination of solvent residues or solvent of crystallisation should form part of specifications. Normally drying processes do not always remove solvent quantitatively from the solid, and GLC (e.g. methanol in *Streptomycin Sulphate*) and sometimes infrared absorption (Chapter 10) are alternative methods. Compounds capable of geometrical or optical isomerism can in certain instances be examined to determine the relative proportions of each form, e.g. *Tranylcypromine Sulphate*. Generally this type of examination requires the formation of suitable derivatives. For example, optical isomers cannot be separated directly on a GLC column except as diastereoisomers. The determination of enantiometric purity therefore requires the formation of diastereomeric derivatives (e.g. the *N*-trifluoroacetyl-L-prolyl-*n*-butyl ester derivatives of amino acids) or the use of a chiral stationary phase (e.g. *N*-*n*-lauroyl-*N*-L-valine-*t*-butylamide for the separation of amino acid enantiomers as their *N*-trifluoroacetyl methyl esters).

The text by Jack (1984) provides an excellent review of applications of gas chromatography in drug analysis. Some applications are shown in Table 4.12.

Table 4.12 Some pharmacopoeial applications of gas chromatography

Substance	Tested for	Column	Temp.	Internal standard
Ampicillin sodium	Dichloromethane (0.2%)	Diglycerol (20%)	60°	1,2-Dichloro-ethane (0.02%)
Atropine sulphate formulations (Silyl derivatives)	$C_{17}H_{23}NO_3$	OV 17 (3%)	220°	Homatropine hydrobromide (0.5%)
Cephaloridine	Residual solvent	Macrogol (10%) 1000	120°	Butan-2-one (0.25%) and dimethylform-amide (0.375%)
Clioquinol (as silyl derivative)	C_9H_5ClINO (≮90.0%)	Silicone gum rubber (SE 30, 5%)	220°	Octadecane (0.2%)
Clofibrate	Volatile related substances and free phenols	Silicone gum rubber (SE 30 30%)	185°	–
Colchicine	Ethyl acetate or chloroform	Macrogol 1000 (10%)	75°	Ethanol (0.1%) or Ethanol (0.02%)
Doxycycline hydrochloride	Ethanol (4.3–6.0% w/w)	Porapak Q	135°	Propan-1-ol (0.05%)
Ethanol (96%)	Other alcohols	Porapak Q	130°	Butan-2-one (0.02%)
Fenfluramine Hydrochloride Tablets	Foreign substances	Carbowax 20M (10%) Potassium hydroxide (2%)	135°	N,N-Diethyl-aniline (0.01%)
Halothane	Volatile related compounds	Polyethylene glycol 400 (30%, 1.8 m) followed by dinonyl phthalate (30%, 1 m)	50°	1,1,2-Trichloro-1,2,2-trifluoro-ethane (0.005%)
Lincomycin Hydrochloride Capsules Injection (Silyl derivative)	Lincomycin (≮82.5%) Lincomycin B (≯5% of area of Lincomycin peak)	Silicone gum rubber (SE 30, 3%)	260°	Tetraphenylcyclo-pentadienone
Methoxyflurane	Volatile related compounds	Polyethylene glycol (30%, 1.8 m) followed by dinonyl phthalate (30%, 1 m)	80°	Chloroform (0.1%)
Orciprenaline Sulphate	Methanol	Porapak Q	140°	Ethanol (0.5%)
Primidone	Ethyl phenyl malondiamide (1%)	OV 17 (3%)	170°	Octadecan-1-ol (0.05%)

Table 4.12 (*cont*)

Substance	Tested for	Column	Temp.	Internal standard
Tetracosactrin	Acetic acid (8–13% calc. with reference to peptide content)	Porapak Q	150°	Dioxan (0.1%)
Tranylcypromine Sulphate Tablets	*Cis*-isomer	OV 225 (3%)	170°	4-chloroaniline
Warfarin Sodium (clathrate)	Propan-2-ol (4.3–8.3% w/w)	Polyethene glycol 1500 (10%)	70°	Propan-1-ol (0.5%)
Wool Alcohols ⎱ Wool Fat ⎰	Antioxidants	Silicone gum rubber (SE 30, 10%) (Use pre-column of silanised glass wool)	150°	Methyl *n*-decanoate (0.005%)

Combination of gas chromatography with other techniques

The identification of fractions in gas chromatography is essentially comparative, in that the characteristics of the unknown are compared with those of known compounds. By correct choice of column, the fractions consist of single substances only, so that, if each is examined by other methods for identification, a powerful analytical tool becomes available. This may be done in several ways, and gas chromatography is now used in conjunction with infrared spectrophotometry (Chapter 10) and with mass spectrometry (Chapter 12).

The interfacing of a gas chromatograph and a mass spectrometer provides one of the most specific and sensitive quantitative techniques available. The mass spectrometer is set to monitor the abundance of preselected fragment ions (selective ion monitoring; SIM) and, in effect, functions as a specific gas chromatographic detector. The emergence of a substance from the column, which yields the monitored fragment, generates a signal in proportion to the concentration of the substance. Only those substances yielding a fragment with a mass to charge ratio of precisely that selected are detected and the sensitivity is such that picogram (10^{-12} g) and even femtogram (10^{-15} g) quantities can be determined quantitatively. The technique of GC–MS is of particular benefit in the identification and quantitation of drugs, drug metabolites and endogenous substances in samples of biological origin.

Capillary gas chromatography

Although open tubular (capillary) columns with a narrow internal diameter were first described by Golay in 1956, their use during the next

20 years was confined mainly to research, rather than routine, applications. This was due principally to the fragility of the narrow glass columns, difficulties in their manufacture, patents restricting their commercial production, and poor precision associated with the injection of the sample into the column. Advances in capillary technology during the 1970s, in particular the introduction of flexible and robust fused-silica columns and improvements in injector design, have resulted in the acceptance of capillary gas chromatography as a routine technique which is capable of fast and excellent resolution of extremely complex mixtures of volatile or volatilisable substances.

Columns

Capillary columns are thin columns with a maximum internal diameter of 1 mm. There are three types in common use:

(a) Wall-coated open tubular (WCOT) columns in which the liquid stationary phase is coated as a thin film on the inside wall of a borosilicate glass or fused-silica column. These columns generally are 10–100 m long and 0.1–0.5 mm internal diameter, although wide-bore (0.7–1 mm) columns are also available. The thickness of the layer of stationary phase is 0.1–1 μm. The fused-silica WCOT columns are extremely inert and do not give rise to the problems of residual adsorption which were exhibited by the earlier borosilicate glass WCOT columns.

(b) Support coated open tubular (SCOT) columns in which the liquid stationary phase is coated onto very fine particles of support adhering to the inner wall of the column. Like WCOT columns, they are 10–100 m in length and approximately 0.5 mm in internal diameter.

(c) Micropacked columns, which are smaller versions of ordinary packed columns. Their lengths vary from 1 to 6 m and they are 0.7–1 mm in internal diameter.

Only a limited number of stationary phases are used in capillary columns. This is due to the excellent resolving power of these columns, which avoids the need for a multiplicity of stationary phases. The stationary phases are, or are similar to, those preferred liquids listed in Table 4.11.

Carrier gases

The choice of carrier gas influences the efficiency (p.142) and speed of analysis. Generally, a high molecular weight gas such as nitrogen produces a smaller height equivalent to a theoretical plate, i.e. greater efficiency, than a lower molecular weight gas such as helium, because of the lower diffusivity of the solutes in the former gas. In contrast, the fastest separations are obtained with a low molecular weight carrier gas. For example, the separation obtained using nitrogen may be achieved up to eight times more quickly using hydrogen. However, the explosive nature of hydrogen tends to limit its use and helium has proven to be an acceptable compromise which gives adequate efficiency and speed for most analytical purposes. Typical flow rates of these gases are: hydrogen 2 ml min^{-1}, helium 1 ml min^{-1}, nitrogen 0.5 ml min^{-1}; compare this with 15–60 ml min^{-1} for these gases when packed columns are used.

Injector systems

A major difference between capillary columns and packed columns is the loading capacity. Capillary columns require a much smaller sample size than, and would be overloaded at the sample quantities used in, packed columns. Typical sample loadings are 10–100 ng for WCOT columns, 100–1000 ng for SCOT columns and 0.5–100 μg for packed analytical columns. The reduction in the amount of sample injected onto capillary columns is achieved by the use of:

(a) Inlet splitters which split the sample volume of a few μl by means of a needle valve into two unequal flows (e.g. 50 : 1, 100 : 1) and allow the smaller flow to pass onto the column.

(b) Splitless injectors; these injectors allow all of the sample to pass onto the column. However, the sample must be very dilute to avoid overloading the column, and a high capacity (e.g. SCOT or heavily coated WCOT) column should be used.

(c) 'Grob' splitless injectors, which are based upon the 'solvent effect'. The sample at a high dilution in a suitable solvent is injected onto a hot injection port, where the solvent and volatile components are quickly vaporised. The column is at a temperature at least 25° lower than the boiling point of the solvent, and this causes the solvent to condense in the first few cm of the column. The condensed solvent acts as a new liquid stationary phase into which the components of the sample partition and re-concentrate as a narrow band. A temperature programme (p.141) is then started which vaporises the components again and allows the chromatographic separation to proceed.

The detailed description of the design and operation of these injectors is outside the scope of this chapter, and the reader is referred to the text by Jennings (1978) or to manufacturers' literature.

Performance

The increase in efficiency of a properly installed capillary column over a conventional packed column is of the order of 100-fold. Typical column efficiencies, in terms of the number of theoretical plates (p.142) are 1000–2000 plates/m for WCOT columns, 500–1000 plates/m for SCOT columns and 500–1500 plates/m for packed columns. Thus, a 50 m capillary column with an efficiency of 1000 plates/m is 100 times more efficient than a 1 m packed column with 500 plates/m.

An example of a gas–liquid chromatogram showing the resolution obtained in the separation of 15 methylated barbiturates is shown in Fig. 4.23.

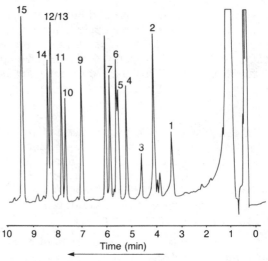

Fig. 4.23. Chromatogram of 15 methylated barbiturates. Temperature programme: initial temperature at 100°, rate 8° min^{-1} to final temperature of 180°. 1. Allobarbitone; 2. Aprobarbitone; 3. Secbutobarbitone; 4. Amylobarbitone; 5. Pentobarbitone; 6. Vinylbitone; 7. Vinbarbitone; 8. Quinalbarbitone; 9. Cyclopentobarbitone; 10. Phenylmethylbarbituric acid; 11, Hexobarbitone; 12. Methylphenobarbitone; 13. Phenobarbitone; 14. Cyclobarbitone; 15. Heptabarbitone.
Reprinted by permission of *J. Forens. Sci. Soc.*

Practical experiments

Experiment 17   *To determine the efficiency of a column*

Method Select a sample appropriate to the column to be tested, e.g. ethyl acetate or propionate for a diethylene glycol succinate column, and prepare a 100 mg ml^{-1} solution of the ester in cyclohexane or other low boiling point non-polar solvent. With the chromatograph at about 50° and a gas flow of about 40 ml min^{-1}, inject 0.2–0.4 μl of the solution. After the components are eluted, repeat the injection: the result should be identical with that from the first injection with regard to the time required for elution of the components. Measure the distance (t_R) and the peak width (W) (Fig. 4.22) and calculate the number of theoretical plates (n) from the formula

$$n = 16 \frac{t_R{}^2}{W^2}$$

where t_R and W are expressed in the same units.

Alternatively, measure the peak width at half height ($W_{1/2}$; Fig. 4.22) and use the formula

$$n = 5.54 \left(\frac{t_R}{W_{1/2}} \right)^2$$

Experiment 18   *To determine the optimum flow rate of carrier gas*

Method As for Experiment 17 but with flow rates of 10, 20, 30, 40, 50, 60, 70 and 80 ml min^{-1}. Plot a graph of flow rate against height equivalent to a theoretical plate (HETP) where

$$\text{HETP} = \frac{\text{length of column}}{\text{number of theoretical plates}}$$

An optimum gas flow exists for each substance and the Van Deemter equation relates band broadening to three factors. In a simplified form the equation is

$$\text{HETP} = A + \frac{B}{u} + Cu$$

where

$$u = \text{linear gas velocity in cm s}^{-1}$$

$$= \frac{\text{length of column (cm)}}{\text{retention time for air peak}}$$

A = eddy diffusion term caused by different rates of flow around different sized particles. Normally its value is small

B = molecular diffusion term caused by diffusion of sample in gas phase

C = mass transfer term caused by a finite time being required for exchange of solutes between gas and liquid phases

Therefore a minimum in the experimental curve is to be expected, corresponding to the flow rate producing the smallest HETP, i.e. highest efficiency.

Experiment 19   *To examine a homologous series of compounds and the relationship between retention time and number of carbon atoms*

Method Prepare a mixture of approximately equal amounts of one of the following series in a low-boiling organic solvent:

methyl esters of lower fatty acids	(diethylene glycol succinate)
hydrocarbons	(Apiezon L)
ethers	(diethylene glycol succinate)
dialkylaminoethanols	(Carbowax 10%; KOH 5%)

Inject the sample (0.2–0.3 µl) onto a suitable column as indicated in brackets above and measure the adjusted retention time (i.e. adjusted from the emergence of the solvent front) for each component in a series. Plot the logarithm of the adjusted retention time against number of carbon atoms in the molecule.

The components are eluted in order of molecular weight and there is no need to inject samples of individual compounds for identification of the peaks. The graph should indicate a straight line relationship between carbon number and the logarithm of the adjusted retention time.

The examples chosen are largely artificial in that for these it is a simple matter to identify any particular member of a series by means of an authentic sample. The real value of the linear relationship shown in the experiment lies in the examination of complex mixtures such as normal and branched chain hydrocarbons obtained from plant waxes or from decomposition of quaternary ammonium compounds such as *Cetri-mide*, or of fatty acids from fixed oils (Experiment 23). Examination of the hydrocarbons, for example, may reveal compounds of carbon number C_{20} to C_{35} or more. Moreover, branched alkanes as well as straight-chain alkanes may be present. Providing a few reference samples are available, the line corresponding to the straight-chain alkanes, and that for the branched alkanes, may be drawn, and identification of the individual peaks carried out with a minimum of reference material.

Experiment 20   *To determine the composition of a mixture of t-butanol and butan-2-ol*

Conditions Column: Porapak Q
            Temperature: 180°

Method Prepare a series of mixtures of the alcohols to contain from 25% w/w *t*-butanol with 75% w/w butan-2-ol to 50% by weight of each component.
   Inject each mixture (0.1 μl) in turn into the column and measure the areas of the peaks for the two alcohols. Plot a graph of

$$\frac{A_1}{A_1 + A_2} \text{ against } \frac{wt_1}{wt_1 + wt_2}$$

where A_1 = area of peak for *t*-butanol; A_2 = area of peak for butan-2-ol; wt_1 = weight of *t*-butanol; and wt_2 = weight of butan-2-ol.
   Determine A_1 and A_2 for the unknown mixture and read off the value of $wt_1/(wt_1 + wt_2)$ from the graph. Calculate the percentage composition of the mixture.

Experiment 21   *To determine the percentage v/v of ethanol in Stramonium Tincture*

Conditions Column: Porous polymer beads (Porapak Q or Chromosorb
            101; 100–120 mesh), 1.5 m (Note 1).
            Temperature: 160°

Method Prepare a standard solution containing dehydrated ethanol (4% v/v) and propan-1-ol (5% v/v). Dilute the sample (8 ml) in a 100 ml volumetric flask with propan-1-ol (5 ml) and *water*. Inject the standard solution and sample solution (0.5 μl) alternately (3 of each) and calculate the ratio

$$\frac{\text{height of ethanol peak}}{\text{height of propan-1-ol peak}}$$

for standard and sample.
   Average the results for each and calculate by proportion the percentage of ethanol in the tincture (Note 2).
   Note 1 A column 1 m long is adequate for separating all the alcohols.
   Note 2 For preparations containing industrial methylated spirit the presence of methanol should be confirmed and the amount determined.

Extensions.
   (a) To confirm that a proportional relationship exists between the peak height ratio and the concentration of ethanol, prepare a calibration series of five solutions containing 5% v/v propan–1-ol and 1, 2, 3, 4 and 5% v/v ethanol respectively. Construct a calibration graph and calculate the linear regression equation (p.280).

(b) As an aid to understanding the reasons for the use of an internal standard in quantitative gas chromatography, inject the sample solution ten times. Calculate the relative standard deviation (RSD) of (i) the heights of the ethanol peak, (ii) the heights of the propan-1-ol peak and (iii) the peak height ratios. The precision of the individual peak heights is expected to be poor (RSD > 5%), because of the difficulty of injecting 0.5 μl precisely, whereas the peak height ratios should show a much better precision (RSD < 2%).

COGNATE EXPERIMENTS

Chloroform in aqueous pharmaceutical preparations Many of these preparations contain 0.1 to 0.2% of chloroform and dilution is therefore not feasible.

Less difficulty is experienced with preparations such as Aconite, Belladonna and Chloroform Liniment, where the chloroform is extracted with carbon tetrachloride containing 1.0% v/v of dichloromethane as internal standard and the solution is compared directly with a 1.66% solution of chloroform in carbon tetrachloride with 1.0% dichloromethane as internal standard. To check for possible interference by constituents of the liniment with the peak for the internal standard, a separate sample is extracted with carbon tetrachloride alone and examined.

A suitable column is silicone gum rubber SE 30 (10% w/w) on white diatomaceous earth (80–120 mesh) at 100° with flame ionisation detector.

Experiment 22   *To determine the percentage of menthol in Peppermint Oil*

Conditions Column: Carbowax 1000 (10%)
              Temperature: 130°

Method Prepare standard solutions of menthol in ethanol to contain 5, 10, 15 and 20 mg ml^{-1}, including in each 20 mg ml^{-1} of camphor as internal standard. Dilute the Peppermint Oil with ethanol so that the expected menthol content of the dilution is about 15–20 mg ml^{-1}, and include 20 mg ml^{-1} of camphor. Inject 1 μl aliquots of each solution and calculate the menthol content (w/w) of the oil by reference to a calibration curve prepared in the normal way (Experiment 21).

The menthol content determined by this method is generally below that determined chemically.

COGNATE EXPERIMENTS

Camphor Liniments. The presence of fixed oil is a disadvantage if the dilution of the sample is not sufficiently great. Therefore, prepare dilutions to contain 1, 2, 3 and 4 mg ml^{-1} of camphor with 3 mg ml^{-1} of menthol as internal standard.

In the absence of fixed oil, the camphor is conveniently extracted with carbon tetrachloride, nitrobenzene being used as internal standard.

A direct comparison of the standard solution containing 4 mg ml^{-1} of camphor and 4 mg ml^{-1} nitrobenzene in carbon tetrachloride is made with an extract of the liniment obtained by extraction with 4 mg ml^{-1} nitrobenzene in carbon tetrachloride. To check that miscellaneous extracted material does not appear with the nitrobenzene peak, a third solution consisting of an extract of the liniment with carbon tetrachloride is also examined. If a peak appears in the same position as that for nitrobenzene, allowance is made for this in the final calculation.

The column suggested for determination of camphor in Aconite Liniment; Aconite, Belladonna and Chloroform Liniment; Ammoniated Belladonna and Camphor Liniment, is silicone gum rubber 5% w/w SE 30 on 60–80 mesh white diatomaceous earth. A temperature of 140° and a flame ionisation detector are satisfactory.

Methyl Salicylate in Liniment and Ointments The standard solution contains 10 mg ml^{-1} of benzyl alcohol (internal standard) and 10 mg ml^{-1} of methyl salicylate in light petroleum (b.p. 80–100°). The sample is diluted with light petroleum to give a concentration of methyl salicylate of about 10 mg ml^{-1} and 10 mg ml^{-1} of benzyl alcohol is included. The third solution required for reasons given above (for camphor) is one of the sample but without the internal standard.

Suitable conditions are 10% w/w of polyethylene glycol 1500 on white diatomaceous earth (60–80 mesh) at 110° with flame ionisation detector.

Experiment 23   *To determine the composition of the fatty acids of Arachis Oil*

Earlier GLC procedures for the determination of the fatty acid composition of oils involved the saponification of the triglycerides with ethanolic potassium hydroxide, several solvent extraction stages, evaporation of the organic solvent and then conversion of the free fatty acids to their methyl ester derivatives with methanolic boron trifluoride. A simple procedure by Christopherson and Glass (1969), which involves the dilution of the oil with petroleum ether and shaking with methanolic potassium hydroxide, combines the saponification and esterification stages and gives results that are comparable to the earlier methods.

Method Dissolve the oil (1 g) in petroleum ether (b.p. 60°–80°; 20 ml) and add 2M potassium hydroxide in methanol (2 ml). Shake the solution for 20–30 s and allow to stand for a few minutes to let the glycerol separate as the lower layer. Inject onto the column 1 μl of the petroleum ether solution containing the methyl esters of the fatty acids.

The conditions for separation of the esters are:

Columns: (i) Apiezon L 0.5%
           (ii) Diethylene glycol succinate 10%

Temperature: 150° for column (i)
               190° for column (ii)

Detector: Flame ionisation

Measure the area of the peaks and calculate the proportion that each peak bears to the sum of all the areas (normalisation). Identify the components by the use of authentic samples of methyl esters of lauric, myristic, palmitic, stearic, oleic and linoleic acids.

When Apiezon L is used as the stationary phase the esters emerge in order of molecular weight, but with the DEGS column unsaturated esters emerge after the corresponding saturated ester, e.g. methyl oleate emerges after methyl stearate.

In determining the composition of the esters the assumption is made that the peaks obtained represent all the esters present in the sample injected.

Extensions

(a) If a sufficient number of authentic methyl esters are available, construct graphs of adjusted retention time *vs* carbon-number (Experiment 19) for series of (i) saturated fatty acids, (ii) fatty acids with one double bond and (iii) fatty acids with two double bonds. Use the linear plots obtained to identify the fatty acids in the sample whose retention times do not coincide exactly with those of authentic standards.

(b) Compare the resolution of the esters, obtained by using the packed columns described above with that obtained by using a suitable capillary column (e.g. DEGS or Carbowax 20M).

(c) Select a suitable internal standard (e.g. a methyl ester of a fatty acid absent in the sample) and develop a quantitative method for the absolute, rather than relative, concentrations of the fatty acids.

(d) Other oils for which the procedure above is suitable are linseed oil (comprising mainly γ-linolenic, oleic, linoleic and palmitic acid triglycerides), cod-liver oil (palmitic, palmitoleic, oleic and myristic acid triglycerides) and olive oil (oleic, palmitic, linoleic, stearic and myristic acid triglycerides).

Experiment 24   *Assay of caffeine and cyclizine hydrochloride in a tablet preparation*

A commercial tablet formulation used in the treatment of migraine is:

Caffeine hydrate: 100 mg
Cyclizine hydrochloride: 50 mg
Ergotamine tartrate: 2 mg

The high solubility of caffeine and cyclizine hydrochloride in chloroform permits a rapid extraction of these drugs and their determination by GLC.

Method Weigh and powder 20 tablets. Transfer an accurately weighed quantity of powder equivalent to one tablet to a 25 ml flask. Shake the powder with 25 ml of a solution of diphenhydramine hydrochloride (internal standard) in chloroform (2 mg ml^{-1}). Allow the insoluble excipients to settle and inject $1 \mu l$ aliquots of the extract onto a packed column of 3% OV-17 on Gas Chrom Q (2 m) at 220°. Use as a reference solution a solution of caffeine hydrate (4 mg ml^{-1}), cyclizine hydrochloride (2 mg ml^{-1}) and diphenhydramine hydrochloride (2 mg ml^{-1}) in chloroform.

The substances are eluted in the order, diphenhydramine : caffeine : cyclizine.

Extensions

(a) Confirm the lack of interference from ergotamine tartrate by shaking 8 mg with 100 ml of chloroform, allowing any undissolved material to settle and injecting the solution onto the column.

(b) Modify the procedure above to determine the uniformity of dose. Powder individual tablets (10) in separate flasks and shake the powder with the internal standard solution (25 ml).

Experiment 25   *To determine volatile bases in urine*

Conditions Column: 10% Carbowax 6000 and 5% KOH
Temperature: 140°
Detector: Flame ionisation

Method Pipette urine (2–5 ml) into a glass-stoppered centrifuge tube, neutralise with dilute hydrochloric acid or sodium hydroxide solution, as appropriate, and add 0.1 ml 5M hydrochloric acid. Extract the urine with freshly distilled ether, centrifuge and reject the ether layer. Add 5M sodium hydroxide (0.5 ml) to the urine and extract with ether (3 × 2.5 ml), centrifuging between each extraction. Transfer the ether extracts to a 15 ml test tube, the base of which is drawn out to a fine taper. Add *N,N*-dimethylaniline solution (5 μg base/ml, 1 ml) and concentrate the solution on a water-bath at 40° to about 50 μl volume. Inject 2 μl into the column. Calculate the ratio of peak heights for *N,N*-dimethylaniline to those of the bases and read off the concentration of the bases from a calibration curve prepared by treating known quantities of the appropriate base in the same way.

The method was developed by Beckett and Rowland (1965) with the object of determining amphetamine in urine. Methylamphetamine and β-phenylethylamine are separated easily from amphetamine and the internal standard. The presence or absence of amphetamine in a treated patient's urine is linked closely with the pH of the urine, so that a negative result is not conclusive evidence that amphetamine has not been taken. The method, however, introduces a confirmatory test for amphetamine when a peak corresponding to amphetamine appears in the trace—add acetone to the residue and inject the solution. Amphetamine forms an acetonide which appears from the column several minutes later than amphetamine itself.

The inclusion of potassium hydroxide in the column is essential when examining amines, as considerable tailing will otherwise occur.

Experiment 26   *To determine the concentration of adrenaline in Adrenaline Injection*

The method is that described by Boon and Mace (1969) who prepared the tri-*O*-trimethylsilyl ether of adrenaline for examination by gas–liquid chromatography. In the preliminary isolation of adrenaline from the injection (or any of its formulations) use is made of ion-pair formation with di(2-ethylhexyl) phosphoric acid at pH 7.4, thus avoiding possible oxidation such as is likely to occur at higher pH values if normal isolation techniques are used.

Reagents. Phosphate buffer solution pH 7.4 made by adding 0.2M sodium hydroxide to 0.2M potassium dihydrogen phosphate (250 ml) until the pH reaches 7.4 and diluting to 500 ml.

Di-(2-ethylhexyl)phosphoric acid (1%) in chloroform.

Silylating reagent N,O-bis(trimethylsilyl) acetamide (1 vol.) in dry pyridine (1 vol.). The solution should contain methyl myristate ($5 \, mg \, ml^{-1}$) as internal standard.

Column 5% *OV*-17, 1.5 m, temperature programmed from 190° to 250° at $8° \, min^{-1}$. In the absence of a temperature programmer a temperature of about 210–220° should be selected.

Method To the injection (10 ml) add phosphate buffer pH 7.4 (15 ml) and extract with the di-(2-ethylhexyl) phosphoric acid solution (4 × 10 ml). Filter each extract through a small plug of cotton wool supporting a little sodium sulphate, bulk the chloroform extracts and remove the solvent by evaporating, preferably by use of a rotary evaporator, until all chloroform is removed (Note 1). To the residue add silylating reagent (2 ml) and allow the mixture to stand for 2 h (Note 2). Inject the solution (1–2 μl) and compare the ratio of peak heights for the sample with that for a standard treated in the same way.

Note 1 It is essential to remove suspended water which reacts with the silylating agent. A residue of adrenaline in di-(2-ethylhexyl) phosphoric acid remains.

Note 2 The silylating agent contains the internal standard and must therefore be accurately measured. The reaction rate is increased by the di(2-ethylhexyl) phosphoric acid and the standard must be treated in the same way; even so, 2 h should be allowed.

High performance liquid chromatography

The technique of high performance liquid chromatography (HPLC) was developed in the late 1960s and early 1970s from a knowledge of the theoretical principles that already had been established for the earlier chromatographic techniques, in particular for column chromatography (pp.86–106), and from advances made in column packing materials and in the design of chromatographic equipment. The technique is based on the same modes of separation as classical column chromatography, i.e. adsorption, partition (including reversed-phase partition), ion exchange and gel permeation, but it differs from column chromatography in that the mobile phase is pumped through the packed column under high pressure. The principal advantages of HPLC compared to classical (gravity-feed) column chromatography are improved resolution of the separated substances, faster separation times and the increased accuracy, precision and sensitivity with which the separated substances may be quantified.

Apparatus

The basic system is illustrated diagrammatically in Fig. 4.24.

The mode of operation of this system is **isocratic**, i.e. one particular solvent or mixture is pumped throughout the analysis. For some determinations the solvent composition may be altered gradually to give **gradient elution**. This may be achieved by a number of techniques and for a detailed description of these the reader is referred to the text by Huber (1978).

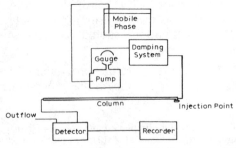

Fig. 4.24. Basic HPLC

Pumps

Pumps are required to deliver a constant flow of mobile phase at pressures ranging from 1 to 550 bar (0.1 to 55 MPa; 14.6 to 8000 psi). Pumps capable of pressures up to 8000 psi provide a wide range of flow rates of mobile phase, typically from 0.01 to 10 ml min^{-1}. Low flow rates (10–100 μl min^{-1}) are used with microbore columns, intermediate flow rates (0.5–2 ml min^{-1}) are used with conventional analytical HPLC columns, and fast flow rates are used for preparative or semi-preparative columns and for slurry packing techniques.

Mechanical pumps of the reciprocating piston type give a pulsating supply of mobile phase. A damping device is therefore required to smooth out the pulses so that excessive noise at high levels of sensitivity or low pressures does not detract from detection of small quantities of sample. This type of pump is extremely useful, however, in that a constant volume of liquid is delivered, the actual value being set by adjustment of piston stroke. This means that the pressure shown on a gauge acts as an indicator of working conditions. Thus, if the column becomes partially blocked, a rise in pressure occurs until ultimately the relief valve (essential in this type of pump) operates. Similarly, leakage from column connections or pump valves shows up as lower pressures. In both cases suitable maintenance measures can be put into operation immediately.

Dual-piston reciprocating pumps produce an almost pulse-free flow because the two pistons are carefully phased so that as one is filling the other is pumping. These pumps are more expensive than single-piston pumps but are of benefit when using a flow-sensitive detector such as an ultraviolet or refractive index detector.

Injection systems

Injection ports are of two basic types, (a) those in which the sample is injected directly into the column and (b) those in which the sample is deposited before the column inlet and then swept by a valving action into the column by the mobile phase.

On-column injection involves the injection of the sample by means of a syringe through a septum into the centre of the packing material. The volume of the sample, which is dependent on the dimesions of the column and the capacity of the packing material, is typically 5-25 μl for analytical columns. High-pressure syringes, that can be used at pressures up to 650 atmospheres, allow the injection of the sample while the mobile phase is flowing. Alternatively, if a low-pressure syringe is used, the flow must be stopped during the injection. These injection methods are not as reproducible as the valve injectors and generally are used in older or simple HPLC apparatus.

Modern injectors are based on injection valves which allow the sample at atmospheric pressure to be transferred to the high-pressure mobile phase immediately before the column inlet. The design of different valves varies widely but a typical arrangement is shown in Fig. 4.25. With the injector in the LOAD position, the sample is injected

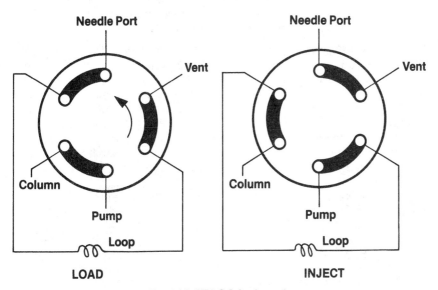

Fig. 4.25. HPLC Injection valve

from a syringe through a needle port into the loop. The valve lever is then turned through 60° to the INJECT position and the sample is swept into the flowing mobile phase. If an excess of sample is flushed through the loop in the LOAD position, the volume injected is the volume of the loop, which is typically 10-20 μl for analytical separations and 0.1-1 ml for semi-preparative or preparative separations. This **complete filling** procedure offers the analyst the highest reproducibility, and is capable of giving relative standard deviations of less than 0.2%. Precision of this order generally avoids the need for an internal standard (see experiment 28). Many of the injection valves also allow a **partial filling** procedure in which any selected volume less than the volume of the loop is injected into the loop. The precision of this technique depends on the precision of the syringe and, with care, relative standard deviations of about 1% can be achieved.

Detectors

The detection of the separated components in the eluate from the column is based upon the **bulk property** of the eluate (e.g. its refractive index) or the **solute property** of the individual components (e.g. their ultraviolet absorption, fluorescence or electrochemical activity). Generally, a detector is selected that will respond to a particular property of the substances being separated, and ideally it should be sensitive to at least 10^{-8} g ml^{-1} and give a linear response over a wide concentration range. It should also have a low dead volume to reduce further band-broadening of the components in the detector and good stability to prevent fluctuations of the response. The most commonly used detectors in the HPLC analysis of pharmaceutical substances are described below.

Photometric detectors. These normally operate in the ultraviolet region of the spectrum and are the most extensively used detectors in pharmaceutical analysis. They comprise essentially a light source, a dispersing device to select an appropriate wavelength for measurement, a flow cell in which the absorbance of the eluate is measured, and a photomultiplier tube or diode to measure the intensity of transmitted light. Commercial spectrophotometers serve as excellent detectors and the only modification required is the installation of the small-volume flow cell in the cell compartment. Photometric detectors are of five principal types:

(a) **Single wavelength detectors** equipped with a low-pressure mercury discharge lamp. The absorbance is measured at the most intense resonance wavelength of mercury at 254 nm.

(b) **Multi-wavelength detectors** employ mercury and other discharge sources which, when used in combination with interference filters (p.266), allow a number of monochromatic wavelengths to be selected, e.g. 206, 226, 280, 313, 340 or 365 nm.

(c) **Variable wavelength detectors** use a deuterium light source and a grating monochromator to allow selection of any wavelength in the deuterium continuum (190–360 nm).

(d) **Programmable detectors** allow the automatic change of wavelength between and during chromatographic analyses.

(e) **Diode array detectors** are microprocessor-controlled photodiode array spectrophotometers in which light from an ultraviolet source passes through the flow cell into a polychromator which disperses the beam so that the full spectrum falls on the array of diodes. Each diode (p.271) detects light at a discrete wavelength and the array provides an almost instantaneous absorption spectrum of the solute in the eluate. Alternatively, if the output of only one diode is selected for measurement then the detector functions as a variable wavelength detector.

Single and multi-wavelength detectors are highly stable and are satisfactory for substances with sufficient absorptivity at the wavelengths available. For substances that are not highly absorbing at these wavelengths, increased sensitivity is obtained by using a variable wavelength detector set at the λ_{max} of the substance in the eluting solvent. Programmable detectors are necessary only when no single wavelength is suitable for all the components of the sample or when trace quantities of substances with significantly different λ_{max} values are being assayed.

The choice of solvents and their purity must be considered in relation to the wavelength of measurement. The detector should not be operated below the ultraviolet cut-off of the solvent (p.308) and the solvents should be of spectral grade quality to ensure the absence of absorbing impurities.

Fluorescence detectors. These are essentially filter fluorimeters or spectrofluorimeters equipped with grating monochromators (Fig. 9.1), and micro flow cells. Their sensitivity depends on the fluorescence properties of the components in the eluate. For substances that are fluorescent, fluorescence detectors are generally more sensitive than photometric detectors, particularly if the excitation and emission monochromators are set at the wavelengths of maximum excitation and fluorescence of the compounds.

Refractive index detectors. These are differential refractometers which respond to the change in the bulk property of the refractive index of the solution of the component in the mobile solvent system. The refractive index detector is the closest approach to the universal detector in that some solvent is usually available in which the sample gives rise to a measurable difference in refractive index between solvent and solution. Although the sensitivity of the refractive index detector is much less than that of specific solute property detectors, they are useful for the detection of substances (e.g. carbohydrates and alcohols) which do not exhibit other properties that can be used as the basis for specific detection.

Electrochemical detectors. These are based on standard electrochemical principles involving amperometry, voltammetry and polarography. The design and operation of electrochemical detectors are described in Chapter 5. These detectors are very sensitive for substances that are electroactive, i.e. those that undergo oxidation or reduction at a suitable potential, and they have found particular application in the assay of low levels of endogenous catecholamines in biological tissues, pesticides, tryptophan derivatives and many drugs.

Columns

HPLC columns are made of high quality stainless steel, polished internally to a mirror finish. Standard analytical columns are 4–5 mm internal diameter and 10–30 cm in length. Shorter columns (3–6 cm in length) containing a small particle size packing material (3 or 5 μm) produce similar or better efficiencies, in terms of the number of theoretical plates (about 7000), that those of 20 cm columns containing 10 μm irregular particles and are used when short analysis times and high throughput of samples are required. Microbore columns of 1–2 mm internal diameter and 10–25 cm in length have certain advantages of lower detection limits and lower consumption of solvent, the latter being important if expensive HPLC-grade solvents are used. HPLC may also be carried out on the semi-preparative or preparative scales by using columns of 7–10 mm or 20–40 mm internal diameter respectively.

Packing materials and mobile phases

Many of the packing materials used in HPLC are based on materials used in classical column chromatography, while others have been developed especially for HPLC. The greatest impact on the efficiency of separation and consequently on separation performance and on the time required for an analysis was the introduction in the mid-1970s of microparticulate (3–10 μm diameter) packing materials. Earlier packing materials of larger particle size (30–70 μm) are now rarely used except in guard columns (Experiment 17). The commonly used packing materials and their associated solvent systems in each of the chromatographic modes are described below.

Adsorption HPLC. Of the various adsorbents used in classical column chromatography (Table 4.1), unmodified silica has proved to be the most widely used in HPLC. It offers high efficiency and a high permeability which allows normal operating pressures of less than 2000 psi to be used. It also has a relatively high surface area, which results in a typical capacity for solutes of 1–5 mg g^{-1}. HPLC-grade silica consists of totally porous microparticles with a spherical or irregular shape and a mean diameter of 3, 5, 7 or 10 μm. The functional group responsible for adsorption is the silanol (Si—OH) group, which interacts with the sample solutes by hydrogen bonding. There is therefore increasing retention of solutes with increasing solute polarity. Alumina is used as an adsorbent less frequently than silica, although for some separations, in particular of aromatic substances and of structural isomers, greater selectivity is obtained with alumina.

The solvents used in adsorption HPLC are almost entirely organic, and a single binary mixture of miscible solvents of the correct polarity often permits elution of all the solutes in the sample within a reasonable time. The eluting power of the solvents increases with increasing polarity. An **elutropic series** of solvents showing the increase in eluting power is given in Table 4.1.

Partition HPLC. Packing materials based on silica are also used in partition chromatography. Early applications of partition HPLC involved coating the silica (in this case acting as a support) with a polar liquid stationary phase, e.g. ethane-1,2-diol, and using as the mobile phase a less polar solvent or mixture of solvents saturated with the stationary phase to avoid the loss of the stationary phase by dissolution in the mobile phase. Such packing materials have now been replaced by silica to which polar phases are chemically bonded. Examples of the functional groups in these chemically bonded partition HPLC packing materials are:

$$Si—(CH_2)_3—O—CH_2—CH—CH_2 \text{ e.g. LiChrosorb Diol}$$
$$\quad\quad\quad\quad\quad\quad\quad\quad | \quad\ |$$
$$\quad\quad\quad\quad\quad\quad\quad\quad OH \ \ OH$$

$$Si—(CH_2)_3—CN \quad\quad \text{e.g. } \mu \text{ Bondpak CN}$$
$$Si—(CH_2)_3—NH_2 \quad\quad \text{e.g. Polygosil NH}_2$$

The order of increasing polarity of these is diol : cyano : amino. This type of chromatography, in which the mobile phase is less polar than the stationary phase, is called **normal-phase partition chromatography**. As in adsorption chromatography, substances are eluted in the order of their increasing polarities, i.e. the least polar substance is eluted first.

In **reversed-phase partition HPLC** the relative polarities of the stationary and mobile phases are the opposite to those in normal-phase HPLC, i.e. the stationary phase is less polar than the mobile phase, and consequently the solutes are eluted in the order of their decreasing polarities. The stationary phase is silica, chemically bonded through a siloxane (Si—O—Si—C) linkage to a low polar functional group. These phases are prepared by treating the surface silanol groups of silica with an organochlorosilane reagent:

$$\quad\quad\quad\quad CH_3 \quad\quad\quad\quad\quad\quad\quad\quad\quad\quad CH_3$$
$$\quad\quad\quad\quad\ | \quad\quad\quad\quad\quad\quad\quad\quad\quad\quad\quad\ |$$
$$—Si—OH + Cl—Si—R \longrightarrow —Si—O—Si—R + HCl$$
$$\quad\quad\quad\quad\ | \quad\quad\quad\quad\quad\quad\quad\quad\quad\quad\quad\ |$$
$$\quad\quad\quad\quad CH_3 \quad\quad\quad\quad\quad\quad\quad\quad\quad\quad CH_3$$

where $R = C_6H_{13}$ (hexyl), C_8H_{17} (octyl) or $C_{18}H_{37}$ (octadecyl). Untreated silanol groups may be 'capped' by treatment with trimethylchlorosilane to eliminate adsorption effects.

The mobile phase in reversed-phase HPLC generally comprises water and a less polar organic solvent modifier, e.g. methanol or acetonitrile. Separations in these systems are considered to be due to different degrees of hydrophobicity of the solutes, the less polar solutes partitioning to a greater extent into the non-polar stationary phase and consequently being retained on the column longer than the more polar solutes. The rate of elution of the components is controlled by the polarity of the organic modifier and its proportion in the mobile phase. The rate of elution is increased by reducing the polarity, e.g. by increasing the proportion of the organic solvent or by using acetonitrile instead of methanol. The simple alteration of the composition of the mobile phase or of the flow rate allows the rate of elution of the solutes to be adjusted to an optimum value and permits the separation of a wide range of chemical types.

Further selectivity in the separation of ionisable substances may be obtained by altering the pH of the mobile phase to cause **ion suppression**. Substances that are ionised at the pH of the mobile phase are polar and are poorly retained on a reversed-phase column. Adjustment of the pH to a value where a greater proportion of the substance is in the non-ionised form reduces its polarity and increases its retention. For example, the retention time of carboxylic acids is increased by reducing the pH with phosphoric acid, acetic acid or buffers. Conversely, the retention of nitrogenous bases is increased by raising the pH. As the pH limits for reversed-phase columns are pH2–7.5, the overall retention of substances with pK_a values near the extremes of this range is made by adjusting both pH and solvent polarity.

The partition coefficients of ionic and ionisable species between the non-polar stationary phase and the polar mobile phase can also be markedly affected by the presence of certain oppositely charged ions (counter ions) in the mobile phase. This effect is used in **ion-pair reversed-phase HPLC**. The principle, which is the same as that of the acid-dye technique in spectrophotometric analysis (p.304), is that a counter ion is added to the mobile phase, which forms an ion-pair association with the solute ions. The reduced charge of the ion-pair renders it more hydrophobic and results in an increased retention of the solute. Tetramethylammonium or tetrabutylammonium ions are used as counter ions for organic acids, and heptane sulphonate or dodecyl sulphonate ($C_{12}H_{25}SO_3^-$) ions are used as counter ions for nitrogenous bases.

Ion exchange HPLC. Packings for ion exchange HPLC are based on the cross-linked polystyrene-divinylbenzene resins described on p.96 or on ion exchange residues chemically bonded to silica. The materials that have the widest general application are the strong cation exchangers containing sulphonate (e.g. sulphopropyl) groups or strong anion exchangers containing tetraalkylammonium (e.g. *N*-propyl-*N*-*N*-*N*-trimethylammonium) groups. The amino bonded phases used in normal phase HPLC may be used as weak anion exchangers.

Most ion exchange HPLC is carried out in aqueous media although the bonded phases can be used with mixtures of aqueous buffers and organic solvents if the solutes show poor water solubility. The retention of the ions on the column is controlled by the ionic strength and pH of the mobile phase. When the ionic strength of the solvent system is increased, greater competition between the ionic solutes and solvent ions occurs and this leads to reduced retention of the solute ions.

Size exclusion HPLC. Cross-linked polystyrene-divinylbenzene resins or silica microspheres (5–15 μm in diameter) are used to fractionate materials of high molecular weight. Sephadex, used in gravity-feed gel chromatography, cannot withstand the pressures required for HPLC. The higher resolution given by the small particle size materials allows the separation of substances of closer molecular weight than classical gel chromatography. Usually only a single aqueous or organic solvent is used as the mobile phase and the desired retention of solutes on the column is achieved by the choice of the appropriate grade of the packing material for the molecular weight range of the solutes in the sample.

Special techniques

Chemical derivatisation

Substances that contain a polar functional group may be treated with a suitable reagent to improve the selectivity and/or sensitivity of the analysis. Pre-column derivatisation, in which the reaction is carried out before the chromatographic separation, is used to improve the chromatographic performance (resolution and/or peak symmetry), to increase the stability of labile solutes or to alter the retention times of the solutes. Post-column derivatisation is used primarily to improve the sensitivity of the detection of the solutes. An example is the conversion of carboxylic acids, phenols, amines or carbonyl compounds to their dansyl (5-N,N-dimethylamino-naphthalene-1-sulphonyl) derivatives for their sensitive detection by photometric or fluorimetric detectors.

Peak identification and purity

An absorption spectrum of the component(s) in the flow cell of a photometric detector may be obtained by briefly stopping the flow of the mobile phase and scanning over a wavelength range or alternatively by using a diode array detector. Characterisation of the eluted solutes in this way often allows their identity to be confirmed if both the retention time and absorption spectrum are identical to those of an authentic reference substance. Also, evidence of the co-elution of a substance with one or more other solutes is obtained if the absorption spectrum fails to match that of the authentic substance.

Retention terminology and column performance

The nomenclature for reporting the **retention** of solutes on an HPLC column, the assessment of peak shape in terms of **symmetry factor** and the evaluation of the performance of the column for a particular mixture of solutes in terms of the **number of theoretical plates and resolution** are the same as in gas–liquid chromatography and have been described earlier in this chapter (pp.138–143).

Applications

HPLC ranks among the most widely used techniques in pharmaceutical analysis. This is due to several reasons.

(a) The wide variety of packing materials allows the separation of most chemical species. The phases that are most extensively used for drug substances of low molecular weight (<1000) and their decomposition products or metabolites are the adsorption systems based on silica and the reversed-phase systems based on octyl silyl or octadecyl silyl bonded on silica. Ion exchange and size exclusion packing materials are limited in their general applications in pharmaceutical analysis.

(b) The different types of detectors available permit the sensitive detection of most chemical types, and the accuracy and precision with which eluted substances may be quantified give analytical data of the highest calibre.

(c) Microparticulate packing materials give excellent separation of similar substances. The number of theoretical plates given by a standard analytical column is of the order of 5000–10000 and this gives adequate resolution of the vast majority of mixtures which are likely to require separation.

(d) The short columns (3–10 cm) in routine use allow fast separations to take place, and often a complete separation of a complex mixture can be achieved within a few minutes. Furthermore, the use of automatic samplers and injectors enables large numbers of samples to be analysed unattended.

(e) A combination of HPLC and spectrometric techniques (e.g. ultraviolet, infrared, mass spectrometry) allows the almost simultaneous quantitation and identification of solutes as they elute from the column.

The principal areas of pharmaceutical analysis in which HPLC is used routinely are the quality control testing of drugs and medicines for compliance with specifications, stability studies, therapeutic monitoring, drug metabolism studies, and pharmacokinetic studies. It is especially useful in providing compound specific assays (e.g. *Oestradiol Benzoate Injection*) and in the separation and control of impurities (e.g. related substances in *Flurbiprofen*). Thus HPLC assays are being used increasingly for the assay of medicines containing natural products such as antibiotics and hormones to reduce dependence on biological assays, which, if still regarded as the essential basis for control, are reserved more and more for the parent substances rather than for their dosage forms. HPLC methods are also effective in the separation of geometrical isomers and racemates, as in the control of *E*-isomer in *Tamoxifen Citrate*, and in the limitation of RS- and SR-enantiomers in *Fenoterol Hydrobromide*, which is primarily a mixture of the RR- and SS-enantiomers. Some examples of applications abstracted from the British Pharmacopoeia, manufacturers' catalogues and the scientific literature are given in Table 4.13.

Practical experiments*

The object of the experiments is to illustrate the development of methods for formulations and biological fluids. Most of the experiments can be carried out using standard HPLC equipment with a 10 cm × 4.6 mm column and a fixed wavelength (254 nm) photometric detector.

*The assistance of Dr R. H. Pryce-Jones in developing some of the experiments is gratefully acknowledged.

Table 4.13 Examples of HPLC analyses

Sample	Column	Mobile phase	Detector	Reference
Oestradiol Benzoate Injection	μ Porasil	Cyclohexane: 1,4-dioxan (9 : 1)	Photometric (A_{254})	BP
Idoxuridine and Ear Drops	μ Bondapak C_{18}	Water: methanol (87 : 13)	Photometric (A_{254})	BP
Clomiphene E and Z isomers	10 μm Silica	Chloroform: methanol: triethanol-amine (98 : 2 : 0.05)	Photometric (A_{254})	BP
Catecholamines	RP–18	0.1M Sodium acetate pH 4: methanol (190 : 10)	Electrochemical (1.45 V)	Pye Unicam
Opium Alkaloids	Partisil-10	Methanol:1M NH_4NO_3 (80 : 20)	Photometric (A_{254})	Pye Unicam
Aspirin, Caffeine, Paracetamol Tablets	3μm ODS Hypersil	Gradient	Photometric (A_{240})	Hewlett-Packard
Barbiturates	Supelcosil LC–18	Water: methanol (50 : 50)	Photometric (A_{220})	Supelco
Sulphasalazine + Metabolites	Nucleosil 10 C_{18}	Methanol: 0.1% NR_4Br (50 : 50)	Fluorescence ($\lambda_{ex} = 310$; $\lambda_{em} = 340$)	Fischer and Klotz (1979)
Verapamil + Metabolites	Nucleosil C_{18}		Fluorescence	Kuwada *et al.* (1981)

Particles should be removed from the solvent system by filtering through a sintered glass filter, and dissolved air should be removed by applying a vacuum or by warming.

Experiment 27 *To prepare a guard column with packing material of particle size about 40 μm*

Guard columns are sometimes used in HPLC to protect analytical columns from contaminating substances that may be present in the sample or solvent system. The irreversible retention of the substances eventually reduces the efficiency of the analytical column. Guard columns are usually packed with the same material as, but of larger particle size than, that in the analytical column and are connected between the injector and the analytical column using low dead-volume tubing.

Material of particle size 30 μm or over can be dry-packed but to ensure satisfactory results more time must be allowed than for GLC columns. Owing to the small particle size there is little point in applying a vacuum to one end, and gentle tapping of the column is sufficient. A mechanical filler which automatically lifts and drops the column about 12 mm is very useful for preparation of reproducible and efficient columns.

Method Rinse the column (50 × 4.6 mm) in turn with nitric acid (25%), water, methanol and ether and dry at 100° (Note 1). Attach a fitting incorporating a 2 μm frit to one end and add enough material to fill about 2 mm of the column. Tap the column vertically on a hard surface for about 1 min with an occasional tap on the side (Note 2). Continue in the same way until the column is filled to the top and then close with a fitting containing a 2 μm frit. Attach the column to the injector and pump the mobile phase (40 ml) through the column, which should be unconnected at the outlet at this stage. Attach the outlet of the column to the connection tube to the analytical column.

Note 1 This is a cleaning process to remove any salts, dust or grime from the tubing.

Note 2 This method is reported to give more efficient columns than that adopted for GLC columns, where a vibrator is often used.

Dry particles of packing material less than 20 μm tend to agglomerate owing to electrostatic forces and hence they cannot be packed efficiently by the dry-packing procedure described above. A variety of 'wet slurry' methods have been described for microparticulate columns. The details of these procedures are beyond the scope of this chapter and the reader is referred to the review by Martin and Guiochon (1977).

Experiment 28   *To assess the precision of quantitative measurements*

In its simplest form an HPLC assay involves the following stages:

(a) sampling of material under investigation
(b) preparation of a sample solution
(c) preparation of a standard solution of the analyte (ideally at a concentration close to that of the sample)
(d) injection of standard and sample solutions onto the HPLC column
(e) measurement of peak heights of the analyte in the standard and sample chromatograms
(f) calculation of the concentration of the analyte in the sample based on the assumption that the peak height is proportional to the concentration of the analyte
(g) reporting the results.

The reliability of the analytical data obtained depends on (among many factors outside the scope of this chapter) the precision with which the standard and sample solutions may be injected onto the column. The procedure to be adopted to investigate the precision of the injection stage is as follows.

Method Select a suitable substance from among the following experiments and make repeated injections of a solution using the chromatographic conditions suggested for that substance. At least ten replicate injections should be made. Measure the peak height of each peak (and, if an integrator is available, the peak areas). Calculate the relative standard deviation (RSD) (coefficient of variation) of the results.

$$\text{RSD} = \frac{\sqrt{\dfrac{\Sigma(x - \bar{x})^2}{(n - 1)}}}{\bar{x}} \times 100$$

where x is the individual peak heights (or areas); $\bar{x}$ is the mean peak height (or area); and n is the number of replicate injections.

Values of RSD below 1% indicate precision that is satisfactory for most analytical purposes. Values in excess of 2% indicate that individual measurements of peak height (or area) may be subject to considerable error. The analyst must consider the reliability of values between 1% and 2% in relation to the purpose of the assay.

If the precision of the injection is poor, investigate the effect of incorporating an internal standard in the solution and calculate the precision of the peak height ratios (p.143) or peak area ratios. In some experiments one or more possible internal standards is suggested. The rationale for using an internal standard is also discussed on p.153 for GLC and, as for GLC, it is expected that there should be a considerable improvement in the precision of measurements when an internal standard is used.

Experiment 29   *Assay of Hydrocortisone in a cream or ointment formulation*

The assay is based on that of the British Pharmacopoeia for Hydrocortisone and Clioquinol Ointment. The mode of separation is reversed-phase partition HPLC.

Standard solution Hydrocortisone (100 μg ml^{-1}) in methanol : *water* (80 : 20).

Sample solution Transfer an accurately weighed quantity of the sample containing about 10 mg of hydrocortisone to a beaker (100 ml), add 30 ml of 2,2,4-trimethylpentane and warm on a water-bath until melted. Transfer the mixture to a separating funnel (100 ml) and extract with 30 ml, 20 ml and 20 ml of a mixture of methanol and *water* (80 : 20). Combine the extracts, allow to cool and dilute to 100 ml with methanol : *water* (80 : 20).

Chromatographic conditions

Column: 5 μm-Spherisorb ODS or 5 μm-Hypersil ODS (10 cm or 25 cm × 4.6 mm).

Mobile phase: Methanol : *water* 65 : 35 (25 cm column), 55 : 45 (10 cm column).

Flow rate: 1 to 1.5 ml min^{-1}.

Detection: Fixed wavelength ultraviolet (254 nm).

Volume of injection (valve or syringe): 20 μl.

Extensions

(a) Assess the precision of the injection as described in Experiment 28. If it is not satisfactory, inject solutions of cortisone, prednisolone, prednisone and hydrocortisone acetate (each 100 μg ml^{-1}) and bromobenzene (400 μg ml^{-1}) in methanol : water (80 : 20) to ascertain which is the most suitable as an internal standard. If an internal standardisation procedure is adopted, a sample solution devoid of internal standard should also be injected to confirm that no sample component co-elutes with the internal standard.

(b) If a variable wavelength detector is available repeat the assay at the wavelength of maximum absorption of hydrocortisone about 242 nm, and note the increase in sensitivity.

(c) Confirm that a proportional relationship exists between the measured value (peak heights or peak height ratios) and concentration by preparing a series of five standard solutions containing 20, 40, 60, 80 and 100 μg ml^{-1} of hydrocortisone (and if necessary a constant concentration of internal standard).

(d) Compare the separation of a mixture of corticosteroids listed in (a) above with that obtained for the same mixture using an adsorption column (e.g. LiChrosorb Si 60; 5 μm) and dichloromethane : methanol : water (100 : 2 : 0.1) as the mobile phase.

Experiment 30 *The assay of theophylline in Aminophylline Tablets or Suppositories*

Standard solutions A series of standard solutions of theophylline in the mobile phase (below), 50, 100, 150, 200 and 250 μg ml^{-1}.
Sample solution Weigh and powder 20 tablets. Add 80 ml of pH 9 phosphate buffer (below) to an accurately weighed quantity of the powder containing about 80 mg of theophylline and stir mechanically for 30 min. Dilute to 100 ml with *water*, mix and filter through Whatman's No. 1 filter paper. Dilute 5 ml of the filtrate to 20 ml with the mobile phase, thus preparing a sample solution containing approximately 200 μg ml^{-1} theophylline.
When extracting suppositories, disperse a weight of suppository equivalent to about 80 mg theophylline in 20 ml of chloroform. Extract with three 25 ml aliquots of pH 9 buffer, combining the extracts and diluting to 100 ml with *water*. Dilute 5 ml to 20 ml with the mobile phase.
Acetate buffer pH 4.8 Glacial acetic acid (2 ml) in *water* (70 ml) adjusted to pH 4.8 with 1M sodium hydroxide (about 17 ml required) and diluted to 100 ml with *water*.
Phosphate buffer pH 9 Dissolve 2.84 g anhydrous disodium hydrogen orthophosphate in water (600 ml), adjust to pH 9 with 1M sodium hydroxide and dilute to 1 l with *water*.
Chromatographic conditions.
Column: 5 μm Hypersil ODS or 5 μm Spherisorb ODS (10 cm × 4.6 mm).
Mobile phase: Methanol : acetate buffer pH 4.8 (15 : 85).
Flow rate: 1.5 ml min^{-1}.
Detection: 254 nm or 272 nm.
Volume of injection: 20 μl.
Internal standard (if required) Hydroxypropyltheophylline (or hydroxyethyltheophylline) added to standard and sample solutions to give a final concentration of about 400 μg ml^{-1}.

Experiment 31 *Separation and identification of the xanthine derivatives in tea or coffee*

Standard solutions Theobromine, theophylline, caffeine and hydroxypropyltheophylline each 200 μg ml^{-1} in the mobile phase, prepared separately and in admixture.
Sample solution Weigh about 300 mg of the sample. Dissolve in or extract with 20 ml of hot *water*, cool, filter into a separating funnel and acidify with 1M hydrochloric acid. Extract the acidic solution with three 20 ml volumes of chloroform. Combine the chloroform extracts and filter through a sintered glass filter overlaid with a filter paper and anhydrous sodium sulphate (to dry) and evaporate the filtrate to dryness on a water-bath. Dissolve the residue in 5 ml of mobile phase, filter and dilute as necessary with the mobile phase.
Identify the xanthine derivatives in the sample solution by comparison of the retention times of the authentic standards using one or more of the following chromatographic conditions.
Chromatographic conditions
(a) Column: 5 μm Hypersil ODS or 5 μm Spherisorb ODS (10 cm × 4.6 mm).
   Mobile phase: Methanol : acetate buffer pH 4.8 (8 : 92) (see Experiment 30).
   Flow rate: 1.5 ml min^{-1}.
   Detection: 272 nm or 254 nm.
   Volume of injection: 20 μl.

(b) Column: Partisil 10-ODS (25 cm × 4.6 mm).
  Mobile phase: Methanol : phosphate buffer pH 2.3 (40 : 60).
  Flow rates: 1 ml min^{-1}.
(c) Column: Nucleosil 5 C$_8$ (15 cm × 4.6 mm).
  Mobile phase: Methanol : 0.01M phosphate buffer pH 7 (20 : 80).

Experiment 32   *Assay of chlorpromazine and its sulphoxide in a syrup formulation*

Chlorpromazine standard solution Dilute a stock solution of chlorpromazine hydrochloride (400 μg ml^{-1} in *water*) 1 ml to 10 ml with the mobile phase.
Chlorpromazine sulphoxide standard solution Transfer 20 ml of the stock solution of chlorpromazine hydrochloride (400 μg ml^{-1}) to a separating funnel (100 ml), add 2 ml of oxidising reagent (Experiment 14, p.333) and allow to stand for 5 min to oxidise chlorpromazine to its sulphoxide. Basify the solution by the careful addition of 5M sodium hydroxide (about 7 ml required) and extract the sulphoxide with three volumes of chloroform (15 ml). Re-extract the combined chloroform extracts with three × 30 ml 0.05M hydrochloric acid and dilute to 100 ml with *water*. Dilute 1 ml to 10 ml with the mobile phase.
Sample Measure the weight/ml of the syrup. Transfer an accurately weighed quantity of the syrup containing about 20 mg chlorpromazine hydrochloride to a 50 ml volumetric flask and dilute to volume with *water*. Dilute 1 ml to 10 ml with the mobile phase.
Chromatographic conditions
Column: Spherisorb (or Hypersil)-5CN (150 × 4.6 mm).
Mobile phase: Acetonitrile : 0.15M acetate buffer pH 6.5 (90 : 10).
Flow rate: 1.4 ml min^{-1}.
Detection: 254 nm.
Volume of injection: 20 μl.
Internal standard (if required): promazine hydrochloride at a concentration of 40 μg ml^{-1}
  in the final standard and sample solutions.

Extension The procedure may also be applied to a suspension containing chlorpromazine embonate. Dilute an accurately weighed quantity of the suspension containing about 20 mg chlorpromazine embonate and dilute to 50 ml with the mobile phase. Shake vigorously for a few minutes. Centrifuge for 5 min and dilute 1 ml of the supernatant solution to 10 ml. Attempt to quantify the separately eluted embonic acid by comparison of its peak height with that of a standard solution of embonic acid. The presence of ascorbic acid in some samples may also be confirmed by comparison of retention times with that of an authentic standard of ascorbic acid.

Experiment 33   *The separation of penicillins and the quantitative assay of ampicillin and cloxacillin in a powder for injection*

Standard solutions of penicillins Solutions of benzylpenicillin, phenoxymethylpenicillin, ampicillin and cloxacillin (500 μg ml^{-1}) alone and in combination.
Ampicillin and Cloxacillin Injection Weigh accurately about 60 mg of the powder for injection and dilute to 100 ml with methanol.
Chromatographic conditions
Column: Hypersil 5-ODS or Spherisorb 5-ODS (250 × 4.6 mm).
Mobile phase: Methanol : ammonium carbonate (0.05M) solution in *water* (35 : 65).

Flow rate: 1 ml min^{-1}.
Detection: 254 nm.
Volume injected: 20 μl.
Internal standard (if necessary): benzylpenicillin (300 μg ml^{-1}).

Extension The hydrolysis of a penicillin to its penicilloic acid may be studied as follows. Choose a penicillin with a retention time of 5 to 12 min and dissolve in a mixture of methanol : buffer (30 : 70). Buffers of different pH values, e.g. pH 8.0, 9.0 and 10.0, should be used to investigate the effect of pH on the rate of hydrolysis, and different temperatures, e.g. 20°, 30° and 40°, should be used to investigate the effect of temperature. The concentration of the penicillin should be about 1 mg ml^{-1}. Immediately dilute 1 ml with 0.1M hydrochloric acid (zero time) and take further samples diluted (1+1) with 0.1M hydrochloric acid at various time intervals, e.g. 30 min intervals up to 6 h. Chromatography should show the decreasing peak heights of the penicillin with time and an increasing peak due to the penicilloic acid. The first order rate constants may be calculated by standard techniques.

Experiment 34   *The assay of methimazole in plasma*

The levels of methimazole in the plasma of patients treated with 60 mg methimazole or carbimazole (of which methimazole is the active metabolite) are normally in the range 1 to 2.5 μg ml^{-1} or 0.2 to 1.0 μg ml^{-1} respectively. In this experiment plasma from treated patients or plasma spiked with methimazole (Aldrich) may be used. The method is based on that of Skellern *et al* (1976).

Sample Plasma containing methimazole 0.2–2.5 μg ml^{-1}.
Standards Accurately weigh about 20 mg of methimazole into a volumetric flask (100 ml) and dilute to volume with *water*. Dilute 25 ml to 100 ml with *water* (50 μg ml^{-1}). Prepare a calibration series containing 0.5 to 2.5 μg ml^{-1} by diluting 1, 2, 3, 4 and 5 ml respectively of the 50 μg ml^{-1} solution to 100 ml with *water*. Use benzamide (18 μg ml^{-1}) in methanol as internal standard.
Extraction procedure Place 1 ml of each of the plasma samples and calibration standards in stoppered test tubes, add 0.5 ml of the internal standard solution and 0.5 ml of water. Extract with chloroform containing 0.25% octan-1-ol (5 ml) mechanically for 15 min. Centrifuge and transfer 4 ml of each chloroform extract to a tapered test tube and evaporate the chloroform on a water-bath at 30° under a stream of oxygen-free nitrogen. Redissolve the residue by vortexing with 50 μl of chloroform.
Chromatographic conditions
Column: Spherisorb Si 60 (10 μm) (100 × 4.6 mm).
Mobile phase: 1% v/v methanol in chloroform.
Flow rate: 1.2 ml min^{-1}.
Detection: 254 nm.
Volume injected: 20 μl (valve or syringe).

Experiment 35   *The assay of paracetamol in plasma*

Paracetamol is rapidly absorbed from the gastrointestinal tract and maximum plasma levels are reached within 0.5 to 2 h after ingestion.

Typical therapeutic levels in plasma are 10–20 μg ml^{-1}. The following procedure is based on that of O'Connell and Zurzola (1982).

Standards Plasma from untreated subjects containing 5, 10, 15, 20 and 25 μg ml^{-1} of paracetamol (Note).

Samples Plasma obtained from subjects 0.5 to 2 h after ingesting 0.5–1.0 g of paracetamol.

Method To 1 ml of plasma samples and standards in small glass vials add 1 ml of 0.6M barium hydroxide solution to denature the proteins. Vortex for 2 min to ensure thorough mixing and then add 1 ml of zinc sulphate solution (50 mg ml^{-1}) to precipitate the proteins and vortex for 1 min. Spin at a high speed for 10 min using a bench-top centrifuge. Filter through glass wool contained in the neck of a Pasteur pipette.

Chromatographic conditions

Column: μBondapak C$_{18}$ (300 × 4.6 mm) and a silica-ODS guard column (50 × 4.6 mm).

Mobile phase: Methanol : *water* (15 : 85).

Flow rate: 1 ml min^{-1}.

Detection: Photometric detection at 240 nm.

Volume injected: 20 μl (valve or syringe).

Note The authors showed that the recovery of paracetamol from plasma was 94% of that from water. If blank plasma is not available for the standards use aqueous solutions of paracetamol as standards and multiply the sample peak heights by 1.06.

Experiment 36   *Assay of adrenaline in Adrenaline Injection by reversed-phase ion-pair HPLC*

Selection of internal standard Prepare standard solutions of the following catecholamines in 0.001M hydrochloric acid (Note 1) and determine their retention times: noradrenaline hydrochloride (0.2 mg ml^{-1}) adrenaline hydrogen tartrate (0.2 mg ml^{-1}), α-methylnoradrenaline (0.05 mg ml^{-1}), 6-hydroxydopamine HBr (0.05 mg ml^{-1}), dopamine (0.05 mg ml^{-1}), methyldopamine (0.1 mg ml^{-1}), isoprenaline hydrochloride (1 mg ml^{-1}). If necessary dilute the solutions with 0.001M hydrochloric acid to obtain suitable peak heights.

Calibration standards Prepare a series of standard solutions in 0.001M hydrochloric acid of adrenaline hydrogen tartrate containing 0.04 to 0.24 mg ml^{-1} and a suitable internal standard from the list above, at a concentration giving a peak height similar to that of the 0.2 mg ml^{-1} solution of adrenaline hydrogen tartrate.

Sample solution Dilute a suitable volume of the injection (Note 2) with 0.001M hydrochloric acid to give a concentration of adrenaline hydrogen tartrate of 0.15 to 0.2 mg ml^{-1} and include the internal standard at the same concentration that is present in the standard solutions.

Chromatographic conditions

Column: Hypersil (or Spherisorb) 5-ODS (250 × 4.6 mm).

Mobile phase: Methanol : sulphonic acid buffer (below), (10 : 90) (Note 3).

Flow rate: 1.2 to 1.4 ml ml^{-1}.

Detection: Ultraviolet (285 nm) (Note 4).

Volume injected: 20 μl

Sulphonic acid buffer Dissolve citric acid (2.802 g), anhydrous disodium hydrogen orthophosphate (0.9464 g), ethylenediaminetetraacetic acid disodium salt (0.0233 g) in 820 ml of 0.005M 1-heptane sulphonic acid and dilute to 1 l with *water*.

Note 1 0.001M hydrochloric acid gives a pH of about 3 and prevents the oxidation of catecholamines which occurs rapidly at pH values in excess of 7.

Note 2 The British Pharmacopoeia formulation contains 1.8 mg ml^{-1} adrenaline hydrogen tartrate.

Note 3 The effect of varying the composition of the mobile phase from 7.5 : 92.5 to 20 : 80 methanol : sulphonic acid buffer should be investigated. The theory of reversed-phase ion-pair HPLC is discussed on p.164.

Note 4 If an electrochemical detector is available repeat the assay using more dilute solutions than those given above for photometric detection and use a potential of +0.72 V.

References

Bailey, F., Barlow, F.S. and Holbrook, A. (1963) *J. Pharm. Pharmacol.* **15**, Suppl. 232T.
Beckett, A.H. and Rowland, M. (1965) *J. Pharm. Pharmacol.* **17**, 59.
Betts, T.J. (1964) *J. Pharm. Pharmacol.* **16**, Suppl. 131T.
Boon, P.F.G. and Mace, A.W. (1969) *J. Pharm. Pharmacol.* **21**, Suppl. 49S.
Caddy, B., Fish, F. and Scott, D. (1973) *Chromatographia* **6**, 293.
Chatten, L.G. and Morrison, J.C. (1965) *J. Pharm. Pharmacol.* **17**, 655.
Christopherson, S.W. and Glass, R.L. (1969) *J. Dairy Sci.* **52**, 1289.
Consden, R., Gordon, A.H. and Martin, A.J.P. (1944) *Biochem. J..* **38**, 224.
Cowley, P.S. and Rowson, J.M. (1963) *J. Pharm. Pharmacol.* **15**, Suppl. 119T.
Craig, L.C. (1944) *J. Biol. Chem.* **55**, 519.
Foster, J.S. and Murfin, J.W. (1965) *Analyst* **90**, 118.
Foster, G.E., MacDonald, J. and Jones, E.S.G. (1949) *J. Pharm. Pharmacol.* **1**, 802.
Garratt, D.C. (1964) *Quantitative Analysis of Drugs*, 3rd edn, Chapman and Hall, London.
Hollis, O.L. (1966) *Anal. Chem.* **38**, 309.
Huber, J.F.K. (ed.) (1978) *Instrumentation for High-Performance Liquid Chromatography*, Elsevier, Amsterdam.
Jack, D.B. (1984) *Drug Analysis by Gas Chromatography*, Academic Press, Orlando.
James, A.T. and Martin, A.J.P. (1952) *Analyst*, **77**, 915; *Biochem. J.* (1952) **50**, 679.
Janâk, J. (1967) *Chem. and Ind.*, 1137.
Jennings, W. (1978) *Gas Chromatography with Glass Capillary Columns*, Academic Press, New York.
Kováts, E. (1958) *Helv. Chim. Acta*, **41**, 1915.
Kováts, E. (1965) *Advances in Chromatography*, **1**, 232
Kuhn, R. and Lederer, E. (1931) *Naturwissenschaften*, **19**, 306; *Ber.* (1931) **64**, 1349.
Lederer, E. and Lederer, M. (1957) *Chromatography*, 108, Elsevier.
Martin, A.J.P. (1949) *Biochem. Soc. Symposium.* No. 3, Cambridge University Press, London.
Martin, A.J.P. and Synge, R.L.M. (1941) *Biochem. J.* **35**, 1358.
Martin, M. and Guiochon, G. (1977) *Chromatagraphia*, **10**, 194.
Mikes, O. (ed.) (1966) *Chromatographic Methods*, p.35, Van Nostrand, London.
Moffat, A.C. (1975) *J. Chromatogr.* **113**, 69.
Nicholson, J.D. (1978) *Analyst*, **103**, 1222.
O'Connell, S.E. and Zurzola, F.J. (1982) *J. Pharm. Sci.* **71**, 1291.
Purdy, S.J. and Truter, E.V. (1962) *Analyst*, **87**, 802.
Savidge, R.A. and Wragg, J.S. (1965) *J. Pharm. Pharmacol.* **17**, Suppl. 60S.
Singer, D.D. (1966) *Analyst*, **91**, 127.
Skellern, G.G., Knight, B.I. and Stenlake, J.B. (1976) *J. Chromatogr.* **124**, 405.
Strain, H.H. (1942) *Chromatographic Adsorption Analysis*, Interscience, New York.
Tiselius, A. (1941) *Arkiv. Kemi. Mineral. Geol.* **14B**, No. 22.
Tswett, M. (1906) *Ber dtsch. botan. Ges.* **24**, 316, 384; translation, Strain, H.H. and Sherma, J. (1967) *J. chem. Educ.* **44**, 238.

5
Electrochemical methods

J.R. JOHNSON

Introduction

In electrochemical methods of analysis one or more electrically related parameters, e.g. voltage, current or charge, are measured and related to the state of the system generating or carrying the charge. In addition these methods can be divided into those relating to systems in equilibrium, e.g. measurement of pH potentiometrically, and those dependent on a transient perturbation being applied to the system before the measurement is made, e.g. voltammetry.

Conductimetric titrations

Theory

The simplest electrochemical method by which electrolyte solutions may be investigated is conductimetry, the theory of which is based simply on Ohm's Law:

$$V = iR$$

where V, i and R represent the applied electromotive force (e.m.f.), the current, and the resistance of the solution respectively. Units of e.m.f. are volts (V; $J\,A^{-1}\,s^{-1}$), of current amps (A), and of resistance ohms (Ω; $kg\,m^2\,s^{-3}\,A^{-2}$). The conductance, G, of the solution is the variable which is measured in conductivity experiments, and is defined as the reciprocal of the resistance and expressed in siemens (S; Ω^{-1}). Thus it can be seen that:

$$G = \frac{1}{R} = \frac{\kappa}{l/a} = \frac{1}{\rho(l/a)}$$

where κ, a and l are conductivity ($\Omega^{-1}\,m^{-1}$; reciprocal of resistivity, ρ), cross-sectional area, and length of the conductor respectively. The conductivity represents the current flowing across unit area of conductor per unit potential gradient. Conductivity depends on ionic concentration and tends to zero as the solution is diluted, whereas the molar conductivity (Λ; $\Omega^{-1}\,m^2\,mol^{-1}$), which represents the conductivity of a solution at a concentration of $1\,mol\,m^{-3}$, reaches its maximum value,

Λ°, at infinite dilution because of the complete elimination of ionic interaction only at this concentration. Molar conductivity is related to conductivity by

$$\Lambda = \frac{\kappa}{c} = \frac{\kappa}{1000c'}$$

where c and c' are the electrolyte concentrations in $mol\,m^{-3}$ and $mol\,dm^{-3}$ ($mol\,l^{-1}$) respectively. Thus Λ° is the limit of this equation when c approaches zero.

In conductimetric titrations the change in conductance on addition of titrant is usually required rather than an absolute value, but if the latter were needed the conductivity cell would have to be calibrated to find the cell constant (l/a) defined above. This is usually done by measuring the conductance of a potassium chloride solution of precisely known concentration and high purity. Care should be taken to exclude atmospheric carbon dioxide. It is recommended that the potassium chloride should be recrystallised repeatedly from high purity conductivity water (resistivity $\geqslant 2 \times 10^5\ \Omega m$) and fused in a platinum crucible. Under these conditions of purity the molar conductivity in $\Omega^{-1}\,m^2\,mol^{-1}$ at 298 K for aqueous potassium chloride solutions of concentration c $mol\,m^{-3}$ is given by

$$= 149.93 \times 10^{-4} - 2.992 \times 10^{-4}c^{1/2} +$$
$$58.74 \times 10^{-7}c \log c + 22.18 \times 10^{-7}c$$

Solutions of potassium chloride of 0.1 or 0.01M are usually used for cell calibration, and, provided the exact concentration is known, Λ, κ, and (l/a) can be calculated from the equations above.

When very low concentrations of electrolyte are investigated the conductivity of the water solvent may contribute significantly to the measured value; this can be accounted for by use of the equation:

$$\kappa \text{ (measured)} = \kappa \text{ (solute)} + \kappa \text{ (solvent)}$$

Ionic conductivities

Comparison of the molar conductivities of pairs of salts having an ion in common, such as potassium chloride–sodium chloride and potassium nitrate–sodium nitrate shows that replacement of one ion by another (i.e. replacement of K^+ by Na^+) leads to a constant difference in the conductance values. Therefore, each ion makes a certain definite contribution to the total conductance of the solution, and this is independent of other ions present in the solution. Hence for any salt:

$$\Lambda^\circ = v_+\Lambda^\circ_+ + v_-\Lambda^\circ_-$$

where v_+ and v_- are the numbers of cations and anions respectively formed on dissociation of one molecule of electrolyte, and Λ°_+ and Λ°_- are the appropriate molar ionic conductivities at infinite dilution.

In conductivity experiments it is important to control the temperature because the molar ionic conductivity of most ions is increased at least 2% per °C rise in temperature.

Table 5.1 Approximate molar ionic conductivity of ions in water at infinite dilution (25°)

Cation	$10^4\Lambda^{\circ}_+/\Omega^{-1}\ m^2\ mol^{-1}$	Anion	$10^4\Lambda^{\circ}_-/\Omega^{-1}\ m^2\ mol^{-1}$
H^+	350	OH^-	199
Li^+	39	Cl^-	76
Na^+	50	Br^-	78
K^+	74	I^-	77
NH_4^+	74	NO_3^-	71
Ag^+	62	HCO_3^-	45
$\frac{1}{2}Mg^{2+}$	53	CH_3COO^-	41
$\frac{1}{2}Ca^{2+}$	60	$\frac{1}{2}CO_3^{2-}$	69

Application of conductimetric titrations

During a titration, changes in the conductivity of the solution usually occur, depending on the relative mobility of the ions added or removed from the solution, and use is made of this to determine, for example, the equivalence point of reactions involving neutralisation or precipitation. The conductance of the solution is measured after each addition of titrant and the results expressed graphically, and, as in all electrometric methods of titration, the 'end point' is found from a series of independent observations rather than from one end point measurement as in indicator methods. Ideally, two straight lines are obtained which intersect at the equivalence point, the accuracy under suitable conditions being better than 0.5%. The change in volume throughout the course of a titration should be kept small by employing a titrant which is at least 100 times more concentrated than the solution being titrated. When concentration differences are less than this, a correction for the dilution effect may be applied by multiplying each conductance reading by the ratio total volume to initial volume.

Measurements of conductance obtained in the region of the equivalence point are of little value due to hydrolysis, dissociation or solubility of the reaction product. For this reason, the graph often shows curvature rather than the ideal clear intersection of two straight lines.

Neutralisation reactions

In the titration of strong acids and strong bases there is always an initial decrease in conductance followed, beyond the equivalence point, by a rise (Fig. 5.1). In the titration of sodium hydroxide with hydrochloric acid, for example, the decrease in conductance is due to the replacement of hydroxyl groups with a molar ionic conductivity of 199 (Table 5.1) by chloride anions with a molar ionic conductivity of 76. Beyond the equivalence point, the increase in conductivity is due to

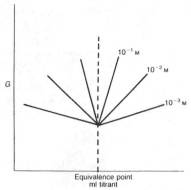

Fig. 5.1 Conductimetric titration of a strong acid with a strong base, showing the effect of dilution

the very high ionic molar conductivity (350) of the excess hydrogen ions.

The titration of hydrochloric acid with sodium hydroxide is similarly explained: the initial fall in conductance is due to replacement of the hydrogen ions (350) with sodium ions (50) and the final rise is due to excess hydroxyl ions (199) being present. The stronger the solutions used, the sharper and more definite is the appearance of the equivalence point, which is obtained graphically by the intersection of two straight lines (Fig. 5.1).

Titrations of a very weak acid with a strong base or a very weak base with a strong acid result in a small initial conductance which increases as the concentration of the salt, which is ionised, increases. In the region of the equivalence point, pronounced hydrolysis often occurs and this is shown by the marked curvature of the titration curve in this region. The equivalence point must be located, therefore, by extrapolating the straight portions of the curve until they intersect (Fig. 5.2).

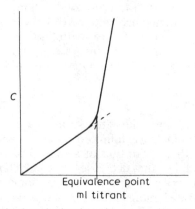

Fig. 5.2 Conductimetric titration of a weak acid with a strong base

Conductimetric titrations can be applied to the determination of phenols, weak acids whose salts are coloured, alkaloids, amines, certain dyes, and mixtures of a strong with a weak acid. Phenolsulphonic acid, for example, may be titrated first followed by the phenolic group. Some authorities claim that conductimetric titration is preferred to potentiometric titrations of weak acids and bases, and for titrating at very low concentrations.

Precipitation and complex formation reactions

Conductimetric titrations of this type may be performed satisfactorily if the reaction product is sparingly soluble or is a stable complex. In order to minimise errors, the intersection of the two branches of the titration curve should be made as acute as possible by suitable selection of titrant. The precipitate should be formed fairly rapidly and should not have strongly adsorbent properties.

Experimental methods

Measurement of conductance

Commercially available integrated or individual component conductance bridges can be used for these measurements (Fig. 5.3). For precise work both the resistance and capacitance of the solution in the conductance cell should be balanced out in the adjustment of the apparatus as each measurement is taken. A typical commercially available apparatus which possesses these facilities is the Wayne Kerr B642 autobalance ratio arm bridge, which can be continuously balanced automatically or manually. All simple conductance apparatus works on the Wheatstone Bridge principle of applying an alternating voltage (ca. 1 kHz) which avoids electrolyte reaction and polarisation to the conductance cell and, by resistance and capacitance adjustment, balancing to zero any current flowing in the circuit. This condition may be found by minimum noise in a headphone detector, by a magic-eye indicator,

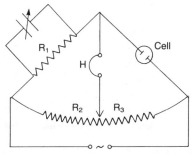

Fig. 5.3 Schematic diagram of a conductivity bridge

or by modulation on a sensitive meter. Instruments operating on the autobalance principle require only minimal operator control.

Conductance cells made of glass may be of the commonly used 'dip type', which are immersed in the solution sufficiently to cover the electrodes completely and without entrapment of air bubbles, or the 'bottle type', which require filling, always to the same level, with the solution to be investigated. For the most precise absolute measurements, cells should be allowed to 'age' for some months to allow the cell constant to stabilise. In all cases the platinised platinum electrodes should be quite rigidly fixed in place, as alteration of their relative positions will change the cell constant.

Conductimetric titration apparatus

In its simplest form, this consists of the conductance bridge, detector, and cell which is in contact with the solution. A 'dip type' cell is most suitable for this as it can be placed in the stirred beaker of solution which acts as the titration vessel (Fig. 5.4). This should preferably be immersed in a thermostatically controlled (±0.1°) bath.

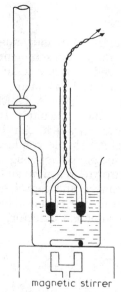

magnetic stirrer

Fig. 5.4 Electrode system and apparatus for conductimetric titrations

Platinisation of electrodes

Black platinum electrodes occasionally become scratched when the cell is being cleaned and should then be recoated by the following general method.

(1) Clean the surface of the platinum wire or foil by careful immersion, for a few seconds, in hot aqua regia, fuming nitric or chromic acid, followed immediately by a

thorough rinse with water.

(2) Immerse the electrodes in a solution of chloroplatinic acid (3%) containing lead acetate (0.025%) and connect them, via a reversing switch, milliammeter and rheostat, to a battery as shown in Fig. 5.5.

Adjust the current density to about 10 mA cm^{-2} electrode surface and reverse the current through the electrodes every 30 s until a thin black velvety coat is obtained (10–15 min). A thick coating results in an electrode with a very sluggish response.

(3) Remove the electrodes from the solution and rinse with water. Do not throw away or contaminate the platinising solution, which can be used repeatedly.

(4) Connect the two electrodes together to the cathode (−) and use another platinum electrode as the anode (+).

(5) Transfer the electrodes to dilute sulphuric acid, the platinised electrodes still being connected to the cathode, and electrolyse for 5–10 min. This process removes the last traces of impurities (chlorine and platinising liquid) adsorbed onto the platinum black and rapidly saturates the electrode surface with electrolytic hydrogen.

(6) Wash the electrodes thoroughly and always store them in water. The platinised electrodes will keep for long periods provided the surface is not allowed to become dry or damaged by scratches.

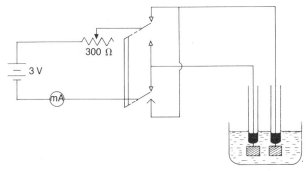

Fig. 5.5 Apparatus for platinising platinum electrodes

Conductimetric titration method

(1) Set up the apparatus (e.g. Fig. 5.3) and switch on the oscillator. After the readings have stabilised, zero the scales on the commercial instrument (if used) according to the manufacturer's instructions.

The platinum electrodes should be freshly platinised and stored in water when not in use. The surface must not be damaged by scratches or allowed to become dry.

(2) Pipette the required amount of the solution to be titrated into the flask and add sufficient water to cover the electrodes.

(3) Balance the bridge to give minimum headphone noise or zero reading on the appropriate meter by adjusting the variable resistance or slide wire on an individual component bridge, or by adjusting the resistance and capacitance controls on a commercial instrument. Adjust the coarse (large value) control first. Ensure that a suitably sensitive scale range is selected.

(4) Add appropriate portions of titrant, stir well, and find the balance point again. Repeat until about 10 ml of titrant has been added in excess of the expected end point.

(5) If necessary, calculate the conductance from the measured resistance readings. Capacitance adjustments serve only to refine the balance point: their values are not used in the conductance calculations.

(6) Plot a graph of conductance G against burette readings (ml) and read the end point from the graph.

If necessary multiply all the values of conductance by the ratio of total volume/initial volume in order to correct for volume change.

Experiments 1 to 10

A representative selection of titrations is given in Table 5.2. The equations representing each reaction should be written down and reference made to Table 5.1. This should enable you to explain the reason for the shape obtained for each graph.

Table 5.2 Conductimetric titration experiments

Experiment	Solution	Strength	Volume (ml)	Titrant	Strength	Titration increments (ml)
1	Hydrochloric acid	0.01M	50	Sodium hydroxide	0.1M	1.0
2	Hydrochloric acid	0.0001M	50	Sodium hydroxide	0.001M	1.0
3	Acetic acid	0.001M	50	Sodium hydroxide	0.1M	0.1
4	Acetic acid	0.1M	25	Piperidine	0.5M	1.0
5	Phosphoric acid	0.5M	50	Sodium hydroxide	2.0M	0.2
6	Acetic acid	0.1M	50	Sodium hydroxide	2.0M	0.2
	+ ammonium hydroxide	1.0M	4			
7	Hydrochloric acid	0.1M	10	Sodium hydroxide	1.0M	0.2
	+ acetic acid	0.1M	40			
8	Sodium acetate	0.1M	50	Hydrochloric acid	2.0M	0.2
9	Strychnine chloride	0.01M	25	Sodium hydroxide	0.1M	0.2
	+ hydrochloric acid	0.01M	25			
	+ ethanol	–	50			
10	Silver nitrate	0.001M	50	Potassium chloride	0.1M	0.1

High-frequency titrations

Introduction

Many difficulties are encountered in electrical methods of titration. Conductimetric measurements are complicated by polarisation, difficulty in wetting electrodes with small amounts of liquid, and corrosion and adsorption at electrode surfaces. Potentiometric and other galvanic methods are usually restricted to ionised solutions and are often impossible where non-aqueous solvents are used, especially if these are poor ionisation media.

By using the field of a high frequency (1–300 MHz) oscillator it is possible to produce ionic or dipole motion without introducing electrodes into the solution. If such an oscillator is placed in an insulated titration vessel, coupling takes place across the walls. The energy required to produce this ionic or dipole motion causes changes in loading of the oscillator. It is also possible to rectify the radiofrequency current bypassed by the solution and measure the resulting direct current with a microammeter. Such an instrument responds only to changes in conductivity of the solution, and is therefore suitable only for the titration of electrolytes. Titrations are carried out in the same way as in the conductimetric method, and similar curve shapes are obtained. The measured impedence represents complicated phenomena and does not lead to a simple explanation of the data.

Applications

The applications of high frequency titration have not yet been fully explored but are nevertheless extensive, and high frequency titrimeters are commercially available. The method may be used for most determinations but is particularly valuable where chemical methods and other electrometric methods fail.

The method is of value in alkaloidal assays and acid–base titrations in highly coloured solutions where visual indicators cannot be used. It may also be used for non-aqueous titrations, which are often limited by the lack of suitable visual indicators and by the lack of adequate electrode systems for potentiometric determinations.

It is possible to analyse complex mixtures by setting up calibration curves. Thus binary and ternary systems such as *o*-xylene-*p*-xylene and water–benzene–butanone may be evaluated, and by simple modification the technique may be applied to automatic process control. Another application is the detection of zones and the analysis of the eluate during chromatographic separations.

High frequency analysis is an elegant and accurate method for following rates of reaction and with the aid of automatic current recorders may be applied to very rapid reactions. It has made possible the direct determination of the products of alkaline hydrolysis of the esters of the chloracetic acids, which have half-lives of a few seconds, thus preventing the use of classical pipetting methods.

Potentiometry

Introduction

Construction of a suitable electrochemical cell can allow the concentration of a wide range of solutes to be measured. All electrochemical cells consist of two electrodes: a reference electrode, whose voltage should be independent of the nature and composition of the solutions into which it is placed, and an indicator electrode, whose voltage should depend on the concentration or, more correctly, the thermodynamic activity, of one component in the solution. When the two electrodes are placed together in the solution an electrochemical cell is constructed; its voltage can be determined by connection to a suitable potentiometer or millivoltmeter which can read to $\pm0.2\,\text{mV}$ or better, and which has a high impedence input of at least 10^{12} ohms. Under these conditions when a negligibly small current (<5 pA typically) is drawn from the electrodes, the e.m.f. of the cell is defined by

$$E_{cell} = E_+ - E_- + E_j$$

where E_j is the liquid junction potential, which in practice is usually eliminated by the use of an appropriate salt bridge integral with the reference electrode. Each electrode assumes a voltage close to its standard electrode potential and so will adopt a polarity ($E\pm$) dependent on its potential relative to the electrode with which it is paired. Thus, when the electrode or half-cell is coupled to another electrode through a common solution and through the external electrical connection of the measuring apparatus, a spontaneous chemical reaction occurs with the passage of electrons through the external circuit from the more negative electrode (or cathode) to the more positive (anode). Reduction, or loss of electrons from the chemical system, takes place at the cathode, and oxidation at the anode. Recent advances in electrode construction technology have resulted in the commercial availability of many ion-selective electrodes, gas sensing probes, and facilities for construction of enzyme electrodes, coated-wire electrodes and other biosensory devices, all of which can be used in combination with a reference electrode to form a concentration measurement system.

Theory

The Nernst equation is the basis for the relationship between the voltage generated by an electrochemical cell as a result of the two half-cell reactions (together giving a complete cell) and the relevant concentration at each electrode. The Nernst equation can be simplified so that it applies to one electrode, or half-cell, only: this is a more convenient approach for analysis, because the reference electrode's behaviour can be regarded as constant and the cell voltage as dependent only on the behaviour of the indicator electrode.

The Nernst equation can be derived by considering the free energy change involved in the complete cell reaction, i.e. by summing the two half-cell reactions, and using the values of chemical potential for all the species involved. After some simplification an equation can be developed which relates either to a cationic electrode (sensitive to a cation concentration):

$$E = E^{\ominus}_{M^{z+},M} + \frac{RT}{zF} \ln a_{M^{z+}}$$

to an anionic electrode;

$$E = E^{\ominus}_{X^{z-},X} - \frac{RT}{zF} \ln a_{X^{z-}}$$

or to a redox electrode;

$$E = E^{\ominus}_{ox,red} + \frac{RT}{zF} \ln \frac{a_{ox}}{a_{red}}$$

In the equations, $E^{\ominus}$ represents the standard electrode potential (SEP), or reduction potential, of the half-cell involved, and a represents the thermodynamic activity of the ion to which the electrode is sensitive. The constants R, T and F represent the gas constant (8.314 J K^{-1} mol^{-1}), absolute temperature (K), and the Faraday (96500 C/mole of electrons) respectively, and z is the number of electrons involved in the electrode reaction. Standard electrode potentials are collected together as the electrochemical series which is based on the arbitrary standard of zero potential for the standard hydrogen gas electrode at 298.15 K with H^+ ion activity of unity, and a hydrogen fugacity (thermodynamic pressure) of 1 atmosphere. The SEP is the electrode e.m.f. when all the species involved in the Nernst equation are at unit activity.

The relationship between activity and concentration is made through the activity coefficient y, as follows

$$a = cy$$

The Nernst equation is normally simplified to use the molar concentration c, so y is the molar activity coefficient. This is often taken as the same as the molal activity coefficient γ, for which there is an analogous equation

$$a = m\gamma$$

but there is in any case very little numerical difference between the two coefficients. If the product cy is used in the Nernst equation, which is then expanded, we obtain

$$E = E^{\ominus}_{ox,red} + \frac{RT}{zF} \ln \frac{c_{ox}}{c_{red}} + \frac{RT}{zF} \ln \frac{y_{ox}}{y_{red}}$$

from which it can be seen that $E^{\ominus}$ will retain its identity only if the two activity coefficients are equal. This qualification is even more limited if only one species exists in solution, as the activity coefficient must then be unity for $E^{\ominus}$ to have the same value in the equation using activities as in that using concentrations. To allow for this, $E^{\ominus}$ is redefined as E', the formal potential, which is numerically the same as $E^{\ominus}$ when the activity term is zero, so in general

$$E = E'_{ox,red} + \frac{RT}{zF} \ln \frac{c_{ox}}{c_{red}}$$

In practice the variable of interest to the analyst is molar concentration, not activity, so it is often necessary to add an inert electrolyte to increase the ionic strength of the solution under investigation and to make the activity coefficients constant over the concentration range of interest. Under these conditions the activity coeffient might have a value of 0.5 or less, but this does not concern the analyst provided it remains constant.

Ionic strength I is defined by

$$I = 0.5 \sum_n c_i\, z_i^2$$

for a solution containing n types of ionic species i whose molar ionic concentration and ionic charge are given by c_i and z_i respectively. Manufacturers of ion-selective electrodes often specify the nature of the ionic strength adjusting solution to be used in relevant analyses.

Measurement of pH

The activity of hydrogen ions in solution is a variable in the Nernst equation for an electrode reversible to these ions, and therefore such an electrode can produce an e.m.f. related to solution pH because of the definitions

$$\mathrm{pH} = -\log_{10} a_{H_3O^+}$$

and

$$\mathrm{pOH} = -\log_{10} a_{OH^-}$$

as even the most highly purified water possesses a small but definite conductivity due to ionisation:

$$2H_2O \rightleftharpoons H_3O^+ + OH^-$$

The hydrogen ion exists in water as the oxonium ion, H_3O^+, but for simplicity the following, and more familiar, equation is used:

$$H_2O \rightleftharpoons H^+ + OH^-$$

The assumption that the ionic activity coefficient is unity and that therefore activities become equivalent to concentrations is an accepted simplification for our purposes, so $a_{H_3O^+}$ becomes $[H^+]$ and a_{OH^-} becomes $[OH^-]$, i.e. the relevant molar ionic concentrations.

Although the potential of an H^+ ion indicating electrode depends on $a_{H_3O^+}$, there are compromises in these equations that must be appreciated. The first is that activity coefficients for individual ions are impossible to measure, so these variables are really mean ionic activity coefficients taken over the anion and the cation in the electrolyte, and the second is that a single electrode potential cannot be measured in isolation, so the modern definition of pH is an operational one involving e.m.f. measurements on a complete cell. This is discussed further in the paragraph on calibration.

By application of the law of mass action to the simplified equilibrium for water above we obtain

$$\frac{[H^+][OH^-]}{[H_2O]} = \text{constant}$$

In pure water or dilute aqueous solution, $[H_2O]$ is a constant; hence

$$[H^+][OH^-] = K_w$$

where K_w is the ionic product for water with an accepted value of 1.008×10^{-14} at 25°. In pure water or in neutral solution when the concentrations of hydrogen and hydroxyl ions are equal:

$$[H^+] = [OH^-] = \sqrt{K_w} = 10^{-7} \text{ (at 25°)}$$

This may also be expressed as:

$$[H^+][OH^-] = K_w = 10^{-14}$$

or

$$\log [H^+] + \log [OH^-] = \log K_w = -14$$

hence

$$pH + pOH = pK_w = 14$$

At 35°, $pK_w = 13.68$, at 60° the value is 13.02, and for a neutral solution pH $= pK_w/2$. Acid and alkaline solutions have pH values less and more than $pK_w/2$ respectively.

Application of the simplified Nernst equation to the hydrogen electrode gives

$$E = E^{\ominus}_{H^+,H_2} + 0.0592 \log a_{H^+}$$

or

$$E = E^{\ominus}_{H^+,H_2} - 0.0592 \text{ pH}$$

but as $E^{\ominus}_{H^+,H_2}$ is defined as zero, forming the basis for the electrochemical series, this becomes

$$E = -0.0592 \text{ pH}$$

(At any absolute temperature T the figure 0.0592 should be replaced (in this and similar expressions) by the product $0.000198T$.)

pH and millivolt meters

A linear relationship exists between the pH of a solution, at a given temperature, and E, the e.m.f. of a cell containing a reference and a suitable indicator electrode, since

$$E = k - 0.0592 \text{ pH (at 25°)}$$

i.e.

$$\Delta E/\Delta pH = -0.0592$$

Thus, a calibration in mV may be converted into pH units by dividing by 0.0592. In practice, however, pH meter scales are calibrated in pH units and in mV and the appropriate scale and range is selected by a simple control. Some multipurpose meters for ion-selective electrode use are fitted with direct reading concentration scales, which enable the meter to be calibrated in concentration units and measurements to be made without a calibration graph conversion step. These scales presume a linear relationship between potential and log concentration as predicted by the Nernst equation, and care must be taken that this is the case. For ion-selective electrodes, rather than pH electrodes, the slope is often below theoretical, i.e. $<0.0592/z$ mV at 25°, but can still be used provided it is linear. The term 'sub-Nernstian' is used for linear slopes $<0.055/z$, but such electrode systems usually prove unreliable.

Most commercial meters can measure with a precision of about ± 0.01 pH units or ± 1 mV, but this can be improved if a scale expansion facility is available. A temperature compensation device such as a variable resistance (adjusted manually) or an automatic temperature compensator (ATC) probe is usually fitted. The latter is immersed in the solution under investigation and has an electrical connection to the pH meter: when in use, care must be taken that the ATC comes to equlibrium before measurements are taken.

pH standard solutions

The most recent British Standard 1647 defines a range of buffer solutions best suited for glass electrode system calibration. Concentrations are now defined in molality (mol/kg of solvent), a unit whose values are independent of temperature. The primary pH standard is 0.05 molal potassium hydrogen phthalate, but in normal use the change from molarity to molality will make no practical difference to the user, as the preparation of the standard solutions is described in mass/mass and mass/volume terms. The British Pharmacopoeia uses molar concentrations in describing standard buffers and pH values so these specifications will change slightly with the adoption of a molal standard. All the buffers in this Appendix, except for the potassium dihydrogen citrate, are included in the latest British Standard but with their concentrations now expressed in molality and their pH values therefore altering slightly from those given in the B.P. table. Most of the new values, if rounded off to 2 decimal places, are identical with the old ones.

Meter calibration

The British Standards 2568:1979 and 3145:1978 refer to the operational definition of pH as mentioned above, and give specifications for suitable glass electrodes and pH meters. Such pH meters should have, in addition to a simple meter zeroing control, both a slope control referring to the Nernst slope RT/zF and a set buffer control for calibration based on the buffer solution standard selected. The two controls interact so the recommended procedure is to use 0.025 molal standard phosphate buffer initially (pH = 6.873 at 20°), and then check with a second recommended buffer of pH near that likely to be encountered in the measurements. The asymmetry control can then be adjusted until the meter again reads the pH of the selected buffer. The asymmetry adjustment setting will apply only for perhaps 24 hours and to the particular glass/reference electrode combination being used. The pivotal point of the slope adjustment is around pH7, which is why this pH should be set first. This and the operational definition of pH can be illustrated by

$$pH_x = pH_s + \frac{F(E_x - E_s)}{2.303RT}$$

which is obtained from the Nernst equation on consideration of the e.m.f. (E) given by a suitable measuring system immersed in buffer solutions x or s whose pH is given by pH_x and pH_s respectively. If solution s is the primary standard and x an additional calibration buffer, the asymmetry adjustment would in effect alter the value of $F/2.303RT$.

Reference electrodes

Calomel electrode

The calomel half-cell is the most widely used reference electrode, owing to the constancy of its potential and ease of preparation. The electrode consists of pure mercury in contact with a mixture of mercury and calomel, and a solution of potassium chloride (Fig.5.6). The potential of the calomel electrode depends upon the concentration of potassium chloride in the cell. The usual concentrations of potassium chloride used are 0.1M, M and saturated, the latter being the most convenient. The saturated calomel electrode (SCE) has, however, a high temperature coefficient (-0.76 mV per degree) which is of significance in accurate work. The potentials for the calomel electrode (*vs* the normal hydrogen electrode) in saturated, M and 0.1M potassium chloride are +250, 286 and 338 mV at 20°, and +246, 285 and 338 mV at 25°, respectively.

Silver–silver chloride electrode

The silver–silver chloride reference electrode is more difficult to prepare but is as convenient to use as a calomel electrode. It consists of a silver wire, coated electrolytically with silver chloride, dipping into a solution of potassium chloride of definite strength. The potential of the silver–silver chloride electrode (*vs* the normal hydrogen electrode) in saturated, M and 0.1M potassium chloride is +200, 235.5 and 288 mV respectively at 25°. The internal element of a broken glass electrode can be used as such an electrode when immersed in a potassium chloride solution.

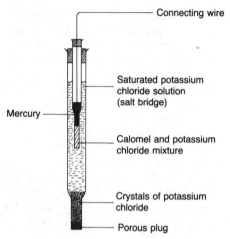

Fig. 5.6 Saturated calomel electrode

Mercury (I) sulphate electrode

This electrode is similar in construction to the calomel electrode but utilises sulphuric acid (0.05M) saturated with mercury (I) sulphate. It is used, for example, in solutions where silver or lead ions are present, and has a potential of 682 mV.

Salt bridges

A salt bridge of saturated potassium chloride, potassium nitrate or ammonium nitrate is used to prevent possible contamination of the reference electrodes with the test solutions. Sometimes the salt bridges are designed as a part of the reference electrode (Fig. 5.6) but are often solidified with a small quantity (3%) of agar. In general, when two solutions of electrolyte are brought into contact, a potential difference is set up between them due to the transference of ions across the boundary. This potential difference is known as a diffusion or liquid junction potential. The salt bridge reduces these potentials almost to zero and they become insignificant.

Indicator electrodes

Hydrogen electrode

The hydrogen electrode consists of a small piece of platinum foil, coated electrolytically with platinum black, over which hydrogen gas is passing. The platinum black surface exhibits a strong adsorptive power towards hydrogen and, provided that the metal surface remains in continuous contact with the gas, the electrode will act as if it were an electrode of metallic hydrogen. In use, therefore, only a part of the foil is immersed in the solution, the remainder being surrounded by pure hydrogen. The potential of the hydrogen electrode is used as a reference zero in the electrochemical series as indicated above.

Glass electrode

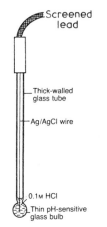

Screened lead

Thick-walled glass tube

Ag/AgCl wire

0.1M HCl

Thin pH-sensitive glass bulb

Fig. 5.7 Glass pH electrode

A glass electrode consists of a very thin bulb or membrane of specially prepared, pH-responsive glass fused on to a piece of comparatively thick, high resistance glass tube (Fig. 5.7). In contact with the thin membrane is a suitable solution such as 0.1M hydrochloric acid. Electrical contact with this solution is usually made with a silver wire coated with silver chloride, which acts as an internal reference electrode (i.e. is unresponsive to pH change).

The potential of the glass electrode, when immersed in a solution, is given by the expression:

$$E = K + 0.0592 \ (\mathrm{pH}_1 - \mathrm{pH}_2) \ (\text{at } 25°)$$

where K is a constant, pH_1 is the pH of the solution in the bulb and pH_2 is the pH of the test solution. Now, pH_1 is constant for a given electrode, hence:

$$E = k - 0.0592 \ \mathrm{pH}_2$$

where k is a constant, known as the asymmetry potential, which depends on several factors such as the existence of strains in the glass, the thickness of the glass bulb, and the composition of the solution within.

The advantages of a glass electrode are its rapid response and the fact that it is unaffected by the presence of oxidising or reducing agents, dissolved gases, highly coloured liquids, or moderate concentrations of many salts, with the main exception of sodium salts. The use of modern lithium silica glasses enables pH measurements to be valid over practically the entire pH range, but high concentrations of alkali can cause errors.

The main disadvantage of the glass electrode is its fragility, although modern devices are available which are especially rugged or resistant to boiling. Small imperfections on the glass bulb, such as scratches, and the presence of dehydrating agents, colloids, and surface deposits can cause interference in the measurements. Glass electrodes generally have a very high internal resistance and thus would never be used with simple potentiometers. The availability of glass electrodes combined with a reference electrode to produce a single unit in a variety of sizes and designs has ensured that the glass electrode remains the most versatile indicator electrode for pH measurements.

Rejuvenation of glass electrodes

After a long period of use, the surface of the glass electrode may deteriorate so that one or more of the following symptoms may appear:

(a) slow electrode response
(b) undue sensitivity of the pH reading to physical movement of the electrode
(c) failure of the electrode to check against a pair of buffer solutions
(d) inability to standardise in the range of the meter's asymmetry potential control.

Under favourable circumstances the electrode can be rejuvenated by momentarily immersing the bulb in 0.1M hydrochloric acid or by alternating immersion in acid and alkaline solutions to reduce residue sodium ion effects. If these methods fail to recondition the electrode, then immersion of the bulb in 20% ammonium fluoride solution for 3 min or in 10% hydrofluoric acid for 15 s should be tried. After this treatment the electrode should be thoroughly rinsed in a stream of tap water and dipped momentarily in 5M hydrochloric acid to remove fluorides. After a final rinse in purified water the electrode should be stored in acidified distilled water.

Redox electrodes

All electrodes are really redox electrodes in that both oxidised and reduced species are present, but the usual anionic and cationic electrodes have one associated chemical species which is insoluble or very slightly soluble. The term 'redox electrode' is therefore confined to an inert platinum plate in contact with the readily soluble oxidised and reduced forms of an electrode couple in the cell solution. The redox potential is the potential adopted by the electrode as a result of the equilibrium:

$$\text{ox} + ze \leftrightharpoons \text{red}$$

The more positive the redox potential of a system, the better is that system as an oxidising agent.

Ion-selective electrodes

A wide range of ion-selective electrodes (ISE) is now available from various established manufacturers. In addition, electrodes can be constructed for special requirements by the analyst using commercially available parts. It must be realised that all electrodes described as ion-selective are not selective exclusively for one ion, but respond to some degree to many ions, most particularly those of the same charge type as, and with a similar chemical nature to, the selected one. Possible interference must therefore be avoided by exclusion of interfering ions or by using selectivity coefficients, described later, in the Nernst equation describing the e.m.f. of the electrode. Ion-selective electrodes are of course used in combination with a reference electrode to make an ion measuring system, but the choice of reference electrode is important because of leakage of bridge electrolyte which could contain an interfering ion from the porous connection with the external solution; choice of a reference electrode with an easily removable outer sleeve so that the salt bridge solution can be readily changed is therefore desirable. Double junction reference electrodes are well suited to this.

ISEs consist, in general, of a thin layer of an electrically conducting material, called the membrane, which separates two solutions and across which a potential develops. The internal solution acts as a reference, and the outer solution is the one being measured. ISEs are classified by IUPAC according to membrane type, these being solid state, heterogeneous, liquid ion exchanger, and glass. The first two types are generally the most stable, long lived, reliable, and selective. The solid state homogeneous type is typified by the fluoride electrode, whose membrane consists of a crystal of lanthanum fluoride and which responds to F^- between 10^{-1} and 10^{-6} mol l^{-1}. Heterogeneous membrane electrodes require an inert matrix to support the active principle, e.g. an insoluble metal salt, to form the membrane: typical electrodes use silver halide or sulphide for detection of Ag^+ and the corresponding anion. Glass electrodes are typified by the pH-responsive electrode, which is obviously the best known ISE. However, glass electrodes for Na^+, K^+ and NH_4^+ are available but show poor selectivity in the presence of other metal ions.

Ion exchanger electrodes are the most versatile and provide a pattern to which special electrodes are constructed. In this case the ion of interest is sometimes incorporated in solution in an organic solvent (hence the name 'liquid ion exchanger'), or more commonly the ion exchanger is incorporated in a solid inert matrix such as PVC, polythene or silicone rubber. A typical liquid cationic exchange electrode (Fig. 5.8) is that for calcium, which is made using calcium bis(di-*n*-decyl) phosphate dissolved in di-*n*-octylphenylphosphonate.

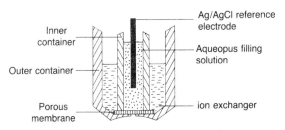

Fig. 5.8 Liquid ion exchange electrode

When choosing, or constructing, ion exchange electrodes, the ease of replacement of ion exchange solution, its toxicity, and the operational lifetime should be carefully considered. Specialised tests and manufacturer's literature should be consulted before this is undertaken. The potentiometric determination of nicotine with a liquid membrane electrode has been reported (*Anal.Chim.Acta*. (1981) **127**, 173). The liquid ion exchanger was nicotine hydrogen tetra(*m*-chlorophenyl)borate dissolved in *o*-nitrotoluene, and the electrode was constructed from a commercially available liquid membrane electrode barrel.

Coated-wire ISEs have no internal reference solution and consist of a PVC matrix sensor system coated onto a metal wire (platinum, silver, copper). Although they have been described as 'improbable devices', they nevertheless have the practical advantage that even microelectrodes of this kind are easily made and their responses are near-Nernstian. Coated-wire ISEs for detection of various drug compounds have been reported; for example (*Anal.Chim.Acta*. (1982) **138**, 97), electrodes based on dinonylnaphthalenesulphonic acid and dioctylphthalate in PVC have been used for analysis of methadone, cocaine, protriptyline and methylamphetamine in their protonated form, and detection limits of about 10^{-6} mol l^{-1} with near-Nernstian responses were observed. Typically, it was noted that calibration curves were found to be reasonably reproducible from day to day, provided the electrodes were soaked in acetate buffer (10^{-2} mol l^{-1}, pH 4.0) between calibrations. Full experimental details of the preparation of a nitrate-coated wire ISE are given in *J.Chem.Educ*. (1980) **57**, 512.

ISE characteristics and usage

(1) Response. In theory, ISEs follow the Nernst equation in responding to the activity of the selected ion, and the earlier discussion concerning activity and concentration is relevant here. Manufacturers often recommend the use of a total ionic strength adjustment buffer (TISAB or ISA) to optimise pH and ionic strength for particular analyses. A commercial fluoride electrode carries the recommendation that the following TISAB should be used to dilute all standards and samples to equalise activity coefficients at all concentrations: NaCl (1 mol l^{-1}), CH_3COOH (0.25 mol l^{-1}), CH_3COONa (0.75 mol l^{-1}), sodium citrate (0.001 mol l^{-1}).

(2) Limits of detection. There are many ways of defining this, but the most useful is perhaps the Nernstian limit below which the response of the electrode is no longer linear. An IUPAC recommendation is that the limit be defined as the concentration at which the electrode response deviates by $18/z$ mV from the extrapolated Nernst line (see Fig. 5.9), where z is the number of electrons involved in the electrode reaction. Orion Research, who manufacture ISEs, use the expression 'mud level' to define an apparent concentration at which the same reading is obtained as from a zero concentration solution. The reason for non-linearity at low concentrations is that the membrane material possesses slight water solubility, thus contributing significantly to the detected ion concentration. Provided reproducibility is achieved and frequent calibration takes place, measurements can be made over non-linear portions of the calibration curve.

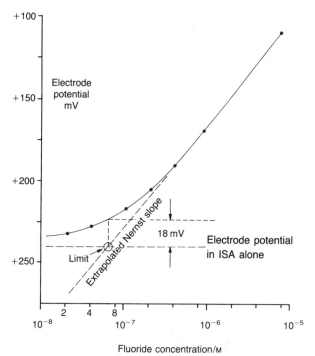

Fig. 5.9 Typical ISE response. (Redrawn from *Orion Research Inc. Newsletter* (1971) **III**(1), 1)

(3) Interference. Direct concentration measurements are not possible when either electrode or method interferences are encountered. Electrode interferences are caused when ions are 'mistaken' by the electrode for the ion being measured, e.g. see the shaded area to the right

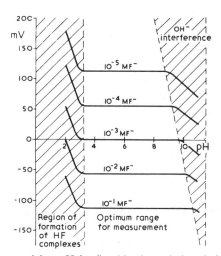

Fig. 5.10 Electrode potential *vs* pH for fluoride electrode in solutions of varying pH

of Fig. 5.10, where a fluoride ISE 'mistakes' OH^- for F^- in alkaline solution. Method interferences are problems in solution chemistry which lead to incorrect concentration values, e.g. see the shaded area on the left of Fig. 5.10 where the fluoride ion forms complexes in acid solutions, thus resulting in a lower level of the free fluoride species detected by the electrode.

Interfering ions for individual electrodes are detailed in manufacturer's literature: e.g. Philips indicate that their iodide electrode also responds to varying extents to CN^-, CrO_4^{2-}, $S_2O_3^{2-}$, CO_3^{2-}, Br^- and Cl^-. These ions should therefore be excluded by careful planning of the analysis or by complexation. An example of the latter is the substitution of 1,2-diaminocyclohexane-N,N,N',N'-tetraacetic acid for sodium citrate in the fluoride TISAB described above if appreciable iron or aluminium is present in the analysis sample. If exclusion of the interferent is not possible, allowance can be made for the e.m.f. contribution by the interferent by extending the Nernst equation to include selectivity coefficients, thus:

$$E = E^{\ominus} \pm \frac{2.303RT}{z_iF} \log [a_i + \sum_n K_{ij} \, a_j^{z_i/z_j}]$$

The extension of the log term includes contributions from each of n types of ion interferents, identified by subscript j, to the e.m.f. of the electrode due to the specific ion i. The selectivity coefficient K_{ij} is given in manufacturer's literature, or can be experimentally determined, and z_i/z_j is the ratio of the charge on ion i to that on ion j.

Direct electrode measurements are best achieved when using pure solutions of the ion being measured. When other ions are present, maximal accuracy is obtained when the standardising solutions also contain these ions at appropriate concentrations. These levels should be reasonably constant between samples.

(4) pH effects. Manufacturers state the optimum pH range for operation of their electrodes: this only considers the resistance of the membrane and electrode body to chemical attack or interference in response, so the specific chemical effects of pH on the system under analysis must be considered individually. For example, one fluoride electrode suffers method interference at pH <4, due to formation of HF, and selectivity interference by OH^- ions at high pH, as the selectivity coefficient $KF^-/OH^- \simeq 0.1$, which is a relatively high value. A similar response to H^+ ions is shown by a calcium electrode (Fig. 5.11), and in pH >10 hydroxide ion complexes a portion of the calcium or magnesium ion, thus reducing the amount of free divalent cations in solution. The electrode does not respond to bound or complex ions, e.g. cations bound to citrates, polyphosphates, carbonate or hydroxide ions, or some proteins. This electrode, like other ISEs, responds only to the activity of free unassociated ions.

(5) Electrode lifetime. This depends on the structure of the electrode, the solid membrane type having a longer lifetime (approximately 2–3 years) than the impregnated support membrane type (3–12 months).

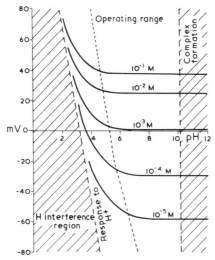

Fig. 5.11 Electrode potential *vs* pH for calcium or magnesium chloride solutions

The former can be repolished if the detection surface is damaged, and the latter can often be refilled with ion exchanger. Hydrolysis of adhesive used to locate the membrane ultimately results in irreparable damage, so careful storage, out of solution, is often advised.

(6) Methods of measurement. There are five common measurement methods used with ISEs:

  (a) direct reading
  (b) standard addition
  (c) standard subtraction
  (d) potentiometric titration
  (e) Gran's plot.

In all but the potentiometric titration method the electrode system must be standardised with one or more solutions of known concentration.

The direct reading technique is the most widely used method for using pH electrodes: a direct measurement of test solution is taken after a two-point standardisation as described above. A full calibration graph of e.m.f. *vs* log *c* should be obtained before direct readings with other solutions are undertaken, the readings being taken on progressively decreasing concentrations of sample with careful washing and rinsing between measurements.

Standard addition and subtraction, and sample addition and subtraction, are routine analytical techniques which will not be described in detail here. In each case an initial measurement is made either on sample or standard in appropriate TISAB, followed by a further measurement after addition either of the agent 'subtracting' the analyte from solution, or of standard or sample as appropriate. Equations for computing the sample concentration can be derived by writing the

Nernst equations, in terms of concentrations, for the two measurements and solving as simultaneous equations. Changes in volume of the solution on addition must be taken into account when the concentrations are calculated, and simplified equations incorporating this factor are available in manufacturer's literature, e.g. Orion's Analytical Methods Guide. The equations require that the slope of the electrode response is known, and ideally this should be determined from a calibration graph beforehand.

Gran's plots were devised in 1952, and only with the recent increase in use of ISEs have they become appreciated. They are used for converting sigmoid potentiometric titration curves or multiple standard addition results to a linear form from which the results are more easily calculated. The method can also be used with complexometric and precipitation titrations whose end points are sometimes difficult to locate with conventional titration readings. In addition, linearised data reduce the emphasis on the accuracy of the data around the end point and allow other data points to make a more significant contribution to the experimental results.

The Gran's plot technique can be illustrated, in varying degrees of complexity, by considering the Nernst equations relating to the analysis being performed. For potentiometric titrations with an ISE responding to an uncomplexed univalent cation, c

$$E = E^\ominus + S \log_{10} a_c$$

where the symbols have been previously defined and S is the electrode response slope. Rearranging gives

$$\text{antilog } (E/S) = \text{antilog } (E^\ominus/S).a_c$$

which is a straight line equation showing that antilog (E/S), containing the measured variable E, is proportional to a_c, and if plotted against the volume of titrant added during a titration will give a linear plot. When this graph, which is a Gran's plot, is extrapolated to the $a_c=0$ ($=$ antilog (E/S)) line it must intersect the volume added axis at the equivalent volume point. Another way of illustrating this is by considering the same type of equation applied to a standard addition based analysis. Moody and Thomas (*Sel.Ann.Rev.Analyt.Sci.* (1973)**3**, 59) showed that the equation below applies:

$$(V_o + V)10^{E/S} = xf10^{k/S}(V_oC_o + VC_s)$$

when volume V of standard solution of concentration C_s is added to V_o of sample solution of concentration C_o to give a potential E from an ISE whose slope is S. The parameter k is a constant, and f and x represent the activity coefficient and the fraction of primary free ions respectively, and should be kept constant by usage of TISAB if multiple standard additions are carried out. A plot of $(V_o + V).10^{E/S}$ against V gives the Gran's plot straight line which, when extrapolated to the abscissa, gives the volume value V_e. From the equation, applied to this point, it is clear that $C_oV_o = -C_sV_e$, from which C_o can be calculated. As in general analytical practice, a blank solution containing no analyte should be analysed and volumes subtracted if the intercept is not exactly at the origin of the graph; sufficient points should be plotted, both for the blank and in the experiment itself, for the straight line graph to be statistically acceptable.

An obvious disadvantage of the Gran's plot method is the need for tedious and repetitive calculation. This is reduced by the availability of microcomputers for processing the data and by the use of Gran's plot paper for direct plotting. One source of this paper is Orion Research who make available both volume corrected and uncorrected paper, the latter being used when no volume changes occur during a titration, e.g. when a reagent is generated coulometrically. Generally, 10% volume corrected paper is used which allows for concentration changes occurring on addition of titrant in potentiometric titrations but does not allow for changes in activitity coefficient, which should be kept constant throughout by the use of appropriate TISAB in the solvent of both sample and titrant. Gran's plot paper is antilog-linear with electrode e.m.f. being plotted on the vertical axis and, for 10% corrected paper, the number of ml (maximum 10) of added titrant per 100 ml of sample. As both axes are calibrated, care must be taken in the design of experiment and choice of solution concentrations that sufficient points are obtained within the ranges covered by the axes. A typical Gran's plot is shown in Fig. 5.12.

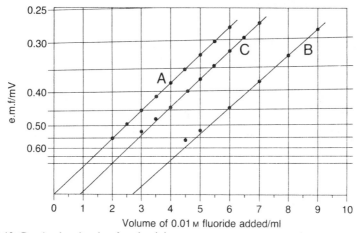

Fig. 5.12 Gran's plot titration for aluminium.
Curve A: blank (0.02M acetate buffer, 0.1M sodium perchlorate). Curve B: 1×10^{-4}M Al^{3+}.
Curve C: 3×10^{-5}M Al^{3+}. Fluoride electrode. (Redrawn from Orion Research Newsletter (1971) **III** (1), 6)

Potentiometric titrations

The end point of most titrations is detected by the use of a visual indicator but the method can be inaccurate in very dilute or coloured solutions. However, under the same conditions, a potentiometric method for the detection of the end point can yield accurate results without difficulty. The electrical apparatus required consists of a potentiometer or pH meter with a suitable indicator and reference electrode. The other apparatus consists of a burette, beaker and stirrer.

The potential of the reference electrode need not be known accurately for most purposes and usually any electrode may be used provided its potential remains constant throughout the titration. The indicator electrode must be suitable for the particular type of titration (i.e. a glass electrode for acid-base reactions and a platinum electrode for redox titrations), and should reach equilibrium rapidly. The electrodes are immersed in the solution to be titrated and the potential difference between the electrodes is measured. Measured volumes of titrant are added, with thorough stirring, and the corresponding values of e.m.f. or pH recorded. Small increments in volume should be added near the equivalence point, which is found graphically by noting the burette reading corresponding to the maximum change of e.m.f. or pH per unit change of volume (Fig. 5.13). When the slope of the curve is more gradual it is not always easy to locate the equivalence point by this method. However, if small increments (0.1 ml or less) of titrant are added near the end point of the titration and a curve of change of e.m.f. or pH per unit volume against volume of titrant is plotted, a differential curve is obtained in which the equivalence point is indicated by a peak (Fig. 5.14).

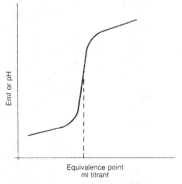

Fig. 5.13 Typical titration curve

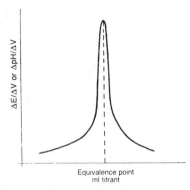

Fig. 5.14 Typical first derivative titration curve

Application of ISEs other than the pH-responsive glass electrode is limited in potentiometric titrations because of the relatively long response times of the electrodes. As previously mentioned, response time depends on electrode type, but concentration of analyte will also have an effect, with readings taking perhaps 15 s at the start of a titration when concentrations are high, but several minutes near the end point, the titration as a whole taking an unacceptably long time. The Gran's plot technique is more satisfactory in these cases, as fewer data need be obtained for a reliable result and the emphasis that potentiometric titrations make on the experimentally most difficult readings around the end point does not apply because of the linearisation of the data.

Automatic equipment, using a constant flow burette and the pH meter connected to a suitable chart recorder, can be made quite easily and is also commercially available.

Neutralisation reactions

Any pH-responsive indicator electrode may be used, but a glass electrode is usually preferable. The potential at the equivalence point is given by the expression:

$$E = k - 0.0592 \text{ pH } (25°)$$

where k, the asymmetry potential, depends on the electrode system used.

Dibasic and tribasic acids may be titrated with alkali to the intermediate equivalence points, provided the dissociation constants of each stage are sufficiently far apart; similarly, mixtures of acids may be titrated satisfactorily (e.g. acetic and sulphuric acids). A sufficiently large inflection is obtained when the difference in pK values exceeds 2.7 pK units. In the titration of a mixture of acids, the first inflection in the titration curve occurs when the stronger acid has been neutralised, and the second when neutralisation is complete.

Redox titrations

The indicator electrode most commonly used is a platinum wire or foil. The potential of the indicator electrode is a function of the ratio of the concentrations of oxidised and reduced forms of an ion as previously discussed. As before, the equivalence point in a redox titration is indicated by a marked inflection in the titration curve, if $E^{\ominus}_{\text{ox,red}}$ of the two chemical systems are sufficiently far apart.

Precipitation reactions

The solubility product of the almost insoluble material formed during a precipitation reaction determines the ionic concentration at the equivalence point. The indicator electrode must readily come into equilibrium with one of the ions. For example, a silver electrode is used for the titration of halides with silver nitrate, and its potential is given by the expression for cationic electrodes discussed earlier.

Experiment 11   *Determination of pH using a glass electrode*

It will be assumed that a commercial potentiometric type pH meter is used, set to read pH units.

(a) Connect the meter to its mains supply and switch on. Modern transistor based instruments may be used almost immediately, but ideally should be left in 'Standby' mode to be ready for instant operation.

(b) Connect the glass and calomel electrodes to the meter. New electrodes should be soaked in pH ≈ 7 buffer for at least eight hours before use. Ensure that the reference electrode is filled with the salt bridge solution and air bubbles excluded. The automatic temperature compensator should be correctly installed and immersed, with the electrodes, in the solution to be measured.

(c) Two-point calibration as described under 'Standardisation' above should be carried out using the pH ≈ 7 buffer first. If the meter has an 'Expand' scale this can be used for more precise adjustment of the reading, which is carried out with the 'Calibration control'. The reading for the second buffer is adjusted with the 'Slope' or 'Temp' control, depending on the meter.

(d) Electrodes should be rinsed with distilled water after removal from a solution, and then rinsed with a sample of the next solution before a measurement is made. Wiping dry is not a recommended practice.

(e) Take the pH reading of the unknown solution.

(f) Switch the meter to 'Standby' and leave the electrodes in buffer of pH ≈ 7 if no further readings are to be taken. Standardisation should be carried out each day.

Experiment 12   *Determination of the linearity of response of a glass electrode*

(a) Prepare 20 ml portions of McIlvaine buffers of pH 2.2, 3.0, 4.0, 5.0, 6.0, 7.0 and 8.0 (Table 5.3).

(b) Set up the pH meter to read potential (mV).

(c) Use the electrode system as in Experiment 11 and measure the potential difference between the glass and calomel electrodes for each of the buffer solutions. Wash well with water before each measurement.

(d) Plot graphically the potential of the glass electrode (ordinate) against the pH of the buffer solution (abscissa) and measure the slope [$dE/d(pH)$] of the line, noting that:

$$E = k - 0.0592 \, pH$$

and

$$\frac{dE}{d(pH)} = -0.0592 \text{ (at 25°)}$$

Table 5.3 McIlvaine Universal Buffer—(pH range 2.2 to 8.0). 20 ml quantities are prepared by mixing X ml disodium hydrogenphosphate (0.2M) with Y ml citric acid (0.1M).

pH	X	Y	pH	X	Y	pH	X	Y
2.2	0.40	19.60	4.2	8.28	11.72	6.2	13.22	6.78
2.4	1.24	18.76	4.4	8.82	11.18	6.4	13.85	6.15
2.6	2.18	17.82	4.6	9.35	10.65	6.6	14.55	5.45
2.8	3.17	16.83	4.8	9.86	10.14	6.8	15.45	4.55
3.0	4.11	15.89	5.0	10.30	9.70	7.0	16.47	3.53
3.2	4.94	15.06	5.2	10.72	9.28	7.2	17.39	2.61
3.4	5.70	14.30	5.4	11.15	8.85	7.4	18.17	1.83
3.6	6.44	13.56	5.6	11.60	8.40	7.6	18.73	1.27
3.8	7.10	12.90	5.8	12.09	7.91	7.8	19.15	0.85
4.0	7.71	12.29	6.0	12.63	7.37	8.0	19.45	0.55

Note: (i) Sodium ion concentration increases with alkalinity. (ii) Temperature coefficient is small. (iii) Buffer solutions should be freshly prepared using recently boiled and cooled water.

Potentiometric titration experiments

General method

(a) Switch on the instrument as before.

(b) Connect the saturated calomel electrode to the positive terminal or socket and the glass electrode to the negative. Use a beaker, as in Experiment 11 and 12, but a mechanical rather than a magnetic stirrer as great care is needed to avoid damage to the glass electrode. Much accidental damage may be prevented by arranging the calomel electrode and the paddle of the stirrer to be at a lower level than the glass electrode, which should be placed between them (Fig. 5.15).

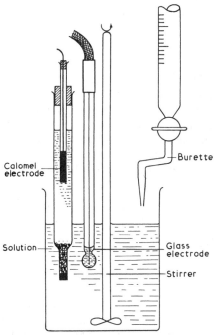

Calomel
electrode

Burette

Solution

Glass
electrode

Stirrer

Fig. 5.15 pH potentiometric titration apparatus

(c) Adjust the meter to zero and standardise the electrode system as above. As subsequent readings are to be taken with the stirrer motor running, the motor should be switched on for this operation (Note).

(d) Rinse the beaker and electrodes thoroughly with water, then with the solution to be titrated, and place a measured volume in the beaker. Add sufficient water to cover the bulb of the glass electrode adequately.

(e) Switch on the stirrer and measure the pH of the solution.

(f) Add about 2 ml of titrant, allow sufficient time to mix and measure the pH of the resultant solution.

(g) Add further quantities of titrant and, as the end point is approached, reduce to 0.1 ml increments and measure the corresponding values of pH.

(h) Obtain further readings of pH for about 5 ml beyond the equivalence point.

(i) Switch the meter to 'Standby', and turn off the stirrer. Wash the beaker and electrodes thoroughly with distilled water.

(j) Plot a graph of pH *vs* ml titrant added. Read off the equivalence point from the graph and calculate any required data from this value. If the titration is of a weak acid against a base, calculate the pK_a value of the acid from the Henderson Equation, which is valid over the range pH 4 to pH 10:

$$pH = pK_a + \log \frac{[salt]}{[acid]}$$

Percentage neutralisation	5	10	25	33.3	50	66.6	75	90	95
Value of log [salt]/[acid]	-1.28	-0.95	-0.48	-0.30	0.00	$+0.30$	$+0.48$	$+0.95$	$+1.28$

Note The stirrer motor should not be switched on or off while the instrument is switched to the 'measure pH' position.

Calculation of titration curves

If the pK_a value of an acid or base is known, theoretical titration curves can be calculated. Usually the pK_a value is used for both acids and bases, and because of the Bronsted–Lowry theory one set of equations only need be used. This theory maintains that weak acids have conjugate weak bases, and weak bases have conjugate weak acids, to which they are respectively converted during titration. The Henderson equation in its most general form then becomes

$$pH = pK_a + \log \frac{[base]}{[acid]}$$

where either the base or the acid form of the weak electrolyte is the conjugate component. The pK_a of a weak base is defined from the acid dissociation constant of its conjugate acid. Thus the equation

$$pH = \tfrac{1}{2}(pK_a - \log c)$$

applies where a weak acid is present alone either at the start of a weak acid–strong base titration or at the end of a weak base–strong acid titration. In the equation, c represents the concentration in $mol\, l^{-1}$ of acid in the solution and must take into account any dilution as a result either of titration or of simple addition of water that may have occurred.

During the titration, when both normal and conjugate forms of the weak electrolyte are present in the solution, the pH may be calculated from the Henderson Equation as above, while at the end point, when completely neutralised, the solution is one of a salt of a weak and a strong electrolyte. The equation:

$$pH = \tfrac{1}{2}(pK_a + pK_w + \log c)$$

applies where a weak base is present alone in solution, as at the end point of a weak acid–strong base titration or at the start of a weak base–strong acid one. The value of pK_w is 14, and c is the weak base concentration; this may have to be corrected for volume changes.

In the titration of polybasic acids, the pH at the intermediate equivalence points is given by the expression:

$$pH = \tfrac{1}{2}(pK_1 + pK_2)$$

The pH beyond the equivalence point can be calculated directly from the excess alkali or acid added, using the formula:

$$pH = pK_w - pOH$$
$$= 14 + \log [OH^-]$$

or the definition of pH as appropriate.

Experiments 13 to 20   *Titrations with sodium hydroxide*

Take 20 ml quantities of the following solutions (0.5M) and titrate with sodium hydroxide (0.5M) using the general method outlined above:

13 Hydrochloric acid
14 Acetic acid
15 Hydrochloric acid–acetic acid mixture
16 Boric acid
17 As Experiment 16 but in the presence of 50 ml glycerol
18 Phosphoric acid (0.15M)
19 Aniline hydrochloride
20 Procaine hydrochloride (0.015M *vs* 0.015M NaOH)

Experiments 21 to 25   *Titrations with hydrochloric acid*

Take 20 ml quantities of the following solutions and titrate with hydrochloric acid (0.5M)
21 Sodium carbonate (0.25M)
22 Sodium nitrate (0.5M)
23 Sodium borate (0.25M)
24 Triethanolamine (0.5M)
25 Triethylamine (0.5M)

In addition, the Gran's plot method should be tried in experiments 19, 20, 24 and 25. Deviations from perfect linearity may occur, and in such cases extrapolation of the linear portion only of the line will yield a satisfactory equivalent volume.

Experiments 26 to 29   *Titrations with hydrochloric acid followed by back titration with sodium hydroxide*

Take 20 ml quantities of the following solutions (all 0.5M) and titrate with hydrochloric acid (0.5M) to beyond the equivalence point (e.g. 25 ml). Back titrate with sodium hydroxide (0.5M) until excess alkali is present (about 50 ml).
26 Pyridine
27 Piperidine
28 Glycine
29 *p*-Aminobenzoic acid

Experiment 30   *Adjustment of a solution to a given pH*

The stability of many solutions of pharmaceutical substances is dependent on the pH of the medium. To obtain maximum stability, it may be necessary to adjust the solution to the required pH by the addition of a suitable quantity of acid or alkali.

(a) Measure the total volume of the solution to be adjusted and take a convenient quantity, say 50 ml, and place in the titration cell.

(b) Measure the pH (glass and calomel electrodes) and add sufficient standard acid or alkali until the correct pH is obtained.

(c) From the burette reading calculate how much acid or alkali should be added to the remainder of the solution. Add the desired amount and mix well.

(d) Again measure the pH and, if necessary, add small quantities of acid or alkali to the bulk solution, with mixing and sampling, until the correct pH is obtained.

Redox titrations

These titrations are performed using a bright platinum wire or foil indicator electrode and either a silver–silver chloride or saturated calomel reference electrode. The method is identical with the general method described above except that the meter is set to read millivolts and not pH units. Electrode standardisation with buffer solutions is not necessary, of course.

Experiments 31 to 35

General method. Place the solution of the substance to be titrated (0.1M: 20 ml) in a 250 ml beaker, add sufficient water to immerse the platinum electrode completely and titrate with the solutions recommended.

31 Titrate iron(II) ammonium sulphate, in the presence of dilute sulphuric acid, with potassium permanganate solution (0.02M).

32 Titrate iron(II) ammonium sulphate, in the presence of dilute sulphuric acid, with potassium dichromate solution (0.02M).

33 Determine the percentage of antimony in antimony sodium tartrate by titration with potassium bromate (0.03M) in hydrochloric acid solution.

34 Titrate iron(II) sulphate (0.1M) with 0.1M ceric sulphate in the presence of sulphuric acid (3M).

Experiment 35   *Assay of iron(II) succinate in Ferromyn S tablets*

Each tablet is stated to contain 113 mg iron(II) succinate. Ten tablets should be weighed and thoroughly pulverised. Titrate an accurately weighed amount of powder equivalent to 50 mg iron(II) succinate dissolved in 75 ml dilute sulphuric acid with standardised 0.002M potassium permanganate, taking readings every 1 ml until several ml past the end point, but reducing their interval for readings around the end point.

Precipitation reactions

Experiments 36 to 42

The electrode system consists of a polished silver wire and a reference electrode (either a silver–silver chloride electrode connected to the solution by means of a potassium nitrate salt bridge, or a mercury(II) sulphate electrode). A glass electrode may also be used as a reference electrode for this type of titration, since the hydrogen ion activity remains practically constant throughout the titration.

36 Place 20 ml of a solution of potassium chloride (0.1M) in the beaker and add sufficient nitric acid (0.05M) to cover the electrodes. Titrate with silver nitrate solution (0.1M), record the e.m.f. of the cell as before, and plot a graph of e.m.f. *vs* burette reading.

37 Titrate 0.1M potassium bromide with silver nitrate.

38 Titrate 0.1M potassium iodide with silver nitrate.

39 Titrate 20 ml of a mixture of equal parts of the 0.1M potassium chloride, bromide and iodide solutions as in Experiment 36.

40 Titrate 20 ml of a mixture of equal parts of potassium thiocyanate (0.1M) and potassium chloride (0.1M) as in Experiment 36.

Specific ion electrodes

Experiments 41 to 44

Specific ion electrodes are used as described above with a reference electrode, and connected to a suitable meter in a similar manner to electrodes used for pH measurements. A conventional pH meter may be used if it has an expanded mV/pH scale, but it is necessary to obtain a suitable calibration curve as described earlier. However, specific ion meters may be used whereby the concentration is obtained directly from the instrumental readings when ion electrodes are used.

It is recommended that the operation of commercially available electrodes is investigated before the construction of ion exchange devices or coated-wire electrodes for drug compounds is attempted.

These experiments use a divalent calcium cation activity electrode suitable for direct measurement as the indicator electrode.

Experiment 41 *Calibration*

Prepare an electrode potential *vs* concentration calibration curve using standard solutions of calcium chloride. Use 0.1M calcium chloride standard solution with dilutions (2% v/v, 4M potassium chloride) of 1 : 5, 1 : 10, 1 : 50 and 1 : 100, or other dilutions according to the expected measurement range.

Experiment 42 *Estimation of calcium chloride*

Determine the concentration of calcium chloride in 'unknown' calcium chloride solutions by measuring the potential obtained when the electrodes are immersed in these solutions and by using the calibration curve already prepared.

Experiment 43 *Electrode response to various ions*

Use 0.1M solutions of various ions at pH 6. Compare the potentials obtained in each case and roughly assess the selectivity coefficient for each using the definition and equation given earlier. Solutions for this experiment should contain Zn^{2+}, Cs^{2+}, Li^+, Mg^{2+}, Na^+. (Comprehensive experiments for selectivity coefficient determination can be found in *Analysis with Ion-Selective Electrodes* by P.L. Bailey, Heyden, 1976.)

Experiment 44   *Determination of calcium by titration*

A suitable electrode can be used as the end point detector in titrations with EDTA or other chelators. The end point is where the potential becomes steady on further addition of EDTA. Determine the concentration of 'unknown' calcium chloride solution by titrating with 10^{-2}M NaEDTA. The calcium chloride solutions should be adjusted to pH 10.

Experiment 45

A considerable number of reports in the literature relate to the use of the fluoride ISE, and its capabilities and conditions of use. One report (*Anal.Abs.* (1984) **46**, 4D58) describes its use for the determination of F^- in urine and proposes the method for inclusion in the Pharmacopoeia of the German Democratic Republic.

Experiment 46   *Determination of fluoride in tablets and solutions (e.g. vitamin supplements or caries preventative) by fluoride ISE*

This method originates from a collaborative FDA study (*J.Assoc. Off.Anal.Chem.* (1984) **67**, 682). Beakers should ideally be made of plastic to reduce any losses of F^- by reaction with glassware. The TISAB to be used should be prepared from 57 ml glacial acetic acid, 58 g sodium chloride, and 4 g 1,2-diaminocyclohexane-N,N,N',N'-tetraacetic acid dissolved in about 500 ml of water, then cooled and adjusted to pH 5.0–5.5 with 5M sodium hydroxide before dilution to 1 litre. Standard solutions of 10^{-2}, 10^{-3}, and 10^{-4}M sodium fluoride solutions should be prepared in water.

Solutions should be prepared by weighing and crushing at least ten tablets, and dissolving powder equivalent to $10 \, mg \, F^-$ in 400 ml of water on a steam bath before cooling and diluting to exactly 500 ml. Titration solutions should be prepared by diluting an aliquot containing $1–2 \, mg \, F^-$ to 100 ml with water.

Measurements should be made under constant temperature conditions on 20 ml of standard or sample diluted with 20 ml TISAB. Two minutes may be required before readings stabilise in the lowest concentration solution. A calibration graph of e.m.f. *vs* log fluoride concentration can be used for determination of the fluoride concentration in each sample.

Voltammetry and coulometry

Introduction

Voltammetric methods of analysis involve the measurement of the current which is caused to flow as a result of the controlled application of a potential to an electroactive system incorporating the species to be analysed. The form of voltammetry most frequently used in pharmaceutical analysis is polarography, which was originally devised by Heyrovsky in 1922. Although it is a fundamentally simple technique, the practical problems of the polarographic apparatus deterred many analysts from investigating the scope of the method until the advent of reliable commercially available polarographs employing the more refined and sensitive techniques which owe their origin mainly to the work carried out in the late 1950s by Barker and co-workers from Harwell.

Polarography

Polarography is the specialised form of voltammetry which investigates solution composition by reducing or oxidising metals, organic compounds or ions at a dropping mercury electrode (dme). The sample, in suitable solution, is placed in a special electrolytic cell which contains, in its simplest form, a polarisable and a non-polarisable electrode. A linear voltage ramp progressing typically at $2–5\,mV\,s^{-1}$ is applied to the cell, the corresponding current is measured and the current–voltage (CV) data recorded graphically (Fig. 5.16). This CV curve, which has a characteristic S-shape and may be used for the qualitative and quantitative analysis of electroactive species, is known as a polarogram and the associated electrical apparatus as a polarograph. Since the electrolysis current is small (usually $<50\,\mu A$) and is physically limited by diffusion of the electroactive species up to the mercury drop, the total amount of electroreaction is negligible, and the same solution may be repeatedly analysed without significant change in the CV curve.

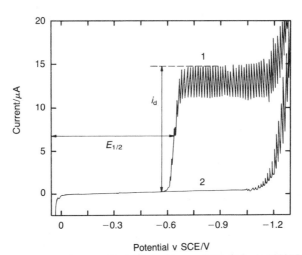

Fig. 5.16 Polarograms of (1) 1M hydrochloric acid containing 0.5×10^{-3}M Cd^{2+}, and (2) 1M hydrochloric acid alone. (Redrawn from Application Note 151, Princeton Applied Research Corporation.)

Apparatus

The basic apparatus used in polarography is very simple, but modern variations in the technique use more refined electronics and apparatus. The simplest d.c. polarograph requires a potentiometer connected, via a microammeter, to two electrodes, one of which is the dme consisting of a mercury reservoir connected by a flexible polythene tube to a short length of very fine (bore diameter 20–80 μm) capillary tubing, and the other a mercury pool anode. However, as the latter has no defined characteristic voltage, a third, reference, electrode should be included in the cell, thereby fixing the relative voltages applied between the electrodes. Such a polarograph is shown in Fig. 5.17.

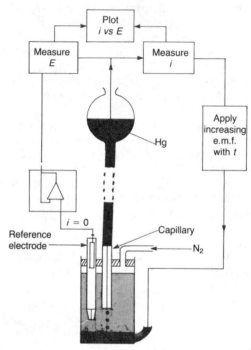

Fig. 5.17 A simple d.c. polarograph in diagrammatic form. (Redrawn from W.J. Albery, *Electrode Kinetics*, Oxford University Press.)

Theoretical considerations

A typical CV curve is shown in Fig. 5.16 and a similar curve but without the superimposed oscillation caused by the mercury drop growth cycle is shown in Fig. 5.18. In this diagram i_r is the residual current, i the current at any point on the wave measured with respect to the line AB produced, or (more correctly) with respect to the residual current curve, and i_d is the diffusion current. The diffusion current is proportional to the concentration of electroreducible ions present, and thus forms the basis of quantitative polarography. $E_{1/2}$, the half-wave potential which produces a current (at C) midway along the steeply rising portion of the curve, is characteristic of the particular system being reduced and thus enables qualitative analyses to be performed.

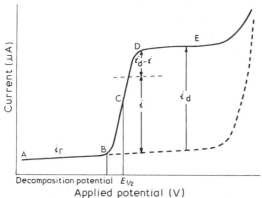

Fig. 5.18 A polarogram illustrating the currents discussed in the text

The gradual rise in current, known as the residual current i_r, is the sum of a relatively large capacitance current i_c and a very small faradaic current i_f. The capacitance current is produced when mercury drops, from the dme, become charged at the mercury–solution interface due to the formation of a Helmholtz double layer of positively and negatively charged ions. This is the cause of the oscillations, superimposed on a d.c. polarogram, which follow the mercury drop growth cycle, showing a steep rise as the drop area rapidly increases at the start of its lifetime but becoming less steep towards the end of the drop life. The residual faradaic current is due to traces of impurities in the solution being reduced. For example, it is very difficult to remove the last traces of oxygen even after bubbling nitrogen through a solution: ordinary distilled water often contains traces of copper and, when solutions are deoxygenated in the presence of the pool anode, mercury ions sometimes go into solution with the formation of hydrogen peroxide.

The residual current should always be subtracted from the total observed current, as shown in Fig. 5.18, in order to obtain the diffusion current i_d. The minimum detectable concentration of electroreducible ions depends, to a very large extent, on the accuracy with which this correction is measured. During an experiment a suitable solution of the electroactive substance under examination is freed from dissolved oxygen by bubbling O_2-free nitrogen through for several minutes and blanketing the solution with nitrogen, because oxygen is electroreducible and this could interfere with the electroreaction being observed. After the voltage scan has been made and the polarogram obtained, a further polarogram should be recorded in the absence of the electroreducible species to give a residual current curve (broken line in Fig. 5.18).

At the point B in Fig. 5.18 the decomposition potential is reached and electrolysis commences with the discharged ions depositing on or amalgamating with the mercury, e.g.

$$Zn^{2+} + 2e^- \leftrightarrows Zn$$

Often only a part of the charge is neutralised and the product is also capable of electroreduction at a higher applied potential, thus giving a two-step polarogram, as for example:

$$Cr^{3+} + e^- \leftrightarrows Cr^{2+}$$
$$Cr^{2+} + 2e^- \leftrightarrows Cr$$

At point C the corresponding voltage is the half-wave potential $E_{1/2}$, and the concentrations of the oxidised and reduced forms at the electrode surface are equal, i.e.

$$[Zn^{2+}] = [Zn]$$

The potential at this point is characteristic of the nature of the reacting material and is independent of the electrode characteristics. Thus, in principle, the half-wave potential may be used for the identification of an unknown substance.

The straight portion of the curve, DE, is known as the limiting current i_d, and is measured in μA as the vertical height between AB produced and the top of the drop oscillations along DE (Fig. 5.18). When the limiting current is reached, the electroreducible substance is reacting as rapidly as diffusion transports it to the electrode surface, and its concentration at the mercury–solution interface remains constant at a value which is negligibly small compared with the concentration in the body of the solution. Under these conditions the current is independent (within certain limits) of the applied potential and in this state the electrode is said to be concentration polarised. The extent of polarisation depends on the surface area of the electrode and the current. In general, electroactive ions may be supplied to the depleted region at the electrode surface by three more or less independent forces:

(a) a convection current, which is minimised by using a quiet solution and uniform solution temperature
(b) a diffusive force, proportional to the concentration gradient at the electrode surface, causing the diffusion current i_d
(c) an electrical force, proportional to the electrical potential gradients at the electrode surface–solution interface, causing a migration current i_m.

Forces (a) and (c) must in practice be minimised because polarographic theory requires that diffusion, which is directly related to concentration, should be the sole means of supply of electroactive material to the electrode surface, and, although the current through an electrolytic solution is carried impartially by all the ions present, regardless of whether or not they take part in the electrode reaction, the fraction of the total current carried by any particular species of ion depends primarily on its relative concentration in the solution. Thus, when a large excess of supporting electrolyte is present, the current through the solution will be almost entirely carried by the large excess of indifferent ions and the proportion of the current carried by the reducible ions will decrease practically to zero. (A supporting or indifferent electrolyte is a salt, whose ions do not participate in the electrode reaction, which is added to increase the conductance of the solution and to minimise effects of electrical migration.) Thus

$$i_m \simeq 0$$

and

$$i_l = i_d$$

and the limiting current is equated with i_d. Therefore, under normal polarographic conditions, when at least a 50-fold excess of supporting electrolyte is present, the limiting current is almost solely a diffusion current.

Ilkovic equation

Ilkovic, on examining the various factors which govern the diffusion current in the presence of a large excess of supporting electrolyte, derived the following equation:

$$i_{max} = 706nD^{1/2}Cm^{2/3}t^{1/6}$$

where i_{max} is the maximum diffusion current in microamperes (μA) during the life of a mercury drop, n is the number of electrons involved in the reduction of one molecule of reducible substance, D is the diffusion coefficient ($cm^2 s^{-1}$), C is the concentration (millimoles per litre), m is the weight of mercury flowing through the capillary ($mg\,s^{-1}$) and t is the lifetime (seconds) of one drop of mercury, normally between 2 and 7 seconds. The value 706 is a combination of numerical constants, one of which is the density of mercury.

If the average current through one drop lifetime were to be measured with a meter which was not capable of following accurately or quickly enough the fluctuations which occur, the average diffusion current i_d would be given by the expression referred to in future as the Ilkovic equation:

$$i_d = 607nD^{1/2}Cm^{2/3}t^{1/6}$$

from which it can be seen that

$$i_d \approx \tfrac{6}{7}i_{max}$$

The Ilkovic equation shows that the diffusion current is proportional to the concentration:

$$i_d = kC$$

where k is a constant defined by the Ilkovic equation, i_d has been corrected for the residual current and the other factors are constant.

This linear relationship may fail when the drop time is too short, owing to the stirring effect disturbing the diffusion layer and producing an abnormally large current. However, the addition to the solution of a small quantity of gelatin, which increases the viscosity, often counteracts this effect.

The term $m^{2/3}t^{1/6}$ in the Ilkovic equation satisfactorily describes the capillary characteristic provided that the drop time of the capillary is within the usual limits. It will be appreciated that capillaries used in various laboratories will be different with respect to bore, length and reservoir height, and so knowledge of m and t will enable comparisons to be made between different capillaries, since:

$$i_d = k'm^{2/3}t^{1/6}$$

and so

$$i_d/m^{2/3}t^{1/6} = k'$$

Factors affecting the variables in the Ilkovic equation

The factors m and t depend upon the dimensions of the dme and on the pressure due to the mercury column. Drop time depends on the interfacial tension at the mercury–solution interface, the nature of the solution and the applied potential. The rate of flow of mercury also depends on the temperature and, to some extent, on the interfacial tension. It can be shown, by consideration of the rate of flow of liquid through a capillary, that at constant temperature:

$$P \propto m$$

and thus

$$P/m = k''$$

where P is the difference in hydrostatic pressure between the two ends of the tube. The ratio P/m is known as the capillary constant.

The height of the mercury column causes changes in the diffusion current, the square of the diffusion current being proportional to the height, after correction for back pressure, of the mercury column above the electrode tip. It can be shown that the value of the back pressure, h_{back}, to be deducted is given by the equation:

$$h_{back} = \frac{3.1}{m^{2/3}t^{1/6}} \text{ cm of mercury}$$

The potential applied to the polarographic cell influences the interfacial tension at the mercury–solution interface. As the applied potential is increased, so the interfacial tension σ increases, initially to a maximum at about -0.56 V *vs* Normal Calomel Electrode (NCE) (in the absence of capillary active substances) and then decreases again (Fig. 5.19). This maximum is known as the electrocapillary maximum, the electrocapillary zero or the isoelectric point, and at this potential the mercury is uncharged. The position of the electrocapillary zero varies when capillary active ions (e.g. potassium iodide, potassium bromide) are adsorbed onto the surface of the mercury drop.   Since it can be shown that drop time is directly proportional to the surface tension, it is to be expected that a drop time *vs* potential curve will be shaped similarly to the electrocapillary curve (Fig. 5.19).

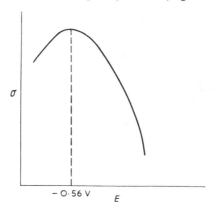

Fig. 5.19 Electrocapillary curve

Every term in the Ilkovic equation, with the exception of n, varies with temperature and the overall effect is quite complex. The factor most affected by temperature is the diffusion coefficient and, in practice, it is found necessary to control the temperature to $\pm 0.5°$ in order to keep variations in i_d, due to temperature, within $\pm 1\%$.

The solvent viscosity affects the diffusion coefficient to a large extent (the latter is inversely proportional to the viscosity) and the factors m and t to a relatively small extent.

Polarographic maxima

Reproducible maxima often occur in CV curves unless eliminated by the addition of a suitable maximum suppressor, such as methylcelullose or gelatin. Figure 5.20 shows the reduction of oxygen where curve a is

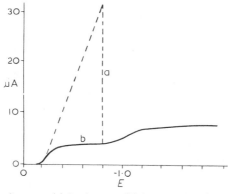

Fig. 5.20 Reduction of oxygen (a) in absence, (b) in presence of maximum suppressor

the unsuppresed oxygen maximum and b is the oxygen wave in the presence of gelatin. The height of each maximum depends largely on the concentration of electroreducible substance in the solution. Maxima are often absent in very low concentration solutions and become more pronounced as the ion concentration is increased.

No entirely satisfactory explanation for the occurrence of maxima is available, although several theories have been advanced.

Study of the polarographic wave

Consider the generalised reversible redox reaction:

$$ox + ze^- \leftrightarrows red$$

e.g. $$Fe^{3+} + e^- \leftrightarrows Fe^{2+}$$

to which the Nernst equation

$$E = E_{ox,red}^{\ominus} + \frac{RT}{zF} \ln \frac{a_{ox}}{a_{red}}$$

applies. In this equation $E_{ox,red}^{\ominus}$ is the standard redox potential of the system concerned, and a_{ox} for example, represents the thermodynamic activity of the oxidised species in the redox system and can be defined as the product of the species' concentration and activity coefficient, i.e. $a = c\gamma$. As there is a direct relationship between concentration c and the diffusion limited polarographic current i_d, the Nernst equation can be rewritten using the currents shown in Fig. 5.18 to represent the concentrations, and thus activities, of the two forms of the electroactive species concerned. In this procedure it is assumed that the electroactive species is being reduced at the dme. The inclusion in the Nernst equation of the diffusion coefficients, D_r and D_o, of the reduced and oxidised species respectively gives

$$E = E_{ox,red}^{\ominus} + \frac{RT}{zF} \ln \frac{(i_d - i)}{i} \left(\frac{D_r}{D_o}\right)^{1/2}$$

which describes the shape of the CV curve. At the half-wave potential, the log term would become zero so that $E^{\ominus} = E_{1/2}$, and if $D_r = D_o$ it can be seen that

$$E_{applied} = E_{1/2} + \frac{0.0592}{z} \log \frac{(i_d - i)}{i}$$

A graph of log $(i_d - i)/i$ plotted against $E_{applied}$ will give a straight line with a slope of $z/0.0592$. If the line is not straight then the reaction is not truly reversible (Fig. 5.21). A 1, 2 or 3 electron change therefore gives a reciprocal slope of 59, 30 or 20 mV (at 25°) respectively.

Three electrode polarographic cell

The potential (in V) of the polarised dme at the mercury–solution interface, E_m, is given by the expression:

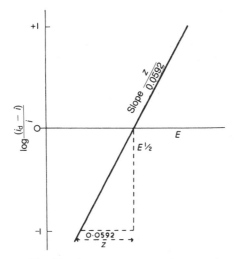

Fig. 5.21 Graph of log $(i_d - i)/i$ vs E

$$E_m = E_{ref} - (V - iR)$$

where E_{ref} is the potential of the reference electrode (0.242 V in the case of a Saturated Calomel Electrode (SCE)), V is the applied voltage, i is the current flowing through the cell, and R is its resistance (note that by convention, for cathodic reduction, V and i are positive). The iR term, known as the ohmic drop, is the potential necessary to overcome the solution resistance, which is quite small in aqueous solutions containing supporting electrolyte. However, at high current levels or in poorly conducting media this resistance causes distortion of the polarogram. This problem is solved in modern polarography by the use of a three electrode cell and an electronic potentiostat. The working electrode, which is the dme in polarography, is the electrode where the electrochemical reaction occurs, while the reference electrode, through which no current should flow, provides a constant reference voltage against which any changes are measured and provides a constant potential difference across the reference electrode–solution interface. The cell is completed by a counter, or auxiliary, electrode in the form of a gold or platinum wire or a mercury pool as in Fig. 5.17 through which current can pass freely and which can adopt any potential applied to it by the potentiostat.

The usual reference electrodes are the SCE mentioned above or the silver–silver chloride electrode with a salt bridge containing saturated silver chloride and making contact with the solution through a porous plug or frit. The voltage generated by this electrode is +0.222 V with respect to the standard hydrogen electrode at 25°. It is normal practice in voltammetry to quote the applied potential and the reference electrode used, e.g. −1.25 V *vs* SCE (or silver–silver chloride).

Miscellaneous phenomena

Hydrogen overvoltage. According to Heyrovsky, hydrogen is reduced at the dme according to the following equations:

$$H + H^+ \leftrightarrows H_2^+$$
$$H_2^+ + e \leftrightarrows H_2$$

i.e. it is assumed that molecular hydrogen is formed by the combination of a hydrogen atom, formed by electrolysis, with a hydrogen ion.

The theoretical potential E at which hydrogen is reduced at the dme at 25° is given by the equation:

$$E = E_{H^\circ} + 0.0592 \log [H^+]$$

where E_{H° is the standard potential of the normal hydrogen electrode and $[H^+]$ is the concentration of hydrogen ions. In practice, a value E', greater than E, is required before hydrogen is evolved. The difference between these two values is known as the hydrogen overvoltage, and its magnitude in voltammetry depends on the electrode type and physical condition, the physical state of the substance deposited, the current density, the hydrogen ion concentration, the temperature and, most importantly in polarography, the presence of other ions in the solution. The value of the dme as an electrode is to a great extent attributable to its high hydrogen overvoltage, which extends its voltage range to about $-2.0\,V$ with respect to SCE if the pH and supporting electrolyte are carefully chosen.

Catalytic hydrogen waves. Herasymenko and Slendyk (1930) found that the hydrogen overvoltage was greatly increased by cations which deposited at potentials more negative than that of hydrogen (e.g. K^+). Certain cations (e.g. Pt^{4+}) and organic substances had the reverse effect.

Quinine, quinidine, cinchonine and cinchonidine in ammonium chloride and also cystine and cysteine in ammoniacal cobalt buffer solution, are found to give well defined catalytic hydrogen waves. Some amino acids (e.g. phenyl-β-alanine) are, however, able to suppress the catalytic cysteine wave.

Oxygen waves. Oxygen is reduced at the dme in two stages, either

$$O_2 + 2H^+ + 2e \leftrightarrows H_2O_2$$

(acid conditions)

$$H_2O_2 + 2H^+ + 2e \leftrightarrows 2H_2O$$

or

$$O_2 + 2H_2O + 2e \leftrightarrows H_2O_2 + 2OH^-$$

(basic conditions)

$$H_2O_2 + 2e \leftrightarrows 2OH^-$$

Oxygen, therefore, must be removed from solutions by displacement with O_2-free nitrogen or hydrogen, otherwise the oxygen wave will be superimposed on the polarograms.

Biological media, e.g. blood serum, sometimes froth badly on deoxygenation with an inert gas. Rapid deoxygenation can be achieved by the addition of glucose, glucose oxidase and catalase.

The removal of oxygen may also be achieved by using carbon dioxide when the solution is acid, or sodium sulphite when it is neutral or alkaline:

$$2SO_3^{2-} + O_2 \leftrightharpoons 2SO_4^{2-}$$

but, as many electrode reactions are strongly influenced by pH, care must be taken in the selection of the deoxygenation method.

Capacitance current

The Ilkovic equation takes no account of the charging or capacitance current, which is due to the adsorption of anions or cations at the electrode surface to form a double layer. A significant current is required to charge this double layer capacitance to the applied potential, and this current, which is an inherent characteristic of the dme, effectively limits the sensitivity of classical d.c. polarography to concentrations of $\geqslant 5 \times 10^{-5}$M. The current oscillations, superimposed on the polarogram, reflect the charging current associated with growth of the mercury drop, the current being very large during the initial life of the drop and decreasing with its growth, in contrast to the faradaic current, which increases with drop life according to the Ilkovic equation. The development of many of the more modern polarographic techniques has been based on the fact that the faradaic current, which is analyte concentration related, is high, but the capacitance current is low towards the end of the drop life.

Care of the dropping mercury electrode

(1) When setting up the dme for the first time it is essential to use clean, dry and dust-free tubing.

(2) Only pure, double or triple distilled mercury should be employed.

(3) The conventional dme should be mounted within $\pm 5°$ of the vertical or else the drop time will be erratic.

(4) The capillary assembly should be mounted on a heavy stand and on a bench free from vibrations, to prevent premature dislodgement of the mercury drops.

(5) Traces of dust cause erratic behaviour of the capillary and should be excluded by keeping the tip of the capillary immersed in water when not in use.

(6) The mercury reservoir must not be lowered unless the capillary tip has been thoroughly washed and immersed in water. When the reservoir is lowered, it should be to a position such that the mercury just stops flowing.

(7) As each mercury drop falls from the capillary, the mercury thread momentarily retracts slightly into the lumen before the succeeding drop begins to form. This pumping of solution in and out at the end of the capillary tends to make it dirty after long use. The tip can

be cleaned by immersing periodically in 50% v/v nitric acid with the mercury flowing and then washing thoroughly with a jet of water. A disassembled capillary can be cleaned by aspirating concentrated hydrochloric acid through it for half a minute, following this with methanol and then air until dry.

Advantages of the dropping mercury electrode

(1) It has a smooth and continually renewable surface exposed to the solution being analysed.

(2) Each drop formed is unaffected by the reactions which occurred at the surface of earlier drops.

(3) Mercury amalgamates readily with most metals.

(4) The high hydrogen overvoltage of mercury enables analyses to be carried out in acid solutions.

(5) The diffusion equilibrium at the mercury–solution interface is rapidly attained.

Disadvantages of the dropping mercury electrode

(1) Mercury has a limited application in the more positive potential range (i.e. when used for anodic polarography), since anodic dissolution of mercury takes place at about +0.5 V.

(2) The surface area of the drop is never constant.

(3) Changes in the applied voltage produce changes in the surface tension of mercury and, therefore, changes in drop size.

(4) The addition of surface active agents produces changes in drop size, and adsorption of surface active agents can interfere with electrode reactions.

(5) Mercury may be toxic in certain biological studies.

Basic principles of polarographic instrumentation

Potentiometric manual polarograph

A means of measuring current used by Kolthoff and Lingane was to pass it through a standard resistance and measure the potential difference set up. Figure 5.22 shows the simplified circuit of such a polarograph which is also capable of measuring accurately the potential difference across the cell electrodes. The primary circuit consists of a battery, switch, and uncalibrated potentiometer.

The resistance R is at least $10\,\mathrm{k}\,\Omega$ and its actual value should be known accurately to within $\pm0.1\%$. The calibrated potentiometer P_2 is shown schematically. Any type of potentiometer is suitable provided it can measure potentials correctly to $\pm1\,\mathrm{mV}$. The galvanometer associated with P_2 should have a period of about 10–20 s so as to minimise oscillations of the needle. In use, S_2/S_3 is put in position V (voltage) and P_2 is set to the desired polarising voltage. P_1 is adjusted until the galvanometer of the calibrated potentiometer P_2 reads zero. S_2/S_3 is then put in position 1 (current) and P_2 adjusted until its galvanometer reads zero. The current flowing through the polarographic cell is calculated from Ohm's Law, $i = V/R$, where V is the potential reading of P at the balance point.

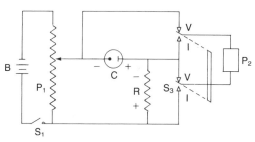

Fig. 5.22 Circuit diagram of a simple potentiometric manual polarograph

In early polarographs of this type the potentiometer for supplying the polarising potential and the recording chart are both driven in unison, usually by means of synchronous motors. The records of current flowing *vs* potential are obtained either photographically or by means of a pen recorder. Recent commercially available instruments use a suitable X–Y recorder to plot a normalised voltage output from the polarograph for both current and applied voltage axes.

The electronic potentiostat

The three electrode cell discussed above allows the true working potential to be applied between the reference and working electrodes while the current between the working and auxiliary electrodes is measured. The potentiostat, which is the central part of the electronics of a modern d.c. or pulse polarograph, allows any increased resistance in the solution to be compensated for, thus maintaining the set potential to within a few millivolts. Its rapid response to resistance changes and its extremely low current drain through the reference electrode (ca. 10^{-12} A) ensures that applied potentials and measured currents are controlled as precisely as possible.

A typical commercial potentiostat circuit using a three electrode cell in which the working electrode is always at ground potential is shown in Fig. 5.23.

Operational amplifier 1 supplies the input signals to the cell, while amplifier 2 works in conjunction with 1 to achieve the condition of stability in the amplifier loop circuit such that the output from 2 is equal in magnitude but opposite in sign to the input voltage of 1. The result is that the voltage relationship between the reference and working electrode is always that programmed, although the absolute voltage of the working electrode will always be that of ground. The third amplifier is connected to the working electrode, which it maintains at ground potential, and produces a voltage output proportional to the polargraphic current to drive the Y-axis of the chart recorder.

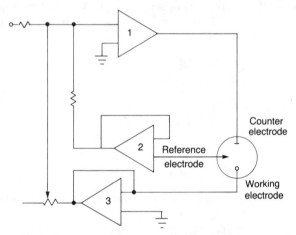

Fig. 5.23 Schematic circuit for controlled potential operation. (Redrawn from Application Note 151, Princeton Applied Research Corporation.)

Compensation circuits

Electrical compensation can be arranged by applying a counter e.m.f. or counter current to reduce to zero electrically an interfering wave or residual current.

Figure 5.24 shows the polarogram of two substances, A and B, where the concentration of A is many times greater than that of B. It is difficult to measure accurately the wave height of B unless wave A is electrically reduced to zero, thus allowing wave B to be examined at a much higher sensitivity.

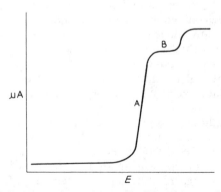

Fig. 5.24 Polarogram of two electroreducible substances in a mixture

In early polarographs, compensation circuitry to offset unwanted background or additional signal was included optionally, but modern instruments include output offset controls which can be used to null any unwanted signal and, in combination with the chart recorder controls, allow scale expansion over a specific range of interest.

Derivative or differential polarography

Before the development of modern integrated circuit potentiostats and current sampling techniques required for the latest polarographic methods, other ingenious approaches were made to improve the performance of polarographs. Some of these approaches are still used, in modified form, in present day methods; one such is that of derivative polarography in which a peak shaped trace is obtained with its obvious advantages of $E_{1/2}$ location and height measurement. The method devised by Heyrovsky for performing derivative polarography was to use two dme's in the same cell working at slightly different voltages. Use of a special galvanometer allowed Δi to be obtained while ΔE between the two dme's was always constant. Thus $\Delta i/\Delta E$ *vs* applied voltage could be recorded. Two cell operation is also possible with some modern commercial polarographs. In this case signals produced by one cell, operating in any mode, are subtracted electronically from those from the other cell, thus allowing baseline correction and large waves appearing before the wave of interest to be nulled.

The static mercury drop electrode

The static mercury drop electrode (smde) has been developed and presented in a commercial form by the Princeton Applied Research Corporation. When used with a polarographic analyser incorporating a potentiostat and drop timing and circuit sampling functions, the smde can perform sampled d.c., pulse, and differential pulse polarographic techniques which rely on sophisticated electronics and precise coordination of drop life, voltage application, and current sampling.

In the smde a capillary tube of larger bore (0.015 cm) than usual is used and a solenoid operated valve assembly at the top of the capillary controls mercury flow. Drops thus grow very quickly but then remain static for about 17 ms, during which the current is sampled, before being mechanically dislodged by the drop timer, and the cycle is then repeated. By this means the interference of capacitance current, described above, in continuous measurement polarography, is almost completely avoided and the sensitivity of the technique increased from about 10^{-5} mol l^{-1} to about 10^{-8} mol l^{-1}. Drop size can be varied by altering the period for which the solenoid valve is open: by using larger drops, greater sensitivity, at the expense of greater noise and susceptibility to accidental dislodgement, is obtained.

Pulse polarographic methods

Pulse polarography was invented by Barker and his co-workers in the late 1950s; modifications of the original method have resulted in pulse techniques being the most widely used analytical polarographic procedures.

In normal pulse polarography (see Fig. 5.25) a series of pulses of increasing amplitude are applied for 50–60 ms towards the end of the drop lifetime. The charging or capacitance current dies away rapidly in comparison with the faradaic current, which is sampled in the last 17 ms of the pulse and recorded in the usual way to give a sigmoid curve as in d.c. polarography, but without charging current oscillations. The mercury drop is then mechanically dislodged and the cycle repeated.

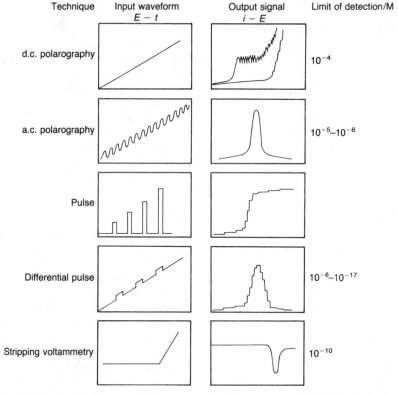

Technique	Input waveform $E - t$	Output signal $i - E$	Limit of detection/M
d.c. polarography			10^{-4}
a.c. polarography			10^{-5}–10^{-6}
Pulse			
Differential pulse			10^{-6}–10^{-17}
Stripping voltammetry			10^{-10}

Fig. 5.25 Applied voltage waveforms, current voltage curves and limits of detection (mol l^{-1}) of modern voltammetric methods. (Redrawn from B. Fleet and N. B. Fouzder, in *Polarography of Molecules of Biological Significance,* ed. W. Franklin Smyth, Academic Press, 1979.)

Diffusion limited currents for normal pulse and d.c. polarography can be compared by the equation:

$$\frac{i_{\text{pulse}}}{i_{\text{d.c.}}} = \left(\frac{3t_p}{7t_d}\right)^{1/2} \approx 5\text{--}10$$

where t_p is the time after application of the pulse and t_d is the d.c. drop lifetime. This equation does not consider the improvement in sensitivity afforded by the reduction in capacitance current interference which improves the performance of current sampled techniques by at least 10^{-2} mol l^{-1}. The value of pulse polarography is that it can respond to both reversible and irreversible processes and that, because the voltage is applied for a very short duration, very little electroreaction actually occurs and only extremely small amounts of product deposit on the electrode.

Differential pulse polarography (dpp) is the most widely used of the modern polarographic developments. The output waveform (see Fig. 5.25) is a peak and is easier to analyse than a sigmoid curve. The voltage pulses of about 60 ms duration in this case are superimposed on a conventional d.c. ramp voltage, and the current is sampled twice in the drop cycle, once immediately before the pulse application and once at the end of the pulse application period, the sampling periods being of about 17 ms duration. The current output is the difference between the two current samples in each drop lifetime. The magnitude of the output is a complicated function of the amplitude, or height, of the applied pulse (ΔE) which is generally between 5 and 50 mV, the larger pulse yielding a bigger signal and greater sensitivities while the smaller pulse allows better resolution of any fine detail of the polarographic wave. The presentation of the output as a peak also allows separate resolution of responses from electroactive substances in a mixture where $E_{1/2}$ potentials are closer together than would be possible with techniques giving sigmoid outputs. The potential E_p at which the current peak occurs is in principle coincident with the half-way potential $E_{1/2}$ derived from d.c. polarography, but the relationship

$$E_p = E_{1/2} - \Delta E/2$$

shows that a deviation dependent on the pulse amplitude (ΔE) obtains and this is most noticeable with large ΔE. At very low concentrations of analyte, the improved signal-to-noise ratios of the differential pulse technique gives this method the advantage over d.c. and normal pulse methods so that it would be the method of choice in pharmaceutical analyses.

Related methods

Alternating current polarography, cyclic voltammetry, and stripping voltammetry each require different voltage waveforms to be applied to the electroactive species with the aim of exploiting some property of the analyte to make the analysis or investigation more specific, e.g. large organic molecules and surfactants can be analysed by a.c. methods because of their adsorptive properties. Stripping voltammetry is the most sensitive of all the techniques mentioned here, because it involves a concentrating step in which some of the electroactive material is first deposited on a stationary electrode from a carefully stirred solution using a controlled and reproducible electrolysis time and electrode location in the cell. The deposited species is then stripped from the electrode, usually by application of a linear voltage sweep of 100–200 mV s^{-1} from cathodic to anodic potential. The application of these techniques in pharmaceutical analyses is fairly limited and is often related to the elucidation of a reaction mechanism or some chemical or physical interaction.

Solid electrodes

Current–voltage curves which are peak shaped in unstirred solutions can be generated in the same way as in polarography when a solid working electrode of 5–50 mm^2 area is used. The advantages of the dme are lost, of course, in particular the production of a fresh surface with every drop, but different reactions can be studied and a wider voltage range in the positive (oxidative) direction can be scanned; in polarography this is limited by reaction of the mercury at about +0.2 V *vs* SCE. Cells with solid electrodes can be used as voltammetric, coulometric or amperometric detectors in liquid chromatography, where specificity and sensitivity improvements are achieved over general purpose ultraviolet or refractive index detectors. Practically, solid electrodes present difficulties in the maintenance of a uniform working surface, and frequently have to be cleaned and polished with silica or diamond abrasive to remove adsorbed products of reaction or surface film contamination and thus maintain consistent and noise-free performance. Most solid electrodes, including platinum, gold, and glassy, pyrolitic, and wax impregnated graphite types show positive (anodic) voltage ranges up to about +1.5 V in aqueous media or +2.0 V in organic solvents, but the carbon types show a good negative (cathodic) range up to −1.0 V compared with platinum (−0.4 V) and gold (−0.7 V). The principles of hydrodynamic voltammetry apply when the electrode itself moves, e.g. by rotation, or the solution is stirred or flows over the electrode. Various cell designs for flow-through systems have been developed with the aims of enhancing mass transport, keeping dead volume as small as possible, and keeping the electrodes close together to minimise resistance effects while retaining an easily cleaned and maintained design. Electrochemical detectors in HPLC are discussed later.

Interference by formulation excipients in polarographic analyses

Multi-component systems can often be analysed by polarography without a prior separation step if only one or two components are electroactive and have $E_{1/2}$ values separated by at least 150 mV. However, non-electroactive compounds can interfere with analyses by altering the

$E_{1/2}$ postion, usually to more negative potentials, and/or altering the limiting current values for a given concentration of electroactive species. An extended study of these interfences in pharmaceutical analyses has been made by Lannigan and co-workers (1984). The movement of $E_{1/2}$ voltages to more negative values implies a resistance to the reduction process being induced by the interfering component, and this is often accompanied by a reduction in peak current as diffusion of the electroactive species may be physically impeded by the interferent. In the unusual case of glyceryl trinitrate being analysed in the presence of polyvinylpyrrolidone (PVP), an increase in i_d and a positive shift of $E_{1/2}$ has been observed. This has been attributed to a synergistic effect of PVP in the transport of glyceryl trinitrate at the mercury–solution interface, an observation which correlates with the stabilising effect of PVP in glyceryl trinitrate tablet formulations.

Binding of non-electroactive species with electroactive molecules can be studied by polarography provided that no surface adsorption effects are shown by the binding species, and no electroactivity is shown by the bound molecule. This would have to be confirmed in practice before an extended series of experiments was initiated.

Applications of polarography

Voltammetry is one of the most versatile analytical techniques available and, although it was formerly considered to be applicable only to metals, there are now many non-metals, anions and organic substances which can be readily analysed by voltammetry. The obvious requirement of the species to be analysed is that it is reducible or oxidisable within the operating range of the dme (i.e. +0.2 V to −2.0 V *vs* SCE but dependent on supporting electrolyte). Provided that the $E_{1/2}$ value of the substance of interest is separated from any other responses by about 100 mV or more, a measurement can be considered. Experimentation with different pH buffers, supporting electrolytes and, in the case of organics, solvent systems, can optimise the response. Table 5.4 gives typical conditions for the analysis of some metal ions.

With organic species, problems of solubility can be overcome by the use of organic solvents, either mixed with water, e.g. aqueous alcoholic mixtures, or with the addition of salts to confer conductivity to the solvent. Such solvents include DMF, DMSO, ethanol and methanol, and suitable supporting electrolyte salts are the perchlorates or fluoroborates of the tetraalkylammonium series or of lithium. However, the usual choice for the analysis of simple organic pharmaceuticals would be an aqueous alcoholic (e.g. 80% buffer–20% methanol) solution containing a simple salt as supporting electrolyte (e.g. at least 0.01M potassium chloride or lithium chloride); there is less chance of solvent or salt impurities interfering with the analysis in these simple systems. Simple buffer solutions are effective over narrow pH ranges, but mixed buffer systems, e.g. Britton–Robinson buffer, are more versatile when the polarographic behaviour of an organic species is to be fully investigated. Care must be taken to ensure that no ions in the electrolyte interact with the substance being determined and that pH changes do not affect its stability, and that the use of aqueous/alcoholic solvents does not cause changes in the $E_{1/2}$ of the electroactive species, the limiting current values, or the reduction mechanism involved, which might lead to erroneous conclusions in an analysis.

Table 5.4 Typical solvents for inorganic polarographic analyses. (Adapted from Application Note 151, Princeton Applied Research Corporation.)

Metal	Supporting electrolyte	$E_{1/2}/V$
As(III)	1M HCl	−0.43/−0.67
Bi(III)	1M HCl	−0.09
Cd(II)	0.2M NH$_4$ citrate, pH 3	−0.62
Co(II)	1M NH$_3$–1M NH$_4$Cl	−1.22
Cr(III)	0.2M KSCN, pH 3 with HOAc	−0.85
Cr(VI)	1M NaOH	−0.85
Cu(II)	0.2M NH$_4$ citrate, pH 3	−0.07
Fe(III)	0.2M TEA–0.2M NaOH	−1.0
Mn(II)	1M NH$_3$–1M NH$_4$Cl	−1.66
Ni(II)	1M NH$_3$–1M NH$_4$Cl	−1.0
Pb(II)	0.2M NH$_4$ citrate, pH 3	−0.45
Sb(II)	6M HCl	−0.18
Sn(II)	0.2M HOAc–0.2M NaOAc	−0.20/−0.53
Sn(IV)	1M HCl–4M NH$_4$Cl	−0.25/−0.52
Tl(I)	1M HNO$_3$	−0.48
Zn(II)	0.2M NH$_4$ citrate, pH 3	−1.04

The wider negative working voltage range for the dme compared with the positive inevitably means that reduction reactions are more frequently encountered in organic analysis than oxidations. A wide range of functional groups undergo electroreduction, but the products of such processes may not be the same as if the reduction were attempted chemically, as the electrode kinetics limit the progress of multi-stage reactions. Some organic functional groups which can be reduced at the dme are shown in Table 5.5.

Table 5.5 Some functional groups reducible in polarography. (Adapted from Application note 151, Princeton Applied Research Corporation.)

Imines	Dienes
Oximes	Alkynes
Nitriles	Ketones
Diazo compounds	Aldehydes
Diazonium salts	Aromatic carboxylic acids
Nitroso compounds	Halides
Sulphones	Thiocyanates
Sulphonium salts	Heterocycles
Nitro compounds	Organometallics

One of the easiest and most frequently encountered organic reductions is that of the nitro group. In nitrofurans and nitroimidazoles, for example, the reaction is

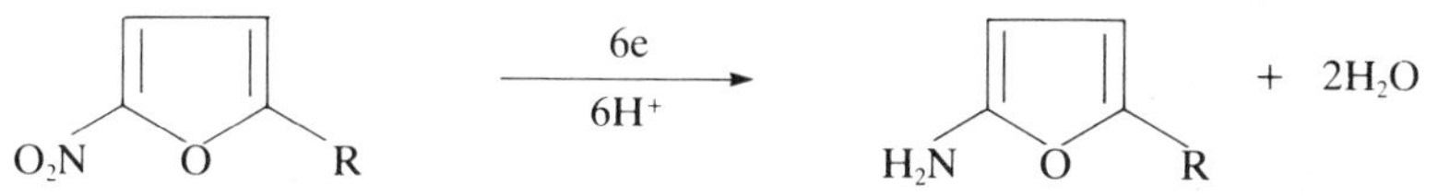

$$\text{(furan with } O_2N, O, R) \xrightarrow[6H^+]{6e} \text{(furan with } H_2N, O, R) \quad + \quad 2H_2O$$

Nitrofurantoin, Metronidazole and *Tinidazole* are examples whose reductions have been shown to follow this route.

The benzodiazepines constitute another class of compounds frequently analysed by polarography. A typical reduction is that of *Nitrazepam* in 20% methanol–0.1M hydrochloric acid, which proceeds by the following mechanism:

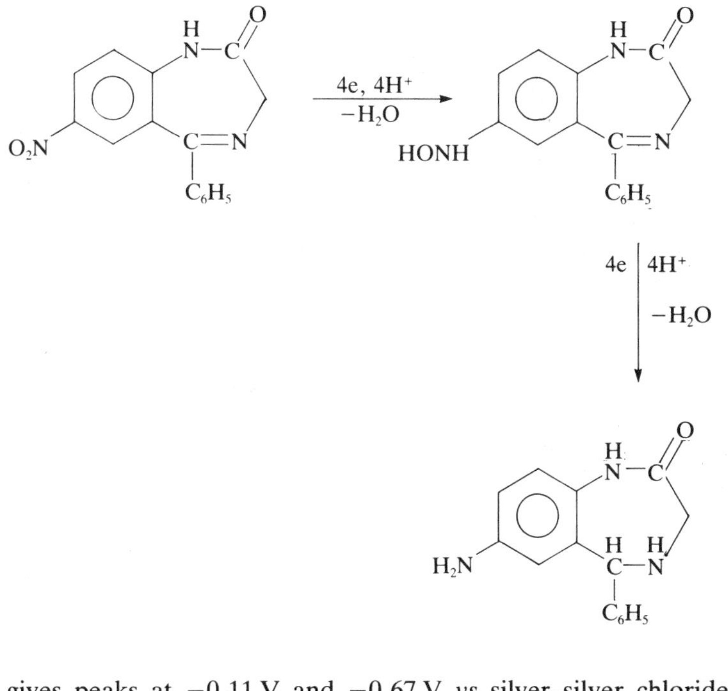

This gives peaks at -0.11 V and -0.67 V *vs* silver–silver chloride. To illustrate further the applicability of polarography in organic analysis and the value of derivitisation, the analysis of atropine by d.c. polarography may be quoted, in which 0.1M tetrabutylammonium perchlorate in acetonitrile can be used as solvent. Nitration of atropine can be effected by reaction at room temperature with 10% potassium nitrate in concentrated sulphuric acid, and a sensitivity of 200 ng atropine ml^{-1} of 1M sodium hydroxide supporting electrolyte with dpp has been claimed by Brooks *et al* (1974).

Polarographic methods of analysis

Direct comparison method

In this method, the diffusion current obtained for the 'test' solution is compared, under identical conditions, with that of a solution of known concentration. Maximum accuracy is obtained when the diffusion currents of both solutions are about equal.

The most important conditions which must be kept constant in the comparison are temperature, concentration of maximum suppressor (if any), composition of the supporting electrolyte and the characteristics of the dme (i.e. constant m and t values).

For complex mixtures it is advisable to keep standard comparison samples of known composition. These comparison samples must approximate closely in content to the samples being analysed.

Use of empirical calibration curves

The dme is calibrated empirically with various known concentrations of the substance in question, and a graph of diffusion current *vs* concentration is plotted. The concentration of the test solution can be read from the graph. It is essential to control the temperature accurately and to check that the capillary characteristics do not vary. This is the most frequently used method, and is satisfactory if standards and test samples are analysed under identical conditions.

Internal standard or pilot ion method

This method is based on the fact that the relative diffusion current constants I are independent of the particular capillary used, provided the nature and concentration of the supporting electrolyte and the temperature are kept constant.

From the Ilkovic equation:

$$i_d = ICm^{2/3}t^{1/6}$$

where

$$I = 607nD^{1/2}$$

For the 'pilot' ion

$$i_{d_1} = I_1 C_1 m^{2/3} t^{1/6}$$

and the 'test' ion

$$i_{d_2} = I_2 C_2 m^{2/3} t^{1/6}$$

hence

$$\frac{i_{d_1}}{i_{d_2}} = \frac{I_1 C_1}{I_2 C_2}$$

The ratio I_1/I_2 is known as the pilot ion ratio (symbol R) and is independent of the capillary characteristics so that

$$\frac{i_{d_1}}{i_{d_2}} = \frac{C_1}{C_2} \cdot R$$

The three methods described above can be used with modern smde polarographs where m and t are not readily defined.

Quasi-absolute method

This method, suggested by Lingane, also makes use of the diffusion current constant. From the Ilkovic equation it can be seen that:

$$C = \frac{i_d}{I m^{2/3} t^{1/6}}$$

with the expression in this form rather than in terms of n and D, evaluation of I is simplified because it is very difficult to determine accurate values of the diffusion coefficient which would otherwise be required. I is constant for a given substance in a given supporting electrolyte at a fixed temperature. It is also independent of the solution and the characteristics of the dme, provided that the drop time exceeds 1.5 seconds.

The absolute method avoids the necessity of time consuming calibration polarograms, the evaluation of $m^{2/3} t^{1/6}$ for a given capillary being much more rapid. However, the method is not as accurate as the more conventional methods of calibration.

Method of standard addition

The polarogram of the test solution, of known volume, is initially recorded, then an accurately measured quantity of a standard solution of the substance in question is added and a second polarogram is obtained. The original concentration of the substance can be calculated from the increase in the diffusion current from the following formula:

$$C_1 = \frac{C_2 i_1 v_2}{v_1(i_2 - i_1) + i_2 v_2}$$

where C, v and i are the concentration, volume and diffusion current, and the subscripts 1 and 2 refer to the test and standard solutions respectively. Maximum precision is obtained when $i_2 \simeq 2i_1$.

Experiments in polarography

Conventional dropping mercury electrode assembly

A vertical capillary, 6–8 cm long and with a bore diameter of about 50 μm, is mounted on a suitable stand. The lower end of the capillary must be cut accurately at right angles to its axis. The upper end of the capillary is connected to a mercury reservoir by means of clean, dry polythene or similar tubing. High grade rubber tubing may be used if traces of surface impurities, such as sulphur, are removed by treating with hot 10% sodium hydroxide (24 h) and thoroughly washing and drying the tube before use.

The reservoir is then filled with pure triple distilled mercury recently filtered through sintered glass or pin-hole filter paper to remove traces of dust. The tip of the capillary should be immersed in water when not in use and, with reasonable care, should remain serviceable for several months.

Polarograph cells

The variety of design of polarographic cells is too large to discuss here. The only essential requirements are to provide access to the cell for the capillary, gas supply, reference electrode, and auxiliary electrode.

Reference electrode

A mercury pool anode can be used but, for more accurate work, a saturated calomel electrode, connected to the solution via a suitable salt bridge, is to be preferred.

Deoxygenation apparatus

Usually O_2-free nitrogen gas is used and it is sometimes necessary to remove last traces of oxygen by passing the gas through vanadium(II) chloride solution before use. In all cases, the gas should be passed through a wash bottle containing the same solvent as is in the polarographic cell, before being bubbled through the solution for analysis. This ensures that the gas is saturated with the solvent vapour and thereby prevents undue concentration of the test solution due to removal of solvent.

Thermostat

For accurate work, the polarographic cell and gas wash bottles should be at the same temperature. A suitable thermostat, set at $25° \pm 0.1°$, should be used for this purpose.

General experimental method

The following selection of experiments illustrates several of the principles discussed in the theoretical sections of this chapter. In all cases, unless otherwise stated, the following general method for a conventional d.c. polarograph should be adopted and full conclusions should be drawn from the results whenever possible.

Method Adjust the reservoir height so as to give a drop time of about 3–4 s with the mercury tip immersed in water. Deoxygenate the test solution by passing nitrogen through it for 5–10 min. Rinse the electrode with a little of the solution and immerse the electrode tip in the solution. If a mercury pool anode is to be used, add 1–2 ml of mercury and bubble gas through the solution for a few minutes longer. Alternatively, a salt bridge, connected to a saturated calomel electrode, can be inserted into the solution before deoxygenation. If damping of the current fluctuations is required, set the appropriate switch to the desired position. Standardise the potentiometer (except on the simplest of manual polarographs, which employ a voltmeter) and set the galvanometer or recording pen to zero.

Connect the electrodes to the polarograph and adjust the galvanometer to the minimum sensitivity, unless the correct setting of the shunt is known. The correct sensitivity is usually found by increasing the polarising voltage to about −1.8 V and adjusting the sensitivity control to give a large but less than full scale deflection.

Plot the polarogram automatically by the following general procedure. After standardising the potentiometer, set it to the desired initial value and the recorder pen to the zero lines. Switch on the synchronous motors used for driving the potentiometer slide wire and recorder chart and, assuming the sensitivity setting is correct, the polarogram will be plotted automatically. Switch the motors off when the applied potential reaches about −2.0 V or the current readings approach full scale deflection.

Wash the tip of the dme well with water before rinsing with the next test solution. When all experiments are complete, wash the dme thoroughly and then immerse it in water. Lower the mercury reservoir slowly to a point where the mercury just ceases to emerge from the capillary.

Use of a typical commercial multimode polarograph Typical commercial polarographs will perform pulse techniques if fitted with a dme whose functions are synchronised with the polarographic signals. An example of such a set-up would be the Princeton Applied Research Model 174A polarograph and Model 303 dme with a suitable X–Y plotter. Full details for use should be obtained from the manufacturer's handbook, but the essential steps are given below.

(1) Fill the cell with approximately 10 ml of solution to be analysed.
(2) Switch on the dme, select N_2 purge time, and initiate purging.
(3) Select mercury drop size.
(4) On the polarographic analyser select the polarographic mode. (It is best to use a conventional valveless dme for classical d.c. work.)
(5) Select starting potential, scan direction, rate, and range to include the peaks of interest.
(6) For dpp select pulse modulation size, e.g. 25 mV.
(7) Select mercury dropping rate (or OFF for d.c. operation).
(8) Select current range and scale (μA or mA) according to expected response and likely concentration of analyte.
(9) If required, adjust output offset and low pass filter to tailor imperfect outputs to requirements.
(10) After purging of the solution is complete, initiate scan on the external cell and record trace on X–Y recorder.
(11) Deoxygenate before repeating a scan.

Quantitative evaluation of the polarogram Of the many methods suggested for the measurement of diffusion current, the following are the most suitable for general use. It will be assumed that two reduction steps, as in the case of many simple mixtures, are present.

(a) *Exact procedure* This method is applicable to well-defined waves whose limiting current plateaux are parallel to the residual current curve of the supporting electrolyte. Fig. 5.26 shows how the measurements are made.

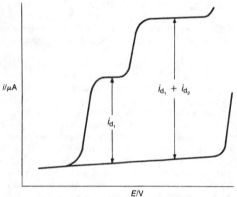

$i/\mu A$

$i_{d_1} + i_{d_2}$

i_{d_1}

E/V

Fig. 5.26 Evaluation of a polarogram by the exact procedure, where i_{d_1} and i_{d_2} represent the diffusion currents of the first and second reducible ions respectively. By this method, values of diffusion current are automatically corrected for i_r.

(b) *Approximate procedure* This method is sufficiently accurate for most purposes and is applicable to well-defined waves only. It has the advantage that the residual current polarogram is not necessary. The near horizontal portions of the curve are produced and the values of i_{d_1} and i_{d_2} are obtained as shown (Fig. 5.27).

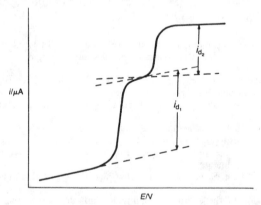

$i/\mu A$

i_{d_2}

i_{d_1}

E/V

Fig. 5.27 Evaluation of a polarogram by the approximate procedure

(c) *Modified approximate procedure* This method is applied when no well-defined plateau separates the two polarographic waves. The value of i_{d_1} is measured by drawing construction lines through the inflection point and parallel to the upper and lower, near horizontal, parts of the polarogram (Fig. 5.28). The values of i_{d_1} and i_{d_2} are then obtained as shown.

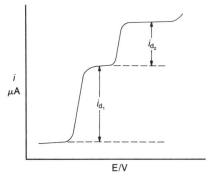

Fig. 5.28 Evaluation of a polarogram by the modified approximate procedure

(d) When dpp is used and peaks are obtained rather than sigmoid curves, the techniques described under ultraviolet spectroscopy (Chapter 7) can be applied to the measurement of peak height. In general, a straight line is drawn across the base of the peak and connecting the lower cusps; the line drawn perpendicularly from the apex of the peak to the constructed 'base' line then represents the peak height. Ideally the 'base' line should be horizontal but in many cases it will not be.

Experiment 47   *Oxygen wave, maximum suppression and residual current*

Method Set up the apparatus and obtain polarograms, by the general method described, as follows:
  (a) Electrolyse about 10 ml of potassium chloride (0.1M) but do not remove dissolved oxygen at this stage. Observe the fluctuations in the current at each applied potential.
  (b) Repeat (a) in the presence of one drop of either gelatin (0.2%) or methylcellulose (1%).
  (c) Repeat (b) in the presence of two, three or more drops of gelatin or methylcellulose in successive experiments until the oxygen maximum is completely eliminated.
  (d) Deoxygenate (c) for about 30 min and obtain the residual current polarogram.
  Note If this experiment is performed carefully it will be seen that the peak height falls with increased concentration of surfactant. This is the basis of an extremely sensitive method of analysis known as adsorption analysis.

It would be to the student's advantage to perform this experiment on the simplest possible (home-made) polarograph and then to repeat on a commercial recording polarograph.

Experiment 48   *Effect of supporting electrolyte on diffusion current*

Method Obtain polarograms of lead chloride (0.002M) by any appropriate technique:
  (a) in the absence of supporting electrolyte and measure i_1, the limiting current,
  (b) in the presence of potassium chloride (0.05M) and measure i_d, the diffusion current.
  From the results the migration current i_m can be found from the relationship:

$$i_d = i_1 - i_m$$

The transport number of the lead cation, t^+, can also be estimated, since:

$$i_m = i_1 t^+$$

and hence

$$t^+ = 1 - i_d/i_1$$

Experiment 49   *Supporting electrolyte and its effect on the half-wave potential*

Method Obtain polarograms by any appropriate technique for a mixture of cadmium chloride (0.01M) and lead chloride (0.008M) in a supporting electrolyte of:
(a) potassium chloride (0.1M).
(b) a mixture of sodium potassium tartrate (0.5M) and sodium hydroxide (0.1M).

Experiment 50   *Diffusion current and concentration*

Method Prepare, from a stock solution (0.1M cadmium sulphate in 0.1M potassium chloride), the following concentrations of cadmium sulphate in potassium chloride (0.1M):

(a) 1×10^{-2}M
(b) 7.5×10^{-3}M
(c) 5×10^{-3}M
(d) 2.5×10^{-3}M
(e) 1×10^{-3}M
(f) 1×10^{-4}M
(g) 1×10^{-5}M
(h) zero, i.e. 0.1M potassium chloride only.

Obtain a polarogram by any appropriate method for solution (a), note the potential range over which the diffusion current plateau is substantially constant and set the potentiometer limit to an intermediate value, say -1.0 V. Obtain and record diffusion currents for solutions (b) to (g) and the residual current, using solution (h), at the fixed polarising potential. The appropriate sensitivity settings for the apparatus should be used in each case. Deduct the residual current observed for solution (h) from all the other values recorded, and thus obtain the corrected diffusion current. Plot a graph of the logarithms of the diffusion current against the logarithms of the concentration. Also plot a calibration curve, for solutions (a) to (e) inclusive, of current against concentration.

Make at least two suitable dilutions of an 'unknown' cadmium solution. Electrolyse at -1.0 V and, after subtraction of the residual current, use the calibration curve to find the concentration of cadmium sulphate in the solution.

Experiment 51   *Capillary characteristics for conventional dme in d.c. mode*

(a) $i_d/m^{2/3}t^{1/6}$ = constant
(b) P/m = constant
(c) i_d^2/P = constant

Method Using four different reservoir heights (say, 50, 60, 70 and 80 cm) and the solution from Experiment 50(a) above, obtain the corresponding diffusion currents and drop times (average of 10 drops) at an applied potential of -1.0 V. Now transfer the electrodes to a 100 ml beaker containing the same solution, but not deoxygenated. At each reservoir height, and at the same applied potential (-1.0 V), collect 10 drops of mercury in a small glass container. Wash the mercury with water and dry by rinsing with acetone followed by a jet of air. Weigh the mercury on an analytical balance and hence calculate *m*. Determine *P* by correcting the reservoir height for back pressure. Conclude whether or not the relationships (a), (b) and (c) above are correct.

Experiment 52   *Analysis of a polarographic wave*

Method
(a) Prepare a solution containing cadmium sulphate (0.01M), potassium chloride (0.1M), and gelatin (0.005%). Place some solution in a polarographic cell which is immersed in a thermostatic bath set at $25° \pm 0.1°$. Deoxygenate and obtain the CV curve from -0.4 to -0.8 *vs* SCE. Use the same current sensitivity throughout this experiment.

(b) Prepare a solution of potassium chloride (0.1M) and gelatin (0.005%) and obtain, at 25°, the residual current curve over the same voltage range. Correct the values of current obtained in (a) by subtraction.

(c) Calculate suitable individual log $(i_d - i)/i$ values and plot against the corresponding applied potentials. Determine the number of electrons involved in the reduction by measuring the slope of the graph (see Fig. 5.21).

Experiment 53   *Limit test for trivalent antimony in Sodium Stibogluconate*

Sodium Stibogluconate contains not more than 0.2% trivalent antimony when determined by the following method.

Method To 0.2 g sample, accurately weighed, in 10 ml of water, add 2 ml 0.1% w/v aqueous solution of gelatin, and 2 ml concentrated hydrochloric acid, and dilute to 20 ml with water. Transfer an aliquot portion to a polarographic cell and bubble nitrogen through the solution for 10 min. Record a polarogram over the range 0 to $+0.5$ V (i.e. an anodic wave) and compare with a standard trivalent antimony calibration curve using, in the above procedure, 0.25 ml of a 0.8% w/v aqueous solution of a trivalent organo-antimony compound. The height of the step at a potential of approximately 0.15 V *vs* SCE is a measure of Sb^{3+}. (A potential of about 0.4 V *vs* SCE is probably better since $E_{1/2} \simeq + 0.15$ V.) The solutions must be examined within 30 min. of preparation as they are unstable.

Experiment 54   *The multimode polarograph in the analysis of Metronidazole*

A multimode polarograph with synchronised dme which can perform sampled d.c., pulse, and differential pulse polarography should be used. The analysis should be performed in each of the three modes and the results compared.

Method Prepare $50 \mu g \, ml^{-1}$ solution of metronidazole in 0.1M hydrochloric acid, and prepare from this six solutions of appropriate concentrations for a calibration graph. Deoxygenate the test solutions for 4 min before each measurement. Choose a suitable voltage scan range advancing at $2 \, mV \, s^{-1}$ to include 0 to -0.5 V w.r.t. silver–silver chloride and start with $10 \mu A$ full scale current range, adjusting as necessary. Use a 1 s drop time with a small mercury drop if this adjustment is available. In differential pulse polarography use a 25 mV pulse amplitude. Run the polarograms in any or all of the available modes, measure the response height and plot a calibration graph. Analyse the data statistically as if it were a straight line.

Compare the data obtained above with those from an identical experiment except for the addition of 10^{-4} mol 1^{-1} polyethylene glycol 4000 (PEG 4000) in each of the test solutions. Analyse the calibration graphs similarly and comment constructively on the results.

Amperometric titrations

Introduction

This technique uses a polarographic or other voltammetric electrode to detect the end point in a titration, thus extending the application of titrimetry to lower concentrations of analyte than would be possible with visual indicators. An abrupt change in limiting current will occur when all the analyte has been removed from the solution by the titrant (Fig. 5.29). No potential scanning is used: a fixed voltage of magnitude sufficient to cause electroreaction of the analyte is applied across the electrodes and the current is monitored as titrant is added. A similar principle is applied in the electrometric end point method for the titration of sulphonamides and in the conventional determination of water by titration with Karl Fischer reagent (q.v.).

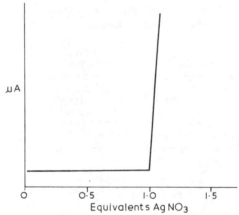

Fig. 5.29 Amperometric titration of potassium chloride with silver nitrate

Amperometric titrations using a dropping mercury electrode

Very dilute solutions may be titrated accurately (ca. ±0.3%) and rapidly. Amperometric titration results are independent of the capillary characteristics and temperature, provided that there is no change during a titration; the reaction need not be reversible, and substances which are not oxidised or reduced may be titrated if the reagent gives a diffusion current. Amperometric titrations involving precipitation may be carried out when the solubility is appreciable and under conditions where potentiometric and indicator methods are inaccurate. Titrations may also be carried out in the presence of large amounts of electrolyte (e.g. potassium chloride) without interference (contrast conductimetric titrations).

Apart from the normal sources of error in volumetric determinations, impurities which give diffusion currents may have to be removed by a preliminary chemical separation, or the conditions must be so chosen that foreign constituents do not contribute to the current. Several types of titration curve may be obtained, as follows:

(1) Consider a lead solution containing an excess of an indifferent electrolyte. The polarogram (Fig. 5.30) has a plateau between points A and B, where the current is practically constant. This solution may therefore be titrated with a solution of sodium oxalate at any fixed applied e.m.f. between the values A and B by measuring the current until the end point is reached, when only a small residual current flows (Fig. 5.31). The slight curvature is due to dilution of the solution with the reagent and is minimised by using a relatively strong titrant (compare conductimetric titrations). In any case, it can be corrected by multiplying the observed values of current by the ratio total volume to initial volume.

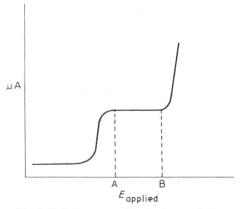

Fig. 5.30 Polarogram of lead nitrate solution

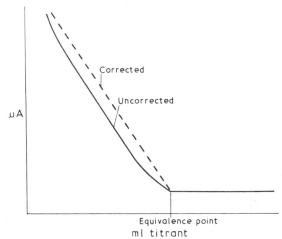

Fig. 5.31 Amperometric titration of lead with sodium oxalate at 1.0 V *vs* SCE

(2) It is possible to titrate a non-reducible substance with a reagent which is electroreducible. Lead nitrate, for example, may be titrated with potassium dichromate, in acetate buffer, at zero applied volts. (Pb^{2+} is not reduced at this voltage.) When the end point is reached, the dichromate ion ($Cr_2O_7^{2-}$) yields a diffusion current (Fig. 5.32).

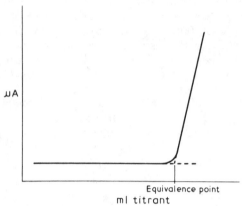

Fig. 5.32 Amperometric titration of lead with potassium dichromate at zero volts *vs* SCE

(3) The substance to be titrated and the reagent used may both be electroreducible. Lead nitrate and potassium dichromate, as above, may be titrated at an applied potential of $-1.0\,V$ *vs* SCE. The diffusion current, due to the lead ions, first falls as lead is precipitated out and rises when excess dichromate ions are present (Fig. 5.33).

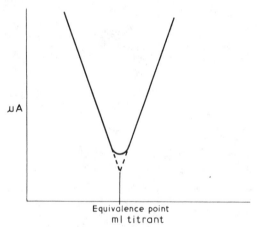

Fig. 5.33 Amperometric titration of lead with potassium dichromate at $-1.0\,V$ *vs* SCE

(4) Amperometric titrations may also be applied to electro-oxidisable substances, provided a suitable electrode is used. Potassium iodide may, for example, be titrated with mercury(II) nitrate:

$$Hg^{2+} + 2e \rightleftharpoons Hg$$
$$2I^- + 2Hg \rightleftharpoons Hg_2I_2 + 2e$$

The titration curve is more or less a straight line and the end point is reached when the current is zero (Fig 5.34). The negative branch of the curve is the anodic diffusion current due to the iodide, and the positive branch is the cathodic diffusion current due to the mercuric ions.

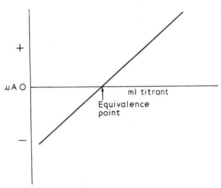

Fig. 5.34 Amperometric titration of iodide ions with mercury(II) nitrate

(5) It is occasionally possible to carry out an amperometric titration when neither the substance titrated nor the reagent yields a polarographic wave. A 'polarographic indicator', which has a diffusion current at the applied e.m.f., may be used to react with excess reagent. For example, aluminium salts may be titrated with a fluoride salt, using an iron(III) salt as indicator. Both aluminium and iron form complexes with fluoride but the former are more stable. During the titration, the diffusion current falls gradually and then very rapidly near the end point which corresponds to zero current.

(6) *Amperometric titration involving chelating compounds.* When a metal ion forms a complex or chelate, the reduction potential is displaced to a more negative value. Use can be made of this fact in amperometric titrations using the dropping mercury electrode. Thus, if the voltage is maintained at a value such that only the metal ions are discharged, the metal ions may be titrated with a complexing agent until the diffusion current has reached a minimum, producing a curve similar to that shown in Fig. 5.31. Since the stability of the complex is usually very dependent on pH, the pH of the solution must be buffered at a suitable value.

If, however, the complexing or chelating agent also undergoes reduction, its reduction potential being less negative than that of the metal complex, it follows that, if the potential is held at a value such that both the metal ion and complexing agent are reduced but not the metal complex, the diffusion current will fall to a minimum, due to the removal of metal ions, and then rise again due to the reduction of the excess chelating agent. The curve obtained is similar to that shown in Fig. 5.33.

An illustration of these procedures is given by the work of Stock, who titrated copper ions (1 ml samples of 1.75×10^{-3}M) with quinoline-8-carboxylic acid (5×10^{-3}M) in a solution buffered at pH 5. At an applied voltage of -0.4 V *vs* SCE, only the copper ions were reduced, producing a curve similar to that shown in Fig. 5.31. When the applied voltage was -1.35 V *vs* SCE, where the chelating agent but not the metal chelate was reduced, a V-shaped curve like that shown in Fig. 5.33 was produced, thus enabling a more accurate determination of the end point to be reached. The advantage of this type of titration is that very low concentrations of metal ions can be determined.

These titration techniques can equally well be used with an appropriate ion selective electrode sensor, and they provide the basis of many published methods of analysis with ISEs.

Apparatus

The apparatus required is essentially the same as for ordinary polarography. The titration cell must have adequate capacity, provision being necessary for a burette, electrodes and gas supply. Stirring may be carried out using a magnetic stirrer or nitrogen bubbling through the solution.

Amperometric titration experiments

General method

Use a general purpose polarographic cell holder, which should have re-sealable apertures in the cover to accommodate the burette. Provided the electrode tips are covered by the solution at all times during the titration, it is better to use a polarographic cell with nitrogen connections etc. *in situ* than to make a special device. Pipette a convenient quantity, e.g. 5 ml, of solution into the cell and add an equal quantity of supporting electrolyte (0.01M). Deoxygenate the solution for about 10 min. Select the required mode of operation if a multimode polarograph is being used and apply the required polarising voltage to the cell and measure the diffusion current. Add 1 ml of titrant, stir the solution (either with magnetic stirrer or with nitrogen) for 20–30 s and measure the diffusion current. Repeat until well past the equivalence point and plot a graph of diffusion current (corrected by proportion for volume change) against the burette reading. The intersection of two straight lines usually marks the position of the equivalence point.

Experiment 55   *Determination of nickel with dimethylglyoxime*

Method

(a) Weigh accurately a sample of nickel salt to give about 0.001M nickel solution. Pipette 25 ml into the titration cell and add an equal quantity of supporting electrolyte (a mixture of ammonium hydroxide (1.0M) and ammonium chloride (0.2M) and 1 or 2 ml gelatin solution (0.2%).

(b) Deoxygenate the solution and set the applied e.m.f. to -1.85 V *vs* SCE.

(c) Measure the diffusion current.

(d) Titrate with dimethylglyoxime solution (0.02M) using the general method. The graph should be V-shaped.

1 ml dimethylglyoxime solution (0.02M) = 0.5869 mg nickel

Experiment 56a   *Determination of lead with potassium dichromate solution*

Method
(a) Weigh accurately a sample of lead salt to give about 0.001M lead solution. Pipette 25 ml into the titration cell and add an equal quantity of potassium nitrate solution (0.01M).
(b) Deoxygenate the solution and set the applied e.m.f. to zero.
(c) Measure the diffusion current.
(d) Titrate with potasium dichromate solution (0.005M) using the general method. The graph should be ⌐-shaped.

1 ml 0.01M potassium dichromate = 2.072 mg lead

Experiment 56b

As Experiment 56a, but at an applied e.m.f. of −1.0 V *vs* SCE. The graph should be V-shaped.

Experiment 57   *Determination of iodide with mercury(II) nitrate*

Method
(a) Weigh accurately a sample of iodide to give about 0.001M iodide solution. Pipette 50 ml of this solution into the titration cell and add an equal quantity of nitric acid (0.2M) containing 0.1% gelatin.
(b) Deoxygenate the solution and set the applied e.m.f. to zero volts.
(c) Measure the diffusion current.
(d) Titrate with mercury(II) nitrate solution (0.01M). The graph is almost a straight line and the equivalence point is when the current reaches zero.

1 ml 0.01M mercury(II) nitrate = 2.538 mg iodide

Coulometric analysis

Introduction

Provided that no extraneous reaction is involved, Faraday's Law states that one equivalent of chemical change occurs at an electrode during electrolysis with the passage of 96 485 coulombs of electricity. If one can assume that there are no competing side reactions (i.e. 100% current efficiency) then:

$$w = QM/96485z$$

where w is the weight of substance transformed during electrolysis, Q is the quantity (coulombs) of electricity, M is the molecular (or atomic) weight of the substance being oxidised or reduced, and z is the number of electrons involved in the electroreaction. It is clear that there is a direct relationship between the weight of material transformed and the quantity of electricity consumed. Therefore, the process is suitable as a method of chemical analysis provided that the variable can be measured accurately.

Coulometric titrations require no standard solutions because the absolute amount of reagent needed to reach a detectable end point is generated in the solution as a result of the precisely measured charge passed. If a constant measured current is used (constant current coulometry) the difficulty of measuring charge is reduced to a problem only of elapsed time measurement, for which suitably accurate (± 0.01 s) devices are available. If applicable, constant current coulometry is the method of choice for checking standard materials because of its absolute nature and lack of dependence on standard solutions.

An inert electrode system and absence of electroreducible impurities (e.g. oxygen) are essential for quantitative applications; microgram quantities are readily determined, the accuracy depending on the precision of measurement of Q and the sensitivity of end point detection.

Under suitable conditions, a large variety of compounds may be analysed, e.g. ascorbic acid, bromoform, carbon tetrachloride, chloroform, cysteine, EDTA, iodoform, methylene blue, oxalic acid, phenol and salicylic acid. Reference to books on coulometry is recommended for a more detailed study; see, e.g., *Coulometry in Analytical Chemistry*, Milner and Phillips, Pergamon, 1967.

Measurement of parameters

In the past, Q has been measured directly by using a silver coulometer and measuring the increase in weight of the cathode. Although this method is theoretically sound, it does present certain practical hazards and it is more usual to determine Q by an indirect method.

Usually, a constant current source i (amps) is available and can be measured accurately, so Q can be determined easily by recording the time t (seconds) accurately, since

$$Q = it$$

Coulometric cell

A variety of designs are available, two of which are shown here (Figs 5.35 and 5.36).

The cell should have two compartments separated by sintered glass or some other suitable membrane. Provision for additional indicator electrodes may be necessary since, in some cases, it is convenient to use an electrochemical method to determine the equivalence point of a coulometric method.

Constant current/constant potential coulometry

Either a constant current d.c. source or a constant potential d.c. supply is used in coulometric analysis. The former method is rapid and is capable of high precision, while the latter method is much slower,

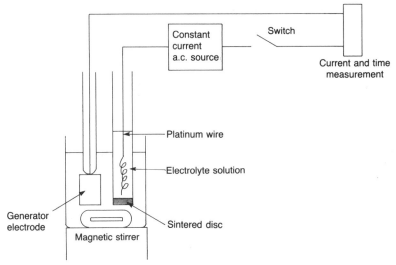

Fig. 5.35 Coulometric titration apparatus. (Redrawn from *Dictionary of Electrochemistry*, D.B. Hibbert and A.M. James, Macmillan, 2nd edn., 1984.)

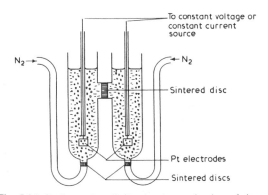

Fig. 5.36 Coulometric cell (for *in situ* production of titrant)

since the current continually falls but has the advantage that optimal conditions can be deduced directly from a polarographic curve by selecting a voltage corresponding to a point on the diffusion current plateau.

Coulometry experiments

The following experiment is intended to illustrate a fundamental approach to coulometric titrations with the 'titrant' prepared *in situ*.

Experiment 58   *Coulometric titration of hydrochloric acid*

Method Fill both electrode compartments of the coulometer cell with 0.01M sodium sulphate solution and pipette 1.00 ml of about 0.1M hydrochloric acid into the cathode compartment, together with a few drops of phenolphthalein. Mix and deoxygenate by bubbling nitrogen through the cell. Connect to a suitable constant current source of about 10 mA and note the time (seconds) for the indicator to change. Assuming 100% current efficiency, calculate the exact concentration of the hydrochloric acid.

Notes

(1) A preliminary run may be necessary to ensure suitable conditions.

(2) If a constant current source is not available, then use a voltage source and keep the current as steady as possible by adjustment of a series rheostat (Fig. 5.37).

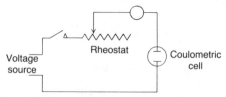

Fig. 5.37 Simple circuit for coulometric titration

(3) The reactions which occur can be represented by the following equations:

Anode compartment   (a) $SO_4{}^{2-} - 2e = SO_4^{2\cdot}$
                                 (b) $2SO_4^{2\cdot} + 2H_2O = 2H_2SO_4{}^- + O_2$

Cathode compartment (a) $Na^+ + e = Na^{\cdot}$
                                 (b) $2Na^{\cdot} + 2H_2O = 2NaOH + H_2$

(4) The anode compartment becomes acidic—this can be checked by addition of a suitable indicator.

Dead-stop end point technique

A small potential difference is applied to a pair of identical platinum electrodes immersed in a solution (Fig. 5.38). Little or no current flows unless the solution is free from polarising substances; this is probably due to adsorbed layers of hydrogen and oxygen on the cathode and anode respectively. Only when both electrodes are depolarised will any current flow. The technique may be applied to any

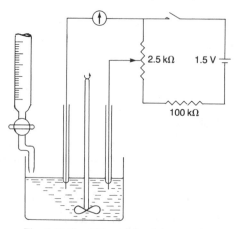

Fig. 5.38 Dead-stop end point apparatus

titration when there is a sharp transition at the equivalence point between at least one polarised electrode and the complete depolarisation of both electrodes. The most important example of the dead-stop technique is in the determination of water content using Karl Fischer reagent.

Automated electrochemical Karl Fischer analysis

Commercially available instruments which perform Karl Fischer (KF) aquametric titrations with greater convenience and accuracy than the classical technique are based on two previously mentioned electrochemical methods, namely coulometry and controlled current potentiometry with two indicator electrodes.

The iodide required for the reaction with water

$$H_2O + I_2 + SO_2 + 3C_5H_5N + CH_3OH$$
$$\rightarrow 2\ C_5H_5N.HI + C_5H_5NH\ SO_4CH_3$$

is generated coulometrically within the titration vessel and the quantitative relationship between charge passed and iodine generated from the reagent is the basis of the analysis. The generation of iodine is automatically stopped when excess is detected by the indicator electrode. This consists of two platinum wires across which an a.c. is applied and, as in the dead-stop end point method, a marked drop in voltage across the electrodes occurs when excess iodine is present. The commercial instrument uses proprietary reagents, of which there is a range with differing properties and composition, for filling the anode and cathode compartments of the coulometric cell. However, normal KF reagents, either with or without pyridine and in one or two component form can be used in the coulometric generator. A notable advantage of this approach to KF analysis is that no calibration is required as the method is absolute, being based on the stoichiometry of the equation above. The commercial instrument will determine amounts of water between $10\ \mu g$ and $10\ mg$ in both solid and liquid samples.

Electrochemical detection in HPLC

Liquid chromatography with electrochemical detection (LCEC) combines the separatory qualities of high performance liquid chromatography (HPLC) with the selectivity and sensitivity of electrochemical analysis. Detection of eluted analyte in HPLC by ultraviolet, fluorescence or refractive index methods has been considered in Chapter 4, and one of their disadvantages is their lack of selectivity. Electrochemical detectors are selective towards compounds possessing electroactive groups, and this selectivity can be improved by the appropriate choice of applied voltage.

In general, LCEC detectors function by applying a fixed d.c. voltage to the working electrode of a flow-through cell, and recording any current arising from oxidation or reduction of analyte as it emerges from the HPLC column after the appropriate elution time. In principle LCEC operates in the same way as the equivalent electrochemical method described earlier in this section. Detectors can be polarographic but are more commonly amperometric or coulometric, with potentiostat, electrode and output requirements essentially identical to those for operation in the conventional way. The cell design must obviously be different, however, to allow for the flow-through requirement of HPLC, and much attention has been given to optimising background noise, sensitivity, electrode accessibility, and to reducing cell volumes. In the familiar three electrode cell arrangement the reference electrode is usually silver–silver chloride in 3M potassium chloride and the auxiliary electrode is usually either glassy carbon, platinum wire, or of a proprietary carbon construction.

Materials used for the working electrode are much more diverse, each possessing virtues for particular applications. The most popular for oxidative detection are carbon paste and glassy carbon, the latter being more robust but giving a higher background current. Carbon paste electrodes prepared by mixing graphite with paraffin oil, ceresin wax or silicon grease, are generally unsuitable for use in non-aqueous solvents, so mixtures containing, for example, >25% v/v methanol or >5% v/v acetonitrile, would be prohibited. Glassy carbon can be readily repolished and carbon paste electrodes readily remade so, if damaged by excessive voltage application, current flow, or solvent degradation, they are easily recoverable. The criteria by which electrodes should be judged are the acceptable range of applied potential, chemical and physical compatibility with the solvent, signal-to-noise performance, long-term stability, and requirement for preconditioning.

Metal working electrodes include platinum or mercury deposited on gold in which the advantage of the high hydrogen overpotential of mercury can be exploited as in polarography in that a wider range in the reductive direction is possible than on many other surfaces. Polarographic mercury drop electrodes are commercially available but lack stability in the flowing stream of eluent from the HPLC column. However, it has been noted that the detection limits (usually in the low ng/ml range) of amperometric detectors made of glassy carbon, carbon paste, and mercury are reduction potential dependent for certain benzodiazepines, and this can in general be expected as many organic compounds react at different rates according to the electrode being used. The highest possible reaction rate is the most desirable condition, because the current is then transport limited rather than rate limited. One of the advantages of the mercury drop in polarography is that contamination is irrelevant because of the renewable electrode surface; however, contamination is reduced in LCEC in comparison with non-chromatographic electrochemistry because the contact time of the contaminant with the solid electrode is minimised by the narrow elution band created by the chromatographic system.

Cell design

Cell design is dependent on the intended mode of operation of the detector, of which there are two. The coulometric mode converts 100% of the electroactive compound passing through the cell and therefore requires a large surface area for the reaction to take place. The amperometric mode converts <10% of the material passing through the cell. The three different cell designs, intended to improve sensitivity or selectivity, are known as tubular, channel, and wall-jet types. In the first type the solution flows through an annulus which is the working electrode, in the second it flows over an electrode plate, and in the third it emerges from a nozzle perpendicularly onto the electrode surface. Each has advantages in mass transfer, resistance to poisoning, ease of replacement of electrode substance, or signal-to-noise ratio.

In coulometric operation a large surface area is required to achieve 100% electroreaction. Two types of cell are available: one consists of an open tube electrode through which the solution passes directly, and in the other the solution passes over a reticulated, cloth or gauze electrode. It can be shown that the smaller the electrode area the higher is the current density, and coulometric cells show that additional current due to area increase is a decreasing function of increased electrode area. In addition, noise and background current are in general linearly related to area, so coulometric cells usually have poorer detection limits than amperometric ones.

Theory

Faraday's Law relates the charge Q in coulombs transferred in the electroreaction to the number of moles N of reactant being processed; thus

$$Q = zFN$$

where z is the number of electrons involved in the reaction of one mole of reactant, and F is Faraday's constant (96 485 coulombs per mole of electrons). In coulometric detection the current flowing is measured and integrated over time to give the charge consumed, which is therefore represented by the peak area on the recorder trace. As 100% conversion occurs in coulometric operation, the peak area will be independent of flow rate u, and thus one calibration curve will apply for all flow rates, but it can be shown that the current i is directly dependent on flow rate as follows:

$$i = zFcu$$

where c is the analyte concentration in the bulk of solution prior to entry into the coulometric detector.

Analogous equations can be written for each geometry of amperometric detector; that for a channel cell in which the eluent passes over the electrode is

$$i = 1.467zFAc \ (D/h)^{2/3}(u/d)^{1/3}$$

where A is the electrode area (cm^2), D is the diffusion coefficient (cm^2s^{-1}), h is the thickness of the channel and d is its width. Both these equations show direct dependence of current on analyte concentration, but the amperometric design is less dependent on volume flow rate. The current, which both detection methods display, describes the rate of analyte conversion while the integrated, coulometric, response reflects the total amount of analyte reacted; a cell design which permits rapid transfer rates is therefore important in both methods of detection.

Operating potential

The choice of applied potential is dependent on the functional groups on the molecule to be analysed. Oxidation or reduction can be undertaken but the voltage range in each direction will be dependent on choice of electrode, as discussed above and on pp. 225–230. Reducible groups are of course the same as described there, and the most frequently encountered oxidisable groups are the aromatic hydroxyl and amine. Less frequently encountered are indoles, phenothiazines, mercaptans and miscellaneous compounds such as ascorbic acid and vitamin A. With LCEC reduction, as in polarography, all traces of oxygen must be excluded from the solvents and no permeable tubing should be used.

Selection of a suitable applied voltage is often done by generating a hydrodynamic voltammogram for the compound after a rough assessment of likely potentials has been made. This allows the potential for maximum signal and selectivity to be found by repetition of the chromatographic run with the detector set at a constant, but successively altered, potential for each run. When insufficient voltage is applied, no response will occur; alteration by 0.1 V steps will eventually lead to a response which is very sensitive to voltage changes. Finally a plateau region will be obtained where diffusion is the rate determining factor for current generation. Further increase in voltage will only reduce selectivity by possibly including more electroactive compounds and functional groups within the voltage range, so an operating potential should be chosen which is sufficiently within the plateau region for maximum sensitivity without decreasing selectivity. A typical hydrodynamic voltammogram is shown in Fig. 5.39. All measurements should be done under controlled and recorded conditions of flow rate, temperature, amplifier gain etc.

Some LCEC coulometric systems have two or more cells operating in series on the eluent. By allowing the application of different potentials at each cell, selective conversion (screening) and analysis of mixtures of electrochemically active substances can be undertaken. Another function of an additional cell is to provide electrochemically clean mobile phase; in this case the cell is connected between the pump and injector. By setting this cell (guard cell) potential at an absolute voltage 50 mV higher than the analytical cell, the cleaning function will be accomplished and background current and noise at the analytical cell reduced.

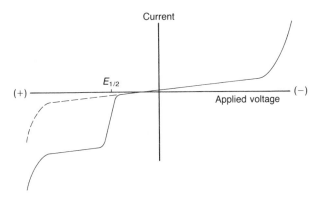

Fig. 5.39 Hydrodynamic voltammogram for mobile phase alone (dotted line) and with electroactive solute with half-wave potential $E_{1/2}$. (Redrawn from Bioanalytical Systems Manual 80–4A, 1980.)

Solvent systems

The mobile phase must be electrically conductive for the same reasons as in polarography. The solvent must therefore permit dissolution and ionisation of between 0.01 and 0.1 mol l^{-1} electrolyte, and the resultant solution should be electrochemically inactive. It will be seen that such mobile phases are employed in the majority of ion exchange and reverse phase HPLC; all solvents used should be of the highest purity, and analytical grade salts and acids should be used for buffer preparation. One authority recommends that deionised water should ideally be filtered through a $0.45 \mu m$ and then a $0.22 \mu m$ low extractable membrane filter before use, and a final filtration through a $0.22 \mu m$ filter should be carried out when all the reagents and additives have been mixed together and before the final degassing process.

Storage, deaeration and delivery of the mobile phase should follow the routines described in Chapter 00. Provided the operator is aware that electrochemical detectors may be up to 100 times more sensitive than ultraviolet or refractive index detectors, and that oxidisable or reducible species, both organic and inorganic, are contaminants to be avoided, then the extent and nature of the precautions to be taken in LCEC follow naturally. The use of prefilters, pulse dampeners and other refinements to HPLC is covered in Chapter 4. Samples should be dissolved in mobile phase when prepared for injection.

Operation

As there are a number of LCEC instruments commercially available, the specific procedures for setting up and running the detector and the chromatograph should be taken from the manufacturer's literature. As the detectors will be measuring nanoamp levels of current, selectable according to analyte concentrations, it is essential that all metal parts are correctly earthed, preferably to a single point to avoid stray currents. Great care must be taken that air is excluded completely from the detector cell when it is assembled and from the mobile phase; checks should be made for leaks along the flow line. Manufacturers always recommend running-in periods for new apparatus, and when a component has been adjusted or the mobile phase altered, because background currents will always vary after change in the state of electrodes, change of mobile phase component, flow rate, temperature etc. These alterations to operating conditions usually require half an hour equilibration time for background current to drop to $<10\,\text{nA}$. Calibration graphs of current *vs* concentration can be drawn as in conventional HPLC, and peak shapes are interpreted in the same way also.

Experiment 59   *Separation and detection of noradrenaline, adrenaline, dopa and dopamine by LCEC*

Method A 250 mm column packed with $5\,\mu\text{m}$ LiChrosorb RP-8 or similar should be used with a glassy carbon LCEC detector set to $+800\,\text{mV}$ w.r.t. silver–silver chloride as the reference. 50 nA is a suitable current range but this can be adjusted according to the sensitivity of the detector. A $20\,\mu\text{l}$ injector loop should be fitted.

The mobile phase should consist of deionised water, acetic acid $(3\,\text{g}\,l^{-1})$, and ammonium sulphate $(5\,\text{g}\,l^{-1})$ of appropriately pure quality. This should be used as solvent for $1\,\text{mg}\,l^{-1}$ each of the analytes.

With a solvent flow of $1\,\text{ml}\,\text{min}^{-1}$ the system should be allowed to equilibrate and the background to fall to a few nanoamps. Injections of samples, either standards or unknowns, can then be carried out according to normal HPLC practice. Ten minutes should be allowed for all components to be eluted separately from the column, and, after annotation of the trace with all the experimental conditions, peak heights should be measured and a calibration graph constructed as in conventional HPLC practice.

Experiment 60   *Determination by LCEC of erythromycin in plasma and urine*

Method A reverse phase $10\,\mu\text{m}$ size μBondapak C_{18} column of 30 cm length should be operated at ambient temperature with a suitable HPLC pump and injector. If possible a dual electrode with the first electrode operating in oxidative screen mode should be used. This electrode, if used, should be set at $+0.7\,\text{V}$ with the detector electrode at $+0.9\,\text{V}$.

The mobile phase should consist of acetonitrile–methanol–0.2M sodium acetate $(40:5:55)$, the sodium acetate solution being carefully pre-adjusted to pH 6.7 with 0.2M acetic acid. The mobile phase components should be of adequate purity and if possible filtered through a $0.22\,\mu\text{m}$ low-extractable membrane.

Flow rate should be $1\,\text{ml}\,\text{min}^{-1}$ at about 2000 p.s.i.

Stock solutions containing 1 and 5 mg ml^{-1} erythromycin A should be prepared in acetonitrile, as should 0.01 and 5 mg ml^{-1} solutions of erythromycin B.

Standard measurements, and unknown samples if required, can be prepared by spiking plasma with erythromycin A solution to give a range from 0.25 to 5 mg ml^{-1}. Urine based solutions can be similarly prepared to give a concentration range from 2.5 to $50 \mu\text{g ml}^{-1}$, these being representative ranges of levels found after a 500 mg oral dose.

Plasma samples should be prepared as follows. A 0.2 ml aliquot should be transferred to a small disposable glass culture tube or similar vessel. An internal standard consisting of $10 \mu\text{l}$ of 0.01 to 0.1 mg ml^{-1} erythromycin B, or erythromycin ethylsuccinate or estolate, in acetonitrile should be added (choice of internal standard depends on whether the sample to be analysed contains it or not). Saturated sodium carbonate $(20 \mu\text{l})$ and diethylether (1 ml) should then be added before vortex mixing. Centrifugation at 800 G for 5 min should be followed by removal of 0.75 ml of the ether layer to another similar vessel. The ether should be allowed to evaporate under reduced pressure for 10–15 min, and the residue reconstituted just prior to analysis with $50 \mu\text{l}$ mobile phase followed by 10 s mixing. After brief centrifugation, $20 \mu\text{l}$ should be injected.

Urine samples are prepared more easily. A 0.1 ml aliquot should be transferred to a culture tube or similar and an appropriate internal standard $(10 \mu\text{l}, 0.2 \text{ mg ml}^{-1})$ added as above together with 0.2 ml acetonitrile for deproteination. After 10 s vortex mixing, the sample should be centrifuged at 800 G for 2 min, and $20 \mu\text{l}$ of supernatant injected onto the column.

Under the conditions described, the maximum retention time (for erythromycin estolate) will be 23 min, with erythromycins A, B and C being eluted after approximately 10, 15 and 7 min respectively.

A calibration graph should be prepared using the internal standards as described in Chapter 4. The concentration of samples, either spiked or of animal origin, can then be determined by comparison.

Further reading and references

General electrochemistry

Bard, A.J. and Faulkner, L.R. (1980). *Electrochemical Methods. Fundamentals and Applications,* Wiley, New York.
Hibbert, D.B. and James, A.M. (1984). *Dictionary of Electrochemistry,* 2nd edn., Macmillan, London.
Skoog, D.A. and West, D.M. (1976). *Fundamentals of Analytical Chemistry,* 3rd edn., Holt, Reinhart and Winston.

Potentiometry and ISEs

Bailey, P.L. (1976). *Analysis with Ion Selective Electrodes,* Heyden.
Cosofret, V.V. (ed.) (1982). *Membrane Electrodes in Drug-Substances Analysis,* Pergamon Press.
Covington, A.K. (1974). *CRC Critical Reviews in Analytical Chemistry,* 355.
Koryta, J. (1982). *Anal.Chim.Acta,* **139**, 1.
Moody, G.J. and Thomas, J.D.R. (1980). *Ion Selective Electrode Reviews,* **2**, 73.
Orion Research (1978). *Analytical Methods Guide,* 9th edn.
Solsky, R.L. (1983). *CRC Critical Reviews in Analytical Chemistry,* **14**, 1.

Voltammetry

Chen, M-L. and Chion, W.L. (1983). *J.Chromatog.Biomedical Applics.*, **278**, 91.

ESA International Inc. (1983). *Model 5100A Coulochem Instruction Manual.*

Fleet, B. and Jee, R.D. (1973). In *Electrochemistry, volume 3, Electroanalytical Chemistry: Voltammetry,* Specialist Periodical Report, Chemical Society, 210.

Franklin Smyth, W. (ed.) (1979). *Polarography of Molecules of Biological Significance,* Academic Press.

Heyrovsky, J. and Zuman, P. (1968). *Practical Polarography,* Academic Press.

Princeton Applied Research (1977). *Application Note* 151.

Volke, J. (1983). *Bioelectrochemistry and Bioenergetics,* **10**, 7.

6
The basis of spectrophotometry

A. G. DAVIDSON

Introduction

Photometric techniques are among the most important instrumental techniques available to the pharmaceutical analyst. Instrumentation ranges from the simple flame photometers, which are used to determine the concentration of certain metallic elements, to the much more expensive spectrometers such as ultraviolet–visible and nuclear magnetic resonance spectrometers which are used in structural and quantitative analysis of molecules.

The basis of all these instrumental techniques is that they measure the interaction of electromagnetic radiation with matter in quantised, i.e. specific, energy levels. The purpose of this chapter is to introduce the various spectrophotometric techniques which are discussed in greater detail in the subsequent chapters, to show their relationship with each other and to describe the theoretical principles involved.

Electromagnetic radiation

White light from an incandescent solid such as the filament of an electric lamp is made up of a large number of individual waves of varying wavelength. This is readily shown by passing a beam of light through a prism when a band of colour, or so-called continuous spectrum, is formed, in which each colour corresponds to waves of a particular wavelength (Fig. 6.1).

The visible spectrum, however, forms only a small part of the complete spectrum of electromagnetic radiation, which extends, as shown in Fig. 6.1, from the ultra-short wave region of the cosmic rays at one end to that of radio waves at the other.

Wavelength is defined as the distance between any two consecutive parts of the wave whose vibrations are in phase, for example from the crest of one wave to that of the next, (A to B or B to C in Fig. 6.2). Its symbol is λ, the Greek letter lambda. The units in which wavelength is commonly expressed are recorded in Table 6.1. By adapting the unit to the appropriate region, the use of cumbersome figures can be avoided.

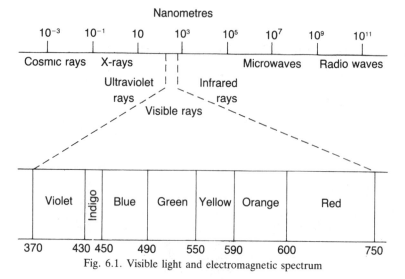

Fig. 6.1. Visible light and electromagnetic spectrum

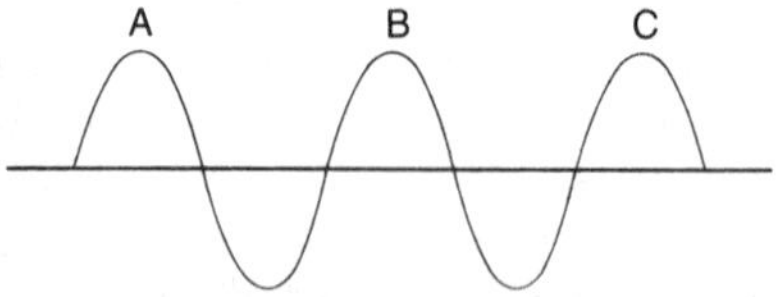

Fig. 6.2. Wavelength

Table 6.1 Wavelength units

Unit	Abbreviation	Metre	Region where used
Angstrom	Å	10^{-10}	Visible and ultraviolet
Nanometre	nm	10^{-9}	Visible and ultraviolet
Micrometre	μm	10^{-6}	Infrared

Older units for micrometre (micron) and for nanometre (millimicron) are now obsolete.

Wavenumber is defined as the reciprocal of the wavelength expressed in cm, i.e. the number of waves per cm. Its units are cm^{-1}.

Frequency is the number of waves passing a point in one second, i.e. the number of cycles per second. Its symbol is ν, the Greek letter nu, and the units are s^{-1} or hertz (Hz).

The interrelationships of these units can be expressed as follows:

$$\frac{1}{\text{wavelength in vacuo (in cm)}} = \text{wavenumber}$$

$$= \frac{\text{frequency}}{\text{speed of light in vacuo (cm s}^{-1})}$$

Atomic spectra

Atomic emission

It has long been known that, when certain compounds are heated, light of characteristic colours is emitted. For example, sodium salts emit yellow light and potassium salts emit lilac light. The work of Kirchhoff and Bunsen in the middle of the 19th century showed that this is due to the emission of light at wavelengths characteristic of the metallic elements in the sample. Flame tests, in which the metallic elements are quickly identified by the characteristic colours imparted to a premixed combustible gas–air flame, remain one of the simplest qualitative analytical procedures.

When a sample containing metallic atoms is heated above 2000°, it first undergoes partial or complete dissociation into free atoms and then volatilises to free gaseous atoms. The electrons of the gaseous atoms exist in discrete quantised energy levels, i.e. they are in orbitals which have specific energy levels that are characteristic of the element. The electrons in the outer orbitals of the atoms may absorb thermal energy and be promoted to one or more higher energy states. The gain in energy of each electron during a transition is a specific quantity corresponding to the difference between the energy levels after and before excitation. Deactivation of the thermally excited atoms to lower energy states occurs very rapidly and photons of light are emitted which have energy (ΔE) equal to the difference between the upper (E_u) and lower (E_l) energy states. The energy of the emitted light is directly proportional to the frequency (v) and inversely proportional to the wavelength (λ). Hence,

$$\Delta E = E_u - E_l = \frac{hc}{\lambda}$$

where h = Planck's constant = 4.132×10^{-15} eV s, and c = speed of light = 3×10^8 m s^{-1} (in vacuo).

$$\Delta E = \frac{1240}{\lambda \text{ (in nm)}}$$

As the atomic energy levels are characteristic of the element, the energies and wavelengths of light emitted are also characteristic of the element. The electronic energy transitions during the thermal excitation of and the emission of light from sodium atoms is shown in Fig. 6.3. The energy levels of the electronic orbitals (on the ordinate) are conventionally given in electronvolts greater than the lowest electronic energy level of the atom, which is given the energy eV = 0.

The wavelengths of light emitted from the sample as a result of thermal excitation may be viewed through a **spectroscope**, which is a simple instrument containing an entrance slit and a dispersing device, e.g. a prism. Alternatively, they may be recorded on photographic film using a **spectrograph**. The emission spectrum of sodium is seen to

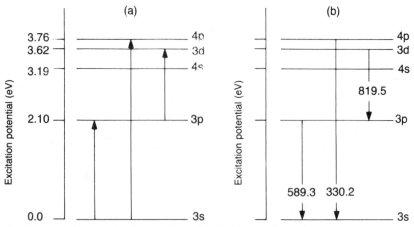

Fig. 6.3. Partial energy levels of sodium (a) during the absorption of thermal energy or electromagnetic radiation and (b) during deactivation with the emission of resonance wavelengths

comprise approximately 20 lines, the most intense of which occurs at 589.3 nm (consisting of a doublet at 589.0 nm and 589.6 nm), with less intense lines at 330.2 nm and 819.5 nm (Fig. 6.4). The line at 589.3 nm (the sodium D-line) is emitted when valence electrons thermally excited from the 3s orbital to the 3p orbital return to the ground state. The difference between the energy levels of the 3p (2.10 eV) and 3s (0 eV) orbitals corresponds with a wavelength of

$$\lambda = \frac{1240}{2.10 - 0} = 589 \text{ nm}$$

The less intense lines at 330.2 nm and 819.5 nm are due to 4p → 3s and 3d → 3p transitions respectively, which are given by fewer atoms. As an exercise you should identify the transition which accounts for the line at 1139 nm in the near infrared region. The intense line at 589.3 nm is yellow light (Fig. 6.1) and accounts for the observation of the yellow colour imparted to a Bunsen flame when sodium salts are introduced. The lilac coloration of a flame in the presence of potassium salts is due to the emission of two lines in the red region of the spectrum at 767 nm (4p → 4s) and 694 nm (4d → 4p) and one line in the violet region at 404 nm (5p → 4s).

The atomic emission techniques of **flame emission spectrometry (flame photometry)** and **emission spectrography** are based on the **measurement of light emitted from thermally excited atoms.**

330.2 nm   589.3 nm   819.5 nm   1139 nm

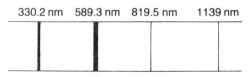

Fig. 6.4. Partial emission spectrum of sodium

Atomic absorption

Eight dark lines (lines A–H) in the continuous solar spectrum were explained by Fraunhofer in 1823 by the phenomenon of atomic absorption. Certain elements in the outer core of the sun absorb radiation at wavelengths characteristic of the elements. For example, absorption by sodium atoms gives rise to the line at 589.3 nm (the sodium D-line, the only line which is still commonly referred to by its original letter).

The excitation of electrons in atomic orbitals to higher energy states may be induced by electromagnetic radiation if the energy of the radiation exactly matches that corresponding to the difference between the upper and lower energy states (Fig. 6.3). Thus, ground state atoms absorb light of exactly the same characteristic resonance wavelengths that they emit after thermal excitation. For example, if a beam of light at 589 nm is passed through a vapour of sodium atoms, a portion of the ground state atoms will absorb radiation promoting the valence electrons to higher energy states.

Atomic absorption spectrophotometry is a technique for the quantitative determination of metallic elements and metalloids, which is based on the **measurement of the absorption of monochromatic light by ground state atoms**.

Molecular spectra

Molecular spectra are characterised by the absorption or emission of light over a much wider range of wavelengths (called spectral bands) than atomic spectra, which consist of sharp resonance lines (Fig. 6.4). This is due to the very large number of transitions which molecules can undergo, in comparison to the relatively few electronic transitions of atoms.

Molecular absorption

The absorption by molecules of electromagnetic radiation of a suitable wavelength can promote:

(a) The energy of electrons to one or more higher energy states (as in atomic absorption). The types of electrons that are responsible for ultraviolet–visible absorption are discussed in Chapter 7.
(b) An increase in the internuclear vibrational energies of the constituent atoms.
(c) An increase in the energy of rotation of the atoms round the bonds joining the atoms.

The total energy of a molecule is the sum of its electronic, vibrational and rotational energies, which are each quantised, i.e. they have specific energies characteristic of the molecular species. The relative energies required to induce electronic, vibrational and rotational transitions between the quantised energy levels are approximately $10\,000 : 100 : 1$ respectively.

At room temperature, the energy of most molecules is the ground electronic (E_{El}) and vibrational (E_{Vib}) states. If polychromatic light, i.e. light comprising a spectrum of a suitable wavelength range, interacts with a large number of identical molecules, some of the molecules absorb a portion of the light. As a result, certain electrons within the molecules (p.316) undergo a transition to an excited energy state and this is normally accompanied by an increase in the vibrational and rotational energies of the molecules. The excited molecules gain different quantities of energy (Fig. 6.5(a)), equivalent to the gain in electronic energy plus a large number of possible vibrational and rotational transitions $(\Delta E_1–\Delta E_7)$ where

$$\Delta E = (E_{El} + E_{Vib} + E_{Rot})_{upper} - (E_{El} + E_{Vib} + E_{Rot})_{lower}$$

It is found that the variation of the gain of energy approximates to a normal or Gaussian distribution, with one particular transition (ΔE_4) given by more molecules than any other transition and with fewer molecules giving the related transitions. The wavelength of light absorbed, corresponding to the total energy gained during each transition, is

$$\lambda \text{ (in nm)} = \frac{1240}{\Delta E}$$

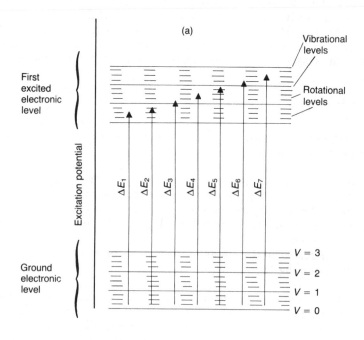

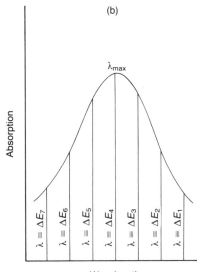

Fig. 6.5. (a) The gain of electronic, vibrational and rotational energy during the molecular absorption of ultraviolet–visible radiation (the relative energies are not drawn to scale) (b) A molecular absorption spectrum in the ultraviolet–visible region

Consequently, a very large number of wavelengths are absorbed, each corresponding to a particular ΔE value, which are so close together that they appear as a continuous band spectrum (Fig. 6.5(b)). The transition ΔE_4, given by the highest proportion of molecules, corresponds with the wavelength of maximum absorption (λ_{max}). Light at longer and shorter wavelengths is absorbed less strongly because fewer molecules give the related transitions.

Vibrational fine structure may be observed in the ultraviolet absorption spectrum of benzene and its simple derivatives (e.g. ephedrine, pethedine) between 240 nm and 270 nm arising from the vibrational sub-levels accompanying the electronic transitions (Fig. 6.6(a)). Chromophoric and auxochromic substitution of benzene destroys the fine structure and such derivatives (e.g. sodium salicylate) show only a broad continuous band spectrum (Fig. 6.6(b)). Fine structure of compounds is also reduced by solute–solvent interactions, particularly in polar solvents.

Molecules absorb electromagnetic radiation only at wavelengths that correspond with the ΔE values between the excited and unexcited states. Short wavelength radiation (X-rays, ultraviolet–visible light) has sufficient energy to induce electronic transitions with the associated vibrational and rotational transitions. As most unsaturated organic molecules contain very few (typically one to three) chromophores (Chapter 7), an ultraviolet–visible absorption spectrum normally comprises one to three absorption bands.

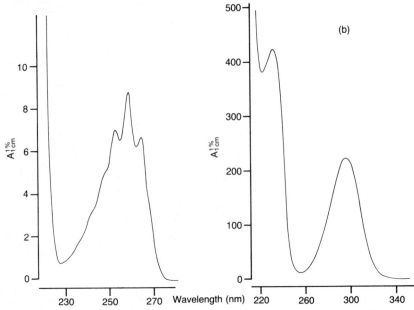

Fig. 6.6. The ultraviolet absorption spectra of (a) ephedrine hydrochloride and (b) sodium salicylate in water

Longer wavelengths have insufficient energy to induce electronic transitions but they are sufficiently energetic to increase the vibrational energy levels, and the associated rotational levels, of many molecular bonds. Consequently an infrared spectrum comprises many absorption bands at wavelengths characteristic of the vibrational energies of the bonds. Microwave radiation (wavelength 10^{-3}–0.67 m) has sufficient energy to induce transitions between the quantised rotational levels only.

In nuclear magnetic resonance spectrometry, electromagnetic radiation in the radiofrequency range (Fig. 6.1) induces transitions between different nuclear energy states in the presence of an external applied magnetic field.

Molecular emission

After the absorption of ultraviolet–visible light, the excited molecular species are extremely short-lived and deactivation occurs due to:

(a) internal collisions (internal conversion)
(b) cleavage of chemical bonds, initiating photochemical reactions
(c) re-emission as light (luminescence).

Re-emission of energy as luminescence occurs from molecules in which the electron system is shielded from normal deactivation processes so that complete deactivation by collisions is discouraged. Molecules on excitation normally possess higher vibrational energy than they had in the ground state. This extra vibrational energy is lost (Fig. 6.7) by collision, after which the molecules return to the ground electronic state with the emission of light as fluorescence. Deactivation as fluorescence is a rapid process occurring within 10^{-6}–10^{-9} seconds of the excitation.

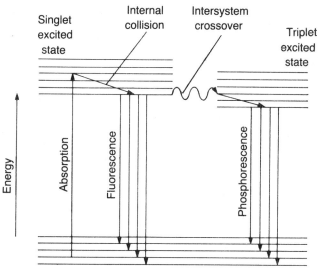

Fig. 6.7. Energy transfer during the absorption, fluorescence and phosphorescence of ultraviolet–visible radiation. The lower set of energy levels represents the ground electronic level and its associated vibrational levels. The rotational sub-levels are not shown. The upper set of energy levels represents the first excited electronic state and its associated vibrational sub-levels. Only one of the many transitions that occur during the absorption of ultraviolet–visible light is shown (see Fig. 6.5(a))

In excited molecules which exhibit fluorescence, the spin of the π electron and that of the π^* electron, which together constitute a pi bond in the chromophore system (Chapter 7), are in opposite directions, i.e. they are antiparallel, and the molecules are in the **singlet state**. Some excited molecules, particularly at low temperatures, may undergo a slow **intersystem crossover** to a state (the **triplet excited state**) in which the spin of the π and π^* electrons are unpaired (parallel). Return from the triplet excited state to the singlet ground state results in the emission of **phosphorescence**. Intersystem crossover is a slower process than fluorescence, and consequently phosphorescence occurs after 10^{-8} s and may be observed even several minutes or hours after the source of excitation is removed.

The differences in the energy levels (ΔE) between the excited and unexcited states during excitation (absorption), fluorescence and phosphorescence are in the order (Fig. 6.7):

$$\Delta E \text{ (Absorption)} > \Delta E \text{ (Fluorescence)} > \Delta E \text{ (Phosphorescence)}$$

As the wavelengths corresponding to the ΔE values are inversely proportional to the energy, the order of λ_{max} are:

$$\lambda\ (\text{Absorption}) < \lambda\ (\text{Fluorescence}) < \lambda\ (\text{Phosphorescence})$$

For example, the wavelengths of maximum excitation, fluorescence and phosphorescence of anthracene are 255 nm, 425 nm, and 680 nm respectively.

The techniques of **spectrofluorimetry** and **phosphorimetry measure the intensity of light emitted from a system that has absorbed radiant energy**.

Instrumentation

The essential components of spectrometers (Fig. 6.8) are:

(a) a source of electromagnetic radiation
(b) a monochromator (to isolate a particular wavelength or range of wavelengths)
(c) a sample compartment
(d) a detector and associated electronics (to measure the intensity of electromagnetic radiation)
(e) a recorder or display.

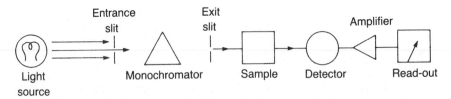

Fig. 6.8. Schematic diagram of a single-beam spectrometer

Light sources

The distribution of energy through a spectrum is mainly a function of temperature; the higher the temperature of the light source the shorter the wavelength of the peak emission. The heating process, however, cannot be carried too far, as changes such as vaporisation would take place with the consequent production of a line spectrum or burning out of a lamp. For example, to obtain ultraviolet light from a tungsten lamp it would have to run at a gross overvoltage, which would shorten its life if not destroy it.

Common energy sources for the various regions are indicated below.

Infrared radiation

Globar and Nernst glowers are common sources of infrared radiation. The globar is an electrically heated rod of silicon carbide, and the

Nernst glower is a small rod of refractory oxides which, when heated to 1200°–1500°, will conduct electricity and thus maintain itself in incandescence. Both these sources operate without a glass envelope, which would absorb infrared radiation of wavelength greater than $2\,\mu$m.

Visible radiation

The tungsten filament lamp is a satisfactory light source for the region 350 to about 2000 nm. It consists of a tungsten filament contained in a glass envelope. Infrared radiation, which may cause stray light effects (p.312), can be removed by suitable filters. The life of the lamp is limited by the evaporation of tungsten, which darkens the inside of the envelope and reduces the energy of incident light.

Tungsten–halogen lamps, which are used in the more expensive instruments, are filled with a halogen gas inside a quartz envelope. The higher operating temperature extends the lower wavelength limit to 310 nm and gives greater light intensity. The halogen prevents evaporation of the tungsten and increases the life of the lamp.

Ultraviolet radiation

The most convenient light source for measurement in the ultraviolet region is a deuterium discharge lamp. It consists of two electrodes contained in a deuterium-filled silica envelope. The passage of a high voltage from a special power supply across the electrodes causes emission of deuterium lines which, at the low pressure inside the lamp, broaden to give a continuous spectrum in the range 185–380 nm. Above 380 nm the emission is not continuous, and the sharp lines at 486.0 nm and 656.1 nm may be used to calibrate the wavelength scale in this region. Below 185 nm the output is reduced by absorption in the silica envelope. Deuterium lamps give approximately five times greater light intensity than hydrogen lamps when operated at the same wattage.

Ultraviolet–visible spectrophotometers normally have both a deuterium lamp and a tungsten (or tungsten–halogen) lamp, and selection of the appropriate lamp is made by moving either the lamp mountings or a mirror to cause the light to fall on the entrance slit of the monochromator.

Monochromators

Filters

Glass filters are pieces of coloured glass which transmit limited wavelength ranges of the spectrum. The colour is produced by incorporating oxides of such metals as vanadium, chromium, manganese, iron, nickel and copper in the glass. The colour absorbed is the complement of the colour of the filter. Thus, a filter absorbing yellow light (575–625 nm) appears blue. The range of wavelengths transmitted (bandwidth) is very wide and may exceed 150 nm. Gelatin filters consisting of a mixture of dyes incorporated in gelatin and sandwiched between glass plates are more selective, with bandwidths about 25 nm.

Interferometric filters have an even narrower bandwidth (about 15 nm) and consist of two parallel glass plates, silvered internally and separated by a thin film of cryolite or other dielectric material. Such filters make use of the interference of light waves rather than absorption to eliminate undesired radiation, and serve as relatively inexpensive monochromators for a specific purpose, e.g. the isolation of calcium radiation at 626 nm from that of sodium at 589.3 nm in the flame photometric method for Na^+ and Ca^{2+} in the same solution. Ordinary filters are incapable of this and cause large errors in measurement of the calcium content.

Prisms

When a beam of monochromatic light passes through a prism, it is bent or refracted. The amount of deviation is dependent on the wavelength, blue light being refracted more than red. If white polychromatic light is substituted for monochromatic radiation, a separation of the different wavelengths leads to the formation of a spectrum (Fig. 6.9) from which the required wavelength may be selected for the spectrometric measurement.

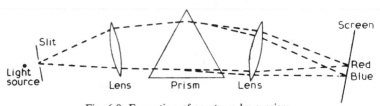

Fig. 6.9. Formation of spectrum by a prism

Prisms are made of quartz for use in the ultraviolet region, since glass absorbs wavelengths shorter than about 330 nm. Glass prisms are preferable for the visible region of the spectrum, as the dispersion is much greater than that obtained with quartz. For the infrared region, the transparent substances usually used for prisms are sodium chloride (2–15 μm), potassium bromide (12–25 μm), lithium fluoride (0.2–6 μm) and caesium bromide (15–38 μm).

Prisms produce a non-linear dispersion, with long wavelengths being less efficiently separated than short wavelengths. The wavelength scales of some (usually older) ultraviolet–visible spectrophotometers show larger divisions between each nm in the ultraviolet region than in the visible region.

Gratings

The dispersing element in the monochromator of most modern ultra-violet, visible and infrared spectrophotometers is the diffraction grating. It consists of a very large number of equispaced lines (200–2000 per mm) ruled on a glass blank coated with a thin film of aluminium. Gratings are now produced by modern holographic techniques. Parallel lines or grooves are chemically etched on a thin layer of photoresist coated on a blank after exposure to the interference fringes produced by the intersection of two beams of light from a laser. They can be used either as transmission gratings, or, when aluminised, as reflection gratings. Rotation of the grating permits appropriate wavelengths of the spectrum to emerge from the exit slit of the monochromator.

The theory of the plane transmission gratings is given below. Fig 6.10(a) shows part of a diffraction grating in which the gaps represent

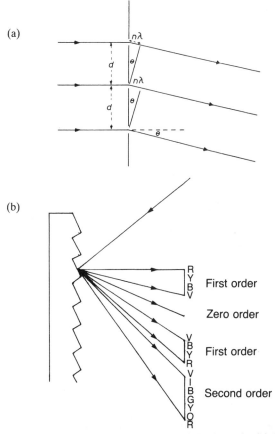

Fig. 6.10. (a) Formation of nth order spectrum: $d \sin \theta = n\lambda$. (b) Dispersion of visible light by a reflection diffraction grating. R, O, Y, G, B, I and V denote the colours of the dispersed light, red, orange, yellow, green, blue, indigo and violet respectively

the transparent spaces. The distance d between consecutive correspond-ing elements is called the grating space. A parallel beam of monochromatic light falls upon the grating at normal incidence. If the path difference for light diffracted through an angle θ from consecutive elements is $n\lambda$, then the various rays will reinforce each other. The values of θ in the equation $d \sin \theta = n\lambda$ ($n = 0, 1, 2, 3$) correspond to the angles of diffraction of different orders. When white light is used instead of monochromatic light, first and successive order images give rise to spectra (Fig 6.10(b)) due to the variation of θ with λ for a given n but the zero order is undispersed. The rulings of a grating can be shaped to concentrate the light in certain orders to give greater efficiency than when the light is spread over many spectra, and this is especially important in the infrared. A grating can be plane or concave, the latter being capable of focussing its own spectrum without the use of lenses or mirrors. However, for reasons of convenience in scanning, plane gratings are most frequently used in monochromators which can be made to cover all spectral regions from 180 nm to 15 μm. To eliminate the effect of overlapping orders, filters have sometimes to be used.

Cells

Samples presented for spectrometric analysis may be in the solid, liquid or gaseous state. The material that contains the sample ideally should be transparent at the wavelength(s) of measurement. For the analysis of liquids and gases in the ultraviolet–visible region above 320 nm, cells (or cuvettes) constructed with optically flat fused glass may be used. Measurements below the ultraviolet cutoff of glass at about 320 nm require the use of more expensive fused silica cells, which are transpa-rent to below 180 nm. The standard pathlength of cells for measure-ments of molecular absorption or fluorescence in the ultraviolet–visible range is 10 mm, although cells of pathlength 1–50 mm are also available for special applications. The most common sample-containing materials in infrared spectrometry (Chapter 10) are sodium chloride (2.5–17 μm) and potassium bromide (2.5–30 μm). In the flame spectroscopic techni-ques of atomic absorption spectrophotometry and flame photometry (Chapter 8), the flame contains the absorbing or emitting atoms of the sample and functions as the sample cell.

Detectors

For the accurate determination of substances by spectrophotometric techniques, precise determinations of the light intensities are necessary. Photoelectric detectors are most frequently used for this purpose. They must be employed in such a way that they give a response linearly proportional to the light input, and they must not suffer from drift or fatigue.

Barrier-layer cells

One of the simplest detectors is the barrier–layer photocell, which has the advantage that it requires no power supply but gives a current which, under suitable conditions, is directly proportional to the light intensity. It consists of a metallic plate, usually copper or iron, upon which is deposited a layer of selenium or sometimes copper(I) oxide. An extremely thin transparent layer of a good conducting metal, e.g. silver, platinum or copper, is formed over the selenium to act as one electrode, the metallic plate acting as the other (Fig. 6.11). Ordinarily, selenium and metallic oxides and sulphides have extremely small electrical conductivities, the electrons being in energy levels where they are not mobile. Light of suitable frequency, however, imparts sufficient energy to the electrons so that they leave the selenium and enter the transparent metal layer. If the two electrodes are now connected through a galvanometer, a current will flow as shown by the deflection of the galvanometer needle. If the resistance in the external circuit is small, the current produced by such a cell is very nearly proportional to the intensity of the light. With higher resistances, leakage back into the selenium layer and hence loss of linearity occur. The current output also depends upon the wavelength of the incident light; when an appropriate correction filter is used the sensitivity is similar to that of the human eye. The useful working range of selenium photocells is 380–780 nm. Their lack of sensitivity, however, compared to phototubes and photomultipliers, restricts their use to the cheapest colorimeters and flame photometers.

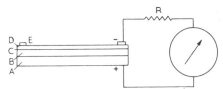

Fig. 6.11. Barrier-layer cell and circuit. A, metal base-plate; B, selenium layer; C, theoretical barrier-layer; D, transparent metal layer; E, collecting ring; R, external resistance

Phototubes

Electrons are liberated when light falls on a metal surface, and, if this is enclosed in an evacuated envelope and kept at a negative voltage, a current which is proportional to the incident light can be drawn from it. The essential feature for a sensitive surface in these cells is that the electrons should be liberated easily from the metal. Elements of high atomic volume, e.g. potassium or caesium, are commonly used and, in order to increase the sensitivity, composite coatings such as caesium/ caesium oxide/silver oxide have proved of more value than the metal alone. The sensitive surface is enclosed in a high vacuum and forms the cathode of the cell (Fig. 6.12). Application of a sufficiently high potential between cathode and anode ensures that all the electrons liberated by the action of light reach the anode. A saturation photocurrent which exhibits a linear relationship with intensity of illumination is then obtained. By using different metals, the cells can be made to respond to different regions of the spectrum. In spectrophotometers two cells are usually used, one being responsive to ultraviolet and visible radiation of wavelengths up to about 620 nm, and the other to radiation of wavelengths 620–1000 nm. A modification of the above type of cell involves the inclusion of an inert gas at a low pressure to give gas-filled cells. The underlying principle is that the liberated electrons cause ionisation of the gas with consequent increase in the photocurrent but, owing to lack of linear response, the cells cannot be used in instruments designed for accurate intensity measurement.

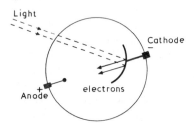

Fig. 6.12. Diagram of phototube

Photomultipliers

In order to obtain greater sensitivity to very weak light intensities, multiplication of the initial photoelectrons by secondary emission is employed. Several anodes at a gradually increasing potential are contained in one bulb (Fig. 6.13). Electrons from the photocathode are attracted to anode 1 and liberate more electrons which travel to anode 2 because of its higher potential relative to anode 1. This process then continues to the last anode, and the result is a final photocurrent 10^6–10^8 times greater than the primary current, which still shows a linear response with increase in the intensity of illumination. Photomultiplier tubes are ideal for measuring weak light intensities such as occur in fluorescence and in the determination of trace elements by their emission spectra.

The current from phototubes and photomultiplier tubes never falls to zero. A small residual current called **dark current** is produced, due to spontaneous discharge at the high voltage of the dynodes or to thermal emission from the cathode. Compensation for dark current is made by external circuitry.

Photodiodes

A full account of the electronic principles of solid-state photodiodes is beyond the scope of this book. In essence, the diode is charged to a pre-set potential by a continuously scanning electron beam. If radiation

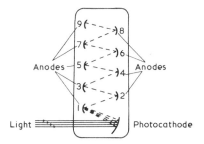

Fig. 6.13. Diagram of photomultiplier tube

impinges on the diode, the charge is depleted in proportion to the intensity of the radiation. When the electron beam scans again, a current in proportion to the light intensity flows to recharge the diode to the pre-set voltage.

Photodiodes, like vacuum phototubes and photomultiplier tubes, are single-channel detectors, i.e. they monitor the total intensity of light and cannot distinguish between different wavelengths. A spectrum is obtained with single-channel detectors by varying the monochromatic light passing through the sample. Recent advances in integrated circuitry and silicon wafer technology have led to the development of linear photodiode arrays containing 256, 512 or 1024 diodes which permit multi-channel detection. When used in combination with a dispersing system (e.g. a grating) each diode can monitor the light intensity at a different wavelength, and the array provides an almost instantaneous spectrum. The multi-channel detector based upon linear diode arrays is of particular advantage in fast reaction kinetics and in monitoring the eluate in HPLC.

Thermocouples

A thermocouple consists of elements of two different metals joined together at one end, the other end being attached to a sensitive galvanometer. When radiant energy impinges on the junction of the metals, a thermoelectromotive force is set up which causes a current to flow. Thermocouples are used in the infrared region, and to assist in the complete absorption of the available energy the 'hot' junction or receiver is usually blackened.

Bolometers

These make use of the increase in resistance of a metal with increase in temperature. For example, if two platinum foils are suitably incorporated into a Wheatstone Bridge, and radiation is allowed to fall on one foil, a change in resistance is produced. This results in an out-of-balance current which is proportional to the incident radiation. Like thermocouples, they are used in the infrared region.

Thermistors

The principle of operation is similar to that described under bolometers but thermistors are constructed of semi-conducting material which has a high negative coefficient of resistivity, i.e. the resistance decreases with increase in temperature.

Golay detector

In this detector, the absorption of infrared radiation causes expansion of an inert gas in a cell chamber. One wall of the chamber consists of a flexible mirror and the resulting distortion varies the intensity of illumination falling on a photocell from a reflected beam of light. The current from the photocell is proportional to the incident radiation.

Readout systems

The signal from the detector is normally proportional to the intensity of light incident on the detector, and after amplification may be displayed as % transmittance (%T) or, after passage through a logarithmic conversion circuit, as absorbance (log $1/T$). There are three common systems for displaying the %T or absorbance, i.e. (a) moving coil meter, (b) digital display or (c) strip-chart recorder.

Spectrophotometers

The brief description of the design and operation of spectrophotometers that follows is restricted to absorption spectrophotometers which measure in the ultraviolet and visible regions of the spectrum. The design of other spectrophotometers is considered in the relevant chapters on these techniques.

Single-beam spectrophotometers

The arrangement of the components in a commercially available single-beam ultraviolet–visible spectrophotometer is shown in Fig. 6.14. The essential characteristic is that the light travels in a single continuous optical path between the light source and the detector.

To make a measurement of absorbance using a manually controlled single-beam instrument, the monochromator is adjusted to the required wavelength and the appropriate lamp and photocell are selected by means of levers or switches. The first step is to close a shutter in the path and adjust a control labelled 'dark current' or 'zero' to offset the dark current from the detector to zero. This sets the scale to 0%T or infinite absorbance. The second step is to open the shutter and place the cell containing only the solvent in the light beam. The scale is set to read 100%T (equivalent to zero absorbance) by a control labelled '100%T' or 'zero absorbance'. The third step is to place the sample cell

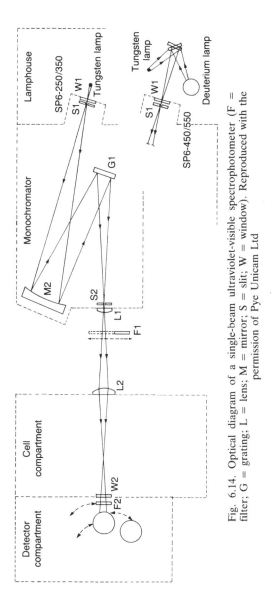

Fig. 6.14. Optical diagram of a single-beam ultraviolet-visible spectrophotometer (F = filter; G = grating; L = lens; M = mirror; S = slit; W = window). Reproduced with the permission of Pye Unicam Ltd

in the light path and to measure the intensity I_T, or its equivalent absorbance, on the scale.

Single-beam instruments are relatively inexpensive and are satisfactory when many samples are being assayed by a simple measurement of absorbance at the same wavelength. A major disadvantage is the need to reset the 100% T at each wavelength to compensate for the large variation of intensity of light from the lamp with wavelength.

Double-beam spectrophotometers

In this type of instrument, the monochromatic light is split by a rapidly rotating beam chopper into two beams which are directed alternately in rapid succession through a cell containing the sample and one containing the solvent only (Fig. 6.15). If there is greater absorption of light in the sample cell than in the reference cell, the recombination of the beams at the detector produces a pulsating current which is converted into two direct current voltages proportional to the light intensities I_0 and I_T transmitted by the reference solution and the sample solution respectively. The ratio of voltages is recorded as a % T ($100I_T/I_0$) or, after logarithmic conversion, as absorbance ($\log I_0/I_T$). Double–beam optics therefore automatically compensate for variation of I_0 with wavelength. Recording spectrophotometers are double-beam instruments equipped with a wavelength scanning device which allows the rapid automatic scanning of spectra.

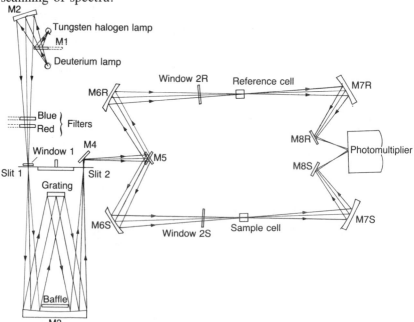

Fig. 6.15. Optical diagram of a double-beam ultraviolet–visible spectrophotometer (M = mirror). Reproduced with the permission of Pye Unicam Ltd

7
Ultraviolet-visible absorption spectrophotometry

A.G.DAVIDSON

Introduction

The technique of ultraviolet–visible spectrophotometry is one of the most frequently employed in pharmaceutical analysis. It involves the measurement of the amount of ultraviolet (190–380 nm) or visible (380–800 nm) radiation absorbed by a substance in solution. Instruments which measure the ratio, or a function of the ratio, of the intensity of two beams of light in the ultraviolet–visible region are called **ultraviolet–visible spectrophotometers**. Absorption of light in both the ultraviolet and visible regions of the electromagnetic spectrum occurs when the energy of the light matches that required to induce in the molecule an electronic transition and its associated vibrational and rotational transitions (p.260). It is thus convenient to consider the techniques of ultraviolet spectrophotometry and visible spectrophotometry together.

Beer–Lambert law

When a beam of light is passed through a transparent cell containing a solution of an absorbing substance, reduction of the intensity of the light may occur (Fig 7.1).

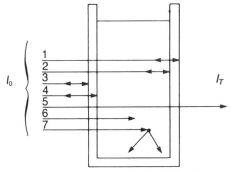

Fig. 7.1 The reduction of the intensity of light by reflection at cell faces (rays 1–4), absorption (ray 6) and scattering by particles (ray 7)

This is due to:

(a) reflections at the inner and outer surfaces of the cell
(b) scatter by particles in the solution
(c) absorption of light by molecules in the solution.

The reflections at the cell surfaces can be compensated by a reference cell containing the solvent only, and scatter may be eliminated by filtration of the solution. The intensity of light absorbed is then given by

$$I_{\text{absorbed}} = I_0 - I_T$$

where I_0 is the original intensity incident on the cell and I_T is the reduced intensity transmitted from the cell.

The **transmittance** (T) is the ratio I_T/I_0 and the % **transmittance** ($\%T$) is given by

$$\%T = \frac{100 I_T}{I_0}$$

In 1760, Lambert investigated the relationship between I_0 and I_T for various thicknesses of substance and found that the rate of decrease in the intensity of light with the thickness, b, of the medium is proportional to the intensity of incident light. Expressed mathematically

$$-\frac{dI}{db} \propto I \text{ or } -\frac{dI}{db} = k' I$$

where k' is a proportionality constant. Therefore,

$$-\frac{db}{dI} = \frac{1}{k'I}$$

Integrating

$$-b = \frac{1}{k'} \ln I_T + C$$

where I_T is the intensity transmitted at thickness b.

When b = 0,

$$C = -\frac{1}{k'} \ln I_0$$

$$\therefore \quad -b = \frac{1}{k'} \ln I_T - \frac{1}{k'} \ln I_0$$

$$\therefore \quad \ln \frac{I_0}{I_T} = k'b$$

On conversion to a common logarithm, the expression becomes

$$\log \frac{I_0}{I_T} = \frac{k'b}{2.303}$$

The quantity $\log I_0/I_T$ is called **absorbance** (A) and is equal to the reciprocal of the common logarithm of transmittance.

$$A = \log_{10}\frac{I_0}{I_T} = \log_{10}\left(\frac{1}{T}\right) = -\log T = 2 -\log(\%T)$$

Older terms for absorbance such as **extinction, optical density** and **absorbancy** are now obsolete.

Lambert's Law is defined as follows: the intensity of a beam of parallel monochromatic radiation decreases exponentially as it passes through a medium of homogeneous thickness. More simply it is stated that the absorbance is proportional to the thickness (pathlength) of the solution.

Beer showed in 1852 that a similar relationship exists between the absorbance and the concentration.

$$\log\frac{I_0}{I_T} = \frac{k''c}{2.303}$$

where k'' is a proportionality constant and c is the concentration.

Beer's Law is defined as follows: the intensity of a beam of parallel monochromatic radiation decreases exponentially with the number of absorbing molecules. More simply it is stated that the absorbance is proportional to the concentration.

A combination of the two laws yields the Beer–Lambert Law:

$$A = \log\frac{I_0}{I_T} = abc$$

in which the proportionality constants $k'/2.303$ and $k''/2.303$ are combined as a single constant called the **absorptivity** (a).

The name and value of a depend on the units of concentration. When c is in moles per litre, the constant is called **molar absorptivity** (formerly the **molar extinction coefficient**) and has the symbol ϵ (the Greek letter epsilon). The equation therefore takes the form

$$A = \epsilon bc$$

The molar absorptivity at a specified wavelength of a substance in solution is the absorbance at that wavelength of a $1\,mol\,l^{-1}$ solution in a $1\,cm$ cell. The units of ϵ are therefore $1\,mol^{-1}\,cm^{-1}$. Expressing the absorptivity in terms of a $1\,mol\,l^{-1}$ solution facilitates the comparison of the light-absorbing abilities of compounds with widely differing molecular weights. Substances that have ϵ values less than 100 are weakly absorbing; those with ϵ values above 10000 are intensely absorbing. Many absorbing drugs have an ϵ value at their wavelength of maximum absorption of $10^{3.5}$–$10^{4.5}$.

Another form of the Beer–Lambert proportionality constant is the **specific absorbance**, which is the absorbance of a specified concentration in a cell of specified pathlength. The most common form in pharmaceutical analysis is the $A(1\%, 1\,cm)$, which is the absorbance of a $1\,g/100\,ml$ (1% w/v) solution in a $1\,cm$ cell. The Beer–Lambert equation therefore takes the form

$$A = A_{1\,cm}^{1\%}bc$$

where c is in g/100 ml and b is in cm. The units of $A(1\%, 1\,\text{cm})$ are $\text{dl g}^{-1}\,\text{cm}^{-1}$. Occasionally, the concentration of liquids in solution is given as % v/v (e.g. in the British Pharmacopoeial assay of methyl salicylate and diethyl phthalate in Surgical Spirit) in which case the specific absorbance is the absorbance of a 1 ml/100 ml solution in a 1 cm cell (see Experiment 12).

A simple easily derived equation allows interconversion of ϵ and $A(1\%, 1\,\text{cm})$ values

$$\epsilon = \frac{A_{1\,\text{cm}}^{1\%} \times \text{molecular weight}}{10}$$

The majority of applications in which spectrophotometric measurements are made rely on the compliance of the absorbing substance in solution with the Beer–Lambert Law at the wavelength of measurement. The absorbance of most substances when correctly measured in a calibrated spectrophotometer (p.325) is normally found to have a proportional relationship with the concentration and pathlength of the solution. Nevertheless, it is essential in the development of new spectrophotometric procedures to confirm that Beer's Law in particular obtains over the range in which the absorbance of the sample is found. A number of instrumental or chemical effects (p.312) may be responsible for deviations from the Beer–Lambert Law.

Quantitative spectrophotometric assay of medicinal substances

The assay of an absorbing substance may be quickly carried out by preparing a solution in a transparent solvent and measuring its absorbance at a suitable wavelength. The wavelength normally selected is a wavelength of maximum absorption (λ_{max}) where small errors in setting the wavelength scale have little effect on the measured absorbance. Ideally, the concentration should be adjusted to give an absorbance of approximately 0.9, around which the accuracy and precision of the measurement are optimal (p.309). The preferred method is to read the absorbance from the instrument display under non-scanning conditions, i.e. with the monochromator set at the analytical wavelength. Alternatively, the absorbance may be read from a recording of the spectrum obtained by using a recording double-beam spectrophotometer. The latter procedure is particularly useful for qualitative purposes (pp.315–325) and in certain assays in which absorbances at more than one wavelength are required. The concentration of the absorbing substance is then calculated from the measured absorbance using one of three principal procedures.

Use of a standard absorptivity value

This procedure is adopted by official compendia, e.g. British Pharmacopoeia, for stable substances such as *Methyltestosterone* that have reasonably broad absorption bands and which are practically unaffected by variation of instrumental parameters, e.g. slit width, scan speed (pp.309–312). The use of standard $A(1\%, 1\,\text{cm})$ or ϵ values avoids the need to prepare a standard solution of the reference substance in order to determine its absorptivity, and is of advantage in situations where it is difficult or expensive to obtain a sample of the reference substance.

Example 1

Calculate the concentration of methyltestosterone in an ethanolic solution of which the absorbance in a 1 cm cell at its λ_{max}, 241 nm, was found to be 0.890. The $A(1\%, 1\,\text{cm})$ in the B.P. monograph of *Methyltestosterone* is given as 540 at 241 nm.

Substituting in the appropriate form of the Beer–Lambert equation:

$$A = A_{1\,\text{cm}}^{1\%} \, bc$$

gives

$$0.890 = 540 \times 1 \times c$$
$$\therefore \quad c = 0.00165 \, \text{g}/100\,\text{ml}$$

Example 2

Calculate the concentration in $\mu\text{g ml}^{-1}$ of a solution of tryptophan (molecular weight 204.2) in 0.1M hydrochloric acid, giving an absorbance at its λ_{max}, 277 nm, of 0.613 in a 4 cm cell. The molar absorptivity at 277 nm is 5432.

Substituting in the appropriate form of the Beer–Lambert equation:

$$A = \epsilon bc$$

gives

$$0.613 = 5432 \times 4 \times c$$
$$\therefore \quad c = 2.82 \times 10^{-5} \, \text{mol} \, \ell^{-1}$$
$$= 2.82 \times 10^{-5} \times 204.2 \, \text{g} \, \ell^{-1}$$
$$= 0.00576 \, \text{g} \, \ell^{-1}$$
$$= 5.76 \, \mu\text{g ml}^{-1}$$

Use of a calibration graph

In this procedure the absorbances of a number (typically 4–6) of standard solutions of the reference substance at concentrations encompassing the sample concentrations are measured and a calibration graph is constructed. The concentration of the analyte in the sample solution is read from the graph as the concentration corresponding to the absorbance of the solution. Calibration data are essential if the absorbance has a non-linear relationship with concentration, if it is necessary to confirm the proportionality of absorbance as a function of concentration, or if the absorbance or linearity is dependent on the assay conditions. In certain visible spectrophotometric assays of colourless substances, based upon conversion to coloured derivatives by heating the substance with one or more reagents, slight variation of assay conditions, e.g. pH, temperature and time of heating, may give rise to a significant variation of absorbance, and experimentally derived calibration data are required for each set of samples.

Statistical treatment of the calibration data, facilitated by micro-computers or pre-programmable calculators, provides a more elegant and accurate determination of the relationship between absorbance and concentration than manually constructed graphs. If the absorbance values and concentrations bear a linear relationship, the regression line $y = \alpha + \beta x$ may be estimated by the **method of least squares**.

$$\alpha = \frac{(\Sigma y)\,(\Sigma x^2) - (\Sigma x)(\Sigma xy)}{N\Sigma x^2 - (\Sigma x)^2}$$

$$\beta = \frac{N\Sigma xy - (\Sigma x)(\Sigma y)}{N\Sigma x^2 - (\Sigma x)^2}$$

where y is the absorbance value at concentration x and N is the number of pairs of values.

Example 3

The absorbance values at 250 nm of five standard solutions, a blank solution and a sample solution of a drug are given in Table 7.1. Calculate, using linear regression analysis, the line of best fit and the concentration of the sample.

Table 7.1 Absorbance Values of Five Standard Solutions, a Blank Solution and a Sample Solution of a Drug

Concentration (μg ml^{-1}) x	A_{250} y
0	0.002
10	0.168
20	0.329
30	0.508
40	0.660
50	0.846
Sample	0.611

Calculation of the slope (β) and intercept (α) using the equations above gives

$$y = 0.01679x - 0.0008$$
$$\therefore \quad \text{Concentration of the sample} = 36.5 \mu\text{g ml}^{-1}$$

Note The data may be further evaluated statistically to confirm that a linear relationship between x and y exists and to provide confidence limits for the slope, intercept and estimated concentration of the sample. For this the reader is referred to standard texts on statistical analysis (e.g. Colquhoun, 1971).

Single-or double-point standardisation

The single-point procedure involves the measurement of the absorbance of a sample solution and of a standard solution of the reference substance. The standard and sample solutions are prepared in a similar manner; ideally, the concentration of the standard solution should be close to that of the sample solution. The concentration of the substance in the sample is calculated from the proportional relationship that exists between absorbance and concentration.

$$c_{\text{test}} = \frac{A_{\text{test}} \times c_{\text{std}}}{A_{\text{std}}}$$

where c_{test} and c_{std} are the concentrations in the sample and standard solutions respectively, and A_{test} and A_{std} are the absorbances of the sample and standard solutions respectively. Since sample and standard solutions are measured under identical conditions, this procedure is the preferred method of assay of substances that obey Beer's Law and for which a reference standard of adequate purity is available. It is the procedure adopted in many spectrophotometric assays of the British Pharmacopoeia and for the majority of spectrophotometric assays of the United States Pharmacopeia.

Occasionally a linear but non-proportional relationship between concentration and absorbance occurs, which is indicated by a significant positive or negative intercept in a Beer's Law plot. A 'two-point bracketing' standardisation is therefore required to determine the concentration of the sample solutions. The concentration of one of the standard solutions is greater than that of the sample while the other standard solution has a lower concentration than the sample. The concentration of the substance in the sample solution is given by the equation (the derivation of which can be attempted as an exercise):

$$c_{\text{test}} = \frac{(A_{\text{test}} - A_{\text{std}_1})(c_{\text{std}_1} - c_{\text{std}_2}) + c_{\text{std}_1}(A_{\text{std}_1} - A_{\text{std}_2})}{A_{\text{std}_1} - A_{\text{std}_2}}$$

where the subscripts std_1 and std_2 refer to the more concentrated standard and less concentrated standard respectively.

Assay of substances in multicomponent samples

The spectrophotometric assay of drugs rarely involves the measurement of absorbance of samples containing only one absorbing component. The pharmaceutical analyst frequently encounters the situation where the concentration of one or more substances is required in samples known to contain other absorbing substances which potentially interfere in the assay. If the recipe of the sample formulation is available to the analyst, the identity and concentration of the interferents are known and the extent of interference in the assay may be determined. Alternatively, interference which is difficult to quantify may arise in the analysis of formulations from manufacturing impurities, decomposition products and formulation excipients. Unwanted absorption from these sources is termed **irrelevant absorption** and, if not removed, imparts a systematic error to the assay of the drug in the sample.

A number of modifications to the simple spectrophotometric procedure described above for single-component samples is available to the analyst, which may eliminate certain sources of interference and permit the accurate determination of one or all of the absorbing components. Each modification of the basic procedure may be applied if certain criteria are satisfied. The correct choice of procedure for a particular analytical problem provides the analyst with an opportunity to demonstrate his/her analytical expertise.

The basis of all the spectrophotometric techniques for multicomponent samples is the property that at all wavelengths:

(a) the absorbance of a solution is the sum of absorbances of the individual components; or
(b) the measured absorbance is the difference between the total absorbance of the solution in the sample cell and that of the solution in the reference (blank) cell.

Assay as a single-component sample

The concentration of a component in a sample which contains other absorbing substances may be determined by a simple spectrophotometric measurement of absorbance as described above, provided that the other components have a sufficiently small absorbance at the wavelength of measurement. This condition is satisfied if the concentration of the interfering substances, their absorptivity or the pathlength of the solution are sufficiently small that their product (i.e. the absorbance) can be ignored. A systematic error of less than 1% would normally be considered to be acceptable. For example, if the contribution to a total absorbance of 1.00 from the interferents is less than 0.01, and if there is no chemical interaction between the components, the sample may be analysed for its principal absorbing component by a simple direct measurement of absorbance at its λ_{max}. An example of this approach is the assay of paracetamol in *Paediatric Paracetamol Elixir*. At the large overall dilution (approximately 3250 times) of the sample the absorbance of the other ultraviolet-absorbing components is negligible.

Assay using absorbance corrected for interference

If the identity, concentration and absorptivity of the absorbing interferents are known, it is possible to calculate their contribution to the total absorbance of a mixture. The concentration of the absorbing component of interest is then calculated from the corrected absorbance (total absorbance minus the absorbance of the interfering substances) in the usual way.

Example 4

The λ_{max} of ephedrine hydrochloride and chlorocresol are 257 nm and 279 nm respectively and the $A(1\%,1 \text{ cm})$ values in 0.1M hydrochloric acid solution are:

ephedrine hydrochloride at 257 nm	=	9.0
ephedrine hydrochloride at 279 nm	=	0
chlorocresol at 257 nm	=	20.0
chlorocresol at 279 nm	=	105.0

Calculate the concentrations of ephedrine hydrochloride and chlorocresol in a batch of *Ephedrine Hydrochloride Injection*, diluted 1 to 25 with *water*, giving the following absorbance values in 1 cm cells.

$$A_{279} = 0.424 \qquad A_{257} = 0.972$$

(a) Since ephedrine does not absorb at 279 nm, calculate the concentration of chlorocresol from the A_{279} of the diluted injection.

$$0.424 = 105 \times 1 \times c$$
$$\therefore \quad c = 0.00404 \text{ g/100 ml}$$
$$\therefore \quad \text{concentration of chlorocresol in the injection}$$
$$= 0.00404 \times 25$$
$$= 0.1010 \text{ g/100 ml}$$
$$= 1.010 \text{ mg/ml}$$

(b) Calculate the absorbance of chlorocresol at 257 nm in the diluted injection.

$$A = 20 \times 1 \times 0.00404$$
$$= 0.081$$

(c) Calculate the concentration of ephedrine hydrochloride from the corrected absorbance at 257 nm

$$\text{Corrected absorbance at 257 nm} = 0.972 - 0.081$$
$$= 0.891$$
$$0.891 = 9.0 \times 1 \times c$$
$$\therefore \quad c = 0.0990 \text{ g/100 ml}$$
$$\therefore \quad \text{concentration of ephedrine hydrochloride in the injection}$$
$$= 0.0990 \times 25$$
$$= 2.475 \text{ g/100 ml}$$
$$= 24.75 \text{ mg/ml}$$

Assay after solvent extraction of the sample

If the interference from other absorbing substances is large or if its contribution to the total absorbance cannot be calculated, it may be possible to separate the absorbing interferents from the analyte by solvent extraction procedures. These are particularly appropriate for acidic or basic drugs whose state of ionisation determines their solvent partitioning behaviour (Part 1, Chapter 9). The judicious choice of pH of the aqueous medium and of immiscible solvent may effect the complete separation of the interferents from the analyte, the concentration of which may be obtained by a simple measurement of absorbance of the extract containing the analyte. An example of this type of assay is the B.P. assay of caffeine in *Aspirin and Caffeine Tablets*.

Simultaneous equations method

If a sample contains two absorbing drugs (X and Y) each of which absorbs at the λ_{max} of the other (Fig.7.2, λ_1 and λ_2), it may be possible to determine both drugs by the technique of simultaneous equations (Vierodt's method) provided that certain criteria apply (see later).

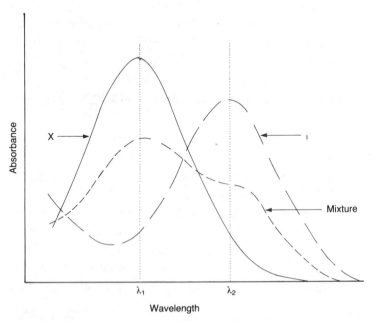

Fig. 7.2 The individual absorption spectra of substances X and Y, showing the wavelengths for the assay of X and Y in admixture by the method of simultaneous equations

The information required is:
(a) the absorptivities of X at λ_1 and λ_2, a_{X_1} and a_{X_2} respectively
(b) the absorptivities of Y at λ_1 and λ_2, a_{Y_1} and a_{Y_2} respectively
(c) the absorbances of the diluted sample at λ_1 and λ_2, A_1 and A_2 respectively.

Let c_X and c_Y be the concentrations of X and Y respectively in the diluted sample.

Two equations are constructed based upon the fact that at λ_1 and λ_2 the absorbance of the mixture is the sum of the individual absorbances of X and Y.

At λ_1

$$A_1 = a_{X_1}bc_X + a_{Y_1}bc_Y \tag{1}$$

At λ_2

$$A_2 = a_{X_2}bc_X + a_{Y_2}bc_Y \tag{2}$$

For measurements in 1 cm cells, $b = 1$.
Rearrange eq.(2).

$$c_Y = \frac{A_2 - a_{X_2}c_X}{a_{Y_2}}$$

Substituting for c_Y in eq.(1) and rearranging gives

$$c_X = \frac{A_2\,a_{Y_1} - A_1\,a_{Y_2}}{a_{X_2}\,a_{Y_1} - a_{X_1}\,a_{Y_2}} \tag{3}$$

and

$$c_Y = \frac{A_1\,a_{X_2} - A_2\,a_{X_1}}{a_{X_2}\,a_{Y_1} - a_{X_1}\,a_{Y_2}} \tag{4}$$

As an exercise you should derive modified equations containing a symbol (b) for pathlength, for application in situations where A_1 and A_2 are measured in cells other than 1 cm pathlength.

Criteria for obtaining maximum precision, based upon absorbance ratios, have been suggested (Glenn, 1960) that place limits on the relative concentrations of the components of the mixture. The criteria are that the ratios

$$\frac{A_2/A_1}{a_{X_2}/a_{X_1}} \quad \text{and} \quad \frac{a_{Y_2}/a_{Y_1}}{A_2/A_1}$$

should lie outside the range 0.1–2.0 for the precise determination of Y and X respectively. These criteria are satisfied only when the λ_{max} of the two components are reasonably dissimilar. An additional criterion is that the two components do not interact chemically, thereby negating the initial assumption that the total absorbance is the sum of the individual absorbances. The additivity of the absorbances should always be confirmed in the development of a new application of this technique. The British Pharmacopoeia assay of quinine-related alkaloids and cinchonine-related alkaloids in *Cinchona Bark* is based upon this technique.

Absorbance ratio method

The absorbance ratio method is a modification of the simultaneous equations procedure. It depends on the property that, for a substance which obeys Beer's Law at all wavelengths, the ratio of absorbances at any two wavelengths is a constant value independent of concentration or pathlength. For example, two different dilutions of the same substance (Fig.7.3) give the same absorbance ratio A_1/A_2, 2.0. In the USP, this ratio is referred to as a **Q value**. The British Pharmacopoeia also uses a ratio of absorbances at specified wavelengths in certain confirmatory tests of identity. For example, *Cyanocobalamin* exhibits three λ_{max}, at 278 nm, 361 nm and 550 nm. The A_{361}/A_{550} is required to be 3.30 ± 0.15 and the A_{361}/A_{278} to be 1.79 ± 0.09.

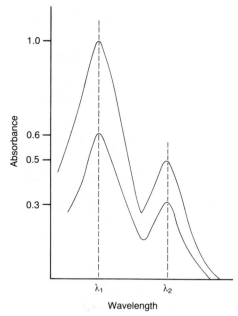

Fig. 7.3 Absorption spectra of two different concentrations of a substance

In the quantitative assay of two components in admixture by the absorbance ratio method, absorbances are measured at two wavelengths (Fig.7.4) one being the λ_{max} of one of the components (λ_2) and the other being a wavelength of equal absorptivity of the two components (λ_1), i.e., an iso-absorptive point (Pernarowski *et al*, 1961). Two equations are constructed as described above for the method of simultaneous

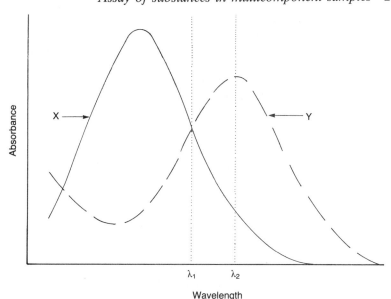

Wavelength

Fig. 7.4 Wavelengths for the assay of substances X and Y in admixture by the method of absorbance ratios

equations (eq.(1) and eq.(2)). Their treatment is somewhat different, however, and uses the relationship $a_X = a_{Y_1}$ at λ_1. Assume $b = 1$ cm.

$$A_1 = a_{X_1}c_X + a_{X_1}C_Y \tag{5}$$

$$\frac{A_2}{A_1} = \frac{a_{X_2}c_X + a_{Y_2}c_Y}{a_{X_1}c_X + a_{X_1}c_Y}$$

Divide each term by $c_X + c_Y$ and let $F_X = c_X/(c_X + c_Y)$ and $F_Y = c_Y/(c_X + c_Y)$ i.e. F_X and F_Y are the fractions of X and Y respectively in the mixture:

$$\frac{A_2}{A_1} = \frac{a_{X_2}F_X + a_{Y_2}F_Y}{A_{X_1}F_X + a_{X_1}F_Y}$$

But $F_Y = 1 - F_X$,

$$\frac{A_2}{A_1} = \frac{F_Xa_{X_2} - F_Xa_{Y_2} + a_{Y_2}}{a_{X_1}}$$

$$\therefore \quad \frac{A_2}{A_1} = \frac{F_Xa_{X_2}}{a_{X_1}} - \frac{F_Xa_{Y_2}}{a_{Y_1}} + \frac{a_{Y_2}}{a_{Y_1}}$$

Let $Q_X = \dfrac{a_{X_2}}{a_{X_1}}$, $Q_Y = \dfrac{a_{Y_2}}{a_{Y_1}}$ and $Q_M = \dfrac{A_2}{A_1}$

$$\therefore \quad Q_M = F_X(Q_X - Q_Y) + Q_Y$$

$$F_X = \frac{Q_M - Q_Y}{Q_X - Q_Y} \tag{6}$$

Equation 6 gives the fraction, rather than the concentration of X (and consequently of Y) in the mixture in terms of absorbance ratios. As these are independent of concentration, only approximate, rather than accurate, dilutions of X, Y and the sample mixture are required to determine Q_X, Q_Y and Q_M respectively.

If the absolute concentrations of X and Y are required, rearrange eq. (5):

$$A_1 = a_{X_1}(c_X + c_Y)$$

$$\therefore \quad c_X + c_Y = \frac{A_1}{a_{X_1}}$$

From eq. (6):

$$\frac{c_X}{f2c_X + c_Y} = \frac{Q_M - Q_Y}{Q_X - Q_Y b7}$$

$$\therefore \quad \frac{c_X}{A_1/a_{X_1}} = \frac{Q_M - Q_Y}{Q_X - Q_Y}$$

$$\therefore \quad c_X = \frac{Q_M - Q_Y}{Q_X - Q_Y} \cdot \frac{A_1}{a_{X_1}} \tag{7}$$

Equation 7 gives the concentration of X in terms of absorbance ratios, the absorbance of the mixture and the absorptivity of the compounds at the iso-absorptive wavelength. Accurate dilutions of the sample solution and of the standard solutions of X and Y are necessary for the accurate measurement of A_1 and a_{X_1} respectively.

This method has been used for the assay of trimethoprim and sulphamethoxazole in Co-trimoxazole Tablets (Ghanem *et al.*, 1979).

Geometric correction method

A number of mathematical correction procedures have been developed which reduce or eliminate the background irrelevant absorption that may be present in samples of biological origin. The simplest of these procedures is the three-point geometric procedure, which may be applied if the irrelevant absorption is linear at the three wavelengths selected. Consider an absorption spectrum (Fig. 7.5(a)) comprising the spectrum of the analyte (Fig. 7.5(b)) and that of the background absorption (Fig. 7.5(c)). If the wavelengths λ_1, λ_2, and λ_3 are selected so that the background absorbances B_1, B_2 and B_3 are linear, then the corrected absorbance, D, of the drug may be calculated from the three absorbances A_1, A_2 and A_3 of the sample solution at λ_1, λ_2 and λ_3 respectively, as follows.

Let vD and wD be the absorbance of the drug alone in the sample solution at λ_1 and λ_3 respectively, i.e. v and w are the absorbance ratios (p.286) vD/D and wD/D respectively.

$$\therefore \quad B_1 = A_1 - vD, \quad B_2 = A_2 - D \quad \text{and} \quad B_3 = A_3 - wD$$

Let y and z be the wavelength intervals $(\lambda_2 - \lambda_1)$ and $(\lambda_3 - \lambda_2)$ respectively.

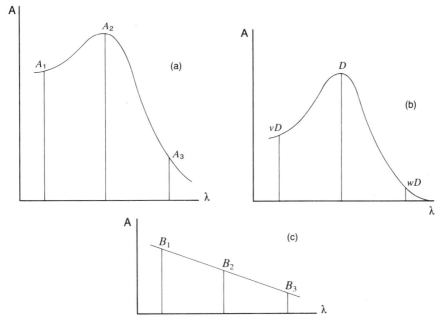

Fig. 7.5(a) The absorption spectrum of a solution of a drug in the presence of linear irrelevant absorption. (b) The individual spectrum of the drug. (c) The individual spectrum of the linear irrelevant absorption

Then

$$\frac{B_1 - B_3}{B_2 - B_3} = \frac{y + z}{z} \text{ (similar triangles)}$$

$$\therefore \quad zB_1 = (y + z)B_2 - yB_3$$

$$\therefore \quad z(A_1 - vD) = (y + z)(A_2 - D) - y(A_3 - vD)$$

Rearranging:

$$D = \frac{y(A_2 - A_3) + z(A_2 - A_1)}{y(1 - w) + z(1 - v)}$$

This is a general equation which may be applied in any situation where A_1, A_2 and A_3 of the sample, the wavelength intervals y and z and the absorbance ratios v and w are known. The values of v and w are determined experimentally using a solution of the drug only. The concentration of the drug is calculated from the corrected absorbance D using any of the normal procedures (p.278–281).

Two special circumstances allow further simplification of the general equation. Firstly, when the wavelengths λ_1, λ_2 and λ_3 are selected to give $v = w = r$, the general equation simplifies to:

$$D = \frac{y(A_2 - A_3) + z(A_2 - A_1)}{(y + z)(1 - r)}$$

The most famous assay of this type is the Morton Stubbs assay of vitamin A in fish liver oils and concentrates. The three wavelengths were selected to give $v = w = 6/7$, conditions given by $\lambda_1 = 313 \, \text{nm}$, $\lambda_2 \, (\lambda_{max}) = 328 \, \text{nm}$ and $\lambda_3 = 338.5 \, \text{nm}$ for solutions of vitamin A acetate in cyclohexane solution. With these values, $y = 15$, $z = 10.5$, $r = 6/7$. The corrected absorbance at 328 nm $(A_{328(corr)})$ is calculated from the simplified equation:

$$A_{328(corr)} = \frac{15(A_{328} - A_{338.5}) + 10.5(A_{328} - A_{313})}{(15 + 10.5)(1 - \frac{6}{7})}$$

$$= 7A_{328} - 4.117A_{338.5} - 2.882A_{313}$$

The second case in which the general equation may be further simplified is when equally spaced wavelengths are selected so that $y = z$. The general equation then simplifies to

$$D = \frac{2A_2 - A_1 - A_3}{2 - v - w}$$

A modified Morton Stubbs equation with three equidistant wavelengths at 316 nm, 328 nm and 340 nm is used to calculate the corrected absorbance at 328 nm of the vitamin A in fish liver oils.

$$A_{328(corr)} = 3.52(2A_{328} - A_{316} - A_{340})$$

The factor 3.52 is derived from $1/(2 - v - w)$ where the absorbance ratios v and w of vitamin A are 0.907 and 0.811 respectively.

It should be noted that the three-point correction procedures are simply algebraic calculations of what the **baseline technique** in infrared spectrophotometry (p.391) does graphically.

Orthogonal polynomial method

The technique of orthogonal polynomials (Glenn, 1963) is another mathematical correction procedure which involves more complex calculations than the three-point correction procedure. The basis of the method is that an absorption spectrum may be represented in terms of orthogonal functions as follows:

$$A(\lambda) = p_0 P_0(\lambda) + p_1 P_1(\lambda) + p_2 P_2(\lambda) \ldots p_n P_n(\lambda)$$

where A denotes the absorbance at wavelength λ belonging to a set of $n + 1$ equally spaced wavelengths at which the orthogonal polynomials, $P_0(\lambda)$, $P_1(\lambda)$, $P_2(\lambda) \ldots P_n(\lambda)$ are each defined. These polynomials represent a series of fundamental shapes (Fig. 7.6) and the contribution that each shape, e.g. P_2 makes to the absorption spectrum is defined by the appropriate coefficient, e.g. p_2 for P_2. The coefficients are proportional to the concentration of the absorbing analyte, and a modified Beer–Lambert equation may be constructed:

$$p_j = \alpha_j bc$$

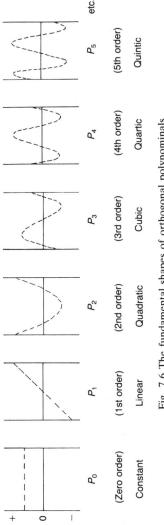

Fig. 7.6 The fundamental shapes of orthogonal polynominals

where p_j is the coefficient of the polynomial and α_j is the proportionality constant analogous to absorptivity. For example, when b is 1 cm and the concentration of the analyte, (c), is in g/dl, α_j is the coefficient of the p_j of the A (1%, 1 cm) of the analyte.

When irrelevant absorption is present in a sample solution, the calculated coefficient (p_j) comprises the coefficients of the analyte and of the irrelevant absorption (Z). Thus

$$p_j = \alpha_j c + p_j(Z)$$

With the correct choice of polynomial, number of wavelengths and the wavelength interval, the contribution from the irrelevant absorption may be negligible. In general, a quadratic (P_2) polynomial eliminates linear or almost linear irrelevant absorption and a cubic (P_3) polynomial eliminates parabolic irrelevant absorption.

Example 5

The individual contributions of an absorbing drug and irrelevant absorption (I.A.) to the total absorbance of a mixture at eight equally spaced wavelengths $(\lambda_1 \ldots \lambda_8)$ and the coefficients of the cubic polynomial for each set of data are given in Table 7.2.

Table 7.2

	λ_1	λ_2	λ_3	λ_4	λ_5	λ_6	λ_7	λ_8	p_3
Drug	0.80	0.65	0.60	0.85	1.01	0.90	0.40	0.17	−5.74
I.A.	0.38	0.31	0.25	0.20	0.16	0.13	0.11	0.10	0
Drug + I.A.	1.18	0.96	0.85	1.05	1.17	1.03	0.51	0.27	−5.74

The segment of the spectrum of the drug between λ_1 and λ_8 shows a minimum around λ_3 and a maximum around λ_5. Its shape may therefore be represented by a cubic (P_3) polynomial. The irrelevant absorption is a simple parabolic curve which does not contain a cubic contribution. The coefficient (p_3) of the cubic polynomial for each set of eight absorbances $(A_1 \ldots A_8)$ is calculated from:

$$p_3 = [(-7)A_1 + (+5)A_2 + (+7)A_3 + (+3)A_4 \\ + (-3)A_5 + (-7)A_6 + (-5)A_7 + (+7)A_8]$$

where the factors are those of an **eight-point cubic polynomial** obtained from standard texts of numerical analysis (e.g. Fischer and Yates, 1953). The contribution of the irrelevant absorption to the coefficient of the polynomial of the sample is eliminated (Table 7.2) by the selection of these parameters, and the concentration of the drug in the sample may be calculated, with reference to a standard solution of the drug, from the proportional relationship that exists between the calculated p_3 value and concentration.

The accuracy of the orthogonal functions procedure depends on the correct choice of the polynomial order and the set of wavelengths. Usually, quadratic or cubic polynomials are selected depending on the shape of the absorption spectra of the drug and the irrelevant absorption. The set of wavelengths is defined by the number of wavelengths, the interval and the mean wavelength of the set (λ_m). Approximately linear irrelevant absorption is normally eliminated using six to eight wavelengths, although many more, up to 20, wavelengths may be required if the irrelevant absorption contains high-frequency components. The wavelength interval and λ_m are best obtained from a **convoluted absorption curve**. This is a plot of the coefficient (p_j) for a specified order of polynomial, a specified number of wavelengths and a specified wavelength interval (on the ordinate) against the λ_m of the set of wavelengths. The optimum set of wavelengths corresponds with a maximum or minimum in the convoluted curve of the analyte and with a coefficient of zero in the convoluted curve of the irrelevant absorption. In favourable circumstances the concentration of an absorbing drug in admixture with another may be calculated if the correct choice of polynomial parameters is made, thereby eliminating the contribution of one drug from the polynomial of the mixture. For example, the selective assay of phenobarbitone, combined with phenytoin in a capsule formulation, using a six-point quadratic polynomial, has been reported (Amer *et al*, 1977).

The determination of the optimum set of wavelengths is readily accomplished with the aid of a microcomputer. A suitable exercise is to write a program to compute and plot the data for a convoluted spectrum.

Difference spectrophotometry

The selectivity and accuracy of spectrophotometric analysis of samples containing absorbing interferents may be markedly improved by the technique of difference spectrophotometry. The essential feature of a difference spectrophotometric assay is that the measured value is the difference absorbance (ΔA) between two equimolar solutions of the analyte in different chemical forms which exhibit different spectral characteristics.

The criteria for applying difference spectrophotometry to the assay of a substance in the presence of other absorbing substances are that:

(a) reproducible changes may be induced in the spectrum of the analyte by the addition of one or more reagents

(b) the absorbance of the interfering substances is not altered by the reagents.

The simplest and most commonly employed technique for altering the spectral properties of the analyte is the adjustment of the pH by means of aqueous solutions of acid, alkali or buffers. The ultraviolet–visible absorption spectra of many substances containing ionisable functional groups, e.g. phenols, aromatic carboxylic acids and amines, are dependent on the state of ionisation of the functional groups and consequently on the pH of the solution.

The absorption spectra of equimolar solutions of *Phenylephrine*, a phenolic sympathomimetic agent, in both 0.1M hydrochloric acid (pH 1) and 0.1M sodium hydroxide (pH 13) are shown in Fig.7.7a. The ionisation of the phenolic group in alkaline solution generates an additional n (non-bonded) electron that interacts with the ring π electrons to produce a bathochromic shift of the λ_{max} from 271 nm in acidic solution to 291 nm and an increase in absorbance at the λ_{max} (hyperchromic effect). The difference absorption spectrum (Fig.7.7b) is a plot of the difference in absorbance between the solution at pH 13 and that at pH 1 against wavelength. It may be generated automatically using a double-beam recording spectrophotometer with the solution at pH 13 in the sample cell and the solution at pH 1 in the reference (blank) cell. At 257 and 278 nm both solutions have identical absorbance and consequently exhibit zero difference absorbance. Such wavelengths of equal absorptivity of the two species are called **isosbestic** or **iso-absorptive points**. Above 278 nm the alkaline solution absorbs more intensely than the acidic solution and the ΔA is therefore positive. Between 257 and 278 nm it has a negative value. The measured value in a quantitative difference spectrophotometric assay is the ΔA at any suitable wavelength measured to the baseline, e.g. ΔA_1 at λ_1 (Fig.7.7b) or amplitude between an adjacent maximum and minimum, e.g. ΔA_2 at λ_2 and λ_1.

$$\text{At } \lambda_1 \quad \Delta A = A_{alk} - A_{acid}$$

where A_{alk} and A_{acid} are the individual absorbances at λ_1 in 0.1M sodium hydroxide and 0.1M hydrochloric acid solution respectively. If the individual absorbances, A_{alk} and A_{acid}, are proportional to the concentration of the analyte and pathlength, the ΔA also obeys the Beer–Lambert Law and a modified equation may be derived

$$\Delta A = \Delta abc$$

where Δa is the difference absorptivity (e.g. $\Delta A_{1\,cm}^{1\%}$ or $\Delta \epsilon$) of the substance at the wavelength of measurement.

If one or more other absorbing substances is present in the sample which at the analytical wavelength has identical absorbance (A_x) in the alkaline and acidic solutions, its interference in the spectrophotometric measurement is eliminated.

$$\Delta A = (A_{alk} + A_x) - (A_{acid} + A_x)$$
$$= A_{alk} - A_{acid}$$

The selectivity of the ΔA procedure depends on the correct choice of the pH values to induce the spectral change of the analyte without altering the absorbance of the interfering components of the sample. The use of 0.1M sodium hydroxide and 0.1M hydrochloric acid to induce the ΔA of the analyte is convenient and satisfactory when the irrelevant absorption arises from pH-insensitive substances. Unwanted absorption

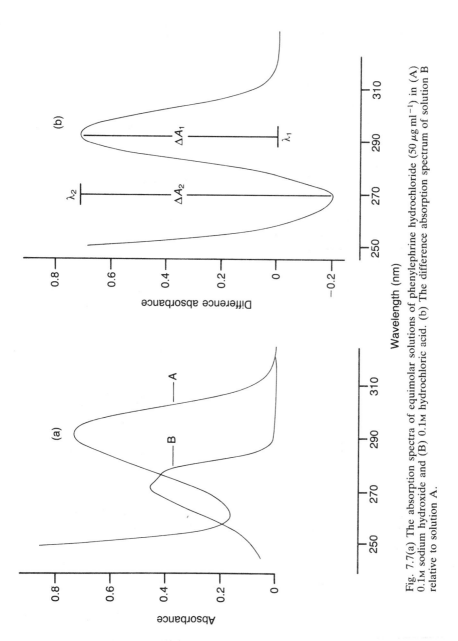

Fig. 7.7(a) The absorption spectra of equimolar solutions of phenylephrine hydrochloride (50 μg ml^{-1}) in (A) 0.1M sodium hydroxide and (B) 0.1M hydrochloric acid. (b) The difference absorption spectrum of solution B relative to solution A.

from pH-sensitive components of the sample may also be eliminated if the pK_a values of the analyte and interferents differ by more than 4. The selection of two buffers, one at a pH 2 units greater than, and the other at a pH 2 units less than, the pK_a of the analyte, ensures that the analyte is 99% in the ionised or molecular states and that the ΔA is almost maximal. Selective assays of *Chlordiazepoxide* (pK_a 4.6) and its major hydrolysis product demoxepam (pK_a 10.5) in degraded formulations of *Chlordiazepoxide*, which may also contain a minor hydrolysis product, 2-amino-5-chlorobenzophenone (pK_a about 1), have been developed, which require the measurement of the ΔA_{269} between equimolar solutions of the sample at pH 8 and pH 3 and the ΔA_{263} between equimolar solutions of the sample at pH 13 and pH 8 respectively (Davidson, 1984). pH-induced difference spectrophotometric assays have also been described for phenols (Wahbi and Farghaly, 1970), amines (Doyle and Fazzari, 1974) and barbiturates (Williams and Zak, 1959).

A substance whose spectrum is unaffected by changes of pH may be determined by a difference spectrophotometric procedure if it can be quantitatively converted by means of a suitable reagent to a chemical species that has different spectral properties to its unreacted parent substance. The ΔA between equimolar solutions of the unreacted substance and its derivative is free of interference if the irrelevant absorption is unaffected by the reagent. Phenothiazine drugs such as *Chlorpromazine* in coloured syrup formulations may be assayed in the presence of their sulphoxide decomposition products, using peracetic acid (Davidson, 1976; see Experiment 14) and steroids with a 4-en-3-one group may be assayed by measuring the A_{241} between a methanolic solution of the steroid and an equimolar solution treated with sodium borohydride, which reduces the 3-carbonyl group to the alcohol and destroys the conjugation necessary for ultraviolet absorption (Görög, 1968).

Derivative spectrophotometry

Derivative spectrophotometry involves the conversion of a normal spectrum to its first, second or higher derivative spectrum. The transformations that occur in the derivative spectra are understood by reference to a Gaussian band which represents an ideal absorption band (Fig. 7.8(a)). In the context of derivative spectrophotometry, the normal absorption spectrum is referred to as the fundamental, zeroth order or D^0 spectrum.

The first derivative (D^1) spectrum (Fig.7.8(b)) is a plot of the rate of change of absorbance with wavelength against wavelength, i.e. a plot of the slope of the fundamental spectrum against wavelength or a plot of $dA/d\lambda$ *vs* λ. At λ_2 and λ_4, the maximum positive and maximum negative slope respectively in the D^0 spectrum correspond with a maximum and a minimum respectively in the D^1 spectrum. The λ_{max} at λ_3 is a wavelength of zero slope and gives $dA/d\lambda = 0$, i.e. a cross-over point, in the D^1 spectrum.

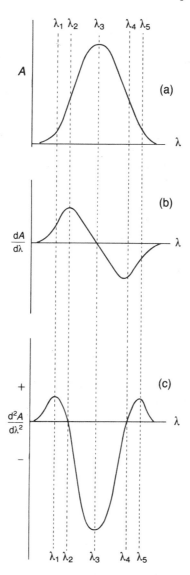

Fig. 7.8 The zeroth (a), first (b) and second (c) derivative spectra of a Gaussian band

The second derivative (D^2) spectrum (Fig.7.8(c)) is a plot of the curvature of the D^0 spectrum against wavelength or a plot of $d^2A/d\lambda^2$ *vs* λ. The maximum negative curvature at λ_3 in the D^0 spectrum gives a minimum in the D^2 spectrum, and at λ_1 and λ_5 the maximum positive curvature in the D^0 spectrum gives two small maxima called 'satellite' bands in the D^2 spectrum. At λ_2 and λ_4 the wavelengths of maximum slope and zero curvature in the D^0 spectrum correspond with cross-over points in the D^2 spectrum.

In summary, the first derivative spectrum of an absorption band is characterised by a maximum, a minimum, and a cross-over point at the λ_{max} of the absorption band. The second derivative spectrum is characterised by two satellite maxima and an inverted band of which the minimum corresponds to the λ_{max} of the fundamental band. As an exercise, you should construct third and fourth derivative spectra (i.e. plots of $d^3A/d\lambda^3$ and $d^4A/d\lambda^4$ respectively against wavelength) of the fundamental spectrum.

These spectral transformations confer two principal advantages on derivative spectrophotometry. Firstly, an even order spectrum is of narrower spectral bandwidth (p.309) than its fundamental spectrum (cf.Figs 7.8(a) and 7.8(c)). A derivative spectrum therefore shows better resolution of overlapping bands than the fundamental spectrum and may permit the accurate determination of the λ_{max} of the individual bands (Fig.7.9). Secondly, derivative spectrophotometry discriminates in favour of substances of narrow spectral bandwidth against broad bandwidth substances. This is because the derivative amplitude (D), i.e. the distance from a maximum to a minimum, is inversely proportional to the fundamental spectral bandwidth (W) raised to the power (n) of the derivative order. Thus,

$$D \ \alpha (1/W)^n$$

Consequently, substances of narrow spectral bandwidth display larger derivative amplitudes than those of broad bandwidth substances.

These advantages of derivative spectrophotometry, enhanced resolution and bandwidth discrimination, permit the selective determination of certain absorbing substances in samples in which non-specific interference may prohibit the application of simple spectrophotometric methods. For example, benzenoid drugs such as *Ephedrine Hydrochloride*, displaying fine structure of narrow spectral bandwidth in the region 240–270 nm, are both weakly absorbing ($A_{1\,cm}^{1\%}$ about 15) and formulated at a relatively low dose in solid dosage preparations (typically 1–50 mg/ unit dose). The high excipient/drug ratio and high sample weight required for the assay may introduce into simple spectrophotometric procedures serious irrelevant absorption from the formulation excipients. Second derivative spectrophotometry discriminates in favour of the narrow bands of the fine structure of the benzenoid drugs and eliminates the broad band absorption of the excipients (Davidson and Hassan, 1984). All the amplitudes in the derivative spectrum are proportional to the concentration of the analyte, provided that Beer's Law is obeyed by the fundamental spectrum. The measured value in a quantitative assay is the largest amplitude that is unaffected by the presence of other absorbing components of the sample (see Experiment 15).

The enhanced resolution and bandwidth discrimination increases with increasing derivative order. However, it is also found that the concomitant increase in electronic noise inherent in the generation of the higher order spectra, and the consequent reduction of the signal-to-noise ratio, place serious practical limitations on the higher order spectra. For

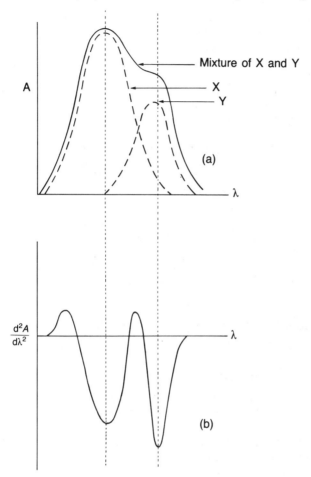

Fig. 7.9(a) The individual spectra of two components X and Y in admixture and their combined spectrum. (b) The second derivative spectrum of the mixture showing the improved resolution of the individual bands. Note the larger amplitude of component Y due to the narrower spectral bandwidth of the fundamental band

quantitative purposes, second and fourth derivative spectra are the most frequently employed derivative orders.

Instrumentation

Derivative spectra may be generated by any of three techniques. The earliest derivative spectra were obtained by modification of the optical system. Spectrophotometers with dual monochromators set a small wavelength interval ($\Delta\lambda$, typically 1–3 nm) apart, or with the facility to oscillate the wavelength over a small range, are required. In either case the photodetector generates a signal with an amplitude proportional to the slope of the spectrum over the wavelength interval. Instruments of this type are expensive and are essentially restricted to the recording of first derivative spectra only.

The second technique to generate derivative spectra is electronic differentiation of the spectrophotometer analog signal. Resistance-capacitance (RC) modules may be incorporated in series between the spectrophotometer and recorder to provide differentiation of the absorbance, not with respect to wavelength, but with respect to time, thereby producing the signal dA/dt. If the wavelength scan rate is constant $(d\lambda/dt = C)$, the derivative with respect to wavelength is given by

$$dA/d\lambda = (dA/dt)/(d\lambda/dt) = (dA/dt)(1/C)$$

Derivative spectra obtained with RC modules are highly dependent on instrumental parameters, in particular the scan speed and the time constant. It is essential, therefore, to employ a standard solution of the analyte to calibrate the measured value under the instrumental conditions selected.

The third technique is based upon microcomputer differentiation. Microcomputers incorporated into or interfaced with the spectrophotometer may be programmed to provide derivative spectra during or after the scan, to measure derivative amplitudes between specified wavelengths and to calculate concentrations and associated statistics from the measured amplitudes.

Chemical derivatisation

Indirect spectrophotometric assays are based on the conversion of the analyte by a chemical reagent to a derivative that has different spectral properties. When an excess of the reagent is used, to ensure complete conversion, the absorbance of the derivative is usually, but not always, proportional to the concentration of the analyte. The majority of indirect spectrophotometric procedures involve the conversion of the analyte to a derivative that has a longer λ_{max} and/or a higher absorptivity. Chemical derivatisation procedures may be adopted for any of several reasons:

(a) If the analyte absorbs weakly in the ultraviolet region, a more sensitive method of assay is obtained by converting the substance to a derivative with a more intensely absorbing chromophore. For example, sugars which do not absorb significantly above 220 nm can be determined spectrophotometrically by heating with anthrone in concentrated sulphuric acid and measuring the absorbance of the coloured derivative at 625 nm.

(b) The interference from irrelevant absorption may be avoided by converting the analyte to a derivative which absorbs in the visible region, where irrelevant absorption is negligible. For example, in the assay of low dosage tablet formulations of 4-en-3-one steroids, which absorb around 240 nm with $A(1\%, 1\,cm)$ values in the range 300–500, the co-extraction of absorbing components from the tablet matrix gives rise to irrelevant absorption in the tablet extract, and consequently falsely high absorbances at 240 nm are measured. The condensation of ketosteroids (e.g. *Methyltestosterone* in *Methyltestosterone Tablets*) with hydrazide reagents (p.303), or the oxidation of the α-ketol group by tetrazolium salts, produces derivatives that absorb in the visible region free of interference from irrelevant absorption.

(c) Indirect spectrophotometric procedures are also used to improve the selectivity of the assay of an ultraviolet-absorbing substance in a sample that contains other ultraviolet-absorbing components. For example, the assay of the low concentrations of adrenaline ($20 \mu g \, ml^{-1}$) in *Procaine and Adrenaline Injection* by a direct measurement of absorbance at the absorption maximum of adrenaline around 279 nm is subject to gross interference from the bactericide chlorocresol and from the high concentration of procaine hydrochloride. Only adrenaline forms the purple derivative in the presence of iron(II) and is measured colorimetrically free of interference from the other ultraviolet-absorbing components.

(d) The adoption of a visible spectrophotometric procedure, instead of an ultraviolet procedure, may be based on cost considerations. In general, single-beam manually adjusted visible spectrophotometers (sometimes called colorimeters) are much cheaper than ultraviolet–visible spectrophotometers.

The analytical literature is replete with examples of indirect spectrophotometric procedures and only a few of the important assays that are based on chemical derivatisation are now discussed.

Diazotisation and coupling of primary aromatic amines

The amine is first diazotised with an aqueous solution of nitrous acid (generated *in situ* by the reaction of hydrochloric acid and sodium nitrite) at 0–5°.

$$Ar{-}NH_2 + HNO_2 \xrightarrow{H^+} Ar{-}N^+{\equiv}N + 2H_2O$$

The colourless diazonium salt is very reactive and when treated with a suitable coupling reagent ($Ar'{-}H$), e.g. a phenol or aromatic amine, undergoes an electrophilic substitution reaction to produce an azo derivative.

$$Ar{-}N^+{\equiv}N + Ar'{-}H \rightarrow Ar{-}N{=}N{-}Ar' + H^+$$

The azo derivatives are coloured and consequently have an absorption maximum in the visible region. The λ_{max} and ϵ_{max} depend on the Ar and Ar' groups. Among the most widely used coupling reagents are 1-naphthol, 2-naphthol and *N*-(1-naphthyl)-ethane-1,2-diammonium dichloride (the Bratton–Marshall reagent) which give high absorptivities. For example, the azo derivative of diazotised sulphadiazine coupled with the Bratton–Marshall reagent is

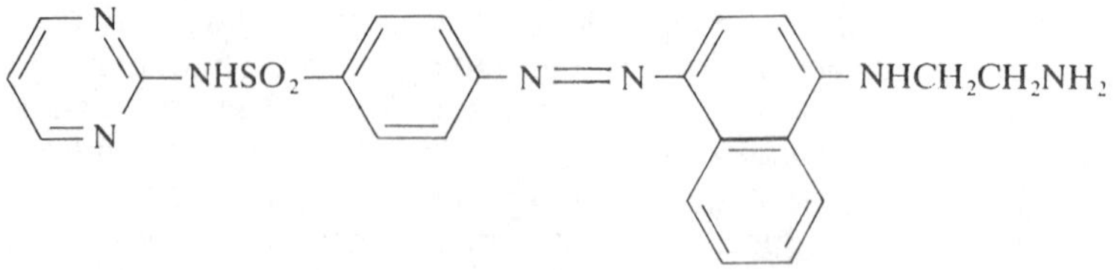

which absorbs intensely around 545 nm owing to its extensive conjugation. Sulphamic acid or ammonium sulphamate is added to the solution of the diazotised amine before the coupling stage to destroy the excess nitrous acid, which inhibits the coupling reaction.

$$HNO_2 + NH_2SO_3H \rightarrow N_2 + H_2SO_4 + H_2O$$

The sensitivity and selectivity of the procedure permit the assay of low concentrations of impurities that contain a primary aromatic amine group in the presence of the parent substance lacking the amine function. British Pharmacopoeia tests for free amine impurities in *Frusemide*, *Iothalamic Acid* and *Iodipamide Meglumine Injection* are based upon diazotisation and coupling. Substances that can be hydrolysed (e.g. Bendrofluazide, Chlorothiazide) or reduced (e.g. Chloramphenicol) to a derivative containing a primary aromatic amine group can also be assayed spectrophotometrically by diazotisation and coupling of the amine derivative.

Condensation reactions

Many colorimetric procedures are based on the rapid reaction that occurs under suitable conditions between amines and carbonyl compounds. The reactions involve the nucleophilic attack by the amine on the carbonyl carbon with the elimination of water. Substances containing a carbonyl group react with a variety of reagents containing an amino group:

$$\begin{matrix} R' \\ R'' \end{matrix}\!\!>C = O + NH_2R''' \longrightarrow \begin{matrix} R' \\ R'' \end{matrix}\!\!>C = NR''' + H_2O$$

When R‴ = Alkyl or aryl, the product is a Schiff's base
= NH_2 (hydrazine), the product is a hydrazone
= $NHCONH_2$ (semicarbazide), the product is a semicarbazone
= OH (hydroxylamine), the product is an oxime

A number of hydrazine and hydrazide reagents have been described for the colorimetric assay of ketosteroids. The selectivity of the reactions for steroids with the keto group in different positions depends on the reagent. Isoniazid (isonicotinic acid hydrazide) reacts with 4-en-3-one and 1,4-dien-3-one steroids in acidic solution to form yellow derivatives with λ_{max} around 400 nm.

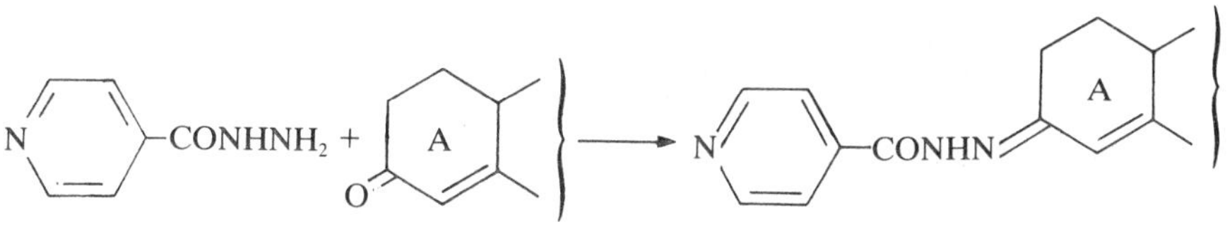

The reagent is used in the assays of *Nandralone Decanoate Injection* and *Betamethasone Sodium Phosphate Injection*. A similar reagent, 2,4-dinitrophenylhydrazine, is used in the assay of *Methyltestosterone Tablets*. The hydrazone formed on condensation of the 3-, 17-, or 20-ketosteroid with (carboxymethyl) trimethylammonium chloride hydrazide (Girard's reagent T) is water-soluble owing to the quaternary ammonium group, and lipid-soluble impurities may be removed by extraction into chloroform. Alternatively, the hydrazone itself may be extracted into dichloroethane as the ion-pair (p.304) with bromothymol blue.

Amino compounds can be assayed spectrophotometrically using a suitable carbonyl reagent. One of the most frequently employed reagents is 4-dimethylaminobenzaldehyde (Ehrlich's reagent) used, for example, in the assay of procaine hydrochloride in *Lymecycline and Procaine Injection*. The condensation product which absorbs at 454 nm is

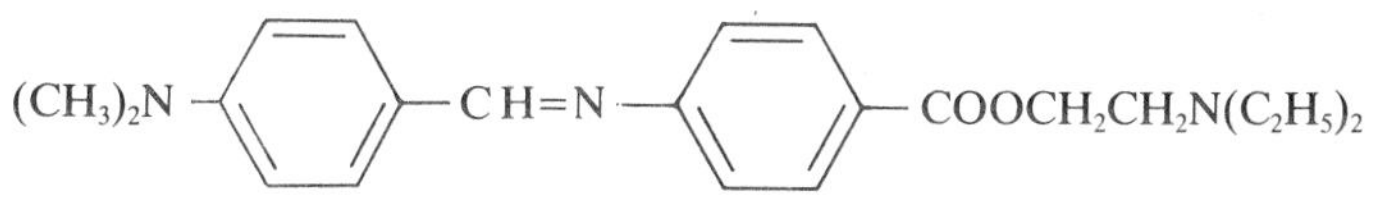

Reduction of tetrazolium salts

In the presence of a steroid with an α-ketol (21-hydroxy-20-keto) side-chain group, tetrazolium salts are reduced to their coloured formazan derivatives. Several formulations containing corticosteroids are assayed using triphenyltetrazolium chloride. The reaction is carried out in an alkaline medium (tetramethylammonium hydroxide) at 30–35° for 1–2 h and the absorbance of the red product is measured around 485 nm. The oxidation of the α-ketol group and the reduction of triphenyltetrazolium chloride to triphenylformazan are shown:

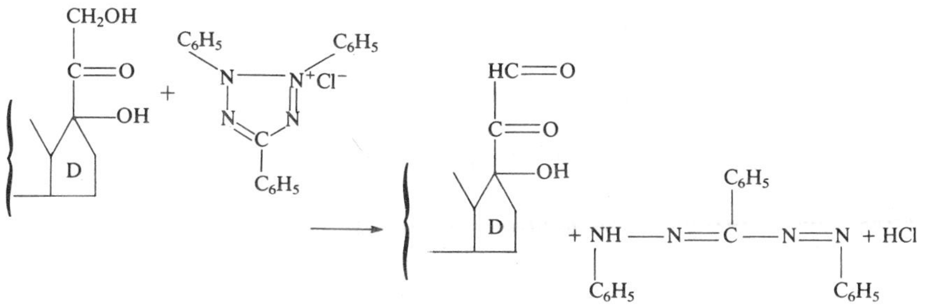

Steroids esterified in the 21-position, e.g. *Hydrocortisone Acetate*, hydrolyse in the alkaline solution to yield the free 21-hydroxysteroids and are also determined by this procedure. Precautions are taken throughout the assay against the effects of light and oxygen.

The acid-dye method

The addition of an amine in its ionised form to an ionised acidic dye, e.g. methyl orange or bromocresol purple, yields a salt (ion-pair) that may be extracted into an organic solvent such as chloroform or dichloromethane. The indicator dye is added in excess and the pH of the aqueous solution is adjusted if necessary to a value where both the amine and dye are in the ionised forms. The ion-pair is separated from the excess indicator by extraction into the organic solvent, and the absorbance is measured at the λ_{max} of the indicator in the solvent. Usually, the most intensely absorbing form of the indicator is measured, with the addition, if required, of acidified or basified ethanol. Alternatively, the absorbance of the indicator may be measured in aqueous solution after back extraction from the organic solvent.

The molar absorptivities of the ion-pairs formed between quaternary ammonium compounds and methyl orange or bromothymol blue are typically 2×10^4–4×10^4. The acid-dye method therefore provides a more sensitive technique for certain amines and quaternary ammonium compounds that absorb weakly in the ultraviolet region, e.g. hyoscine butylbromide ($\epsilon_{257} = 202$).

The correct choice of pH may permit the selective assay of a mixture of an amine and a quaternary ammonium salt. For example, in the assay of a tertiary base and a quaternary ammonium salt both substances are ionic at pH 3, and the resultant absorbance of the ion-pair extracted into the organic solvent measures the total concentration, whereas at pH 9 only the quaternary compound is ionised and forms the extractable ion-pair.

The acid-dye technique is used for the assay of formulations containing certain quaternary ammonium salts or amines, i.e., *Biperidine Lactate Injection, Clonidine Hydrochloride Injection and Tablets, Neostigmine Methylsulphate Injection*, and *Benzhexol Hydrochloride Tablets*, and the reader is referred to the monographs of these formulations for details of the procedure.

Oxidation methods

Oxidation of the side chain of weakly absorbing compounds containing a simple phenyl group produces a carbonyl derivative that has a much greater absorptivity than the parent compound. Commonly used oxidation reagents are alkaline potassium permanganate solution, acidified potassium dichromate solution, or perchlorate solution. The product formed from simple monophenyl compounds, e.g. ephedrine or propanolamine, is the corresponding benzaldehyde derivative, which exhibits intense absorption, at around 240 nm owing to the interaction of the carbonyl π electrons with the ring electrons.

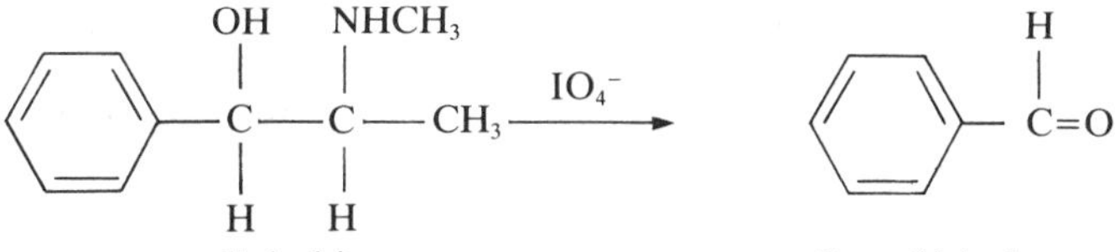

Ephedrine   Benzaldehyde

The assay of *Ephedrine Hydrochloride Elixir* involves the extraction of the benzaldehyde into cyclohexane, and measurement of the absorbance at its λ_{max}, 241 nm.

Compounds with a diphenylmethylidene $[(C_6H_5)_2C<]$ nucleus are oxidised to benzophenone, which has a λ_{max} in hexane solution at 247 nm.

Metal–ligand complexation

Many organic reagents (in this context called ligands) form complexes with metal atoms by the formation of coordinate bonds (in which both electrons are donated by the ligand) and covalent bonds (Part 1, Chapter 8). Ligands with two or more donating groups may share more than one pair of electrons with a single metal atom by coordinating to two or more positions. The chelates formed by these multidentate ligands with metal atoms are often coloured, and consequently their concentration may be determined by visible spectrophotometry. Many examples of the photometric assay of either the concentration of heavy metal ions by the addition of an excess of a chelating reagent or the concentration of organic substances by the addition of an excess of a suitable complexing metal have been described (Cheng *et al.*, 1982).

Characteristic colours which vary in hue with change of pH are given by the reaction of iron(II) ions with phenols that contain two adjacent hydroxyl groups. Thus, on mixing a solution of adrenaline with a buffered solution containing iron(II) sulphate, a purple complex (λ_{max}, 540 nm) is formed that has a maximum intensity at pH 8–8.5.

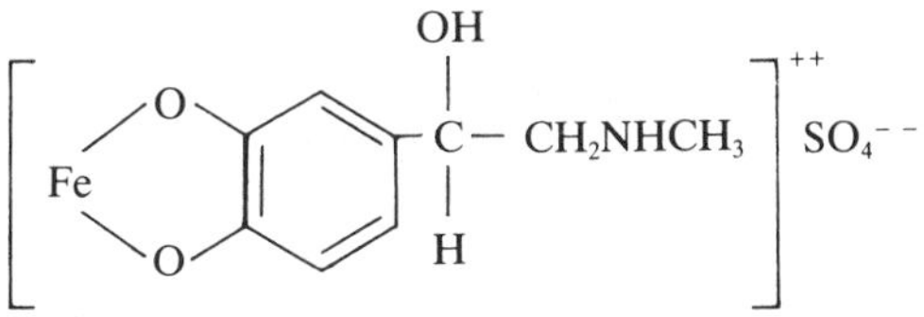

Assays of adrenaline in *Procaine and Adrenaline Injection* and in *Lignocaine and Adrenaline Injection*, of *Methyldopa Tablets*, *Methyldopate Injection* and *Isoprenaline Hydrochloride Injection* are based upon the coloured complexes formed with iron(II) sulphate–citrate reagent, buffered with glycine buffer solution. Alternatively, the concentration of iron(III) ions in a sample may be determined using the chromogenic reagent Tiron (1,2-dihydroxy-benzene-3,5-disulphonic acid, disodium salt), which forms an intense red complex.

One of the first substances to be used as a general reagent for the photometric assay of heavy metal ions such as Cd, Hg, Cu, Pb and Zn was dithizone(1,5-diphenylthiocarbazone). The coloured complex formed, for example with lead:

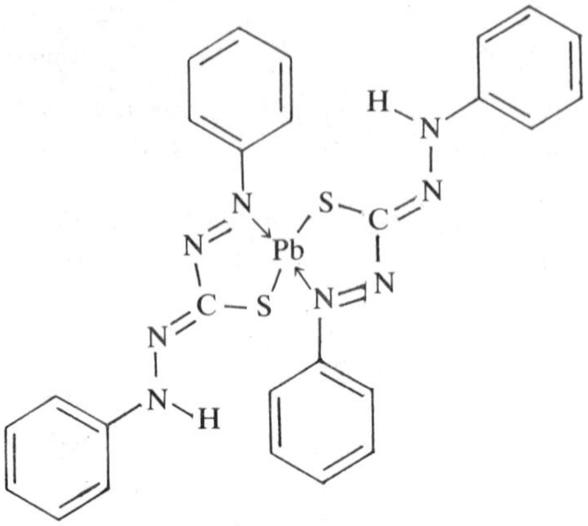

is soluble in organic solvents and may be extracted from aqueous solution into an immiscible organic solvent such as chloroform or carbon tetrachloride. Lead dithizonate has a carmine red colour in carbon tetrachloride and the high absorptivity ($\epsilon_{520} = 7 \times 10^4$) permits the assay of only a few μg of lead. Thio-Michler's reagent [4,4'-bis-(dimethylamino)-thiobenzophenone] is an extremely sensitive chromogenic reagent for Pd ($\epsilon_{520} = 2.1 \times 10^5$) and for Hg ($\epsilon_{560} = 1.2 \times 10^5$).

Body fluids

The determination of drugs and metabolites in body fluids forms part of several major fields of investigation—bioavailability, biochemistry, drug metabolism and toxicology. Pharmaceutical analysts are necessarily concerned in all these aspects because information of this type is required in submissions of data on new drugs to the Committee on Safety of Medicines (Part 1, Chapter 2).

The examination of body fluids is more difficult than that of pharmaceutical preparations in several ways:

(a) A small quantity of drug or metabolite is usually present in a large volume of blood, urine or tissue.

(b) Solvent extraction of body fluids gives rise to an extract that may contain, in addition to the drug, endogenous pigments or compounds which make optical methods of analysis subject to error unless great care is taken in the analytical conditions, e.g. choice of solvent, pH of extraction and subsequent purification methods. Even so, it is often necessary to use chemical methods as an intermediate step in the spectrophotometric method of analysis to introduce some degree of specifity or to increase absorbance.

(c) The drug may occur both free and combined as conjugates e.g. glucuronide or ethereal sulphates, both of which are polar and water-soluble.

(d) Protein-binding of the drug may occur and this leads to poor recoveries unless the protein is denatured during the extraction procedure.

(e) The use of several extractions for quantitative recovery of drug may lead to difficulties, e.g. emulsion formation with plasma samples and large volumes of solvent for evaporation. These difficulties can be mitigated by using extraction with a large solvent : sample ratio, and by carrying out control analyses with normal body fluids to which the drug to be determined has been added.

Direct spectrophotometric procedures, even after purification of the sample by solvent extraction or chromatographic 'clean-up', often lack the sensitivity and selectivity required for the assay of low concentrations of drugs that are found in body fluids after the adminstration of therapeutic doses. However, some of the modified spectrophotometric techniques, e.g. those involving chemical derivatisation, derivative spectrophotometry, or difference spectrophotometry, are sufficiently discriminating and sensitive for the assay of therapeutic levels of certain drugs.

Optimum conditions for spectrophotometric measurements

In developing an analytical method based on ultraviolet–visible spectrophotometry, the analyst is required to select the most appropriate sample and instrumental conditions for the assay with regard to the nature of the sample being analysed and the purpose of the assay. The accuracy and precision of spectrophotometric measurements depend on the intensity of stray light and the choice of sample conditions (solvent, concentration and pathlength) and instrumental parameters (wavelength, slitwidth and scan speed).

Sample conditions

Solvent

The choice of solvent is governed by the solubility of the absorbing substance and by the absorption of the solvent at the analytical wavelength.

The solubility of a substance in polar and non-polar solvents can often be predicted from a consideration of its chemical structure. For example, substances that are essentially hydrocarbon in nature and devoid of polar functional groups (e.g. vitamin A and related compounds, p.317) are lipophilic and are usually soluble in non-polar solvents such as cyclohexane. Substances with several functional groups that confer polarity are normally hydrophilic and soluble in polar solvents such as ethanol or water. Water is the ideal solvent as it is transparent at all wavelengths in the visible and ultraviolet regions, above 180 nm. It is also cheap and may be readily purified by using a laboratory still or de-ioniser. If the analyte can undergo an equilibrium (e.g. acid–base, keto–enol) reaction in which the equilibrium depends on the pH of the solution, it may be necessary to adjust the pH by means of acid, alkali or buffers to a value that maintains the analyte as a single chemical species. Organic solvents show either end-absorption below 220 nm due to n $\rightarrow$ π^* transitions within their functional groups (e.g. ethanol) or specific absorption bands due to n $\rightarrow$ π^* transitions (e.g. acetone) or $\pi \rightarrow \pi^*$ transitions (e.g. toluene) within the unsaturated group (p.316). Organic solvents are therefore restricted to measurements at wavelengths where the solvents are reasonably transparent. Compensation for weak absorption by the solvent in the sample cell may be made by using a reference (blank) cell containing the solvent only. Strong absorption by the solvent cannot be compensated in this way and will result in unacceptable noise in a spectral recording. The requirement in the British Pharmacopoeia is that the absorbance of the reference cell and its contents should not exceed 0.4, and preferably should be less than 0.2 when measured with reference to air. Table 7.3 lists the ultraviolet cut-off of several common solvents. This is the wavelength at which the absorbance in a 1 cm cell is 1 and below which it is extremely inadvisable to attempt to make spectral measurements.

Table 7.3 Ultraviolet cut-off of some common organic solvents

Solvent	Cut-off (nm)	Solvent	Cut-off (nm)
Ethanol	205	Diethyl ether	220
Methanol	210	Chloroform	245
Acetonitrile	210	Carbon tetrachloride	265
Hexane	210	Toluene	280
Cyclohexane	210	Acetone	330

Concentration and pathlength

The intensity of light transmitted from an absorbing solution (I_T) will show small random fluctuations owing to small variations in the light source intensity, detector and amplifier noise etc. Consequently, every absorbance value will have a small random error associated with it. The % relative error is the difference between the true concentration and that calculated from the measured absorbance (Δc) expressed as a % of the true concentration (c):

$$\% \text{ Relative error} = \frac{100 \Delta c}{c}$$

The % relative error is dependent on the magnitude of the measured absorbance and the type of detector used to measure the intensities of light. For modern instruments equipped with a phototube detector (e.g. photomultiplier tube), it may be shown that the relative error is at a minimum when the absorbance is 0.869 ($\%T = 13.5\%$). The optimum accuracy and precision are therefore obtained when the absorbance is around 0.9. However, in practice absorbances in the range 0.3 to 1.5 are sufficiently reliable, and the combination of cell pathlength and concentration of the analyte should be adjusted to give an absorbance within this range. In older instruments equipped with a photo-emissive detector (e.g. a lead sulphide detector) the minimum relative error is given by an absorbance of 0.434, and the combination of pathlength and concentration is adjusted to give absorbances in the range 0.2 to 0.7.

Instrumental parameters

Slitwidth

Most modern spectrophotometers have a control for adjusting the width of the slit through which the monochromatic light from the monochromator passes before it enters the cell compartment. Light from a prism or grating monochromator is never absolutely monochromatic but consists of a narrow range of wavelengths. The light of the greatest intensity occurs at the centre of the range corresponding to the nominal wavelength setting, but other wavelengths of lower intensity which have a triangular slit function are also present. The monochromaticity or spectral purity of the light incident on the sample cell is defined by the **spectral bandwidth**, which is the width of the triangle (in nm) at one half of the peak intensity (Fig.7.10(a)). The spectral bandwidth (SBW) is related to the mechanical width of the slit (SW) and the dispersion (D) of the monochromator

$$\text{SBW} = SW \times D$$

For example, for a slitwidth of 0.1 mm and a dispersion of 3 nm mm^{-1}, the spectral bandwidth is 0.3 nm. An increase in the slitwidth increases the spectral bandwidth and reduces the monochromaticity of the light.

The effect that increasing the slitwidth has on a measured absorbance value depends on the width of the absorption band and on whether the wavelength of measurement corresponds with a maximum or minimum absorbance. The width of the absorption band is defined by its **natural bandwidth**, which is the width of the absorption band at one half the A_{max} (Fig.7.10(b)). It has been shown that, if the spectral bandwidth exceeds one-tenth the natural bandwidth of the substance, the absorbance at the λ_{max} will be less than the true value (Fig.7.10(c)). This is because lower absorbances at wavelengths on the slope of the absorption band, included within a broad spectral bandwidth, contribute to the measured absorbance. Furthermore, as the absorbance increases due to higher concentrations or longer pathlengths, the % error increases,

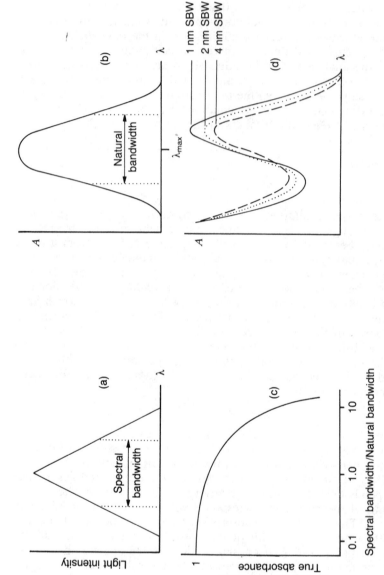

Fig. 7.10(a) The variation of light intensity with wavelength of the light emergent from the exit slit of a monochomator. The spectral bandwidth is the wavelength range in nm at one-half the maximum intensity. (b) An ideal absorption band. The natural bandwidth is the width of the band in nm at one-half the maximum absorbance (A_{max}). (c) The effect of increasing spectral bandwidth on the measured absorbance. (d) The effect of increasing spectral bandwidth on the absorption spectrum of a substance with a narrow natural bandwidth

resulting in a negative deviation from the Beer–Lambert Law. Conversely, when the spectral bandwidth is wide in relation to the natural
bandwidth, falsely high absorbances will be measured at the λ_{min} and
positive curvature in Beer–Lambert Law plots will be observed. It
follows that effects of variation of slitwidth are greater for narrow
bandwidth substances (e.g. benzenoid substances displaying fine structure in the range 240–270 nm) than for substances exhibiting broad
absorption bands (Fig.7.10(d)). If the slitwidth is too narrow, the
resultant reduction of incident energy lowers the signal-to-noise ratio
and decreases the precision of the measurement. The optimum slitwidth
is therefore the widest setting that provides spectral fidelity.

Scanning speed

If the scan speed selected for the automatic recording of the spectrum is
too fast, electronic and mechanical damping of the spectrophotometer's
signal to the recorder may prevent the recorder pen responding quickly
enough to the rapid changes of absorbance. The effects observed in
progressively faster replicate scans of an absorption spectrum (Fig.7.11)
are that the apparent λ_{max} is displaced in the direction of the scan, A_{max}
are decreased, A_{min} are increased and the resolution between adjacent
bands is reduced. The optimum scan speed is normally the fastest speed
that maintains pen fidelity.

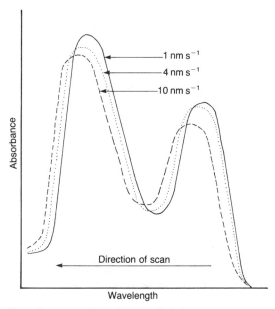

Fig. 7.11 The effect of scan speed on the recorded absorption spectrum of a substance

Stray light

Stray light is any radiation reaching the photodetector other than the narrow range of wavelengths normally transmitted by the monochromator. It arises from scattering and refraction inside the monochromator, due mainly to imperfections on the optical surfaces.

If stray light (SL) comprises wavelengths which the sample does not absorb, it will fall on the photocell during the measurement of the intensity of both the incident light (I_0) and unabsorbed light (I_T). Thus the observed absorbances is given by

$$A_{\text{obs}} = \log \frac{I_0 + SL}{I_T + SL}$$

The proportionately greater effect that the stray light has on I_T than on I_0 results in the measurement of falsely low absorbance values. The effect of stray light is particularly evident at high absorbance, i.e. when I_T is low, where deviation from Beer's Law may be observed. Fig.7.12 shows the effect of increasing levels of stray light on a Beer's Law plot. It should be noted that a spectrophotometer cannot measure absorbances greater than that corresponding with the level of stray light. For example, if the unabsorbed stray light is equivalent to 1% of the intensity of incident light at a particular wavelength, the limiting value of A_{obs} is

$$\log \frac{100 + 1}{0 + 1} = 2 \text{ (approximately)}$$

The effect of stray light may also be seen at wavelengths near the extremes of the usual wavelength range of light sources, where the available energy is low. For example, the intensity of light from a deuterium lamp below 220 nm decreases rapidly with decreasing wavelength. Unabsorbed stray light of higher wavelengths may reduce the absorbance sufficiently for a spurious maximum to be seen (Fig.7.13). These spurious maxima are common at wavelengths below 210 nm and may be detected by marked deviations in Beer's Law when solutions of different concentration are examined. Alternatively, stray light at a particular wavelength may be detected by placing in the sample beam a cut-off (stray light) filter or a solution of a substance that is known to absorb completely at that wavelength. Observed absorbance other than infinite absorbance ($\%T = 0$) confirms the presence of stray light (see Experiment 5).

Deviation from the Beer–Lambert Law

Occasionally a positive deviation (upward curve) or negative deviation (downward curve) may be observed in graphs of absorbance *vs* concentration (Beer's Law plots) or of absorbance *vs* pathlength (Lambert's Law plots). A knowledge of the possible explanations may assist the analyst to alter assay conditions to achieve the desired proportional relationship between absorbance and both concentration and pathlength.

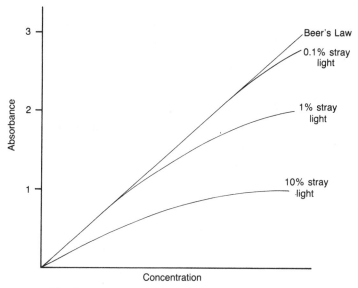

Fig. 7.12 The effect of stray light on Beer's Law plots

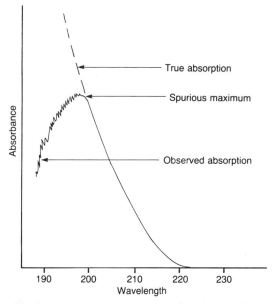

Fig. 7.13 The effect of stray light on end absorption

Non-monochromatic light

A requirement for adherence to the Beer–Lambert Law is that the light incident on the sample is monochromatic (p.309). In normal circumstances, light of spectral bandwidth less than one-tenth the natural bandwidth of the substance gives linear Beer's Law and Lambert's Law graphs if stray light is absent. If, however, the light incident on the sample is insufficiently monochromatic due to the presence of stray light (p.312) or the use of a wide slitwidth (p.309), and if the absorptivity of the substance at these extraneous wavelengths is less than that at the nominal wavelength of measurement, the measured absorbance will be less than the true absorbance and negative deviation from linearity, particularly at high absorbances, will be observed in Beer's Law and Lambert's Law plots. Conversely, if the absorptivity at the extraneous wavelengths is greater than that at the wavelength of measurement, then a positive deviation will be observed.

Chemical effects

Deviation from Beer's Law may occur if the substance undergoes chemical changes (e.g. dissociation, association, polymerisation, complex formation) as a result of the variation of concentration. For example, a solution of benzoic acid of high concentration in a simple unbuffered aqueous solution has a lower pH and contains a higher proportion of the unionised form than a solution of low concentration. The ionised and unionised forms of benzoic acid have different absorption characteristics:

$$C_6H_5COOH \rightleftharpoons C_6H_5COO^- + H^+$$

$$\lambda_{max} = 273 \text{ nm} \qquad \lambda_{max} = 268 \text{ nm}$$

$$\epsilon_{273} = 970 \qquad \epsilon_{268} = 560$$

Therefore, increasing the concentration of benzoic acid produces higher absorptivity values at 273 nm with positive deviation from Beer's Law, and lower absorptivity values at 268 nm with negative deviation from Beer's Law.

Similarly, in unbuffered solutions of potassium dichromate, the following pH-dependent dissociation of the dichromate ions occurs:

$$Cr_2O_7^{2-} + H_2O \rightleftharpoons 2HCrO_4^- \rightleftharpoons 2CrO_4^{2-} + 2H^+$$

$$\lambda_{max} = 350 \text{ nm} \qquad \lambda_{max} = 373 \text{ nm}$$

$$\epsilon_{350} = 3136 \qquad \epsilon_{373} = 4820$$

At pH above 10 the proportion of chromium(VI) as CrO_4^{2-} is greater than 99.9%. In weakly acidic solution, pH approximately 3, the predominant species is $HCrO_4^-$, although a small amount, dependent on the concentration of potassium dichromate, may exist as the dimer $(Cr_2O_7^{2-})$. Different proportions of the ions may be found in unbuffered solutions resulting in non-linear plots of absorbance and concentration. Solutions of potassium dichromate, 0.06 g l^{-1} in 0.005M sulphuric acid or 0.04 g l^{-1} in 0.05M potassium hydroxide, are used extensively in the routine calibration of wavelength and absorbance scales (Experiment 1).

To identify the cause of the deviation in a Beer's Law plot, Lambert's Law should also be tested for the substance over the same absorbance range. If the absorbances of a single solution in cells of different pathlength show that Lambert's Law is obeyed, deviation from Beer's law must be due to chemical factors rather than to a lack of monochromaticity of the light.

Structural analysis

A detailed account of the application of absorption spectra to the determination of the structure of organic compounds is beyond the scope of this book, but elementary principles can be cited to help in understanding the appearance and position of the absorption bands of many pharmacopoeial substances.

Absorption of light in the visible and ultraviolet regions of the spectrum is due to the presence of a **chromophore** in the absorbing molecule. The term 'chromophore' was originally used for unsaturated groups of atoms which were thought to be essential for colour. Now that light absorption studies have been extended into the ultraviolet region, the term includes such multiple bonds as those of -ene, -yne and carbonyl groups in which the electrons are more loosely bound than those in fully saturated compounds. Unfortunately, these simple chromophores absorb in a region inaccessible to ordinary spectrophotometers, and special instruments and light sources are needed to obtain reliable results at wavelengths less than 210 nm. In Table 7.4 are recorded representative examples of simple chromophores and their constants.

Table 7.4 Typical chromophores and their constants

Substance	Chromophore	λ_{max}(nm)	ϵ_{max}
Methane	–	125	–
Oct-3-ene	$C{=}C$	185	8000
Acetone	$C{=}O$	188	900
		279	15
Acetoxime	$C{=}N{-}$	190	5000
Acetic acid	$-C{=}O$ $\mid$ OH	204	40

Absorption of ultraviolet–visible radiation occurs when its energy corresponds to that required to induce electronic transitions in the absorbing molecules (p.259). The electrons involved in the absorption of energy may be considered under three headings:

σ **electrons** occur in fully saturated systems such as the bonds of alkanes, and they require so much energy for excitation that the absorption bands lie in the vacuum ultraviolet region of the spectrum. They do not concern us further except to say that interaction with π electron systems gives rise to hyperconjugation and to the small effects noted on p.316.

n electrons are non-bonding, such as those of the lone pairs in O, N and S, and the transition n → π^*, i.e. to an excited bonding orbital, accounts for the low intensity absorption of the carbonyl group at about 280 nm. Interaction with electrons of conjugated systems gives rise to the effects noted under 'auxochromes' below.

π **electrons** are those of multiple bonds, the so-called 'mobile' electrons. The basic absorption, due to the transition $\pi \to \pi^*$, lies in the 180–200 nm region.

The absorptions of two or more chromophores which are separated by more than one bond are usually additive, but when such chromophores are conjugated, i.e. separated by a single bond, pronounced effects are produced. The maximum absorption is shifted to longer wavelengths, thus bringing it into the working range of spectrophotometers. Such an effect is called a **bathochromic** shift and the increase in ϵ_{max} which often accompanies such a shift is known as a **hyperchromic** effect. The reverse changes are known as **hypsochromic shift** and **hypochromic** effect respectively, and occur quite often when a chromophoric system is changed, e.g. by alteration of pH. In Table 7.5 are recorded typical examples of conjugated chromophores and auxochromes.

Table 7.5 Typical conjugated chromophores and auxochromes

Compound	Chromophore	λ_{max}(nm)	ϵ_{max}
Butadiene	C=C—C=C	217	21 000
Crotonaldehyde	C=C—C=O	217	16 000
Sulphanilamide in NaOH solution	$H\bar{N}O_2S$——NH$_2$	251	16 300
Sulphanilamide in HCl solution	H_2NO_2S——$\overset{+}{N}H_3$	218 / 265	12 700 / 1080

Changes in absorption spectra can also be produced by fully saturated groups attached to a chromophoric system, and these groups are called **auxochromes**. Unlike chromophores, which are covalently unsaturated (e.g. double bonds), auxochromes are covalently saturated. They are of two types :
(a) co-ordinatively unsaturated, e.g. —NH$_2$, —S—, which contain lone pairs of electrons; and (b) co-ordinatively saturated, e.g. —$\overset{+}{N}H_3$. Auxochromes of type (a) are generally more effective in modifying absorption spectra. The influence of alkyl groups is usually very small and merely affects the position of the absorption maximum (Table 7.6).

The correlation of absorption with structure is still largely empirical, as different types of compound may absorb in the same region. Hence, for complete identification of a compound, spectrophotometric measurements must be supplemented by chemical tests and other physical measurements.

Before proceeding to more complicated chromophoric systems, the absorption characteristics of pharmacopoeial substances which contain simple combinations of double bonds can now be discussed by reference to the above effects. Vitamin A_1 is a particularly good example of the effect of conjugation

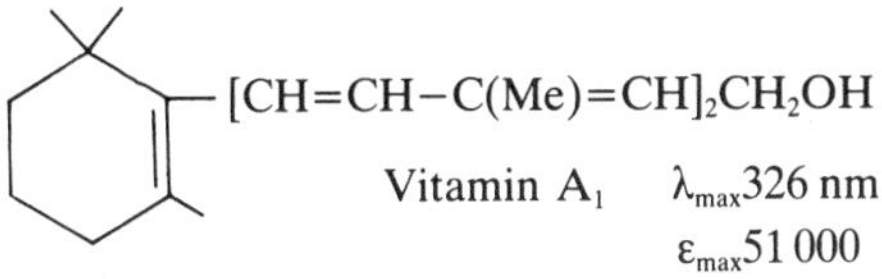

—[CH=CH−C(Me)=CH]$_2$CH$_2$OH

Vitamin A_1 λ_{max}326 nm
ε_{max}51 000

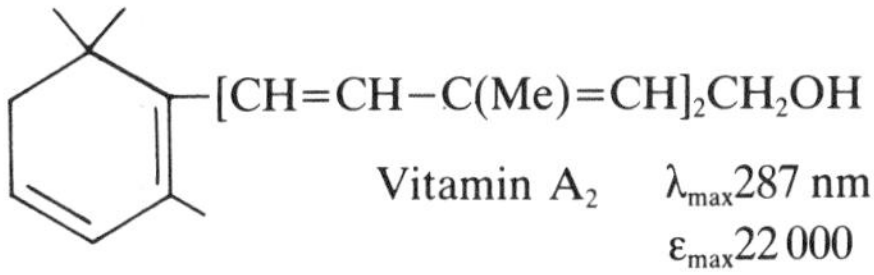

—[CH=CH−C(Me)=CH]$_2$CH$_2$OH

Vitamin A_2 λ_{max}287 nm
ε_{max}22 000
λ_{max}351 nm
ε_{max}41 000

on the position and intensity of the absorption maximum. The additional double bond in vitamin A_2 changes the absorption maximum to 351 nm. This compound is also of interest because, in addition to the absorption caused by the six alkenic bonds, there is evidence of absorption by the partial chromophoric system —[CH=CH—CMe=CH]$_2$ at 287 nm, an effect which is enhanced by the steric effects of the methyl groups in the ring.

Comparison of the absorption of Calciferol with that of its stereo-isomers iso-vitamin D_2 (*all-trans*) and precalciferol illustrates the effects produced by *cis-trans* isomerism and also by steric factors. The shift of maximum from 265 nm in Calciferol to 287 nm in iso-vitamin D_2 is typical of the increase of

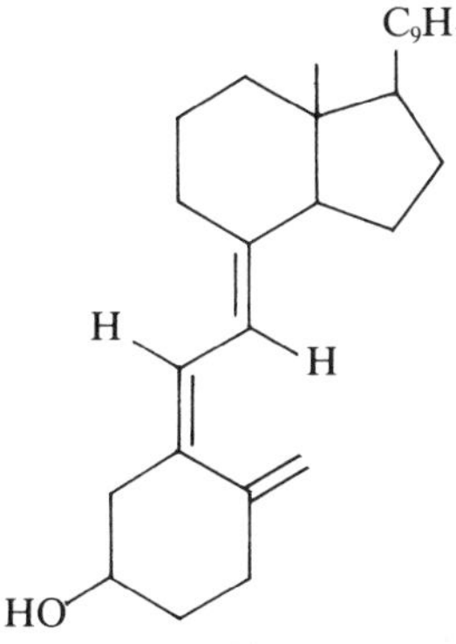

Calciferol
λ_{max}265 nm
ε_{max}18 200

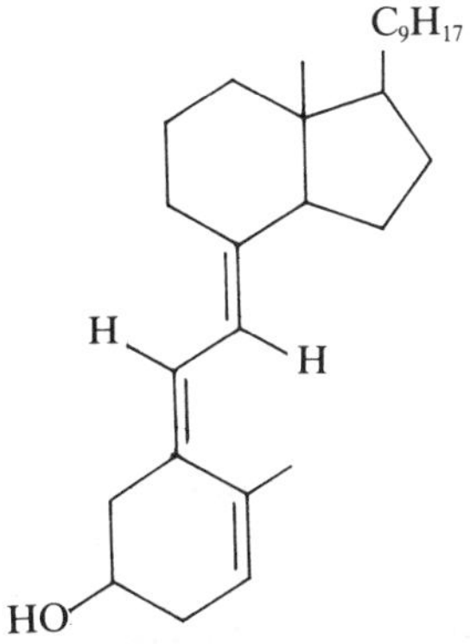

iso-Vitamin D_2 (*all-trans*)
λ_{max}287 nm
ε_{max}44 100

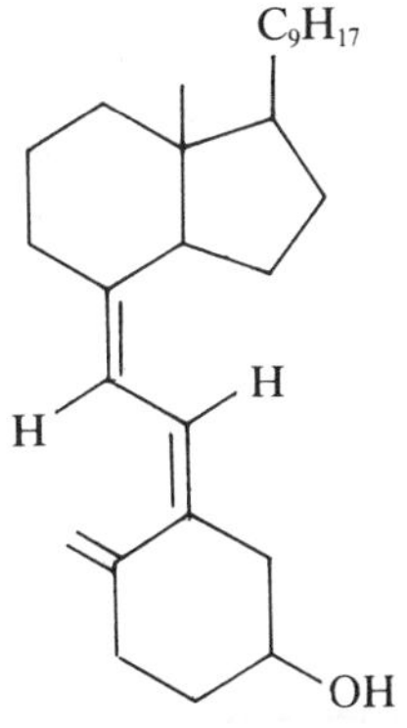

Precalciferol
λ_{max}265 nm
ε_{max}9800

chromophore length in passing from a *cisoid* to a *transoid* diene system. The chromophoric systems in Calciferol and precalciferol are identical but, owing to steric hindrance between the hydrogen atoms on $C_{(7)}$ and $C_{(15)}$ in precalciferol, the intensity of the absorption is decreased.

The intense ultraviolet absorption of many keto-steroids is due entirely to the conjugated system 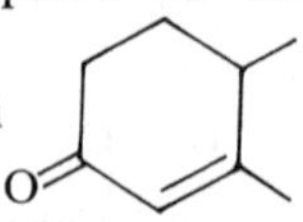 , and even when a third double bond is present in conjugation as in *Prednisolone* and *Predni-sone*, very little change is observed in the position of maximum absorption. These are examples of crossed conjugation, and under these conditions the absorption is always characteristic of the main chromophoric system.

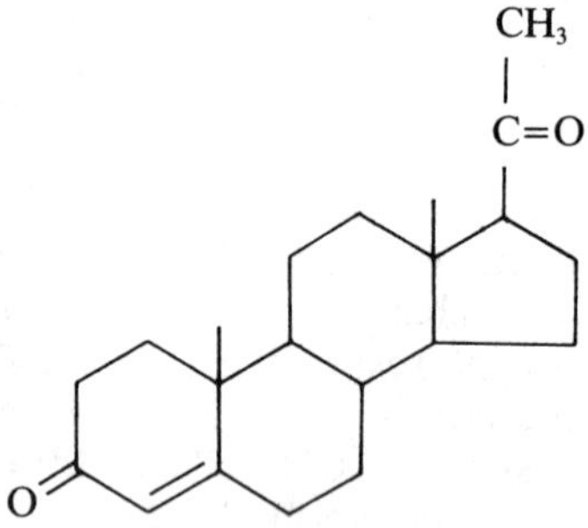

Progesterone
λ_{max} 241 nm

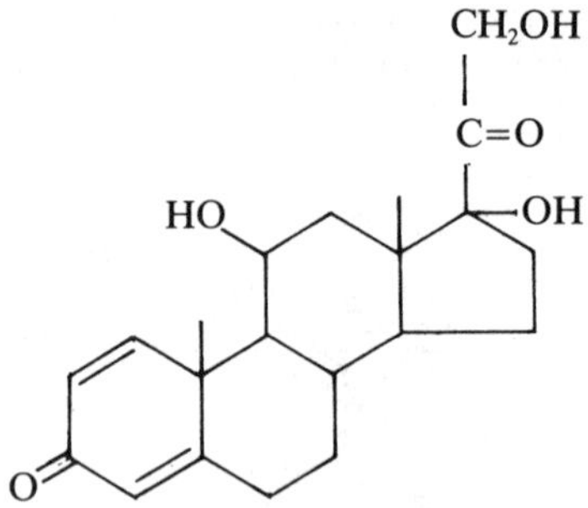

Prednisolone
λ_{max} 241 nm

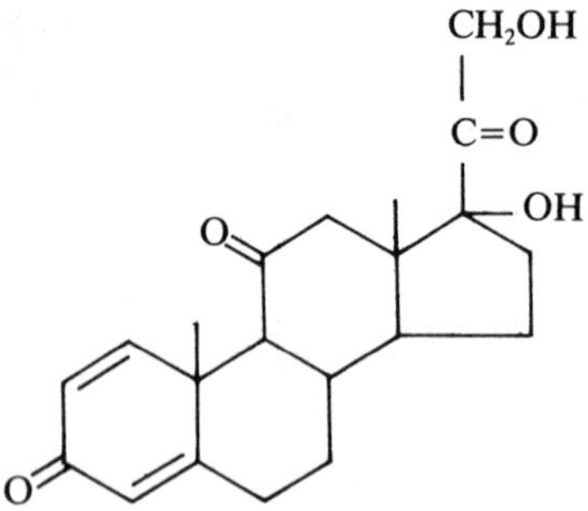

Prednisone
λ_{max} 241 nm

In order to produce a marked change in wavelength, conjugation must extend the chromophoric system as with vitamins A_1 and A_2. The identity of the auxochromes present in the compounds discussed may not be clear at first, but small modifications in the basic absorption of the chromophoric systems are caused by substituents. Increasing alkyl substitution causes bathochromic displacement of the absorption band, and the compounds in Table 7.6 illustrate

Table 7.6 Effect of alkyl substitution on absorption maximum

Compound	λ_{max}	Substituents
Crotonaldehyde	217	1
α-Ionone	228	2
Progesterone	241	3

this point, although other factors such as the position of double bonds relative to ring systems (Woodward, 1941, 1942) also operate in fixing the position of the absorption bands. The effect is best observed with an isolated ethylenic bond. Bladon *et al.* (1952) selected 210 nm (not a maximum) as a convenient wavelength for observation because of the limitation of the instrument used. The bathochromic shift of the end-absorption associated with increasing alkyl substitution caused significant increases in ϵ values sufficient to identify the degree of substitution, e.g. cholest-2-ene (disubstituted, $\epsilon_{210} = 200$); cholest-8-ene (tetrasubstituted, $\epsilon_{210} = 4400$).

Certain arrangements of chromophores appear to act as a unit, as for example in benzene, the ultraviolet absorption spectrum of which is capable of modification by other chromophores and by auxochromes. Benzene itself shows a high intensity absorption band at 200 nm, and a second band at 255 nm of low intensity, which exhibits very fine vibrational structure. Substitution of one of the hydrogens by an alkyl group reduces the fine structure, but it is still evident in all of the following pharmacopoeial compounds, which exhibit typical benzenoid absorption with a maximum at 257 nm (Fig.6.6(a)):

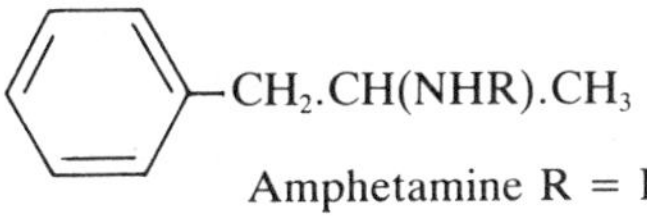

—$CH_2.CH(NHR).CH_3$

Amphetamine R = H
Methylamphetamine R = CH_3

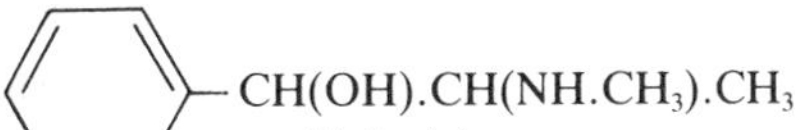

—$CH(OH).CH(NH.CH_3).CH_3$
Ephedrine

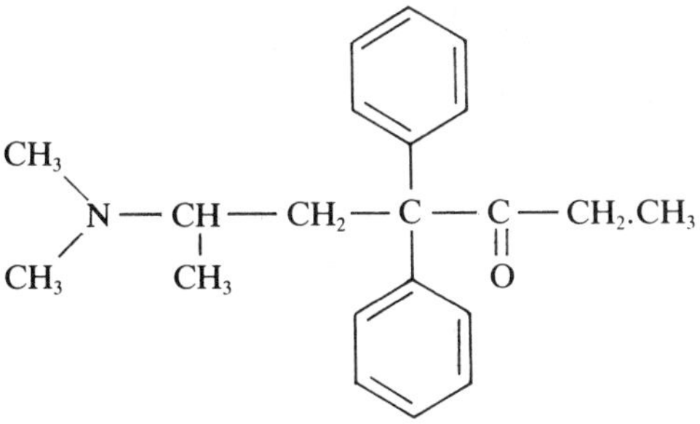

Methadone

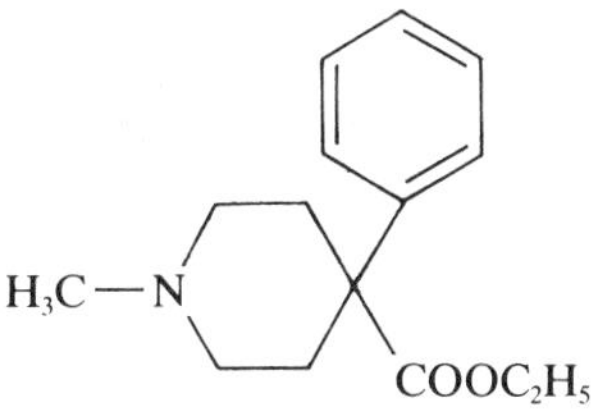

Pethidine

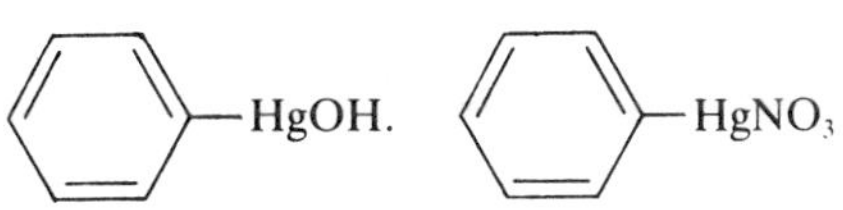

—HgOH.

—$HgNO_3$

Phenylmercuric Nitrate

When phenolic hydroxyl groups are present, a marked bathochromic displacement of the absorption to about 280 nm occurs, and the effect of solvent on the absorption is considerable because, whereas in neutral or acid media the auxochrome is —OH, in alkaline media it becomes —O⁻. For example, the absorption maximum for chlorocresol in acid solution occurs at 279 nm, but in alkaline media the peak is found at 296 nm. Typical examples of phenols are *Morphine Hydrochloride, Nalorphine Hydrobromide* and *Ethinyloestradiol*, whilst *Adrenaline* and *Isoprenaline* are examples of o-dihydroxyphenols.

The amino group is a powerful auxochrome when attached directly to a benzene system, and aniline exhibits high intensity absorption at about 230 nm and typical low intensity benzenoid absorption at 280 nm. If additional conjugation is present, a cumulative effect is obtained. Thus in *Procainamide* the 200 nm high intenstiy absorption of benzene suffers a bathochromic displacement to 280 nm.

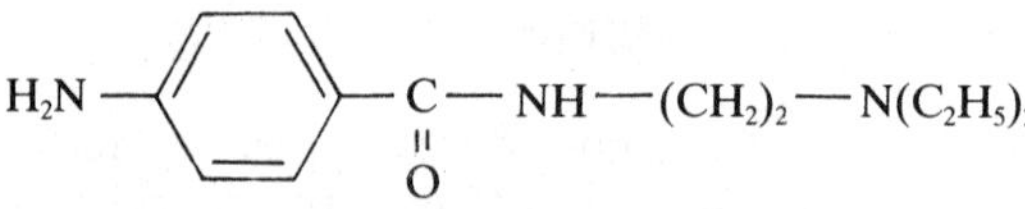

Procainamide

Similarly, the sulphonamide drugs of general formula

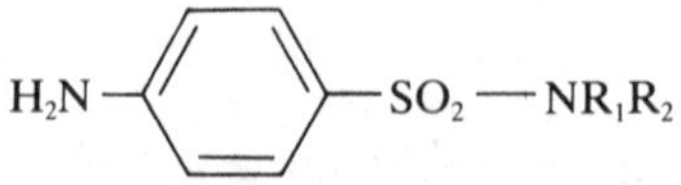

show high intensity absorption at about 251 nm. The effect of change in pH in the solvent used for the ultraviolet measurements is very striking (Fig.7.14). In alkaline solution, the absorbing system is as given above, but in acidic solution the amino —NH_2 group is replaced by —$\overset{+}{N}H_3$ which is considerably less efficient as an auxochrome.

p-Aminobenzoic acid is similar in absorption characteristics to *Procainamide* and *Procaine* in neutral or alkaline media, but in acid solution the absorption curves of all three approach that of benzoic acid.

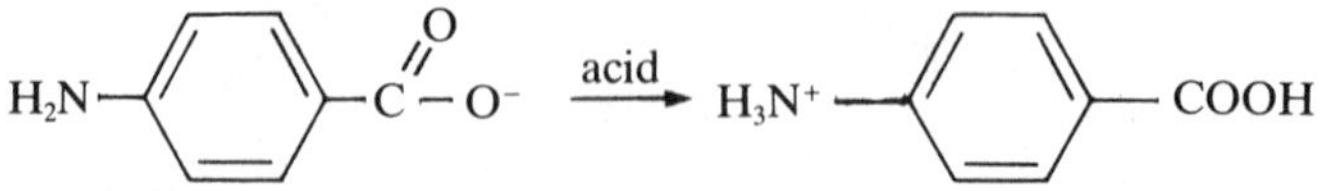

Sodium aminosalicylate in alkaline solution has an additional auxochrome the phenate ion, so that the high intensity absorption noted in

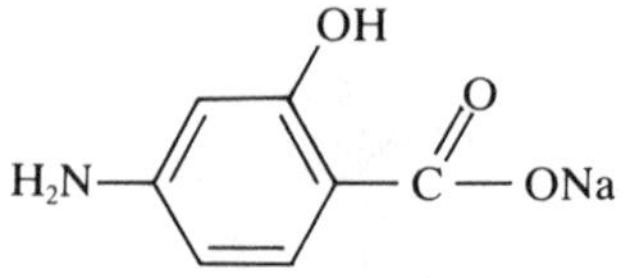

Sodium Aminosalicylate

the *p*-aminobenzoic acid (λ_{max} 280 nm) now appears at 265 nm with the typical phenate absorption of lower intensity at 299 nm. The absorption curves for *p*-aminobenzoic acid, benzoic acid and *p*-aminosalicylic acid are shown in Figs 7.15 and 7.16.

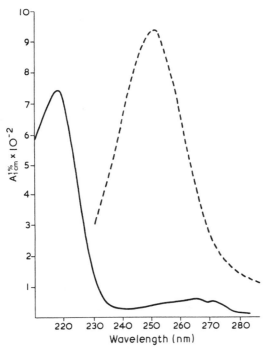

Fig.7.14 Absorption spectra for sulphanilamide in 1M hydrochloric acid (——) and in 1M sodium hydroxide (‒‒‒‒)

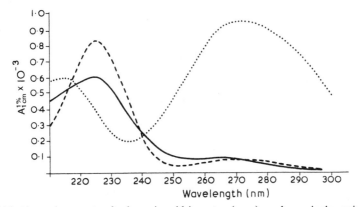

Fig.7.15 Absorption spectra for benzoic acid in water (——), and *p*-aminobenzoic acid in water (. . . .) and in 0.1M hydrochloric acid (‒‒‒‒)

The fusion of two or more benzene rings causes bathochromic displacement of the 200 nm and 255 nm bands, and in general, three regions, distinguished by the intensity of the absorption, can be discerned (Table 7.7). Auxochromes and conjugated chromophores cause modification in the basic absorption in the manner already described for benzene so that quite complex absorption spectra result.

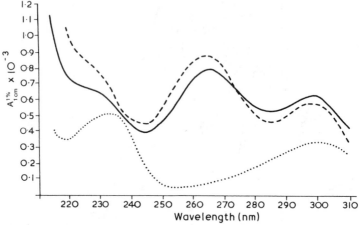

Fig.7.16 Absorption spectra for *p*-aminosalicylic acid in water (———) in 0.1M hydrochloric acid (....), and in 0.1M sodium hydroxide (_ _ _ _)

Table 7.7 Absorption characteristics of fused-ring polycyclic aromatics

Compound	Region					
	I		II		Benzenoid	
	λ_{max}(nm)	log ϵ_{max}	λ_{max}(nm)	logϵ_{max}	λ_{max}(nm)	logϵ_{max}
Benzene	–	–	200	3.65	255	2.35
Naphthalene	220	5.05	275	3.75	314	2.50
Anthracene	250	5.20	380	3.80	–	–
Phenanthrene	250	4.50	295	4.10	330	2.90

The absorption spectra of pyridine and its derivatives are completely analogous to those of benzene and its derivatives, as shown by the following comparison of benzene and benzoic acid with pyridine, nicotinic acid and its derivatives (Table 7.8). Changes in pH will cause changes in the absorption spectra of pyridine derivatives

so that the solvent used should always be specified. The absorption band at 266 nm (in 0.01M hydrochloric acid) is used as an identity test for

Isoniazid. The increase in absorption at about 263 nm in acid conditions as compared with that in alkaline conditions is a valuable property of pyridine compounds as it enables them to be determined in the presence of those compounds for which no such ΔA value is observed, e.g. *Triprolidine* in coloured syrup formulations.

Table 7.8 Absorption characteristics of pyridine derivatives

Substance	Formula	λ_{1max}(nm)	λ_{2max}(nm)	Solvent
Benzene		198	255	Ethanol
Pyridine	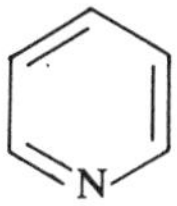	195	250	Hexane
Benzoic acid	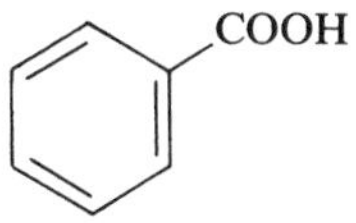 COOH	230	270	Ethanol
Nicotinic acid	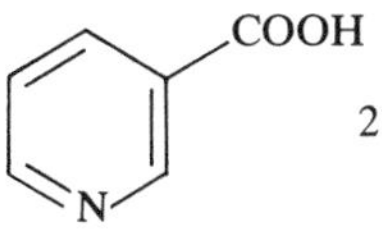 COOH	212	263	0.1M NaOH
Nicotinamide	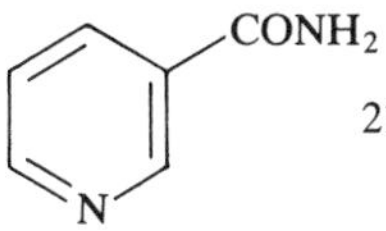 CONH$_2$	212	263	0.1M NaOH
Nikethamide	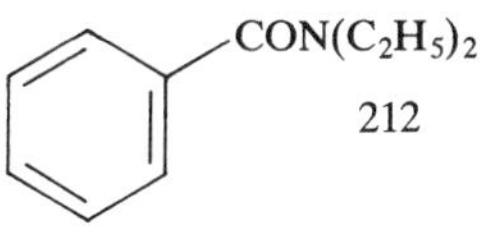 CON(C$_2$H$_5$)$_2$	212	263	0.1M NaOH
Isoniazid	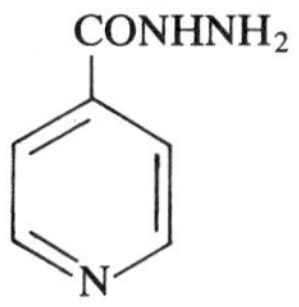 CONHNH$_2$	215	266	0.1M HCl

Comparisons may be drawn in the same way between the quinolines and napthalene, and between acridine and anthracene.

When a compound exists in tautomeric forms it may be possible, by careful selection of pH values for the solution of the substance, to obtain selective absorption in the ultraviolet region of the spectrum. For example *Phenobarbitone* shows strong absorption under alkaline conditions (pH 13) at about 255 nm (Fig. 7.17), due to the —C=N—C=O

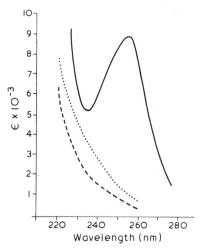

pH 2   pH 10   pH 13

chromophoric system. At pH 10 maximum absorption occurs at about 240–245 nm.

Fig.7.17 Absorption spectra of phenobarbitone in 0.1M hydrochloric acid (----), in water (....) and in 0.1M sodium hydroxide (——)

The purines, *Caffeine* and *Theophylline*, similarly show characteristic absorption because of the chromophoric system shown in heavy type.

Caffeine

Because of the tautomeric nature of nitrosophenols, a pronounced increase in colour can be obtained by making solutions of such compounds alkaline.

This is actually done in the colorimetric assay for morphine in *Camphorated Opium Tincture*. The sensitivity of the method is increased by measuring the absorption of the nitroso-morphine in alkaline solution.

The subject could be expanded greatly but enough has been said to indicate that inspection of the formula of a substance will reveal whether or not interesting features of absorption in the ultraviolet region can be expected.

Ultraviolet–visible spectra rarely provide a complete identification of a substance, owing to the lack of characteristic spectral features. Structural elucidation of new compounds is usually carried out by the examination of infrared, nuclear magnetic reasonance and mass spectrometric data in conjunction with ultraviolet data (Chapter 14). The ultraviolet–visible absorption spectra and spectral characteristics (λ_{max} and $A_{1\,cm}^{1\%}$ or ϵ values) of many drug substances may be obtained from a number of sources (Sharkey et al, 1968; Clarke, 1985; Sunshine, 1981).

Experiments in ultraviolet–visible spectrophotometry

The following exercises are designed to illustrate principles of method and theory and to provide practical examples of spectrophotometric techniques in the assays of drugs in medicines and in body fluids. The variation in the design of spectrophotometers makes it inadvisable to relate experiments to any one particular instrument.

Instrument performance

Experiment 1 *Calibration of absorbance scale*

Prepare a standard solution of potassium dichromate (Note 1) approximately 60 mg (Note 2) diluted to 1 l with 0.005M sulphuric acid (Note 3). Measure the absorbance of the solution in a 1 cm cell using 0.005M sulphuric acid in the reference cell (Note 4) at the two wavelengths of maximum absorption (257 nm and 350 nm) and at the two wavelengths of minimum absorption (235 nm and 313 nm). Alternatively, record the absorption spectrum on the 0–1 absorbance range between 400 nm and 220 nm using a *slow* scan speed (Note 5) and read the absorbance at the two λ_{max} and two λ_{min} values. Calculate the $A(1\%, 1\,cm)$ values using the exact concentration of the standard solution and compare the results with the standard values in Table 7.9 (Note 6).

Note 1 Analytical reagent quality; dried at 110° for 1 h.
Note 2 Accurately weighed ±0.1 mg.
Note 3 The effect of pH on the dissociation of dichromate is discussed on p.314.
Note 4 Before using the solution, a check should be carried out on the cells in use.

Normally these are matched, but it is of interest to note how often the following will indicate differences in the cells. Fill both cells with the solvent being used. Dry the outside of the cells with paper tissues. Do not touch the polished faces with the fingers but handle the cells by the ground glass sides only. Check the absorbance of one cell against the other as blank at the selected wavelength. Note the reading, reject the solvent in one cell and fill again with solvent. Check the reading and repeat the emptying and filling until consistent readings are obtained. Now reject the solvent in the other cell and repeat the above procedure. Any residual difference at this stage can be used to correct absorbances. When the difference becomes large (0.005–0.010) then cleaning of cells must be undertaken.

Note 5 The effect of scan speed is discussed on p.311.

Note 6 The $A(1\%, 1\,cm)$ values are the mean values obtained in a study by the UV Spectrometry Group and the range is based on the variation found in the literature values.

Table 7.9 Recommended $A(1\%, 1\,cm)$ values for acidic potassium dichromate solutions

Wavelength (nm)	$A(1\%, 1\,cm)$
235 (min)	125.1 ± 1.9
257 (max)	145.4 ± 1.5
313 (min)	48.8 ± 0.7
350 (max)	107.1 ± 1.1

Experiment 2 *Calibration of wavelength*

(a) *Holmium filter* For the routine calibration of instruments, a holmium filter is satisfactory. Record the absorption spectrum (0–2 absorbance range) from 500 to 230 nm using the slowest scan speed and the narrowest slit setting. Identify the three fused absorption bands centred around 452.2 nm and the single band at around 360.9 nm. Instruments with accurately calibrated wavelength scales will show λ_{max} at 453.2, 418.4, 360.9, 287.5, 279.4 and 241.5 nm.

(b) *Holmium perchlorate solution* Prepare a solution of holmium(III) perchlorate by dissolving 0.5 g of holmium oxide in 2.4 ml perchloric acid (72%; AR) by warming gently and diluting to 10 ml with *water*. Record the absorption spectrum (0–1 absorbance) as described for the holmium filter. The wavelengths of the principal bands ($A > 0.4$) should be 485.8, 450.8, 416.3, 361.5, 287.1, 278.7 and 241.1 nm.

(c) *Discharge lamps* A low pressure mercury discharge lamp is the most suitable for the high-accuracy calibration of instruments. Record the **transmission** spectrum from 600 nm to 240 nm (i.e. with the instrument in the 'single-beam' or 'energy' mode) of the mercury lamp placed near the entrance to the monochromator, using the minimum slit setting and slowest scan speed. The principal emission lines of mercury are at 579.0, 576.9, 546.1, 435.8, 404.5, 364.9 and 253.7 nm.

Experiment 3 *Detection of stray light*

The effects of stray light on an absorption spectrum are discussed on p.312. Stray light at a particular wavelength may be detected using a cut-off filter which absorbs intensely at that wavelength.

Method Measure the absorbance at 200 nm of a 1.2% solution of potassium chloride in *water* against *water* in the reference cell. Alternatively, measure the absorbance in the range 200–210 nm of a Vycor glass filter against air. Any absorbance less than 2 indicates the presence of stray light in this region.

Instrumental settings

Experiment 4   *Slitwidth*

For this experiment, a spectrophotometer that will allow measurements to be made at various slit settings in the range 0.25–4 nm, is required.

Method Prepare a solution of toluene in hexane (0.02% v/v) and record its absorption spectrum from 300 to 230 nm in a 1 cm cell against hexane in the reference cell, using the slowest scan speed and narrowest slit setting. Repeat the spectral recording several times under identical conditions except for slitwidth, which should be increased, e.g. 0.5, 1, 2, 3 and 4 nm. In each spectrum, measure the absorbance at the λ_{max} at 269 nm and at the λ_{min} at 266.5 nm and note the effect of increasing the slitwidth.

Repeat the experiment with a standard solution of potassium dichromate (0.006%) in 0.005M sulphuric acid and measure the absorbance at the λ_{max} (350 nm and 257 nm) and the λ_{min} (313 nm and 235 nm) values.

Explain the different effects of increasing the slitwidths on the toluene and potassium dichromate spectra in terms of their natural bandwidths (p.309).

Experiment 5   *Scan speed*

A recording spectrophotometer that will permit scanning at a variety of scan speeds, from 0.1 to 10 nm s^{-1} is required.

Method Record the absorption spectrum (0–1 absorbance range) from 290 nm to 230 nm of one of the following solutions which exhibit narrow bandwidth absorption bands:
   0.02% toluene in hexane (experiment 4);
   0.10% ephedrine hydrochloride in *water*;
   0.05% diphenhydramine hydrochloride in *water*;
using the slowest scan speed and narrowest slit setting. Repeat the spectral recording using progressively faster scan speeds. Note the effect on the λ_{max} and λ_{min} values and absorbances at these wavelengths (Fig.7.11).

Solvent effects

Experiment 6   *To show the effect of solvent upon the absorption spectrum of phenol*

Method Prepare a 0.002% w/v solution of phenol in (a) *water* and (b) cyclohexane. Record the absorption spectrum of each solution over the range 230–300 nm using 1 cm cells and the appropriate solvent as a blank. Typical absorption spectra are shown in Fig 7.18.

In cyclohexane, little interaction between solvent and solute occurs and vibrational fine structure is observed. In water, however, solvation of the solute and hydrogen bonding are possible so that the fine structure is almost eliminated, only the band envelope being obtained. Similarly, if the two solvents are aqueous and differ only in their ionic character, e.g. different strengths of the same buffer (salt) solution, it is possible for a 'salt effect' to cause slight differences in the absorption band of a compound.

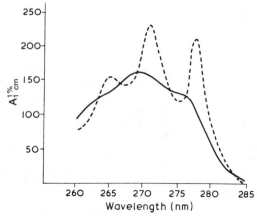

Fig.7.18 Absorption spectra of phenol in cyclohexane (‑ ‑ ‑ ‑) and in water (——)

Experiment 7   *To show the effect of pH upon the absorption spectrum of sulphanilamide*

Method Prepare 100 ml of a 0.1% w/v aqueous solution of sulphanilamide. Dilute 1 ml of this solution to 100 ml using 1M hydrochloric acid and also dilute 1 ml to 100 ml using 1M sodium hydroxide. Record the absorption spectrum of each solution over the range 210–300 nm for the acid solution and 230–300 nm for the alkaline solution, using 1 cm cells and appropriate reference solutions. Typical absorption spectra are shown in Fig.7.14.

In alkaline solution the primary amino group is retained as the auxochrome, but in acid solution quaternisation occurs to a coordinatively saturated auxochrome.

$$H_2N-\left\langle\bigcirc\right\rangle- \overset{O}{\underset{O}{\overset{\uparrow}{S}}}-NH_2 \quad \underset{\text{alkali}}{\overset{\text{acid}}{\rightleftharpoons}} \quad \overset{+}{H_3N}-\left\langle\bigcirc\right\rangle- \overset{O}{\underset{O}{\overset{\uparrow}{S}}}-NH_2$$

This is much less effective in modifying absorption and the characteristic benzenoid absorption is obtained at 265 nm.

Note that solutions of sodium hydroxide absorb radiation below about 230 nm, so that, unless the concentration of alkali is very small, the readings will be unreliable.

Structural effects

Experiment 8   *To show typical benzenoid absorption*

Method Prepare a 0.05% solution in 0.1M HCl of any of the following chemicals:
    Ephedrine hydrochloride:
    Atropine sulphate;
    Phenylmercuric nitrate.
Record the absorption spectrum over the range 220–280 nm using 1 cm cells and 0.1M hydrochloric acid as reference solution.

The absorption curve of ephedrine hydrochloride is shown in Fig.6.6(a) and is typical of all the above compounds because all possess the general formula ⬡—R . Strong auxochromes such as −NHR and −OH, are not directly attached to the benzene ring, and chromophores such as $\overset{\backslash}{\underset{/}{C}}=O$ are not in conjugation so that the absorption corresponds closely to that of toluene.

Experiment 9 *To show typical pyridine absorption*

Method Prepare a 0.002% w/v solution of nicotinic acid (or nicotinamide) in *water* and record the absorption spectrum over the range 230–290 nm.
The absorption spectrum for nicotinamide is shown in Fig.7.19.

The chromophoric system is (pyridine ring with $\overset{O}{\overset{\|}{-C}}-R$) but the absorption caused by the pyridine nucleus occurs at 265 nm. Compare the spectrum with those due to the benzene system, in particular that of benzoic acid (see below).

Experiment 10 *To show the influence of conjugated chromophores and auxochromes on chromophores*

Method 1 The effect of conjugation on the carbonyl chromophore
Prepare a 0.01% solution of cyclohexanone in ethanol, a 0.001% solution of testosterone in ethanol and a 0.001% solution of *p*-dimethylaminobenzaldehyde in ethanol and record the absorption spectra over the range 210–320 nm.

The carbonyl chromophore absorbs strongly at about 185 nm and weakly at about 280 nm, but when conjugated with an alkenic bond as

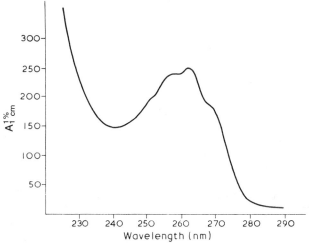

Fig 7.19 Absorption spectrum for nicotinamide in water

in testosterone, the absorptions are shifted to about 240 nm and 300 nm respectively. Note that in the absorption curve of *p*-dimethylaminobenzaldehyde, the benzenoid absorption is shifted to a longer wavelength (about 280 nm). A molecule shows, not only the absorption corresponding to the interacting chromophores, but also that corresponding to an individual chromophore. Where the latter is weak, it is often masked by the strong absorption of the conjugated chromophoric system.

Method 2 The effect of substituents on the benzene chromophore
For this exercise the following solutions are required:

Aniline	0.005% w/v in water
Benzoic Acid	0.001% w/v in water
p-Aminobenzoic Acid	0.0005% w/v in water
Cinnamic Acid	0.0005% w/v in water

Record the absorption spectra over the range 210–320 nm using 1 cm cells and water as reference solution.

The —NH$_2$ auxochrome is comparable with the —COOH chromophore in its effect on the absorption, cf. Figs 7.15 and 7.16. They can readily be distinguished, however, by determining the absorption spectra in acid solution. The absorption spectra of benzoic and cinnamic acids will remain unaffected, whereas those of aniline and *p*-aminobenzoic acid will alter appreciably because of the change in the chromophoric system. Compare the spectra with those of ephedrine hydrochloride etc. (Experiment 8).

Experiment 11 *To determine the isosbestic point of an indicator and to show its use in analysis*

Method Pipette 2.0 ml of bromocresol green solution (0.02%) into each of 8 × 50 ml graduated flasks. Adjust to volume with buffer solutions of pH 3.0, 3.4, 3.8, 4.2, 4.6, 5.0, 5.4 and 6.0 respectively. Mix well and determine the absorption spectrum of each solution over the range 470–700 nm using 1 cm cells. Record each spectrum on the same chart and note the isosbestic point, i.e the point of equal absorbance regardless of pH, and the position of the absorption maximum.

Draw a graph of the ratio

$$\frac{\text{Absorbance at wavelength of maximum absorption at pH 3.0}}{\text{Absorbance at isosbestic point}}$$

at each pH as ordinate against pH as abscissa.

After the completion of the graph, any pH between about 3.1 and 5.2 can be determined by adding a few drops of bromocresol green indicator to the test solution and measuring the absorbance at two wavelengths— one at the isosbestic point and the other at the maximum absorption. Calculation of the ratio and reference to the graph gives the required pH. In this method the quantity of indicator need not be known, because the ratio of the absorbances would be the same regardless of that quantity (p.286).

This principle has been applied to the determination of Thiamine in pharmaceutical products. The vitamin shows marked changes in absorption at about 245 nm with change in pH, but no change in absorbance at about 273 nm—the isosbestic point. If a solution of the vitamin is examined at 245 nm using pH values of 1 and 7 respectively, the difference in absorbances can be related directly to the vitamin content of the solution *provided that irrelevant absorption is absent*. This proviso can be confirmed by measuring the absorbance of both solutions at 273 nm, when no change should be observed. Any change which does occur must be due to irrelevant absorption or change in solvent, and a correction can be applied to the absorbance at 245 nm. The difference in the absorbances at 245 nm then becomes directly proportional to the concentration of the vitamin.

Mixtures

Experiment 12 *Assay for methyl salicylate and diethyl phthalate in surgical spirit.*

The following exercise illustrates the assay of an absorbing component (diethyl phthalate) in admixture with another (methyl salicylate) by calculating its net absorbance at a wavelength (227 nm) where they both absorb. The methyl salicylate is first determined by measuring the absorbance of the diluted sample at a wavelength (306 nm) where the diethyl phthalate does not absorb.

Assay for methyl salicylate Dilute 5 ml of the Surgical Spirit to 100 ml with absolute ethanol. Dilute 5 ml of this solution to 100 ml with ethanol and measure its absorbance at 306 nm in a 1 cm cell using ethanol in the reference cell.
Assay for diethyl phthalate Dilute 10 ml of the final solution for the assay of methyl salicylate to 50 ml with ethanol and measure the absorbance at 227 nm, the λ_{max} of diethyl phthalate, in a 1 cm cell.

Using the following values of $A(1\%$ v/v, 1 cm) of the two substances, calculate the individual concentrations of methyl salicylate and diethyl phthalate in the Surgical Spirit. If necessary refer to the specimen calculations for ephedrine hydrochloride and chlorocresol (p.283).

	$A(1\%$ v/v, 1 cm)	
	227 nm	306 nm
Methyl salicylate	432	335
Diethyl phthalate	419	0

Experiment 13 *Assay of caffeine and sodium benzoate injection by the simultaneous equations method and by the absorbance ratio method*

The composition of the injection is:

Anhydrous caffeine	125 mg
Sodium benzoate	125 mg
Water to	1 ml

Method Prepare a standard solution of sodium benzoate (1 mg ml^{-1}) by dissolving about 100 mg accurately weighed in *water* and diluting to 100 ml. Dilute 5 ml to 250 ml with *water* to give a working standard of 20 μg ml^{-1}. Similarly, prepare a standard solution of anhydrous caffeine 20 μg ml^{-1} (Notes 1–3).

By means of serial dilutions (Note 4), dilute the injection with *water* to give a final concentration 10 μg ml^{-1} each of sodium benzoate and caffeine, based upon the stated concentrations in the injection.

Record on the same chart the absorption spectra (320–210 nm) of the standard and sample solutions in a 1 cm cell (Note 5).

For the simultaneous equations method, measure the absorbance of each solution at the wavelengths of maximum absorption of sodium benzoate and caffeine at around 227 nm and 273 nm respectively. Calculate the concentrations of sodium benzoate and caffeine in the diluted sample and hence in the injection by the method of simultaneous equations (Note 6).

For the absorbance ratio method measure the absorbances of the standard and sample solutions at the wavelength of maximum absorbance of caffeine at around 273 nm and at the iso-absorptive point around 245 nm where the two substances have equal absorptivity. Calculate the concentratons of sodium benzoate and caffeine in the diluted sample and hence in the injection by the absorbance ratio method (Note 6).

Which technique is likely to offer the greater precision? (Note 7). What aspects of the procedure are particularly liable to cause error if not carefully controlled?

Note 1 Dry the caffeine at 80° for 4 h.

Note 2 The caffeine standard may require gentle warming to effect dissolution. Cool to room temperature before completing to 100 ml.

Note 3 The determination of the iso-absorptive point in the absorbance ratios method is facilitated by using identical weights of sodium benzoate and caffeine.

Note 4 It is part of the exercise to calculate the dilution factor and to propose a suitable series of dilutions.

Note 5 Use a wide wavelength scale expansion, e.g. 2–5 nm cm^{-1}, to avoid errors in the measurement of absorbance values at certain wavelengths on the slope of the spectra.

Note 6 The mathematical treatment of the data for the simultaneous equations and absorbance ratio methods is given on pp.284–288. You are encouraged to insert the analytical data (absorbances and absorptivities) directly into equations 1 and 2 (p.285) and to solve these equations for the concentrations of sodium benzoate and caffeine rather than to use the final questions derived for c_X and c_Y.

Note 7 Hint. Consider the ratio of absorbance ratios as discussed on p.285.

Experiment 14   *Assay for chlorpromazine hydrochloride in a syrup formulation by difference spectrophotometry*

A direct spectrophotometric assay for chlorpromazine in a simple aqueous dilution of the syrup is subject to interference from the colouring and flavouring agents and from the decomposition product chlorpromazine-5-sulphoxide, which may occur as a result of aerial oxidation during storage. The difference spectrophotometric assay involves the oxidation of the intact chlorpromazine in an aliquot of the syrup to chlorpromazine sulphoxide by the reagent peroxyacetic acid, which is prepared by adding hydrogen peroxide solution to glacial acetic acid

$$H_2O_2 + CH_3COOH \rightarrow CH_3COOOH + H_2O$$

The absorbance of the oxidised solution is measured using an unoxidised solution of the syrup of the same concentration in the reference cell. The difference absorbance at 343 nm is proportional to the concentration of intact chlorpromazine, since any sulphoxide originally present as a degradation product and other absorbing substances are present in equal concentration in both the test and reference solutions, and their difference absorbances are zero (Davidson, 1976).

Method
Oxidising reagent Dilute hydrogen peroxide (100 volumes, 5 ml) to 500 ml with glacial acetic acid. Allow the solution to stand at room temperature for 16 h or heat the solution at 70° for 1 h. The reagent may be kept for 2 months.

Standard chlorpromazine solutions Dissolve chlorpromazine hydrochloride (about 60 mg accurately weighed into a 100 ml volumetric flask) in *water* and dilute to 100 ml. Pipette a 10 ml aliquot of the solution into two volumetric flasks (100 ml). To one flask add peroxyacetic acid oxidising reagent (5 ml) and dilute the contents of both flasks to 100 ml with *water*. As an aid to understanding the difference technique, record the absorption spectrum of each solution in 1 cm cells on the same chart using water in the reference cell. Now record the difference spectrum of the oxidised solution using the unoxidised solution in the reference cell. Note that the difference absorbance at each wavelength is simply the arithmetical difference in absorbance between the solutions measured separately. From the difference absorption spectrum measure the difference absorbance at the λ_{max} around 343 nm and calculate the $\Delta A_{1\,cm}^{1\%}$ using the modified Beer–Lambert equation:

$$\Delta A = \Delta A_{1\,cm}^{1\%}\, bc$$

where c is the concentration of the chlorpromazine hydrochloride in the reference cell in g/100 ml and b is the pathlength (1 cm).

Syrup solutions Measure the *weight per ml* of the preparation by weighing 10 ml in a dry volumetric flask (10 ml). Weigh an amount of the preparation equivalent to about 30 mg chlorpromazine hydrochloride into a volumetric flask (50 ml) and dilute to volume with *water*. Continue as described for the standard solutions. From the ΔA_{343} calculate the concentration of the chlorpromazine hydrochloride in the test solutions using the Beer–Lambert equation and hence determine the concentration of chlorpromazine hydrochloride in the syrup (Note 1).

Note 1 Evidence of extensive decomposition to the sulphoxide (i.e. greater than 10% of the chlorpromazine hydrochloride concentration) may be seen in the spectrum of the unoxidised solution of the syrup as a shoulder around 343 nm due to the sulphoxide. A difference spectrophotometric assay of the sulphoxide levels in syrups based upon the selective reduction with zinc and hydrochloric acid has been described (Davidson, 1978).

Experiment 15 *Assay for ephedrine hydrochloride in Ephedrine Hydrochloride Elixir by second derivative spectrophotometry*

The composition of the elixir is:

Ephedrine Hydrochloride	0.3 g
Water	6 ml
Glycerol	20 ml
Ethanol (90%)	10 ml
Chloroform Spirit	4 ml
Invert Syrup	20 ml
Syrup sufficient to produce	100 ml

The assay is based on a published procedure (Davidson and Elsheikh, 1982). Before carrying out the assay, carry out the preliminary experiment as an aid to understanding the spectral transformations that occur during the generation of derivative spectra.

Preliminary exercise Record the normal (zero order) absorption spectrum of a holmium filter from 300 to 500 nm (see Experiment 2a) using a suitable scan speed. Also record (on the same chart as the zero order spectrum, if possible) the first and second derivative spectra (Note 1). Examine the spectra and note:

(a) In the zero order spectrum, the isolated absorption band at 360.9 nm and the three fused absorption bands centred at 453.2 nm.

(b) The typical shape of the first derivative spectrum of an isolated band (at 360.9 nm) showing the maximum rate of change of absorbance with wavelength ($dA/d\lambda$), occurring at wavelengths of maximum slope in the zero order spectrum and the cross-over point ($dA/d\lambda = 0$) coinciding with the λ_{max} in the zero order spectrum. Note also the fusion of the first order bands of the triplet in the 453.2 nm region (Note 2).

(c) The typical shape of a second derivative spectrum of an isolated band (at 360.9 nm) showing a negative inverted sharpened band at the wavelength coinciding with the λ_{max} in the zero order spectrum, and *two* satellite bands. The triplet at 453.2 nm shows three inverted peaks and a fusion of satellite bands (Note 2).

Assay of ephedrine hydrochloride elixir The stated concentration of Ephedrine Hydrochloride is 3 mg ml^{-1}. Prepare a 'blank ephedrine elixir' matrix containing all the ingredients except Ephedrine Hydrochloride. Prepare a stock solution of Ephedrine Hydrochloride (3 mg ml^{-1}) and a working standard (0.3 mg ml^{-1}). Dilute the ephedrine elixir and the blank elixir (2 ml to 20 ml, Note 3) with *water*.

Record the zero and second order spectra of the standard solution in a 1 cm cell from 285 nm to 235 nm. Select suitable instrumental parameters (scan speed, absorbance range, recorder voltage) to give a second derivative (D^2) spectrum of adequate sensitivity. Under the same experimental conditions record the D^2 spectra of the diluted blank and sample elixirs.

Note in the D^2 spectra the discrimination in favour of the sharp benzenoid bands of the ephedrine against the broad spectral band of the elixir matrix which is due mainly to colouring agents (if present) and to the breakdown products of sugars e.g. 5-hydroxymethylfurfuraldehyde.

Confirm the proportionality of the D^2 amplitudes with concentration by plotting the five largest amplitudes (measured to the nearest 0.5 mm) against the concentration of standard solutions of ephedrine (0–0.5 mg ml^{-1}). Record each spectrum three times and use the mean value of each amplitude. As the temperature of the solutions affects the measured amplitudes (approximately 1% reduction per 1° rise in temperature; Davidson, 1983), ensure that the solutions are at room temperature.

Confirm the specificity of the procedure by observing the effect of an increasing elixir matrix concentration on the amplitudes of a constant concentration of Ephedrine Hydrochloride as follows. Prepare 4 × 20 ml solutions each containing Ephedrine Hydrochloride (3 mg ml^{-1}, 2 ml) and containing 0, 4, 8 and 12 ml of a blank elixir solution diluted with water. Record each spectrum three times and use the mean value of each amplitude.

Determine the most sensitive amplitude in the D^2 spectrum of ephedrine, which shows a proportional response with concentration and which is unaffected by the presence of the elixir matrix. Use this amplitude to assay the content of ephedrine in the elixir.

Note 1 If necessary obtain the assistance of a demonstrator to optimise the various instrumental settings that affect derivative spectra.

Note 2 Displacement of the true wavelengths by up to 2 nm in the direction of the scan may occur due to scan speed effects (Experiment 5). If necessary correct the displaced spectra.

Note 3 In the absence of a 'to contain' pipette which should be rinsed into the 20 ml flask, it will be necessary to measure the weight per ml and sample the elixir by weight rather than by volume.

Body fluids

In the following examples it must be remembered that an important preliminary part of the investigation has been omitted. It is essential that the compound to be determined is identified in the sample by appropriate techniques, e.g. TLC or GLC. Thus, in Experiment 18 bases other than Diphenhydramine will form the ion-pair and be measured by the procedure.

Experiment 16 *Assay of a barbiturate in plasma or urine by pH-induced difference spectrophotometry*

The assay depends upon the formation of a conjugated chromophore system in the molecule by change by pH, and the subsequent measurement of a ΔA value to overcome the effect of irrelevant absorption. The relevant formulae are given on p.324.

Reagents Boric acid solution: a solution containing boric acid (1M) and potassium chloride (1M). 0.45M sodium hydroxide. 1M sulphuric acid.
Solvent Wash chloroform (10 volumes) with approximately 1M sodium hydroxide (1 volume) and then with water (2 × 1 volume).

Method Transfer plasma or urine (5 ml) (Note 1) to a separator and add chloroform (75 ml). Shake thoroughly and allow to separate. Filter the chloroform extract and transfer an aliquot of the filtrate (50–60 ml accurately measured) to a clean separator. Add 0.45M sodium hydroxide (8 ml) and shake thoroughly. Run off the lower chloroform layer, transfer the alkaline solution to a centrifuge tube and centrifuge until clear. Dilute the clear supernatant solution (2 ml) with (a) 0.45M sodium hydroxide (1 ml), (b) boric acid solution (1 ml) and (c) 1M sulphuric acid (1 ml) in 1 cm cells (Note 2). Record the absorption spectrum of each solution (225–300 nm) against an appropriate blank solution obtained by treating 5 ml water as described for plasma or urine. Record also the difference absorption spectra of (i) the solution in 0.45M sodium hydroxide (pH approximately 13.5) against the equimolar solution in borate buffer (pH 10.5) and (ii) the solution in 0.45 M sodium hydroxide against the equimolar solution in sulphuric acid (pH approximately 0.5).

A positive ΔA around 260 nm and a negative ΔA around 240 nm in the difference spectrum of the pH 13.5 solution against the pH 10.5 solution is reasonable evidence of the presence of a barbiturate, because few other substances exhibit these spectral transformations between pH 13.5 and pH 10.5. For quantitative purposes, the larger ΔA at the maximum around 255 nm in the difference spectrum of the sodium hydroxide solution against the sulphuric acid solution should be measured. The calculation of the concentration of the barbiturate in the plasma or urine (Note 3) is based upon the proportional relationship that exists between the ΔA of the sample solution and that of standard solutions of the appropriate reference barbiturate (5 ml of a 20 μg ml^{-1} solution) carried through the procedure (Note 4).

Note 1 For urine samples, considerable amounts of interfering substances are to be expected in the chloroform extract. The extract should be purified by shaking with a phosphate buffer pH 7.4 (5 ml) before continuing as described.

Note 2 A melting point tube, sealed at both ends, is a convenient stirrer for this purpose.

Note 3 In simulating a case of poisoning, the body fluid should contain about 20 μg of barbiturate/ml.

Note 4 As an extension to this experiment an unknown barbiturate could be identified by TLC or HPLC (Chapter 4).

Experiment 17 *Assay of diphenylhydantoin (phenytoin) in blood or urine*

The determination is based upon the conversion of a weak ultraviolet absorbing system to one which is more highly absorbing, by oxidation as described by Wallace (1966, 1968).

Phenytoin Sodium   Benzophenone
                                    λ_{max}257 nm

Method Place 10 ml of the sample in a separator and adjust the pH to about 6–7 by dropwise addition of 0.5M hydrochloric acid. Add chloroform (100 ml) and shake vigorously (3 min). Filter through a phase-separating paper into a glass-stoppered graduated measuring cylinder and record the volume. Add 1M sodium hydroxide (5 ml) and shake thoroughly (3 min). Allow to separate and transfer the alkaline layer (4 ml) to a 250 ml round-bottomed flask and evaporate to approximately 1 ml by means of a rotary evaporator at 50–60° under reduced pressure (Note 1). Cool, add a solution of potassium permanganate in 1M sodium hydroxide (1%; 20 ml) (Note 2), heptane (5 ml) and reflux the contents, with stirring for 30 min. Cool and record the absorption spectrum of the heptane layer over the region 220–340 nm using a blank of heptane.

Calculate the amount of diphenylhydantoin by reference to a calibration curve (0–20 μg ml^{-1}) prepared by treating aqueous solutions of the drug to identical procedures.

Note 1 It is essential that thoroughly clean glassware is used and a solvent-free residue is obtained, otherwise loss of permanganate and poor oxidation occur.

Note 2 It is essential that the permanganate solution is purple in colour. The solution turns dark green on storage and should not then be used. Caddy *et al* (1973) recommend using separate solutions and mixing when required. An alternative simple method is to weigh approximately 6 g of sodium hydroxide pellets into a sample tube, mark the volume produced, and use as a measure for the quantity of sodium hydroxide in future assays. Potassium permanganate (0.2 g) is weighed for each determination. *Water* (15 ml) should be added to the reaction flask before addition of reagents.

Experiment 18   *Assay of diphenhydramine in plasma by the acid-dye technique*

Spectrophotometric procedures are not sufficiently sensitive for the assay of normal plasma levels of diphenhydramine, which are usually less than 0.2 μg ml^{-1}. Even the high sensitivity offered by the acid-dye technique (p.304) allows the assay of only high concentrations, in excess of 5 μg ml^{-1}, which may occur after overdosage or in cases of poisoning. The assay is based on the formation of the coloured ion-pair when methyl orange is added to an extract of diphenhydramine in dichloroethane. After the excess methyl orange is removed, the remaining indicator's absorption is a measure of the concentration of diphenhydramine.

Reagents Hexane A.R. Chloroform containing 1% v/v of isoamyl alcohol; the presence of the alcohol is essential to avoid adsorption of the diphenhydramine–indicator complex on glass surfaces, otherwise extremely variable results are obtained. Methyl orange solution: store over a layer of the chloroform : isoamyl alcohol solvent. Acid–alcohol. 2% w/v H_2SO_4 in ethanol.

Method To blood plasma (5 ml) add 0.1M sodium hydroxide (5 ml) and extract with hexane (25 ml). Reject the lower aqueous layer and extract the hexane layer with 2M hydrochloric acid (6 ml). Transfer the acid extract to a small separator (Note 1) and add 2M sodium hydroxide (10 ml). Extract with the special chloroform solvent (10 ml) and reject the aqueous phase. Shake the organic layer with methyl orange solution (0.5 ml) and centrifuge the organic layer to remove excess reagent (Note 2). To the clear chloroform extract (5 ml) add acid–alcohol (0.5 ml) and measure the absorbance at 535 nm (Note 3).

Prepare a calibration graph using standard solutions of diphenhydramine hydrochloride $(0–10 \ \mu g \ ml^{-1})$ treated as described for plasma.

Note 1 All the acid extract must be used and washing with small volumes of water will assist in the recovery.

Note 2 The volume of methyl orange reagent is kept small as the presence of more water would cause dissociation of the complex. It is for this reason that centrifugation is employed rather than washing with water.

Note 3 The addition of acid–ethanol changes the yellow colour of the indicator complex to the red colour of the indicator itself.

References

Amer, M.M., Ahmed, A.K.S. and Hassan, S.M. (1977) *J. Pharm. Pharmacol.* **29**, 355.
Bladon, P., Henbest, H.B. and Wood, G.W. (1952) *J. Chem. Soc.* 2737.
Caddy, B., Fish, F., Mullen, P.W. and Tranter, J. (1973) *J. Forens. Sci. Soc.* **13**, 127.
Colquhoun, D. (1971) *Lectures on Biostatistics*, p.222, Clarendon Press, Oxford.
Davidson, A.G. (1976) *J. Pharm. Pharmacol.* **28**, 795.
Davidson, A.G. (1978) *J. Pharm. Pharmacol.* **30**, 410.
Davidson, A.G. (1983) *Analyst* **108**, 728.
Davidson, A.G. (1984) *J. Pharm. Sci.* **73**, 55.
Davidson, A.G. and Elsheikh, H. (1982) *Analyst* **107**, 879.
Davidson, A.G. and Hassan, S.M. (1984) *J. Pharm. Sci.* **73**, 413.
Doyle, T.D. and Fazzari, F.R. (1974) *J. Pharm. Sci.* **63**, 1921.
Fisher, R.A. and Yates, F. (1953) *Statistical Tables for Biological, Agricultural and Medical Research*, 4th edn, p.80, Oliver and Boyd, Edinburgh.
Ghanem, A., Meshali, M. and Foda, A. (1979) *J. Pharm. Pharmacol.* **31**, 122.
Glenn, A.L. (1960) *J. Pharm. Pharmacol.* **12**, 595.
Glenn, A.L. (1963) *J. Pharm. Pharmacol., Suppl.* **15**, 123T.
Görög, S. (1968) *J. Pharm. Sci.* **57**, 1737.
Moffat, A.C. (Editor) (1985) *Clarke's Isolation and Identification of Drugs* (2nd edn) The Pharmaceutical Press, London.
Pernarowski, M., Knevel, A.M. and Christian, J.E. (1960) *J. Pharm. Sci.* **50**, 943.
Sharkey, M.F., Andres, C.N., Snow, S.W., Major, A., Kram, T., Warner, V. and Alexander, T. (1968) *J. Assoc. Offic. Anal. Chem.* **51**, 1124.
Sunshine, I. (1981) *Handbook of Spectrophotometric Data of Drugs*, CRC Press, Boca Raton.
Wahbi, A.M. and Farghaly, A.M. (1970) *J. Pharm. Pharmacol.* **22**, 848.
Wallace, J.E. (1966) *J. Forens. Sci.* **11**, 552.
Wallace, J.E. (1968) *Anal. Chem.* **40**, 978.
Williams, L.A. and Zak, B. (1959) *Clin. Chim. Acta* **4**, 170.
Woodward, R.B. (1941) *J. Am. Chem. Soc.* **63**, 1123.
Woodward, R.B. (1942) *J. Am. Chem. Soc.* **64**, 72, 76.

8
Atomic emission spectrometry and atomic absorption spectrophotometry

A.G. DAVIDSON

Atomic emission spectrometry

Emission spectra

When a solid body is heated to incandescence, radiation is emitted as a continuous spectrum, i.e. one which is uniform and without the appearance of bright or dark lines. On the other hand, excitation of gaseous or vaporised material in the atomic or molecular form causes the emission of line or band spectra respectively at specific wavelengths, characteristic of the material (Chapter 6). The energy necessary to excite the emission of spectral lines varies from element to element, and although, theoretically, it is possible to detect all elements in this way, atomic emission methods are usually limited in practice to the detection and determination of metals, metalloids and certain non-metals, such as silicon.

The energy available from a natural gas–air flame, which reaches a temperature of the order of 1600–1800°, is able to excite only a very limited number of elements to a level where emission will occur. At higher temperatures, in the oxygen–acetylene flame ($T = 3050°$) and in the direct current arc ($T = 4000$–$8000°$), the number of elements which can be detected increases rapidly with temperature, so that, whereas only 30 to 40 emit in the oxygen–acetylene flame, more than 70 can be detected in the direct current arc.

Spectrographic analysis

The technique of **emission spectrography** involves the recording on a photographic film of the individual wavelengths ('lines') of light emitted from a vaporised sample subjected to an electric discharge. The individual lines are separated by a prism or grating (Chapter 6). The elements vaporised into the discharge may then be identified by the position and relative intensity of the lines on the film.

The principal application of the technique is therefore the qualitative analysis of samples for the identification of metallic elements. Although the intensity of certain lines may be measured by a photoelectric receiver, quantitative analysis is more accurately carried out by **flame emission spectrometry** or **atomic absorption spectrophotometry**.

Flame emission spectrometry (flame photometry)

Flame emission spectrometry is the measurement of the intensity of light emitted at a particular wavelength from atoms that are excited thermally in a flame. Thermal excitation of the metallic atoms in the sample is more reproducible and more readily controlled for quantitative purposes by means of a flame than by electric discharge. The sensitivity of the former is lower, however, because the intensity of atomic emission depends on the number of thermally excited atoms, which is highly dependent on temperature.

The sensitivity of flame emission spectrometry for different elements depends on the energy difference (ΔE) between the excited and ground states (p.258). In general, the elements with the lowest ΔE values are the most easily excited in the flame and give a greater proportion of thermally excited atoms than those with higher ΔE values. In practice, only the alkali and alkaline earth metals that have ΔE less than 3 eV, i.e. calcium, sodium, potassium, lithium and barium, produce a sufficient number of thermally excited atoms at the temperature of an air–natural gas flame. Flame photometers that are designed to measure only alkali and alkaline earth metals use a relatively cool (e.g. air–natural gas or air–acetylene (2300°)) flame and consequently only the principal lines are emitted. The principal line of the analyte element is isolated by a filter (p.265) and its intensity is measured by a barrier-layer photocell (p.269).

Less easily excited elements, e.g. cobalt, copper, iron, magnesium and manganese, may be thermally excited with sufficient intensity at the higher temperatures of an oxygen–acetylene or nitrous oxide–acetylene ($T = 3000°$) flame. To isolate the required wavelength, a grating or prism monochromator is used instead of a simple colour filter, and the intensity is measured by a photomultiplier tube rather than by a simple photocell. However, these elements are normally assayed with greater accuracy, precision, sensitivity and selectivity by atomic absorption spectrophotometry, which is the spectrophotometric method of choice for most metallic elements other than the alkali and alkaline earth metals.

Interferences

Interference in the assay of an element may occur in the presence of certain other substances in the sample. It is convenient to classify these interferences as follows.

Cationic interference

Filter instruments give adequate resolution of the principal lines of sodium (589 nm), potassium (767 nm) and lithium (671 nm) to allow the assay of each element free of interference from the others. However, when sodium and calcium are both present in the sample, interference may occur as a result of their relatively close wavelengths of emission. Calcium can combine with the products of combustion of the flame gases to form calcium hydroxide, which gives a broad band molecular emission centred at 554 nm. Some of this emission may be transmitted by a broad band-pass sodium filter and give a falsely high reading for sodium. Conversely, the emission at 589 nm from high concentrations of sodium may interfere in the assay of calcium when measured at its principal wavelength 626 nm, but sodium does not interfere when measured at the less intense line of calcium at 423 nm. The extent of each interference depends upon the relative concentrations of alkali and alkaline earth metals and on the bandwidth of the filters. Cationic interference is eliminated by using an efficient monochromator rather than filters.

Anionic interference

Anions do not emit light but some, particularly polyvalent anions, depress the emission of certain cations by forming less volatile salts in the flame. For example, a solution of calcium ($10 \mu g\, ml^{-1}$) containing phosphate or sulphate ($10 \mu g\, ml^{-1}$) emits less intensely than one containing chloride. This is due to the formation of the less volatile calcium phosphate or calcium sulphate, which give fewer free excited calcium atoms than an equimolar solution of calcium chloride. The problem is avoided by adding a large excess of a **releasing agent** such as lanthanum chloride, which releases the calcium by the competitive complexation of the lanthanum with the phosphate. Alternatively, a **chelating agent** (Part 1, Chapter 9), such as EDTA, is used, which preferentially complexes with the calcium and removes the interference. The lack of interference from chloride allows the use of hydrochloric acid, at a concentration as high as $1\, mol\, l^{-1}$, for the dissolution of the sample.

Physical interference

To ensure reproducible and accurate results, the sample and standard solutions should be aspirated into the flame at the same rate and should give identical drop sizes in the nebuliser chamber. Substances that alter the physical properties of the sample may affect flow rate or drop size and cause erroneous results. For example, substances such as sugars, which increase the viscosity of sample solutions, reduce the flow rate to the flame and lower the emission of the elements. Also, alcohols in high concentration enhance emission. This is because the lower surface tension of the sample solution produces a smaller drop size and consequently a greater efficiency of thermal excitation of the atoms in the flame. These effects are minimised by preparing standard solutions containing the same concentration of the interferent as the sample solution, or by using the 'standard additions' procedure (p.342).

Flame photometer

The simple exercises described below may be carried out on any simple flame photometer. A description of one such instrument is given here because it illustrates the theoretical principles which have been given earlier. A diagrammatic view of the components is shown in Fig. 8.1.

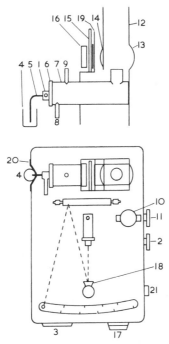

Fig. 8.1. Diagram of the components of the EEL flame photometer: 1, atomiser; 2, control valve; 3, pressure guage; 4, beaker; 5, stainless steel capillary tube; 6, ebonite plug; 7, mixing chamber; 8, draining tube; 9, gas inlet tube; 10, pressure stabilizer; 11, control valve; 12, chimney; 13, reflector; 14, lens; 15, optical filter; 16, photocell; 17, potentiometer; 18, taut-suspension galvanometer unit; 19, window; 20, plate; 21, zero control

Compressed air is supplied to a small annular-type atomiser (1) through a control valve (2) at a pressure of 12 lb per square inch as indicated on the gauge (3) mounted on the front of the instrument. The flow of air through the atomiser draws the sample from the beaker (4) up the stainless steel capillary tube (5) and sprays it as a fine mist through the ebonite plug (6) into the mixing chamber (7). Here the large droplets fall out and flow to waste through the drain tube (8). Gas is introduced into the mixing chamber through the inlet tube (9) from the internal gas pressure stabiliser (10) and control valve (11). The gas–air mixture passes to a multi-jet burner mounted above the mixing chamber, where it burns as a broad flat flame, and the hot gases pass up the well-ventilated chimney (12). The light emitted by the flame is collected by a reflector (13) and focussed by a lens (14) through the interchangeable optical filters (15) onto a barrier-layer photocell (16). The current generated by this cell is taken through a potentiometer (17) to a Tinsley taut-suspension galvanometer unit (18). A glass window (19) is interposed between the lens and filter for cooling purposes.

Operation With the flame burning at constant air and gas pressures, and with the appropriate filter in position, the strongest standard solution is sprayed by moving the beaker containing the solution up the recessed plate (20) on the side of the instrument. This automatically places it in position relative to the capillary tube up which the liquid is then drawn. The sensitivity control is adjusted to give a full scale reading. Water is now sprayed and the zero control (21) used to zero the instrument. The spraying of standard and water and the adjustments are repeated until the full-scale deflection corresponds to the strongest standard solution, for example to $10 \,\mu\text{g ml}^{-1}$ of K. The intermediate dilutions are now sprayed in turn and the readings are noted to prepare a calibration curve. The use of this instrument in the determination of concentrations is described in Experiment 1.

Calibration techniques

Direct calibration

When it is known that the sample is not subject to anionic or physical interference, the method of direct calibration may be used. This involves setting the full-scale and zero readings with the highest standard and blank solutions respectively as described above, and then spraying the other standard solutions (3–5 in number) containing intermediate concentrations of the element. A graph is constructed of emission *vs* concentration (Fig.8.2) and the concentration of the element in the sample is read from the graph, as the concentration corresponding to the emission of the sample. If the graph is linear, the data may be treated statistically as described on p.280. Frequently, calibration graphs show negative curvature, particularly at high concentrations of the element. This is due to the reabsorption of light emitted by the excited atoms in the hot central region of the flame, by the unexcited atoms in the cooler outer regions of the flame (cf. concentration quenching in spectrofluorimetry, p.362). Linearity is improved by using more dilute solutions. If the graph is linear over the concentration range in which the sample concentration occurs, a single-point standardisation (p.281) may be adopted.

Standard additions

This is a useful procedure for overcoming anionic or physical interferences from certain substances in the sample that affect the efficiency of the emission of the element. The method involves preparing a number (usually 4–6) of equal dilutions of the sample, each containing an added

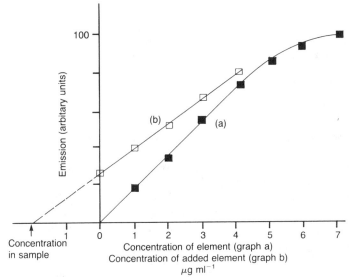

Fig. 8.2. Graphs for direct calibration procedure (graph a) and for the method of standard addition (graph b)

known concentration of the element. Ideally, the concentrations of added element should be approximately 0, 0.5x, x and 2x, where x is the concentration of the element in the sample. The full-scale reading and zero are set with the most concentrated solution and solvent respectively, as described above, and then the other solutions are sprayed. The data are plotted on a graph in which the abscissa is the concentration of added element (Fig.8.2). Extrapolation of the graph to an extended abscissa gives the concentration of element in the sample.

It may be useful to compare the slopes of 'direct calibration' and 'standard additions' data which cover a similar range of concentrations and which are measured on the same sensitivity setting. Non-parallel graphs show the presence of interferents in the sample solution and indicate the need to use the latter procedure to overcome the interference. Parallel graphs provide reasonable, but not conclusive, proof of the absence of anionic or physical interference. It should be noted that the standard additions procedure does not overcome cationic interference.

A number of examples of the application of flame photometry are given in official compendia and include limit tests for sodium in certain potassium salts and *Magnesium Chloride*, in which potassium and calcium are also subject to control. It is also used to determine potassium and sodium in *Haemodialysis Solutions* and sodium in *Intraperitoneal Solutions*.

Practical exercises

Experiment 1   *Determination of the concentration of potassium in a dilute aqueous solution of potassium chloride*

Calibration Curve

Method Prepare a stock solution of potassium chloride (AnalaR) by dissolving the dried salt (0.477 g) in sufficient *water* to produce 500 ml. The concentration of K is 500 μg ml^{-1} (500 ppm). Dilute aliquot portions of this stock solution with water to obtain standard solutions containing 2, 4, 6, 8 and 10 μg ml^{-1}. This series of solutions should be used to prepare the calibration curve.

Solution to be examined If the concentration of potassium is completely unknown, it is possible to avoid the preparation of a large number of dilutions by making use of the potentiometer sensitivity control. This is calibrated with a relative concentration scale which may be used in the following way to determine the approximate dilution required.

Method Spray a standard solution, say the 6 μg ml^{-1} standard, with the potentiometer set at full sensitivity, i.e. 1 on the dial. Note the scale reading. Reduce the sensitivity to zero and readjust while spraying the solution to be examined until the same scale reading is obtained. The reading of the potentiometer will then give the *approximate* dilution required.

Prepare and spray the dilution after standardising the instrument for full-scale deflection with the strongest standard solution in the normal way. Note the reading and calculate the concentration of K by means of the calibration curve.

COGNATE DETERMINATIONS

Haemodialysis Solutions. Determination of sodium and potassium.

Intraperitoneal Dialysis Solutions. Determination of sodium.

Body fluids The normal levels of sodium and potassium in urine are 100–250 mmol l^{-1} (2300–5750 ppm) and 40–120 mmol l^{-1} (1560–4680 ppm) respectively, and in blood serum are 136–148 mmol l^{-1} (3100–3400 ppm) and 3.8–5.0 mmol l^{-1} (147–195 ppm) respectively. Dilute the body fluid to give an emission that falls on the calibration graph for the element, ideally at about 60–80% of the emission of the most concentrated standard solution.

Experiment 2   *To show the effect of (a) ethanol and (b) dextrose on the intensity of the light emitted*

Method

(a) Dilute an aliquot portion of the stock solution of potassium chloride wih ethanol (10% v/v) to give a solution containing 6 μg ml^{-1} of potassium. Compare the reading obtained when spraying this solution with that given by the standard solution of the same strength. When setting the instrument to zero, spray ethanol (10% v/v) to overcome the objection that the increase in the reading may be due to traces of potassium in the ethanol.

(b) Repeat the above, but use a solution of dextrose (10%) in *water* in place of the ethanol (10% v/v).

The concentrations of ethanol and dextrose in the above solutions are grossly in excess of those that would normally be encountered in practice, but the exercise illustrates the importance of including in the standard solutions the same substances that are present in the solutions under examination. The concentration of these other substances should be of the same order in standard and test solutions.

Experiment 3 *Determination of calcium in Magnesium Chloride for Dialysis*

The B.P. limit for calcium is 0.01% whereas that for sodium is 0.5%. There is therefore considerable risk of error by interference from Na emission, unless selection of the calcium emission line at 423 nm is possible. This requires a suitable monochromator and a more elaborate instrument than a simple flame photometer. It is instructive, however, to use the emission at 626 nm isolated by means of an interference filter, and to investigate the effect of sodium on the apparent calcium content when added to standard solutions of calcium. The method suggested in the British Pharmacopoeia is the *standard additions* procedure, which allows for the presence of the sample by adding standard calcium solutions to the sample. It does not, however, allow for interference from sodium, which is the reason for selecting 423 nm if possible.

Method Add the magnesium chloride sample (5 g) to each of four 50 ml volumetric flasks A, B, C and D and add *water* to volume in flask A. To flasks B, C and D add respectively 0.5, 1.5 and 2.5 ml of calcium solution (400 μg ml^{-1}) and dilute to volume with *water*. The concentration of added calcium in flasks B, C and D is therefore 4, 12 and 20 μg ml^{-1}. Use the 20 μg ml^{-1} solution to set the instrument and spray each solution three times, with a *water* wash between each, and plot the average of each three readings against concentration of added standard. Extrapolate the line joining the points to cut the concentration axis. Read off the concentration of the sample solution and calculate the amount of calcium in the sample.

Note that sodium is also assayed in the sample using the standard additions procedure, whereas potassium is assayed by the direct calibration procedure.

COGNATE DETERMINATION

A commercially available Intraperitoneal Dialysis Solution prepared to the following formula provides a challenging exercise for the assay of calcium:

Sodium Lactate	3.924 g
Sodium Chloride	5.786 g
Calcium Chloride (dihydrate)	0.2573 g
Magnesium Chloride (hexahydrate)	0.1017 g
Dextrose (anhydrous)	42.5 g
Water to	1 litre

Investigate, at both 423 nm and 626 nm, the individual effects of (a) sodium, (b) lactate and (c) dextrose on the emission of a standard solution of calcium (40 μg ml^{-1}). The concentration of the interferent added to the solution of calcium should be that present when the sample is diluted to 40 μg ml^{-1} of calcium. The effect of lactate free of interference from sodium may be investigated by neutralising a solution of lactic acid with potassium hydroxide. The exercise may be extended by investigating the effect of lanthanum chloride on lactate interference (see p.355). Develop a flame photometric procedure for calcium, which avoids any interferences observed above. Compare the results with those obtained by using the complexometric titration procedure of the B.P.C 1973.

Atomic absorption spectrophotometry

Flame photometry has met with considerable success in analysis, but, even so, there are limitations, e.g. in the number of elements detectable and also in the number of atoms of an element excited by the flame. In any population of atoms at a temperature T, the ratio:

$$\frac{\text{Number of atoms in first excited state}}{\text{Number of atoms in ground state}} = \frac{N_2}{N_1} = \frac{P_2}{P_1} e^{-(E_2 - E_1)/kT}$$

where P_2 and P_1 are the statistical weights of the respective states and are related to the total quantum number J of the atom by $P = 2(J + 1)$, $(E_2 - E_1)$ is the excitation energy , and k is the Boltzmann constant. The ratio for zinc in a flame at $3100°\,K$ is 1.13×10^{-9} and most of the atoms in the flame are therefore in the ground state.

Atomic absorption spectrophotometry (AAS) uses the flame, into which the solution is sprayed, as if it were a cell containing an absorbing solution, the light source being a hollow cathode lamp of the **element being determined**. The intensity of a selected emission line is measured before and during the spraying of the solution in the flame to give an absorbance as in molecular spectrophotometry (p.277). Thermally excited atoms in the flame will, of course, emit some light of the same wavelength as that of the source, but, even if this interference is quite large, it is eliminated by modulation of the light source and incorporation of a suitably tuned amplifier to respond only to the modulation.

The absorption technique is useful when one element is being determined in the presence of a large amount of another which interferes in flame photometry. If interference is due to actual chemical combination, however, the absorption technique will be affected also. Examples of elements which are difficult to determine by flame photometry, but readily so by atomic absorption spectrophotometry, are magnesium and zinc.

Apparatus

Figure 8.3 shows the basic design of a simple atomic absorption spectrophotometer.

The sequence of events in the determination of an element is as follows. Sample solution is drawn through the atomiser (1) into the

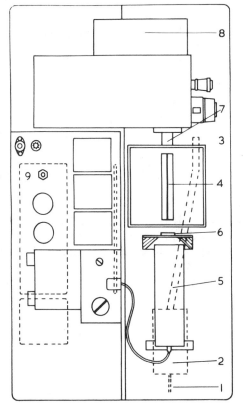

Fig. 8.3. Atomic absorption spectrophotometer

mixing chamber (2) where it mixes with the fuel and the support gas, either air or nitrous oxide. Large droplets run to waste through the drain (3) while the rest of the mixture passes to the burner (4), where the sample's chemical bonds are dissociated.

Emission from the lamp (5) which has a cathode made wholly or partly of the element to be determined, is modulated at 50 Hz from the line supply. The beam is focussed by the lens (6), on the centre of the flame. Absorption occurs. The transmitted remnant of the beam is focussed on to the inlet slit of the Czerny-Turner monochromator (7). The isolated resonance line emerges from the exit slit to energise the photomultiplier (8), the output of which is then amplified.

The amplifier (9), electronically switched in synchronisation with the lamp modulation, acts as a phase-sensitive detector discriminating against all signals originating outside the lamp including emission from thermally excited atoms. The amplified signal therefore represents only the intensity of the resonance line transmission.

Light source

Conventional sources such as tungsten lamps and deuterium or xenon discharge lamps are not satisfactory for atomic absorption, because the bandwidth of the radiation emerging from the monochromator is very broad by comparison with the absorption band of the ground state atom. The absorption line width for the latter may be from 0.001 to 0.01 nm, in comparison with a bandwidth of 1 nm from the monochromator of a conventional spectrophotometer. The light source must supply radiation of a line width less than that of the absorption line width of the element being determined. This is achieved by vapour discharge lamps for certain easily excited elements, e.g. sodium and potassium. Normally, however, *hollow cathode lamps* of the appropriate element are used (Fig. 8.4). When a current flows between the anode and cathode in these lamps, metal atoms are sputtered from the cathode cup, and collisions occur with the filler gas. A number of the metal atoms become excited and give off their characteristic radiation. The choice of the filler gas depends on its emitted spectrum compared with the spectral lines of the element of interest, and very often either argon or neon is used.

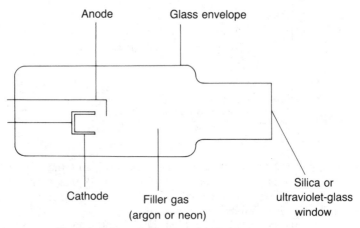

Fig. 8.4. Schematic diagram of hollow cathode lamp

Most commercial types of hollow cathode lamp are low current devices. This is necessary to prevent the heat developed by the discharge from melting the cathode filling, or the filling being lost due to evaporation. Examples are lead, zinc and bismuth, which are highly volatile, and can be run only at low excitation energies, although alloys of elements with low melting points can be run at considerably higher currents than those lamps with pure cathodes. Another reason for running hollow cathode lamps at low currents is their tendency to self-absorb the radiation emitted, thereby giving lower sensitivity than could otherwise be obtained without self-absorption. This effect is clearly demonstrated on lamps for the alkali metals and zinc.

It is possible to obtain multi-element lamps, but care is needed in their use as interference from adjacent lines of the element not being determined could give rise to non-linear calibration curves or spectral interference if the elements are also present in solution. Two-element lamps that are useful, however, are Ca/Mg, Cu/Zn and Na/K, and a three-element lamp is obtainable for Ca/Mg/Zn.

Present day lamps are stable so that although a large initial capital outlay is required for a selection of lamps, a useful life of several years may be expected.

The flame

The optimum conditions for the flame depend upon the physical dimensions of the burner, its position relative to the beam from the light source, temperature and fuel–air supplies. A long pathlength is desirable for adequate sensitivity, and typical dimensions of the burner slot are 100 mm × 0.4 mm.

An air–acetylene mixture gives a temperature of about 2300°, which is satisfactory for most elements. The temperature of the flame using nitrous oxide–acetylene is about 3000°, and this is required for refractory elements, e.g. aluminium. Special precautions are, however, required in using the latter mixture.

The ultimate sensitivity attained with the flame also depends upon the efficiency of the nebuliser unit for aspiration of the sample.

The monochromator

The radiation from the source consists of the line spectrum of the element forming the cathode, along with emission from the gas in the lamp. Isolation of the required radiation by grating monochromators is therefore fairly readily accomplished. However, in view of the presence of other elements in sample solutions, monochromators giving suitable dispersion are required. Dispersion will vary according to the quality of the monochromator and ranges from $1 \, \text{nm} \, \text{mm}^{-1}$ to $10 \, \text{nm} \, \text{mm}^{-1}$. Clearly, for resolution of closely spaced lines, a good monochromator is essential, but for many of the routine determinations less expensive systems operate quite well.

The wavelength scale associated with the monochromator need not be calibrated in subdivisions of nm units, as it is always necessary to adjust the setting manually in conjunction with the response of the galvanometer as described below.

Situation

It is unlikely that toxic levels of metals will occur in the atmosphere on spraying solutions, but it is a useful precaution to have some means of drawing off possibly high local concentrations of, for example, lead and mercury in the vapour above the flame.

Setting-up procedure

The operation of the spectrophotometer will vary from one instrument to another but there are common features that occur in operation.

Fuel and air supplies

The pressures are critical for optimum performance and once the air pressure is fixed the pressure of the fuel supply can be adjusted to give maximum absorbance when spraying a standard solution into the flame.

Selection of emission line

The light source will have several emission lines but one will usually show a much greater sensitivity to the element than the others, even although the intensity of emission may be less. For example, the lead line at 217 nm is more sensitive than that at 283.3 nm, although the latter is about 10 times more intense. On the other hand, the most useful line for copper at 324.8 nm is more intense than the other possible ones in the spectrum. Generally, it is the most sensitive line that is chosen, except where other factors must be considered, e.g. in the determination of lead excessive scatter at 217 nm could well make the selection of 283.3 nm radiation preferable if reasonable levels of lead are present (10–75 ppm).

Difficulty sometimes occurs, particularly with relatively simple monochromators, where the wavelength indicator may be one or two nm different from the true value. The monochromator should be carefully adjusted to give maximum transmission in the region of the known wavelength of the emission line. Immediately check that absorbance occurs when a solution of the particular element is sprayed. In a particular example, the wavelength indicator read 1.5 nm low, and it was quite easy to confuse the lead ion line at 220 nm with the lead line at 217 nm, with the result that no absorption occurred when standard lead solutions were sprayed.

Spraying technique

A standard method should be adopted to ensure reproducibility of results and reduce contamination and corrosion of the nebuliser and burner. It is usual to take three readings for each solution, with a spray of water between each, and to average the readings for plotting or calculation purposes. Immediately the determination is completed, 20 to 30 ml of water should be sprayed through the nebuliser and burner unit to sweep out all traces of previous solutions. Normally, solutions are very dilute but in the determination of trace impurities they may contain large amounts of corrosive salts.

Flameless techniques

Electrothermal atomisers

One of the major advances in instrumentation was the introduction of electrothermal atomisers for the generation of the atomic vapour. The most widely used system is a graphite tube (3–5 cm long by 0.5–1 cm in diameter) which is heated electrically in a furnace containing an inert gas, e.g. argon. The sample (liquid or solid) is inserted through a hole in the centre of the tube, which is then positioned in the light path in place of the flame burner and heated according to a predetermined programme as follows:

Stage 1. Drying, e.g. about 100° to remove water.
Stage 2. Ashing (300–500°) to remove organic constituents.
Stage 3. Atomisation (2000–2900°) to liberate the element as free gaseous atoms.
Stage 4. Removal of remaining inorganic matter at the maximum operating temperature (about 3000°).

The advantages of electrothermal atomisation are:

(a) only a small sample weight or volume is required (typically 10–20 μl or μg)
(b) the atomisation efficiency, and consequently the sensitivity, is greater than that given by flame atomisation, by up to 10 000 times
(c) chemical pretreatment of the sample is not usually required.

The technique is particularly useful for the assay of trace levels of metals and for the assay of metals when only small quantities of sample, e.g. blood, are available. A major disadvantage, however, is the reduction of precision compared to the flame atomisation technique. The latter is preferred when the sample contains sufficient quantities of the element under test.

Cold vapour systems

These are possible for particular elements and that for mercury is described in the practical section.

Hydride generation systems

A number of elements such as arsenic, tellurium and selenium may be reduced by a suitable reagent, e.g. sodium borohydride in acidic solution, to their gaseous hydrides, e.g.

$$6BH_4^- + As^{3+} + 3H^+ \rightarrow 3B_2H_6 + 3H_2 + AsH_3 \uparrow$$

The hydride is swept out of solution into a heated tube positioned in the optical axis of the spectrophotometer, where it dissociates to its constituent atoms.

The hydride generation and cold vapour techniques are many times more sensitive for those elements that can be converted to an atomic vapour than the corresponding flame methods. They also eliminate certain types of interference because the element is separated in a gaseous state from the interfering matrix.

Calibration techniques

The techniques of **direct calibration** and **standard additions** described above for flame photometry are also used in AAS, except that the ordinate scale is **absorbance** (p.277) rather than emission. Most calibration graphs show negative deviation from linearity at absorbance above 0.5, due to stray light (p.312), i.e. unabsorbed resonance lines from the lamp, close to the analytical wavelength. Many modern instruments have a 'curve correction' facility to extend the linear range, which is operated either by the analyst or automatically by the instrument's built-in microprocessor.

Instrument performance

The major criteria by which spectrophotometers, techniques and assay procedures for different elements are assessed are (a) sensitivity and (b) detection limit. The universally agreed definitions of these are as follows.

Sensitivity

This is the concentration of the element in $\mu g \, ml^{-1}$ (ppm) that reduces the intensity of incident monochromatic radiation by 1%, i.e. the concentration giving an absorbance of 0.00436 (p.277). The sensitivity is calculated by simple proportion from the concentration, read from the calibration graph, giving an absorbance of 0.1.

If the sensitivity of the instrument used in the assay of a particular element is known, e.g. from manufacturer's literature, the approximate concentrations to give absorbances in the optimum range (0.1 to 0.8) may be readily calculated.

Detection limit

This is the concentration of the element that gives a reading equal to twice the standard deviation of at least ten readings obtained with a blank solution. It is the smallest concentration of the element that can be distinguished from a zero concentration, with 95% certainty.

Interferences

Atomic absorption spectrophotometry is subject to the same anionic and physical interferences from those substances that alter the number of free gaseous atoms in the flame, as described for emission spectrometry (p.340). The techniques for avoiding these interferences have also been described earlier. Unlike emission spectrometry, however, atomic absorption is almost entirely free of cationic interferences. This is because the absorption of the sharp resonance wavelengths from the cathode of the light source containing a particular element is given only by gaseous atoms of the same element.

Two other types of interference which are generally observed only in absorption spectrometry are **scattering effects** and **ionisation** interference. The former occurs when high concentrations of certain interfering elements or particulate matter in the sample scatter the incident radiation, giving a falsely high absorbance for the analyte. This is a particular problem at low wavelengths. For example, in the assay of lead in urine the high concentrations of urinary sodium cause significant scatter at the wavelength of measurement, 217 nm. One technique for eliminating scatter interference is described in Experiment 5. Another technique for eliminating scatter or background absorption from molecular species is the use of a continuum light source such as a deuterium lamp (p.265) in addition to the hollow cathode lamp. Absorption by the analyte element of the broad bandwidth (relative to the very narrow width of the line source) of the continuum source is negligible, whereas scattering or molecular absorption gives a significant reduction of the light intensity. The corrected absorbance is obtained by subtracting the apparent absorbance measured using the continuum source from the total absorbance measured using the line source.

Ionisation interference may occur in hot flames which provide sufficient thermal energy to cause ionisation of those elements with a low ionisation potential, such as the alkali and alkaline earth metals. Since the wavelengths at which the gaseous ions absorb are different from those absorbed by the free atoms, ionisation reduces the absorbance of the element when it is measured at a wavelength absorbed by the free atoms. Ionisation is an equilibrium reaction and consequently the presence of other easily ionisable elements, e.g. potassium or calcium, will suppress ionisation and increase the absorbance. To avoid the different degrees of suppression that will occur in solutions containing different concentrations of the ionisation suppressant, an excess of a suitable suppressant, e.g. potassium ($2000\,\mu g\,ml^{-1}$) should be added to all solutions of those elements which undergo ionisation.

Practical experiments

Experiment 4 *Determination of total zinc in insulin zinc suspension (100 units ml^{-1})*

Insulin Zinc Suspension is a neutral suspension of insulin in the form of a water-insoluble complex with zinc chloride. Tests are applied to control both total zinc and also zinc in solution, determined on a sample of the supernatant liquid obtained by centrifuging the suspension. The percentage of total zinc and of zinc in solution varies according to the strength (i.e. 40, 80 or 100 units ml^{-1}) of the preparation.

Stock solution of zinc (5000 µg ml^{-1}). Dissolve zinc metal (AnalaR; 2.5 g) in 5M hydrochloric acid (20 ml) and dilute to 500 ml with *water*

Method To the suspension (well shaken; 2 ml) add hydrochloric acid (0.1 M; 1 ml) and dilute with *water* to 200 ml. Spray the solution using the standard procedure and read off the concentration of zinc from a calibration curve prepared with solutions containing 0.5, 1, 2 and 3 µg ml^{-1} of zinc.

Experiment 5   *Determination of lead in calcium carbonate*

There are many examples of pharmaceutical substances which require long, tedious methods for the determination of traces of Pb. For such compounds atomic absorption may offer a considerable advantage in time if they can be brought into solution quickly and easily.

Formerly *Calcium Carbonate* was in this class of compound, and although the method described in the British Pharmacopoeia to control heavy metals at not greater than 20 ppm is now very simple, the application of atomic absorption to the determination of traces of lead is described to illustrate several useful points. These are (a) effect of strong solutions on absorbance, (b) scale expansion and (c) a correction procedure sometimes applicable to solutions which show marked scattering of radiation. The limit for lead in *Calcium Carbonate* was formerly 10 ppm, and the solution of sample must necessarily contain a fairly high concentration of salt. This leads to difficulties such as matrix effects, weakening of radiation and low intensity of absorption by the trace element being determined. The last can be overcome to a certain extent by a scale expansion accessory, which is a valuable addition to an atomic absorption spectrophotometer. The effect of the presence of calcium chloride is to cause scatter and molecular absorption, both of which can be allowed for by making use of the lead ion line at 220 nm. The principle underlying the determination is therefore to obtain a reading at 217 nm (lead + other effects) and at 220 nm (other effects only). The reading at 220 nm is used to correct the absorption at 217 nm and the difference is related to the lead content of the sample. This convenient correction method for lead is based on the reasonable assumption that scatter and molecular absorption are similar at the two *adjacent* wavelengths. It would be quite wrong to use radiation, if any were present, situated 10–20 nm from the line of interest because scatter of radiation decreases with increase in wavelength.

A correction can, of course, be applied if a sample of Calcium Carbonate of known lead content is available for comparison. Alternatively, a deuterium hollow cathode lamp as a continuum source for background correction (p.353) may be used.

Stock solution of Lead Dissolve lead nitrate (3.995 g) in *water* to produce 500 ml of solution (5000 µg ml^{-1}).

Method To Calcium Carbonate (10 g) add *water* (20 ml) and hydrochloric acid (22 ml) *slowly* (Note 1). Boil to remove carbon dioxide, cool and dilute to 50 ml with *water*. Spray the solution adopting the standard procedure and record the response at 217 nm and

220 nm using scale expansion (Note 2). Measure the difference between the readings and calculate the amount of lead in the sample by reference to a calibration curve made at the same time with standard lead solutions containing 0.5, 1.0, 1.5 and 2.0 μg ml^{-1} lead (Note 3).

Note 1 Care should be taken to avoid loss of sample by the vigorous effervescence which occurs.

Note 2 The scale expansion should be such that the response for the sample solution should be almost full scale. It follows that the response at 220 nm will be less, and therefore on scale. It is also important, to illustrate underlying theory, to add some lead solution to the sample solution after completion of the exercise and redetermine the absorption at both wavelengths. The difference between the readings should be greater than that found in the exercise, but the response at 220 nm should be much the same, indicating that the ground state atom does not absorb the radiation characteristic of the lead ion.

Note 3 These values were suitable for a relatively simple instrument and should be well within the capabilities of most atomic absorption spectrophotometers. The volume of hydrochloric acid used in this determination is quite large and allowance should be made in the final calculation for any lead that may be present in the acid.

COGNATE DETERMINATION *Dextrose. Determination of lead* Use 20% solutions and the method of standard addition. Note that irrelevant absorption is absent when spraying this solution, in marked contrast to that shown by the calcium solution.

Experiment 6 *To show (a) the effect of phosphate on the absorption by Ca and (b) the elimination of the phosphate interference by the addition of Lanthanum*

Calcium (100 μg ml^{-1}) Dissolve 0.2497 g of dry calcium carbonate in 5 ml of hydrochloric acid (S.G. 1.18), boil to remove carbon dioxide and dilute to 1 l with *water*.
Phosphate (100 μg ml^{-1} PO$_4^{3-}$) Dissolve 0.2254 g of disodium hydrogen orthophosphate dihydrate (Na$_2$HPO$_4$. 2H$_2$O), or an equivalent weight of another sodium phosphate salt, and dilute to 1 l with *water*.
Lanthanum (10 000 μg ml^{-1}; 1%) Dissolve lanthanum chloride (26.6 g) or lanthanum oxide (11.72 g) in 1M hydrochloric acid and dilute to 1 l.

Method Transfer a 2 ml aliquot of the stock calcium solution to each of five 100 ml volumetric flasks. Add 0, 5, 10, 15 and 20 ml respectively of the phosphate solution to each flask and dilute to 100 ml with *water*. Measure the absorbance of the calcium at 422.7 nm in a 10 cm air–acetylene flame.

Repeat the experiment with the addition of 20 ml of the lanthanum solution to each flask. Plot the absorbance of the calcium against the concentration of phosphate for the two series of solutions.

Experiment 7 *The assay of calcium and magnesium in serum*

The normal serum levels of calcium and magnesium are 2.1–2.6 mmol l^{-1} (85–105 μg ml^{-1}) and 0.75–1.1mmol l^{-1} (18.2–26.7 μg ml^{-1}) respectively. The assay is carried out without prior deproteinisation or extraction by diluting the serum 1:50 or 1:100, depending on the sensitivity of the instrument. Interference in the assay of calcium and magnesium from serum phosphate (2 mEq l^{-1}) is eliminated by the addition of lanthanum to the serum samples.

Solutions Dilute 10 ml of the 100 μg ml^{-1} solution of calcium (Experiment 6) to 100 ml with *water*, giving a standard solution of calcium (10 μg ml^{-1}).

Prepare a standard solution of magnesium ($125\,\mu g\,ml^{-1}$) by dissolving oxide-free magnesium ribbon (125 mg) in 5 ml hydrochloric acid (S.G. 1.18) and diluting to 1 l with *water*. Dilute 5 ml of this solution to 100 ml, giving a standard solution of magnesium ($6.25\,\mu g\,ml^{-1}$).

Method Prepare a series of standard solutions of calcium (0, 0.5, 1.0, 1.5, 2.0 and $2.5\,\mu g\,ml^{-1}$) by appropriate dilutions of the standard solution ($10\,\mu g\,ml^{-1}$). Also prepare a series of solutions of magnesium (0, 0.125, 0.25, 0.375, 0.50 and $0.625\,\mu g\,ml^{-1}$) by appropriate dilutions of the standard solution ($6.25\,\mu g\,ml^{-1}$).

To 1 ml of the serum add 10 ml of 1% lanthanum solution (Experiment 6) and dilute to 50 ml with *water*. Record the absorbances at 422.7 nm of the diluted serum samples and the calibration series of standard solutions of calcium in a 10 cm air–acetylene flame. Similarly record the absorbances at 285.2 nm of the diluted serum sample and the calibration series of standard solutions of magnesium. Calculate the concentrations of calcium and magnesium in the serum with reference to the appropriate calibration graphs.

Experiment 7   *Determination of thiomersal (0.002%) as mercury in a solution for contact lenses*

The direct examination of the solution by atomic absorption is complicated by (a) the relatively low sensitivity of the method for mercury when using a flame, (b) the different response given by Hg^+ and Hg^{2+} and (c) the organic combination of mercury in the bacteriostat (*Thiomersal*). A considerable increase in sensitivity is obtained by converting organically combined mercury to the mercury (II) salt followed by reduction to Hg and sweeping this as vapour through a gas cell with end-windows of quartz. Absorption of 254 nm radiation gives adequate response for less than $0.1\,\mu g$ of mercury. The method is the basis for the determination of the very low levels of mercury encountered in pollution studies.

Apparatus For the Pye-Unicam, EEL 240 and Perkin-Elmer instruments the gas cell may be purchased to fit into the space occupied by the burner, but, if no such accessory is available, it may be easily prepared from glass tubing and two end-windows of quartz or silica. Figure 8.5 illustrates the cell and general arrangement. The pathlength of the cell is about 11 cm with end-windows 22 mm diameter fixed with any convenient resin, which also serves to hold the glass tube for locating the cell in position in the place normally occupied by the burner.

The coarse sinter-glass filter should be fairly large but capable of being covered by 5 ml of solution. A 50 ml Quickfit boiling tube is convenient for holding the sample. The delivery of air should be about $1000\,ml\,min^{-1}$ as supplied by a small pump. The system as described is open-ended but the vapour from the sample may be vented to the atmosphere through dilute HNO_3. A closed-circuit system is also possible but a more elaborate pump is required and all leaks must be eliminated.

Reagent Tin(II) chloride solution made by dissolving tin(II) chloride A.R. (20 g) in hydrochloric acid (20 ml) and water (20 ml), boiling in the presence of tin granules and diluting to 100 ml with *water*.

Method Evaporate the sample (5 ml) to dryness on a boiling water-bath, add sulphuric acid (1 ml) and continue heating with the occasional addition of a few drops of hydrogen

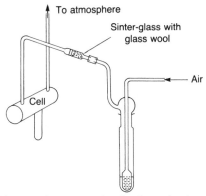

Fig. 8.5 Diagram of apparatus for the determination of mercury

peroxide (100 vol) until a clear solution is obtained (Note 1). Cool and dilute the solution to 200 ml with *water*.

Transfer the solution (1 ml) to the boiling tube, add *water* (4 ml) (Note 2) and tin(II) chloride solution (4 drops). Immediately connect up the apparatus, switch on the pump and record the absorbance as the vapour is swept through the cell (Note 3). Repeat the determination on aliquot portions of a solution of thiomersal treated in the same way.

Note 1 This simple treatment is sufficient for material low in organic matter, otherwise more drastic conditions are necessary, e.g. the normal wet-oxidation with sulphuric and nitric acids as prescribed for *Penicillamine*.

Note 2 It is essential for reproducible results that conditions are constant for each determination. The volume (5 ml) should be maintained and when aliquot portions of a standard solution are used, e.g. 0, 1, 2, 3 and 4 ml, water (5, 4, 3, 2 and 1 ml) should be added.

Note 3 For the conditions and apparatus described, 0.1 μg of mercury in the boiling tube should easily give a response almost full scale on the recorder. Scale expansion should be adjusted to achieve this. As the mercury vapour is swept out of solution, the recorder pen reaches a maximum and then returns to the base line.

COGNATE DETERMINATIONS

Investigate the application of the method to the following using 0.1 g and oxidation with nitric acid and sulphuric acid. Standard solutions of mercury should be prepared with mercury (II) chloride.

Danthron
Penicillamine
Penicillamine Capsules
Penicillamine Tablets
Penicillamine Hydrochloride

Note The limit test of mercuric salts in *Penicillamine* and its preparations is based upon flame atomic absorption spectrophotometry, after complexation of mercury (II) with ammonium pyrrolidinedithiocarbamate and extraction into 4-methyl-pentan-2-one.

9
Spectrofluorimetry

A. G. DAVIDSON

Introduction

The origin of fluorescence and the molecular energy changes that occur during absorption (excitation), fluorescence and phosphorescence are described in Chapter 6 (pp.259–264). Although some information about molecular structure may be derived from excitation and emission spectra, qualitative applications of spectrofluorimetry are rare, and the vast majority of applications in pharmaceutical analysis concern the quantitative assay of drugs, decomposition products and metabolites.

Fluorescence spectra

Instruments that measure the intensity of fluorescence are called **fluorimeters**. Those that measure the fluorescence intensity at variable wavelengths of excitation and emission, and are able to produce fluorescence spectra, are called **spectrofluorimeters**.

In the recording of fluorescence spectra, the limitations of light sources and measuring devices assume real significance. These limitations are variations of the intensity of available energy with wavelength, and variation in the response of the detector to light of different wavelength (Experiment 3). In absorption spectrophotometry both these factors are not immediately evident, because comparison of blank and test solution is carried out under identical conditions and the absorption spectrum recorded is true (within the limitations of the instrument). The excitation and fluorescence spectra may, however, be grossly distorted versions of the true spectra if the instrument is not specially adapted.

Excitation spectra

Before a compound can fluoresce, energy must be absorbed, and with an ideal light source of constant intensity at different wavelengths the most intense fluorescence is produced by radiation corresponding in wavelength to that of the absorption peak of the substance. Therefore, if the intensity of the fluorescence is plotted as a function of the wavelength of the radiation used to excite the fluorescence, an activation or excitation spectrum will result. This will be identical to the absorption spectrum when corrected for instrumental effects, because the fluorescence efficiency is generally independent of wavelength. As the intensity of fluorescence is measured at a particular wavelength, the disadvantage of variation in sensitivity of the detector with wavelength does not appear. However, in practice, the light source is not ideal and the output from the monochromator used to supply exciting radiation will vary according to wavelength. The detector will therefore respond to variations in intensity of fluorescence caused by more (or less) absorption of energy by the sample, and also by more (or less) excitation energy available from the light source. A curve of intensity of exciting light as a function of wavelength can be prepared for the light source and may be used to correct the apparent excitation curve obtained (Experiment 3).

Emission spectra (fluorescence)

When a monochromatic source of constant light intensity is used to irradiate a sample the fluorescence may be analysed in a monochromator at constant slitwidth to give an apparent emission spectrum. The true spectrum is obtained by applying a correction for change in detector sensitivity with wavelength and for changes due to the fluorescence monochromator, i.e. half-band width of emergent light and light losses.

Fluorescence emission spectra arise because of transitions from the first excited state (Fig.6.7) and their shapes are therefore independent of the light used to excite fluorescence. If the substance has one absorption band, the emission spectrum often bears a mirror-image relationship to it when plotted on a **frequency** scale but, if several bands occur this relationship may be highly distorted because of overlapping of absorption and fluorescence bands.

Instrumentation

When both excitation and emission spectra are to be recorded two monochromators are essential, one for the light source (excitation monochromator) and one for the fluorescence (emission monochromator). The light source must provide a high level of ultraviolet and visible radiation and a compact high pressure xenon arc lamp is commonly used.

The production of ozone by the photochemical conversion of atmospheric oxygen in the lamp compartment presents a potential toxic hazard unless the ozone is thermally decomposed or removed by adsorption onto charcoal.

As many experiments will almost certainly entail the measurement of very weak fluorescence, the detector must be a highly sensitive photomultiplier tube of low dark current.

If the main interest lies in fluorescence emission spectra one monochromator may be dispensed with and a suitable light source and filter used instead. The rather poor luminosity associated with the monochromator, even with a xenon arc lamp, is replaced by the much more intense light from a source such as a mercury vapour lamp, from which a suitable activation beam is isolated by means of the filter. This arrangement partially overcomes one of the difficulties inherent in spectrofluorimetry, i.e. that so much of the available light is lost.

The arrangement of a typical instrument is shown in Fig. 9.1.

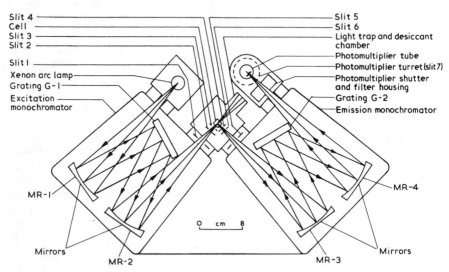

Fig. 9.1. Components of Aminco-Bowman spectrophotofluorimeter

Advantages

High sensitivity

Substances that are reasonably fluorescent may be determined at concentrations up to 1000 times lower than those required for absorption spectrophotometry. In a spectrofluorimetric measurement the photomultiplier tube measures a single light intensity (relative to a zero light intensity) which may be amplified electronically many times without introducing significant noise. In ultraviolet–visible absorption spectrophotometry the photomultiplier measures two intensities I_0 and I_T (p.274) and a reasonable difference, given by $I_T = 0.05I_0$ to $0.5I_0$ ($A = 1.3$ to 0.3), should exist between these intensities for the accurate and precise measurement of absorbance. At very low absorbances the small difference between I_0 and I_T approaches the noise of the signal and cannot be measured with satisfactory precision.

The high sensitivity offered by spectrofluorimetry may be of no advantage if the sample contains a sufficient quantity of the analyte for assay by absorption spectrophotometry, the latter being generally the more precise technique. For example, the highly fluorescent substance quinine sulphate may be assayed with good accuracy and precision in Quinine Sulphate Tablets (300 mg) by measuring the absorbance at 348 nm of a filtered extract of the tablet powder in 0.1M hydrochloric acid. However, low dosage drug formulations containing less than 1 mg per dose unit and biological samples (blood, urine etc.) containing low concentrations of drugs, may require the high sensitivity of spectro-fluorimetry, which is the spectrophotometric method of choice for many hormones, alkaloids and vitamins in formulation or biological samples.

Selectivity

Two factors confer on spectrofluorimetry a greater selectivity than that given by ultraviolet–visible absorption spectrophotometry. First, not all substances that absorb in the ultraviolet–visible region fluoresce. In non-fluorescent molecules, absorbed energy is lost by alternative radiationless pathways, principally by internal conversion (p.263). Molecules require, in addition to a chromophore (p.315), a degree of rigidity in their structure to reduce the dissipation of absorbed energy by internal conversion.

Substances that are fluorescent are characterised by their wavelengths of maximum excitation and maximum emission. Different fluorescent species may show different wavelengths of maximum excitation and/or emission. The facility to vary independently the wavelength of excitation and the wavelength of fluorescence allows the analyst to select the optimum combination of wavelengths for the analyte and to reduce interference from other fluorescing species in the sample.

Quantitative aspects

Many of the quantitative aspects of spectrofluorimetry may be understood by reference to the fundamental equation for the intensity of fluorescence emitted. This equation may be derived from that of the Beer–Lambert Law

$$\frac{I_0}{I_T} = 10^{abc} \qquad \text{(Chapter 7, p.277)}$$

$$\therefore \quad I_T = I_0 \times 10^{-abc}$$

But fluorescence $(F) = (I_0 - I_T)\Phi$ where $\Phi =$ quantum yield of fluorescence

$$\therefore \quad F = (I_0 - I_0 \times 10^{-abc})\Phi$$
$$= I_0(1 - 10^{-abc})\Phi$$
$$= I_0(1 - e^{-2.3abc})\Phi$$

Now

$$e = 1 + x + \frac{x^2}{2!} + \frac{x^3}{3!} + \dots$$

Let

$$x = -2.3abc$$

Then

$$F = I_0[(1 - 1 - x - \frac{x^2}{2!} - \frac{x^3}{3!} - ...]\Phi$$

$$= I_0[-x - \frac{x^2}{2!} - \frac{x^3}{3!} - ...]\Phi$$

$$= I_0[2.3abc - \frac{(-2.3abc)^2}{2!} - \frac{(-2.3abc)^3}{3!} - ...]\Phi \quad (1)$$

But

$$abc = \text{absorbance } (A)$$

$$\therefore \quad F = I_0[2.3A - \frac{(2.3A)^2}{2} + \frac{(2.3A)^3}{6} - ...]\Phi \quad (2)$$

At very low absorbance (< 0.02), only the first term is significant and eq. (1) simplifies to

$$F = 2.3I_0\, abc\Phi \quad (3)$$

For a fixed set of instrumental (I_0 and b) and sample (a and Φ) parameters, the fluorescence is proportional to concentration

$$F = Kc \text{ where } K = 2.3I_0\, ab\Phi$$

Factors affecting fluorescence intensity

Concentration

Equations (2) and (3) show that the fluorescence intensity of a substance is proportional to concentration only when the absorbance in a 1 cm cell is less than 0.02. With increasing absorbance the factorial terms in eq. (1) introduce an increasingly significant error (the **inner filter effect**) and cause negative curvature in calibration graphs. If the concentration of the fluorescent substance is so great that all incident radiation is absorbed, eq. (2) reduces to

$$F = I_0\Phi \quad (4)$$

That is, fluorescence is independent of concentration, and proportional to the intensity of incident radiation only, a property that may be utilised to determine the approximate emission characteristics of a light source (Experiment 3).

A further problem ensues if the emission and excitation spectra overlap, which results in the reabsorption of fluorescence and a negative dependence of fluorescence on concentration. The variation of fluorescence over a wide concentration range is shown in Fig. 9.2.

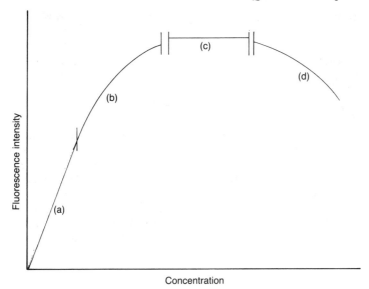

Fig. 9.2. Diagrammatic representation of the variation of fluorescence intensity with concentration. Region (a), Proportional relationship. Region (b), Negative deviation from linearity. Region (c), Fluorescence independent of concentration. Region (d), Reabsorption of fluorescence

Quantum yield of fluorescence (Φ)

This is the ratio

$$\Phi = \frac{\text{Number of photons emitted}}{\text{Number of photons absorbed}}$$

Since some absorbed energy is lost by radiationless pathways, the quantum efficiency is less than 1. Highly fluorescent substances have Φ values near 1, which shows that most of the absorbed energy is re-emitted as fluorescence. For example, fluorescein in 0.1M sodium hydroxide and quinine in 0.05M sulphuric acid have Φ values of 0.85 and 0.54 respectively at 23°. Non-fluorescent substances have $\Phi = 0$.

Intensity of incident light (I_0)

An increase in the intensity of light incident on the sample produces a proportional increase in the fluorescence intensity. The intensity of incident light depends on the intensity of light emitted from the lamp (discussed above; see also Experiment 3), the excitation monochromator transmission properties (which for a particular instrument are constant) and the excitation slitwidth. The intensity of incident light and sensitivity of a fluorescence measurement are increased by increasing the width of the excitation slit. However, wide slit settings introduce problems due to photochemical decomposition (see below) or to spectral overlap, with consequent reduction of selectivity. The choice of the excitation slit-width is therefore a compromise between sensitivity, selectively and photostability.

Pathlength

The symbol for pathlength (b) in eq. (1) does not refer to the dimension of the sample cuvette but to the internal volume of sample solution, where the fluorescence is both generated and detected. The effective pathlength viewed by the detector depends on both the excitation and emission slitwidths (Fig. 9.3). Therefore, the use of microcuvettes does not necessarily reduce the fluorescence. In fact, if inner-filter quenching or self-absorption are significant, the use of microcells may reduce these interferences and actually increase the measured fluorescence.

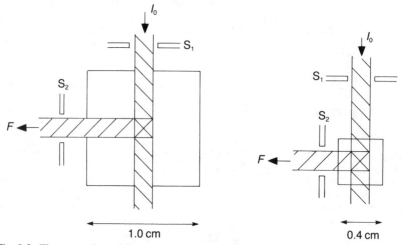

Fig. 9.3. The generation of fluorescence in conventional cuvettes (1 cm) and microcuvettes (0.4 cm), showing the effect of excitation (S_1) and emission (S_2) slitwidths. The diagonal hatching represents the paths of incident light (I_0) and fluorescence (F). The small region of cross-hatching represents the 'effective cell' dimensions

Adsorption

The extreme sensitivity of the method requires very dilute solutions, 10–100 times weaker than those employed in absorption spectrophotometry. Adsorption of the fluorescent substance on the container walls may therefore present a serious problem, and strong stock solutions must be kept and diluted as required. Quinine is a typical example of a substance which is adsorbed on to cell walls.

Oxygen

The presence of oxygen may interfere in two ways: by direct oxidation of the fluorescent substance to non-fluorescent products, or by quenching of fluorescence. It is a useful precaution, therefore, to check a de-aerated solution and compare the result obtained with that from the oxygen-containing solution. Anthracene is well known to be susceptible to the presence of oxygen.

pH

It is to be expected that alteration of the pH of a solution will have a significant effect on fluorescence if the absorption spectrum of the solute is changed. Many phenols, for example, are fluorescent in both dissociated and undissociated forms. Consequently, the fluorescence from a solution of the phenol will show two peaks, one being due to the ionic form; acidic solutions may be necessary to suppress the peak due to the ionic form.

Photodecomposition

In absorption spectrophotometry, the intensity of the radiation passing through solutions is weak by photochemical standards, although adequate for measurements; decomposition of solute is therefore not very likely. Spectrofluorimetry, on the other hand, requires high intensity illumination for irradiation and the risk of photochemical change is thereby increased. Parker and Barnes (1957) have calculated for a hypothetical case involving quantities to be expected in normal practice that an error of 20% could quite easily arise. It may be possible in unfavourable cases to select radiation of a wavelength which is not strongly absorbed so that the extent of photochemical change is reduced, whilst at the same time adequate sensitivity is retained.

Temperature and viscosity

Variations in temperature and viscosity will cause variations in the frequency of collision between molecules. Thus, an increase in temperature or decrease in viscosity is likely to decrease fluorescence by deactivation of the excited molecules by coliision. Similarly, many substances not normally fluorescent at room temperature are capable of emitting light when excited at a low temperature or when in a viscous solvent or glassy matrix. The temperature coefficients of fluorescence are typically about -1% per degree increase in temperature.

Quenchers

Quenching is the reduction of fluorescence intensity by the presence of substances in the sample other than the fluorescent analyte(s). Absorption of incident or emitted radiation quenches fluorescence by the

inner-filter effect (discussed above). **Collisional quenchers** reduce fluorescence by dissipating absorbed energy as heat due to collisions with the quenching species. For example, quinine is highly fluorescent in 0.05M sulphuric acid but non-fluorescent in 0.1M hydrochloric acid due to collisional quenching by the halide ion.

Static quenchers form a chemical complex with the fluorescent substance and alter its fluorescence characteristics. Certain xanthine derivatives, e.g. caffeine, reduce the fluorescence of riboflavin by static quenching. The application of spectrofluorimetry for the study of binding properties (number of binding sites, binding constants, location and mechanism of binding) of drugs and macromolecules, e.g. proteins, is based upon the alteration of fluorescence intensity of the protein, or of the drug, with binding. For a detailed account of fluorescence procedures in drug-binding studies, see the text by Pesce *et al.* (1971).

Scatter

When the excitation and emission monochromators are at the same wavelength, scattered light of the same wavelength as the incident light will be detected by the photomultiplier arising from colloidal particles in the sample (**Tyndall scatter**) and from the molecules (**Rayleigh scatter**). Even when the excitation and emission monochromators are set 20 nm or more apart, a little Rayleigh–Tyndall scatter may be detected. Although it is compensated by using a blank solution, it limits the sensitivity of measurements. Reduction of excitation and emission slit-widths to reduce spectral overlap of excitation and emission spectra will reduce Rayleigh–Tyndall scatter, at the expense of the sensitivity of the measurement.

Raman scatter arises from the conversion of some incident radiation into vibrational and rotational energy by the solvent molecules. The resultant scattered light is of lower energy and, consequently, of longer wavelength. Raman scatter by solvents is usually weak compared to the fluorescence intensity of the sample, but may be significant when the fluorescence of a very dilute solution is measured at a wavelength coinciding with the Raman scatter band. The difference in **wavenumbers** (p.256) between that of the incident radiation and that of Raman scatter is a constant value for each solvent, e.g. about $3.4 \times 10^3\,\mathrm{cm}^{-1}$ for water, $2.9 \times 10^3\,\mathrm{cm}^{-1}$ for ethanol and $0.7 \times 10^3\,\mathrm{cm}^{-1}$ for carbon tetrachloride.

Applications

Compounds which are intrinsically fluorescent are readily determined with simple instruments as the solution for examination is normally obtained by dissolution of the sample in a suitable solvent. Table 9.1 lists some typical examples.

Table 9.1 Fluorescent compounds

Compound	pH	Wavelength (nm)		Minimum concentration required (μg ml^{-1})
		Excitation	Fluorescence	
Adrenaline	1	295	335	0.1
Allylmorphine	1	285	355	0.1
Amylobarbitone	14	265	410	0.1
p-Aminosalicylic acid	11	300	405	0.004
Aureomycin	11	355	445	0.02
Chloroquine	11	335	400	0.05
Chlorpromazine	11	350	480	0.1
Chlorpromazine sulphoxide	7	335	400	0.02
Cinchonidine	1	315	445	0.01
Cinchonine	1	320	420	0.01
Cyanocobalamin	7	275	305	0.003
Ergometrine	1	325	465	0.01
Folic Acid	7	365	450	0.01
Lysergic acid diethylamide	7	325	465	0.002
Menadione		280	320	0.07
Noradrenaline	1	285	325	0.006
Oxychloroquine	11	335	380	0.08
Oxytetracycline	11	390	520	0.05
Pamaquin	11	300,370	530	0.06
Pentobarbitone	13	265	440	0.1
Phenobarbitone	13	265	440	0.5
Procaine	11	275	345	0.01
Procainamide	11	295	385	0.01
Proflavine	1	440	510	0.01
Physostigmine	1	300	360	0.04
Quinine	1	250,350	450	0.002
Rescinnamine	1	310	400	0.008
Reserpine	1	300	375	0.008
Riboflavine	6	444	520	0.01
Salicylic acid	11	310	435	0.01
Thiopentone	13	315	530	0.1
Thymol	7	265	300	0.1
Vitamin A		325	470	0.01

Single substances which are, in themselves, non-fluorescent may be determined as a result of chemical change. This method is useful for both inorganic and organic compounds, and many inorganic compounds form highly fluorescent complexes by combination with organic reagents (Gomez-Hens and Valcárel, 1982). The determination of selenium illustrates the increase in sensitivity which can be obtained with fluorimetry as compared with that for absorptiometry. Thus 0.3 μg of selenium may be determined by measurement of the absorbance of its complex with 3,3'-diaminobenzidine, but, by using the fluorescence of the complex, 0.04 μg of selenium can be measured. The sensitivity is further increased to 0.002 μg of selenium with 2,3-diaminonaphthalene as reagent.

Table 9.2 Compounds Readily Converted to Fluorescent Derivatives

Compound or element	Reagent	Wavelength (nm) Excitation	Wavelength (nm) Emission	Reference
Adrenaline	$K_3Fe(CN)_6$ or I_2 and then NaOH	410	530	(c)
Chlordiazepoxide	Photo-oxidation after hydrolysis	380	480	(e)
Hydrocortisone	70% H_2SO_4 in ethanol	470	520	(b)
5-Hydroxytryptamine	o-Phthalaldehyde	365	495	(d)
Primary amines and amino acids	Fluorescamine	390	485	(f)
Zinc	Rhodamine B	366	580	(a)

References: (a) Haddad, 1977; (b) Mattingly, 1962; (c) Prasad *et al.*, 1973; (d) Maickel and Miller, 1966; (e) Koechlin and D'Arconte, 1963; (f) Udenfriend *et al.*, 1972.

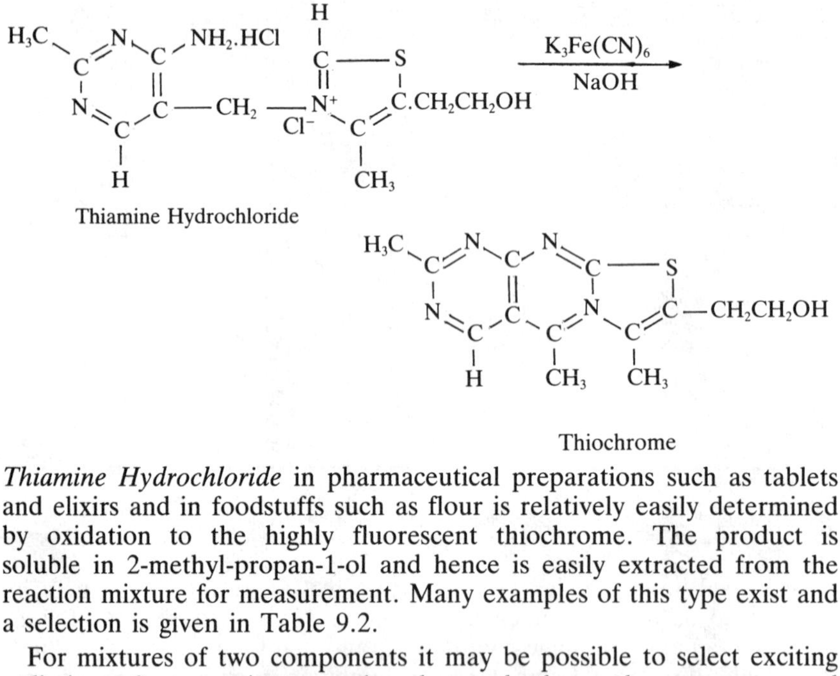

Thiamine Hydrochloride

Thiochrome

Thiamine Hydrochloride in pharmaceutical preparations such as tablets and elixirs and in foodstuffs such as flour is relatively easily determined by oxidation to the highly fluorescent thiochrome. The product is soluble in 2-methyl-propan-1-ol and hence is easily extracted from the reaction mixture for measurement. Many examples of this type exist and a selection is given in Table 9.2.

For mixtures of two components it may be possible to select exciting radiation of appropriate wavelengths, such that only one compound fluoresces at any one time. Even if this is not possible, measurement of fluorescence at two wavelengths may be sufficient to determine the composition of the mixture. Techniques which have proved their worth in absorption spectrophotometry are also of considerable value in spectrofluorimetry, e.g.:

(a) Application of separation methods and determination of each component separately.

(b) Changes in pH with concomitant changes in fluorescence (difference spectrofluorimetry), e.g. both morphine and codeine show fluorescence at 355 nm but only that of morphine is eliminated on making the solution alkaline (Experiment 10; see also experiments 9 and 11).

(c) Acid-dye technique, e.g. atropine forms a fluorescent chloroform-soluble complex with eosin. The technique is therefore useful in solving the many difficult analytical problems that arise when traces of compounds are to be determined in biological materials.

(d) Derivative spectrofluorimetry (cf. derivative absorption spectrophotometry; p.296). Generation of derivative excitation or emission spectra may permit the selective assay of one fluorescent substance in the presence of another, particularly if the spectral bandwidth of the former is smaller than that of the latter. Thus, chlorpromazine sulphoxide, an oxidative decomposition product of chlorpromazine, has been assayed with good accuracy, precision and selectively in formulations of chlorpromazine at concentrations down to 0.1% of that of the intact drug by recording the second derivative excitation spectrum (Davidson and Fadiran, 1984).

(e) Simultaneous equations (see p.284). An assay illustrating an application of simultaneous equations is described in Experiment 8.

Practical experiments

Experiment 1 *To determine the excitation and emission spectra of a substance for which no data are available*

Method Prepare a 1 μg ml^{-1} solution of the substance in *water*, ethanol or cyclohexane, as appropriate, and transfer to the fluorimeter cell. Set the excitation monochromator to an arbitary wavelength in the ultraviolet region (say 300 nm), and using high sensitivity , examine for fluorescence by turning the fluorescence monochromator manually. Note the wavelength for which a maximum response is obtained and set the monochromator to this wavelength, λ_{em}.

Obtain the excitation spectrum of the substance by recording the intensity of fluorescence at the wavelength λ_{em} as the excitation monochromator scans the spectrum from 200 nm. Note the excitation wavelength for which the fluorescence is a maximum and set the excitation monochromator to this wavelength, λ_{ex}.

Obtain the fluorescence emission spectrum by recording the intensity of fluorescence as a function of wavelength as the fluorescence monochromator scans the spectrum from a wavelength shorter than λ_{em}.

The spectra are obtained in a reasonably short space of time but it must be emphasised that, with so many factors involved, considerable manipulation of sensitivity controls may be required before the optimum conditions are found.

Experiment 2 *To determine the fluorescence emission spectrum of quinine sulphate solution (1 μg ml^{-1} in 0.05M sulphuric acid) at two different wavelengths of excitation*

Method Examine the excitation spectrum of quinine and note the two maxima at about 250 and 350 nm respectively. Record the fluorescence emission spectrum under the same conditions of sensitivity using first one and then the other wavelength for excitation.

Fluorescence occurs because of transitions from the first excited state to the ground state (Fig. 6.7). Even when the second and higher excited states are present, reversion to the first excited state occurs before fluorescence is emitted. Therefore, the position of the fluorescence emission spectrum is the same in both tracings, as would be expected. The difference between the spectra is one of intensity caused by the incident intensity of the radiation at about 350 nm being so much greater than that available at 250 nm. This difference is shown in the following experiment.

Experiment 3   *To determine the approximate emission characteristics of the light source and its monochromator*

Method 1 Prepare a 0.001% solution of quinine sulphate in 0.05M sulphuric acid and determine the absorption spectrum of the solution in the normal way (curve 1).

Dilute the solution with 0.05M sulphuric acid to give a $1 \mu g\,ml^{-1}$ solution of quinine sulphate and determine the apparent excitation spectrum over the range 200–400 nm with the fluorescence monochromators set at 450 nm (curve 2). Determine the blank values for 0.05M sulphuric acid and subtract from curve 2. Calculate a factor at each 10 nm interval by dividing the observed fluorescence (corrected curve 2) by the absorbance found in curve 1 at each wavelength. Plot the values against wavelength to give a graph, representing, in relative figures, the intensity of emission from the monochromator at each wavelength.

Method 2 Prepare a 1% solution of quinine sulphate in 1M sulphuric acid and determine the apparent excitation spectrum of the solution over the range 200–400 nm with the fluorescence monochromator set at 450 nm. The sensitivity controls must be adjusted to maintain the response of the photomultiplier at a reasonable level. Plot the readings (after allowing for changes in sensitivity) against wavelength to give a graph similar to that obtained in method 1.

The excitation spectrum is a function of the absorption of radiation by the sample, the fluorescence efficiency of the sample, and the intensity of the light source (eq. 3). Of these factors the fluorescence efficiency is normally constant over a wide range of wavelengths; therefore, a comparison of the absorption and excitation spectra of a well-characterised standard substance should show the variation of the intensity of the light source with wavelength. This is the principle of method 1 and among the standard substances available are:

aluminium chelate of
   2,2'-dihydroxy-1,1'-azonaphthalene-4-sulphonic acid
3-aminophthalimide
4-dimethyl-4'-nitrostilbene
m-nitromethylaniline
quinine sulphate

Argauer and White (1964) discuss the method and substances in detail but note that for accurate results the wavelength scales of the spectrofluorimeter and spectrophotometer must correspond very closely. As the intensity of the light source falls off rapidly at the shorter wavelengths, the possible error is greater at these wavelengths.

The response curve obtained by method 2 is possible because all the light is absorbed and eq.(4) is now relevant. Frontal illumination of the cell is preferable but this cannot be done in the Aminco-Bowman instrument unless the cell holder is removed completely and the cell is carefully adjusted manually to reflect the light into the fluorescence monochromator.

Both methods give figures which are proportional to the **number of quanta**. If, however, a thermopile is used to obtain the relative intensity of the emission of the light source with wavelength, the result is in terms proportional to **energy units**.

To convert from one form to the other use the relationship

$$\text{No. of quanta} = \frac{\text{No. of energy units}}{hv} = \frac{\text{No. of energy units}}{hc} \times \lambda$$

in which the symbols have their usual significance (pp.255–256). There-fore, multiply the figures obtained for the energy curve by the corres-ponding wavelengths to obtain the curve representing intensity in terms proportional to quanta.

Method 3 A spectrofluorimeter that allows scanning of both the excitation and emission monochromators synchronously is required for this experiment. Place a mirror in the cell compartment to reflect the incident radiation directly into the emission monochromator. Adjust both excitation and emission slitwidths to a narrow setting and adjust both monochromators to the same wavelength. To avoid damage to the photomultiplier tube by the high intensity reflected light, a very narrow slit (or other means of reducing the light intensity) should be inserted between the emission monochromator and photomultiplier. Adjust the sensitivity with both monochromators set at the same wavelength around 400 nm where emission from the xenon lamp is maximal and then record the relative spectral output by scanning the excitation and emission monochromators synchronously from 200 to 600 nm. A typical spectrum obtained in this experiment is shown in Fig. 9.4. The resonance lines of the xenon around 450 nm superimposed upon the continuous emission are readily observed.

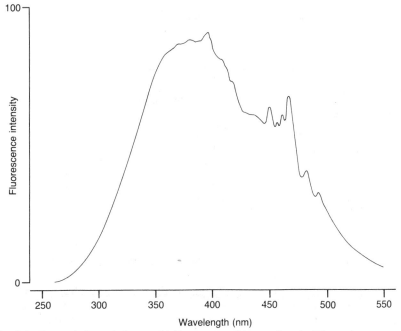

Fig. 9.4. The variation of photomultiplier output with wavelength. The major source of variation is the intensity of light emitted from the xenon lamp

Experiment 4   *To determine the concentration of quinine sulphate in Ferrous Phosphate Syrup with Strychnine and Quinine*

Method Dilute the syrup (about 2 g, accurately weighed) with 0.05M sulphuric acid to 100 ml. Dilute this solution further to give a concentration of quinine sulphate in 0.05M sulphuric acid of about 0.4 μg ml^{-1}. Measure the fluorescence (λ_{ex} = 350 nm; λ_{em} = 450 nm) of this solution and of a series of standard solutions of quinine sulphate (0.0 to 0.5 μg ml^{-1}) in 0.05M sulphuric acid. Read the concentration of quinine sulphate from a calibration graph or calculate it by using linear regression analysis (p.280). Determine the weight per ml of the syrup and calculate the concentration of anhydrous quinine (as % w/v) in the sample.

Experiment 5   *To determine the concentration of quinine in urine*

The assay may be applied to urine spiked with quinine (0.2 to 0.4 μg ml^{-1}) or to a sample of urine collected over 24 h after drinking 100–200 ml of tonic water. The latter sample may require a modification to the procedure, e.g. a further dilution, to cope with concentrations outside the 0.2 to 0.5 μg ml^{-1} range. The method is based upon that of Mulé and Hushin (1971).

Method Transfer 5 ml of urine to a 35 ml stoppered centrifuge tube, adjust the pH to 9–10 with drops of dilute ammonium hydroxide solution and extract with 10 ml of chloroform : propan-2-ol (3:1 v/v) for 1 min. Allow the phases to separate, with centrifugation if necessary. Transfer 5 ml of the organic phase to a dry centrifuge tube, extract with 5 ml of 0.05M sulphuric acid for 1 min and centrifuge. Measure the fluorescence of the acidic extract (λ_{ex} = 350 nm; λ_{em} = 450 nm) and of a standard solution of quinine (0.5 μg ml^{-1}) and blank carried through an identical procedure.

Experiment 6   *To show quenching of fluorescence*

(a) *Inner-filter effect* Dissolve 25 mg quinine sulphate in sufficient 0.05M sulphuric acid to produce 500 ml of solution. By dilution with 0.05M sulphuric acid, prepare two series of solutions of quinine sulphate so that the concentrations are (i) 0.1, 0.2, 0.3, 0.4, and 0.5, and (ii) 10, 20, 30, 40, and 50 μg ml^{-1} respectively. Measure the fluorescence (λ_{ex} = 350 nm; λ_{em} = 450 nm) of the first series setting the sensitivity with the 0.5 μg ml^{-1} solution and of the second series setting the sensitivity with the 50 μg ml^{-1} solution. Plot the measurements against concentration in the normal manner (see Fig. 9.2).

(b) *Collisional quenching* Using the 50 μg ml^{-1} standard solution of quinine sulphate, prepared above, and a solution of 0.1M potassium iodide, prepare solutions of quinine sulphate (0.5 μg ml^{-1}) in 0.05M sulphuric acid containing 0, 0.0005, 0.001, 0.0015, 0.002, 0.0025, 0.005, 0.0075, 0.01 and 0.015M potassium iodide. Plot the fluorescence against concentration of potassium iodide.

The fluorescence of the weak solutions shows a linear relationship with concentration. The stronger solutions show marked quenching effects due in this instance to the inner-filter effect (p.362). It is thus possible for a particular intensity of fluorescence to correspond to two concentrations of the substance. In fluorimetric assays, therefore, it is usual to check to avoid or detect this source of error as follows. Determine the apparent content of the fluorescing substance in the tablet, solution or preparation and then repeat the determination after adding a known amount of the substance being determined. The increase in fluorescence should correspond to the quantity of the substance added. If this is not the case, then quenching effects are present caused either by the substance itself (concentration quenching) or by other substances in solution. The latter effect is observed in the presence of potassium iodide (Experiment b) when the fluorescence values are seen to decrease with increasing iodide concentration due to collisional quenching (p.366).

Experiment 7 *To determine the concentration of proflavine hemisulphate in proflavine cream (0.1% w/w)*

Method Thoroughly mix the sample and transfer about 1 g, accurately weighed, to a separator. Dilute with ether (50 ml) and extract the mixture with 0.1 M hydrochloric acid (4 × 20 ml). Wash each extract with ether (20 ml), transfer to a 500 ml volumetric flask and make up to volume with 0.1 M hydrochloric acid. Dilute 10 ml of this solution to 100 ml with 0.1 M hydrochloric acid. Compare the fluorescence of the solution with that of a standard solution of proflavine hemisulphate (0.2 μg ml^{-1} in 0.1 M hydrochloric acid) using monochromators or filters of maximum transmission at about 440 and 510 nm for excitation and fluorescence respectively.

Extension Confirm that a proportional relationship exists between the fluorescence and concentraton of proflavine hemisulphate.

Experiment 8 *To determine the composition of a mixture of 1- and 2-naphthol*

Method Prepare stock solutions of 1- and 2-naphthol (about 0.001–0.003%) in *water, making both solutions identical in concentration.* Dilute each solution (5 ml) with *water* to 500 ml, including while doing so a buffer solution (pH 7.5, 5 ml). Determine the excitation and fluorescence emission spectra for each solution, and select one excitation wavelength such that a reasonable response to the fluorescence is obtained with each phenol.
  Prepare a solution of the sample of mixed naphthols (about 0.003%) and dilute 5 ml with *water* and buffer as described above. Determine for each solution (standards and sample), the fluorescence emission spectrum, recording each on the same chart. Record also the blank for the diluted buffer solution. Calculate the percentage composition of the mixture as described below.

  The calculation is similar to that for a two-component mixture in spectrophotometry (p.284) and is best illustrated by reference to an actual determination. From a consideration of the excitation and fluorescence emission spectra, 300 nm was selected as the excitation wavelength. With the instrument set at this value the traces shown in Fig. 9.5 were obtained.

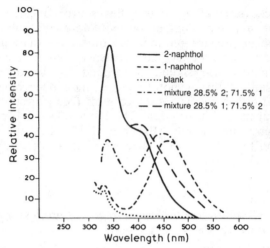

Fig. 9.5. Fluorescence spectra of 1- and 2-naphthols

Data:

Concentrations of standards:	$1 \times 10^{-5}\%$
λ_{max} for 2-naphthol:	350 nm
λ_{max} for 1-naphthol:	465 nm
Reading at 350 nm standard 2-naphthol	76.5 (net)
Reading at 465 nm standard 2-naphthol	36.7 (net)
Reading at 465 nm standard 2-naphthol	10.0 (net)
Reading at 350 nm standard 1-naphthol	2.5 (net)
Reading at 350 nm (mixture)	30.2 (net)
Reading at 465 nm (mixture)	40.5 (net)

Calculation:

Reading at 350 nm $= 76.5\beta + 2.5\alpha$

Reading at 465 nm $= 10\beta + 36.7\alpha$

where α and β are the proportions of 1- and 2-naphthol respectively.

$$\therefore \quad 30.2 = 76.5\beta + 2.5\alpha$$
$$40.5 = 10\beta + 36.7\alpha$$

Solving for α and β

$$\alpha = 1.003$$
$$\beta = 0.367$$
$$\therefore \quad \text{Percentage 1-naphthol} = 26.8$$
$$\text{Percentage 2-naphthol} = 73.2$$

Note that with a mixture in the solid state the actual concentration need not be known, but in a crude product and also when the naphthols are present in solution, the concentration must be obtained before the percentage can be determined. This information is available from the spectra as both naphthols give the same intensity of fluorescence at about 435 nm. By making the concentrations the same for the standards, this point is clearly seen on the recording. Note this reading (26.6 net) and also that for the mixture (36.8 net).

Then,

$$\text{concentration of mixture} = \frac{36.8}{26.6} \times \text{concentrations of standards}$$

$$= 1.38 \times 10^{-5}\%$$

The setting of the excitation monochromator is critical particularly if the scale is shorter than those encountered on absorption spectrophotometers. Therefore, if it is difficult to reproduce the setting with the desirable degree of accuracy, it is better to examine the solutions in the way described rather than to use constants for application to assays carried out after an interval.

Experiment 9 *To determine the content of $C_{20}H_{24}O_2$ in an Ethinyloestradiol tablet of average weight*

Method Weigh and powder 20 tablets. Place an accurately weighed quantity of the powder equivalent to about 0.2 mg of Ethinyloestradiol in a 50 ml graduated flask previously rinsed with methanol (Note). Add water (2 ml), warm gently and add methanol (20 ml). Shake for 5 min and dilute to volume with methanol. Allow to stand for tablet base to settle, and measure the fluorescence of exactly 2 ml of the clear supernatant liquid (λ_{ex} about 290 nm; λ_{em} about 320 nm). Add 1M sodium hydroxide (0.2 ml), mix carefully and again measure the fluorescence. Compare the difference between the two values with that obtained by measuring under the same conditions, the fluorescence of a standard solution of pure Ethinyloestradiol (4 μg ml) in methanol, before and after the addition of 1M sodium hydroxide (0.2 ml).

Note Purify the methanol before use by boiling under reflux over sodium hydroxide, and distillation.

The determination which is an example of difference spectrofluorimetry is based upon the fluorescence of the phenolic portion of the molecule, and the effect of pH is to give increased specificity to the assay.

Experiment 10 *Determination of morphine and codeine in admixture*

The determination depends upon the fact that codeine fluoresces in both acid and alkaline media, whereas morphine fluoresces in acid media only. The method was used by Chalmers and Wadds (1970) for the determination of both compounds in opium. They extended their work to the determination of papaverine and noscapine which were differentiated by the effect of protonation in chloroform solution on the fluorescence spectra.

Standard Solutions 5 μg ml^{-1} solutions of codeine and morphine in 0.1M hydrochloric acid and 0.1M sodium hydroxide.
Method Determine and record the fluorescence emission spectra for both solutions using appropriate blanks. Make a trial dilution of the solution of morphine and codeine to obtain almost full-scale reading in acid solution. When this is achieved make appropriate dilutions in 0.1M hydrochloric acid and 0.1M sodium hydroxide and record the fluorescence emission spectra as for the standard solutions. Calculate the concentration of morphine and codeine.

Experiment 11   *Determination of total alkaloids (calculated as emetine) in Ipecacuanha Tincture*

The determination depends upon the fluorescence of ipecacuanha alkaloids in acid solution and the large dilution of the tincture such that the effect of the colour of the tincture may be neglected.

Method Dilute the tincture (1 ml) to 200 ml with water in a volumetric flask. To the dilution (10 ml), add 0.05M sulphuric acid (10 ml), and dilute to 100 ml with *water*. Compare the fluorescence at about 318 nm (excitation at about 283 nm) with that from a 1 μg ml^{-1} solution of emetine hydrochloride in 0.005M sulphuric acid. (Note). Calculate the percentage of total alkaloids (calculated as emetine) from the relationship.

$$\% \text{alkaloids} = \frac{F_{\text{sample}}}{F_{\text{std}}} \times 0.0001 \times 2000 \times \frac{480.7}{552.7}$$

Note Emetine Hydrochloride BP contains 15–19% water and that of the USP contains 8–14%. It is therefore advisable to dry the standard material *in vacuo* at 100° before use to arrive at material of definite composition.

Extension

(a) Determine the effect of pH on the fluorescence characteristics of emetine and cephaeline. Develop selective spectrofluorimetric assays for emetine and cephaeline (see Davidson and Hassan, 1984).

(b) Determine the effect of light on the fluorescence characteristics of emetine and cephaeline in 0.01M sulphuric acid and 0.1M sodium hydroxide.

COGNATE DETERMINATION

Ipecacuanha Powder Weigh accurately the well-mixed sample in fine powder (0.1 g), add dimethylsulphoxide (5 ml) and warm in a water-bath (15 min). Cool, dilute to 200 ml with 0.005M sulphuric acid, mix well, and filter a portion. Dilute 10 ml to 100 ml with 0.005M sulphuric acid and proceed as described for Ipecacuanha Tincture.

Experiment 12   *Determination of dissolution rate for Digoxin tablets*

The tablets contain either 0.25 mg or 0.0625 mg of Digoxin and a sensitive method of assay is required. Although digoxin contains an α, β-unsaturated lactone ring, the intensity of absorption is not sufficient at the concentrations obtained in practice for ultraviolet absorption to be used. Wells *et al* (1961) developed a spectrofluorimetric method based on the conversion of digoxin to a mono-anhydro compound with the double bond situated at $\Delta^{8(14)}$.

Reagents

Ascorbic acid (0.1%) in methanol (AR).

Hydrogen peroxide 0.009M) prepared from strong hydrogen peroxide solution previously standardised with potassium permanganate.

Hydrochloric Acid AR.

Water, freshly distilled.

Digoxin EPCRS (100 mg) in ethanol (80%, 100 ml) stored at 4° and diluted immediately before use as follows: 10 ml to 100 ml with freshly distilled *water*, and this solution is further diluted (10 ml) to 100 ml with *water*.

Apparatus As described in Part 1, Chapter 14 using a stirrer speed of 120±5 rpm and the basket so positioned that the bottom of the basket is 2 cm from the base of the beaker.

Method Place 6 tablets in the basket and lower into the beaker containing freshly distilled *water* (600 ml) at 36.5 to 37.5°. Set the stirrer in motion and at intervals of 0.5, 1.0, 2.0 and 4.0 hours withdraw a sample (5 ml) from a point midway between the basket and the wall of the beaker and level with the middle of the basket. Replace the sample with water (5 ml).

Filter through a Millipore filter (0.8 μm) and reject the first 1 ml of filtrate. To the filtrate (1 ml) in a 10 ml volumetric flask, add, with thorough mixing, ascorbic acid solution (3 ml), hydrogen peroxide solution (0.2 ml), and dilute to volume with hydrochloric acid. Measure the fluorescence after exactly 2 h, using an excitation wavelength of 360 nm and a fluorescence emission wavelength of 490 nm.

The instrument should be set to zero with *water* and to 100 with the diluted standard digoxin solution (1.0 ml for 0.25 mg tablets, and 0.25 ml + 0.75 ml water for 0.0625 mg tablets) treated in the same way. Calculate the percentage of available digoxin in solution and plot the results. The B.P. specification is that the amount of digoxin per tablet in solution after 1 h is not less than 75% of the stated strength.

Extension Examine and compare the results for tablets obtained from different manufacturers (see Beckett and Cowan, 1973).

Experiment 13 *Determination of 11-hydroxysteroids in plasma*

The determination is based upon the development of fluorescence when such steroids are treated with sulphuric acid. It is not a specific reaction and Stenlake *et al* (1970) have shown that cholesterol and cholesterol esters are some of the major interfering fluorogens. Nevertheless, the determination is routinely done in clinical laboratories by a method essentially that of Mattingly (1962) and studied collaboratively (1971).

Reagents Dichloromethane. Purify by shaking with sulphuric acid ($\frac{1}{10}$th volume, 5 extractions), washing with water ($\frac{1}{10}$th volume) until the washings are neutral. Add anhydrous sodium sulphate, shake well and decant into a thoroughly clean distillation apparatus. Recover the dichloromethane by distillation (Note 1).

Water Glass distilled *water* should be used.

Fluorescence reagent. Add ice-cold sulphuric acid A.R. (7 volumes) to ethanol (for spectroscopy, 3 volumes) *slowly, with cooling in ice*. The mixture should be colourless and used within a few days.

Stock standard Dissolve hydrocortisone (50 mg) in ethanol (20 ml) and dilute 1 ml of the solution to 100 ml with *water* (concentration = 25 μg ml^{-1}). Store at 4°.

Working standard Dilute the stock standard (1 ml) to 100 ml with *water*.

Apparatus Scrupulous cleanliness is essential and all glass apparatus should be cleaned with chromic acid, followed by washing with sodium metabisulphite solution, tap water and finally with purified water.

Extraction of plasma Pipette the plasma (2 ml) into a glass-stoppered centrifuge tube (20 ml) and add dichloromethane (15 ml). Moisten the stopper with water to ensure a good fit and avoid loss of solvent, and shake gently, horizontally, for 2 min. Allow the phases to separate and remove the plasma by aspiration. Duplicate blanks (water, 2 ml) and standard (2 ml) are treated in the same way.

Method To the dichloromethane extract (10 ml) in a 20 ml glass-stoppered tube, add fluorescence reagent (5 ml) and shake vigorously for 20 s. Repeat the procedure with blanks and standards at 1 min intervals (Note 2). Transfer the acid layer to a spectrofluorimeter cell and measure the intensity of fluorescence at 520 nm using an excitation wavelength of 470 nm exactly 12 min after mixing the extracts with fluorescence reagent. Calculate the concentration of plasma corticosteroids using the formula

$$\mu g/100 \, ml = \frac{F_T - F_B}{F_S - F_B} \times 25$$

where F_T = Intensity of fluorescence of sample; F_S = Intensity of fluorescence of standard; and F_B = Intensity of fluorescence of blank.

Note 1 Dichloromethane vapour is toxic and care should be exercised in its use. If further purification is required, the solvent may be refluxed with and distilled from phosphorus pentoxide as described by Spencer-Peet *et al.* (1965).

Note 2 Strict timing is essential and no more than six tubes should be handled in one experiment (2 samples, 2 blanks, 2 standards). The fluorescence intensity for corticosteroids decreases with time, whereas that for cholesterol and its esters increases, an effect which is readily shown if the various compounds in fluorescence reagent are incubated at 60° over a period of time.

References

Argauer, R.J. and White, C.E. (1964) *Anal. Chem.* **36**, 368 (and correction 1022).
Beckett, A.H. and Cowan, D.A. (1973) *Pharm. J.* 111.
Chalmers, R.A. and Wadds, G.A. (1970) *Analyst* **95**, 234.
Collaborative Report, (1971) *Brit. Med. J.* **ii**, 310.
Davidson, A.G. and Fadiran, E.O. (1984) *J. Pharm. Pharmacol.* **36**, Suppl., 15P.
Davidson, A.G. and Hassan, S.M. (1984) *J. Pharm. Biomed. Anal.* **2**, 45.
Gomez-Hens, A. and Valcárel, M. (1982) *Analyst* **107**, 465.
Haddad, P.R. (1977) *Talanta* **24**, 1.
Koechlin, B.A. and D'Arconte, L. (1963) *Analyt. Biochem.* **5**, 195.
Maickel, R.P. and Miller, F.P. (1966) *Anal. Chem.* **38**, 1937.
Mattingly, D. (1962) *J. Clin. Path.* **15** 374.
Mulé, S.J. and Hushin, P.L. (1971) *Anal. Chem.* **43**, 708.
Parker, C.A. and Barnes, W.J. (1957) *Analyst* **82**, 606.
Prasad, V.K., Ricci, R.A., Nunning, B.C. and Granatek, A.P. (1973) *J. Pharm. Sci.* **62**, 1130.
Spencer-Peet, J., Daly, J.R. and Smith, V. (1965) *J. Endocr.* **31**, 235.
Stenlake, J.B., Williams, W.D., Davidson, A.G. and Downie, W.W. (1970) *J. Endocr.* **46**, 209.
Udenfriend, S., Stein, S., Bohlen, P., Darman, W., Leimgruber, W. and Weigele, M. (1972) *Science* **178**, 871.
Wells, D., Katzung, B. and Meyers, F.H. (1961) *J. Pharm. Pharmacol.* **13**, 389.

Reading

Argauer, R.J. and White, C.E. (1970) *Fluorescence Analysis*, Marcel Dekker, New York.
Hercules, D.M. (ed.) (1966) *Fluorescence and Phosphorescence Analysis*, Interscience, New York.
Miller, J.N. (1981) *Standards in Fluorescence Spectrometry*, Chapman and Hall, London.
Parker, C.A. (1968) *Photoluminescence of Solutions*, Elsevier, Amsterdam.
Pesce, A.J., Rosén, C–G. and Pasby, T.L. (1971) *Fluorescence Spectroscopy*, Marcel Dekker, New York.
Udenfriend, S. (1965 and 1969) *Fluorescence Assay in Biology and Medicine*, Academic Press, New York.

10
Infrared Spectrophotometry

A.G. DAVIDSON

Introduction

The infrared (IR) region of the electromagnetic spectrum extends from 0.8 μm (800 nm) to 1000 μm (1 mm) and is subdivided into near infrared (0.8 to 2 μm), middle infrared (2 to 15 μm), and far infrared (15 to 1000 μm).

The **fundamental region** between 2 and 15 μm is the region that provides the greatest information for the elucidation of molecular structure and most IR spectrophotometers are limited to measurements in this region.

Absorption of IR radiation causes changes in vibrational energy in the ground state of the molecule. The transition from vibrational level 0 to vibrational level 1 (Fig 6.5(a)) gives rise to the fundamental absorption of the molecule, and overtones or harmonics are caused by the transitions 0–2, 0–3 and so on, though the intensity of absorption for these overtones is very much less than that for the fundamental frequencies. For energy to be transferred from the light source to the molecule the frequency of vibration of each must coincide and, moreover, must be accompanied by a change in the dipole moment of the molecule. Certain symmetrical molecules, e.g. ethene, show no change in dipole moment during a stretching vibration and such vibrations do not result in the absorption of IR radiation. The C=C stretching vibration of ethene is described as **infrared inactive**.

Just as ultraviolet absorption spectra are strictly caused by changes in electronic, vibrational and rotational energy, so IR absorption spectra are due to changes in vibrational energy accompanied by changes in rotational energy. For simple molecules in the gas phase this leads to results from which information may be obtained on force constants, bond lengths, moments of inertia and, where adequate resolution is available, on isotopes in an easily accessible region of the spectrum. Experiment 1 for hydrochloric acid illustrates some of these points. For non-linear polyatomic molecules the number of normal modes of vibration is given by the expression $3n - 6$ ($3n - 5$ for a linear molecule) where n = number of atoms. All these vibrations occur at the same time and an IR spectrophotometer has been likened to a stroboscope in that it enables the particular frequencies of vibration to be recorded.

Particular groups in the molecule, e.g. hydroxyl, carbonyl and amines also have characteristic absorption frequencies known as **group frequencies**, which are almost independent of the nature of the rest of the molecule and which therefore occur in particular regions of the absorption spectrum. Neighbouring groups and interaction between atoms, e.g. hydrogen bonding, will, however, have some influence on the frequency and so changes of considerable value for use in structural analysis are observed.

Molecular vibrations are classified into **stretching** and **bending** (or **deformation**) vibrations. The former involves changes in the bond length, as if the atoms are connected by a spring vibrating with a particular frequency. Polyatomic molecules may have in-phase (symmetric) or out-of-phase (asymmetric) stretching vibrations as shown for a single group of atoms, e.g. methylene group (CH_2), in Fig. 10.1.

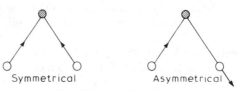

Fig. 10.1 Stretching modes

The different types of bending vibration, which involve a change in bond angles, are shown in Fig. 10.2.

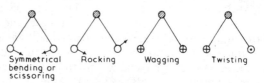

Fig. 10.2. Bending modes. (+) below plane of paper; (·) above plane of paper

Not all the vibrations of a molecule give rise to IR absorption, because certain vibrations are inactive (e.g. in symmetric molecules) or only weakly active and because some of the absorption bands involving similar and even different groups overlap.

Spectra

An IR spectrum is a plot of the absorption of IR light with variation of wavelength (λ) or, alternatively, wavenumber. The equation to inter-convert wavelength and wavenumber is

$$\text{wavenumber (in cm}^{-1}) = \frac{10^4}{(\lambda \text{ in } \mu\text{m})}$$

The ordinate may be linear in either % transmittance (%T) or absorbance (A). Examples of IR spectra in which the ordinate is linear in %T

are given in Figs 10.5–10.12. Unlike ultraviolet–visible spectra, IR spectra are conventionally plotted with the 100%*T* (or zero absorbance) at the top of the spectrum. IR absorption bands are therefore inverted compared to those in ultraviolet–visible spectra.

Factors that affect the quality and reliability of IR spectra, such as scan speed, slitwidth, sample concentration, interference from solvents, and stray light, have already been discussed with reference to ultraviolet–visible spectra (pp.307–312).

Instrumentation

The arrangement of a typical IR spectrophotometer is shown in diagrammatic form in Fig. 10.3.

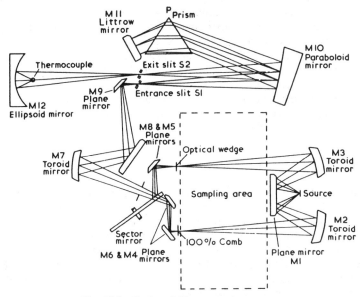

Fig. 10.3. Optics of IR spectrophotometer

The various components of a spectrophotometer, i.e. light source, monochromator, and detector are easily recognised and have been discussed in general terms in Chapter 6. The rotating sector mirror reflects the reference beam from M3 into the apparatus during one half of its rotation and allows the sample beam from M2 through during the other half. Any difference between the two beams causes an out-of-balance signal; at balance the two beams have the same intensity and produce a DC signal, but when unbalanced an AC signal results, which causes movement of the optical wedge into or out of the reference beam until the two beams are balanced. The optical wedge is coupled to a pen and an absorption curve is drawn as the spectrum is scanned. This system is known as the optical null method. Some modern instruments operate on the ratio-recording principle, described for ultraviolet–visible spectrophotometers (p.274). The wavelength of the radiation emerging from the exit slit depends upon the position of the Littrow mirror M11 relative to the prism. This mirror, rotating at a definite speed, moves the spectrum across the exit slit so that an absorption curve covering the region 2.5–15 μm can be obtained. The 100% comb in the sample beam is manually operated to adjust the pen to 100% transmission in the initial setting-up stage.

Prisms of sodium chloride are of use for the whole of the region from 4000–650 cm^{-1} but suffer from the disadvantage of low resolution at 4000–2500 cm^{-1}. The use of potassium bromide prisms extends the range to 400 cm^{-1}. Modern instruments are now usually equipped with a grating monochromator (p.267) which provides a better overall resolution throughout the range 4000–625 cm^{-1}. Typical rulings are 240 lines per mm for the 4000–1500 cm^{-1} region and 120 lines per mm for the 1500–650 cm^{-1} region. The presence of various orders of spectra passed or reflected by gratings makes it necessary to use filters or a fore-prism to ensure that unwanted radiation is eliminated; construction of the instrument is simplified if filters are used. The sample and reference beams are in close proximity to one another so that absorption by atmospheric moisture and carbon dioxide can be neglected unless so much is present that the available energy in certain parts of the region is reduced. Flushing of the instrument with dry nitrogen must then be carried out.

Qualitative uses

The number of absorption bands in an IR spectrum may be considerable, and therein lies its value as a means of identification. Two compounds are one and the same if their spectra agree in all respects, i.e in the position and relative intensity of the bands. **Identification Tests** in the B.P. monographs of many drug substances and their formulations are based upon a comparison of the sample spectrum with an **Authentic Spectrum** published in the B.P. If the spectra are not identical and have been obtained by examination of the sample in the solid state, further tests must be carried out. The pharmacopoeia describes the following additional tests:

(a) Recrystallise the sample from the solvent specified and repeat the determination.
(b) Dissolve the sample in a suitable solvent and measure the absorption spectrum against the solvent as blank. Compare with the authentic substance under the same conditions.

Different absorption spectra, after taking these precautions, indicate non-identity of the compounds. This procedure is necessary because various crystalline forms of the same substance may give rise to different spectra, e.g. *Cortisone Acetate* has been obtained in five different crystalline forms. The preparation of a disc of a hydrohalide salt of an organic base should be carried out using a diluent with the same halide, to prevent exchange of halide and distortion of the spectrum that may occur if a different halide is used. Thus a hydrochloride salt of a base should be examined in a sodium chloride disc and a hydrobromide salt in a potassium bromide disc.

Comparison of spectra in this way is probably the simplest use of IR absorption. Very often, however, it is necessary to identify a completely unknown substance or substances in, perhaps, a single tablet. All the resources of the analyst must be brought to bear on the problem, including chromatography and ultraviolet absorption. As a first step, separation of the active principle from inert tablet base will be necessary and partition between immiscible solvents under controlled conditions of pH is very useful for this purpose. It may be possible to use the residue from an appropriate extract directly for IR analysis from the results of which the presence or absence of functional groups may be inferred. An indication may thus be obtained of the type of compound for which a search should be made in the reference files. The residue may, however, be a mixture and require further treatment. In this connection, gas chromatography or thin-layer chromatography, combined with IR analysis of fractions is a useful technique for the isolation and identification of pure components.

Interpretation of infrared spectra

Interpretation of absorption spectra is considerably simplified by charts or tables which correlate the frequency at which a band occurs with molecular structure, and Fig. 10.4 illustrates such a comprehensive chart. The region between 4000 and 1500 cm^{-1} is probably easier to interpret than that between 1500 and 650 cm^{-1}, because the latter includes many skeletal vibrations which are typical of the molecule as a whole. It is outside the scope of this book to give detailed analyses of the assignments, and recourse must be made to the references at the end of this chapter. The following spectra are therefore presented as an introduction to the interpretation of bands in various regions of the spectrum.

In the discussion of the spectra, note that not all the bands are assigned to particular groups. Such assignment would be impracticable because many skeletal vibrations occur, and the first step is therefore to interpret the region between 3500 and 1500 cm^{-1}. Strong absorption bands in the region 850–700 cm^{-1} are often indicative of aromatic systems, but no indication is given here of the wealth of information afforded by the pattern of the bands on the degree and position of substitution of the benzene nucleus. A further aid to interpretation is to examine the spectra of a series of compounds derived from a simple parent compound (see Table 10.1).

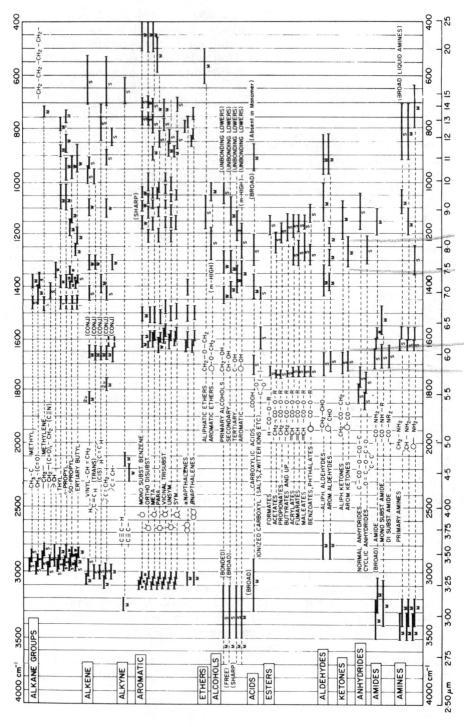

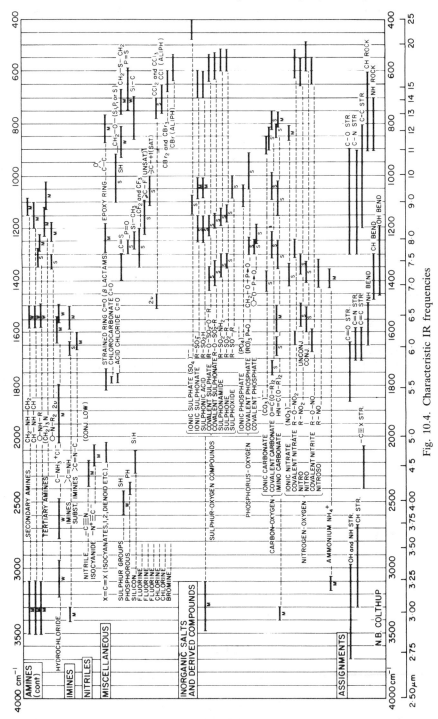

Fig. 10.4. Characteristic IR frequencies

Table 10.1 Chemically related compounds for comparison of IR spectra

Parent compound	Derived compounds
Heptane	Heptenes, methylhexanes, heptanol, heptanal, heptanoic acid and methyl heptanoate, methylcyclohexane
Benzene	Phenol, benzaldehyde, benzoic acid, methyl benzoate, nitrobenzene
Toluene	Benzyl alcohol, phenylacetic acid and methyl phenylacetate

A careful comparison of the individual spectra with those in the same and different series enables the effect of cyclisation and aromatisation to be observed, as well as the appearance of bands typical of group frequencies.

Examples

(a) Allyl alcohol

The absorption spectrum (Fig. 10.5) consists of few bands and the substance is therefore relatively simple. The presence of O—H is indicated by the broad absorption at 3300 (stretching) and at $1020\,cm^{-1}$ (C—OH stretching/deformation). Aromatic absorption is absent in the diagnostic regions (1600, 1500 and $850–700\,cm^{-1}$) but medium intensity absorption at $1650\,cm^{-1}$ may be due to unsaturation (C=C). This is confirmed by the much stronger absorption at 920 and $990\,cm^{-1}$ typical of the system $H_2C=CH—$.

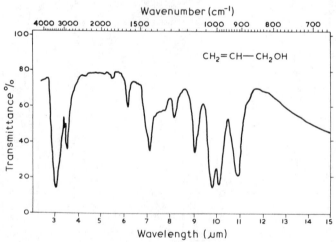

Fig. 10.5. IR spectrum of allyl alcohol

(b) Benzyl alcohol

The presence of the hydroxyl group is indicated immediately by the absorption at 3300 and $1020\,cm^{-1}$ (Fig. 10.6). This region also shows

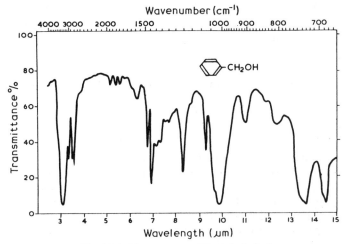

Fig. 10.6. IR spectrum of benzyl alcohol

absorption due to C—H in HC=C and CH_2 at 3010 and 2800 cm^{-1} respectively. That the unsaturation is of an aromatic nature is evident from the characteristics absorption in the regions 1600, 1500 and 850–700 cm^{-1}. The hydroxyl group cannot be attached directly to the aromatic system, otherwise enhancement of the absorption at 1600 cm^{-1} would be noticeable. Aromatic absorption at 695 and 735 cm^{-1} indicates monosubstitution of the benzene ring.

(c) Benzoic acid

The broad absorption band at 3000–2500 cm^{-1} (Fig. 10.7) indicates hydrogen bonded O—H of —COOH. Hydrochlorides of organic bases give somewhat similar absorption, but the carbonyl group is more likely

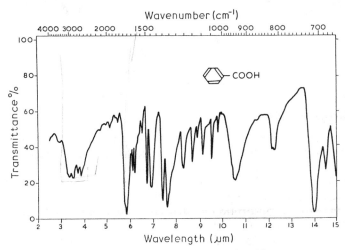

Fig. 10.7. IR spectrum of benzoic acid

in view of the strong carbonyl absorption at 1690 cm^{-1}. Confirmation is given by the bands at 1280 and 1310 cm^{-1} (—C—O—) and by the characteristic broad band at 940 cm^{-1} due to bonded O—H deformation in dimeric acids.

An aromatic system is indicated by the absorption at 1600, 1500, 720 and 695 cm^{-1}, the latter bands being characteristic of monosubstitution in the benzene ring.

(d) Acetylsalicylic acid

The appearance of the absorption in the 3000–2000 cm^{-1} (Fig. 10.8) is typical of O—H stretching in a carboxylic acid dimer. Strong bands at 1690 and 1300 cm^{-1} are also indicative of this group. Other carbonyl absorption is also present at 1750 cm^{-1}, confirmed by the characteristic ester absorption at 1180 and 1220 cm^{-1}.

The compound is aromatic as shown by the bands at 1600, 1490 and 850–700 cm^{-1}. The absorption at 1600 cm^{-1} is very strong for an aromatic compound, but it can be explained as being enhanced by the presence of polar substituents.

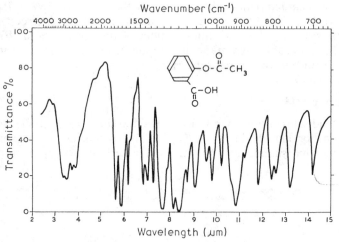

Fig. 10.8. IR spectrum of acetylsalicylic acid

(e) Ethyl-p-aminobenzoate

Aliphatic C—H is present only in small amount because of the very weak absorption at about 3000 cm^{-1} (Fig. 10.9). The band at 3200 cm^{-1} may be due to hydroxyl but is probably (from the intensity and appearance), due to NH stretching. An aromatic system is confirmed by the bands at 1580, 1500 cm^{-1} and by those in the 850–700 cm^{-1} region.

Strong carbonyl absorption at 1680 cm^{-1} might easily be interpreted as amide —C=O with the band at 1580 cm^{-1} as the second peak for amide absorption. This is in agreement with the N—H stretching frequency at 3200 cm^{-1}. However, typical ester absorption (C—O) appears at 1260

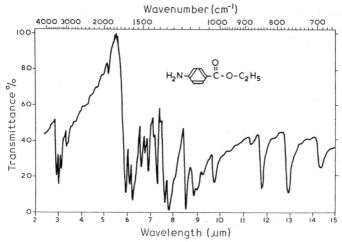

Fig. 10.9. IR spectrum of ethyl *p*-aminobenzoate

and $1280 \, \text{cm}^{-1}$, so that a better interpretation would be the presence of ester and amino groups.

(f) Phenacetin

The absorption at about $3200 \, \text{cm}^{-1}$ (Fig. 10.10) may be due to the NH stretching vibration rather than to —OH; compare the normal appearance of the latter absorption (Figs 10.5 and 10.6) with that in this curve. An aromatic system is probably present because of the low intensity of CH_2 vibrations. This is confirmed by the absorption at 1600, 1500 and $850–700 \, \text{cm}^{-1}$.

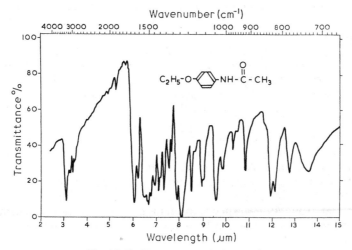

Fig. 10.10. IR spectrum of phenacetin

Carbonyl absorption is most likely to account for the band at $1650\,cm^{-1}$, as double bond systems give rise to much less intense absorption. It is however, rather low in frequency for an aldehyde, ketone or ester. The band at $1550\,cm^{-1}$ might be caused by NH deformation in an amide, by an amine or by a nitro group. When considered with the band at $1650\,cm^{-1}$, the most likely explanation is the presence of an amide group.

The absorption at $1250\,cm^{-1}$ is certainly in the C—O stretching region of an ester group, but aromatic ethers also absorb strongly in this region. Some confirmation for the group is given by the band at $1040\,cm^{-1}$.

Quantitative analysis

The principles underlying quantitative ultraviolet spectrophotometry apply also to IR work. Moreover, IR absorption spectra possess an advantage over those in the ultraviolet region in the greater number of bands present. It may often be possible to select a fairly strong band for each component in a mixture such that little or no interference occurs one with another. A calibration curve of absorbance against concentration may be constructed or, if Beer's Law is shown to be obeyed, a direct comparison of sample and standard absorption may be used. Strictly, the integrated areas of the absorption bands should be compared, but, with sharp bands, peak heights can be used in calculations.

Infrared and ultraviolet spectrophotometry differ, however, in the concentrations used. Ten per cent solutions are common in IR work and such a concentration is necessary because all solvents have some absorption in one part or another of the IR region. This means that very short pathlengths of 0.025 to 0.1 mm must be used in many assays. The high concentration of solute makes the accurate cancellation of solvent absorption very difficult, but errors may be reduced by applying a base-line technique. The assumption is made that absorption due to solvent (or a second component) is constant or varies linearly with wavelength over the region of the absorption band. The method of calculation is illustrated by reference to Fig. 10.11. The band *abc* is the recorded absorption of component A and *def* is the absorption caused by solvent and other components. Draw the line *agc* connecting the two minima *a* and *c* or between two suitable wavelengths on each side of the band. The point *g* is obtained by dropping a line perpendicular to the zero transmittance line to meet *ac* at *g*. The absorbance ($\log I_0/I_T$) is calculated from the distances shown in Fig. 10.11.

Even when bands do interfere, simultaneous equations may be set up as described for ultraviolet work. As many more components may be determined from an IR spectrum than is possible with an ultraviolet absorption spectrum, the work involved in solving a large number of

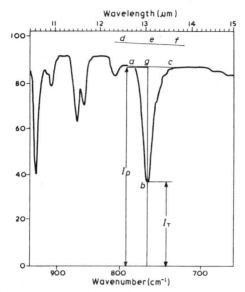

Fig. 10.11 IR spectrum showing application of base-line technique

simultaneous equations is considerable. Computers and matrix methods for solving simultaneous equations are therefore being used increasingly in analytical work. Infrared absorption assays also differ from ultraviolet absorption assays in the initial period of investigation necessary to obtain reproducible conditions. It may take some time to select the right solvent, cell thickness, slitwidth and absorption band.

For the determination of small quantities of substances, present either as impurity or as solute in a preparation, compensation for absorption by the major component or solvent in the preparation may be achieved by introducing that component into the reference beam. A spectrum of the minor component only is thereby obtained. This method has been applied to the determination of β-picoline, 2,6-lutidine and 2-ethylpyridine in γ-picoline and also to solutions of testosterone propionate in arachis oil. Sufficient energy must be available for reliable results and the validity of Beer's Law for appropriate standard solutions will normally confirm this.

Deviations from Beer's Law due to chemical effects (p.314) are encountered more frequently in IR spectrophotometry, as a result of the higher concentration of sample necessary for IR measurements. For example, the O—H stretching and O—H bending vibrations of alcohols and phenols are sensitive to hydrogen bonding. Intermolecular bonding increases as the concentration of the hydroxylic substance increases, and the absorbance at the λ_{max} of the free or unbonded hydroxyl group actually decreases with increasing concentration (See experiment 5).

Practical experiments

Experiment 1   *Determination of the vibration–rotation spectrum of hydrogen chloride, and calculation of the force constant, moment of inertia and bond length for hydrogen chloride*

Method   Place a thin layer of sulphuric acid in a 4 cm quartz or fused silica cell such as is used in ultraviolet spectrophotometry. Add carefully a few crystals of sodium chloride and cover the cell immediately with a polythene cap. Record the absorption spectrum over the region 3500–2500 cm^{-1} by means of a grating instrument using an empty quartz cell in the reference beam, and:

(a) Determine the position (in cm^{-1}) of the Q branch, i.e. the 'gap' in the spectrum (Fig. 10.12) and hence calculate k the force constant.

(b) Count the number of bands in the P and R branches and the range in cm^{-1}. Calculate the average separation in wavenumbers and hence obtain I, the moment of inertia.

(c) Calculate the bond distance from the relationship $I = \mu r^2$ (see below).

Adequate resolution for the complete experiment is obtained with a simple grating instrument or a more expensive large prism machine. If a simple prism instrument is available, the Q branch may still be observed and part (a) of the calculation can be done. The 4 cm quartz cell enables a gas cell to be dispensed with, although the latter is essential if hydrogen iodide or hydrogen bromide are to be examined.

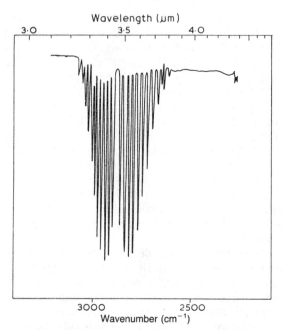

Fig. 10.12. IR spectrum of hydrogen chloride

The change in vibrational energy observed in this experiment is due to the transition $V_0 \rightarrow V_1$ (Fig. 6.5(a)). Changes in rotational energy accompany this transition so that the net result is given by

$$\Delta E = \Delta E_{\text{vibrational}} \pm \Delta E_{\text{rotational}} \qquad (1)$$

The rotational energy of the molecule is given by

$$E_{\text{rotational}} = \frac{h^2}{8\pi^2 I} J(J + 1)$$

where J = rotational quantum number, and I = moment of inertia. When changes occur ΔJ is restricted to ± 1 so that

$$\Delta E_{\text{rotational}} = \frac{h^2}{8\pi^2 I} 2(J + 1)$$

or

$$\overline{v} = \frac{h}{8\pi I c} \times 2(J + 1)$$

$$= 2B(J + 1)$$

where c = velocity of light.

B is a constant and therefore, according to expression (1) above, the spectrum should consist of equally spaced peaks on each side of a central position corresponding to $\Delta E_{\text{vibrational}}$. As a result of the selection rule $\Delta J = \pm 1$, $(J + 1)$ may be ± 1, ± 2, ± 3 . . . and therefore absorption corresponding to $\Delta E_{\text{vibrational}}$ alone will not be observed. This 'gap' in the spectrum is known as the Q branch and shows clearly in the absorption spectrum obtained in an actual experiment and recorded in Fig. 10.12. When $(J + 1)$ is equal to $+1$, $+2$, $+3$ . . . the R branch is observed, and when it is equal to -1, -2, -3 . . . the P branch is recorded.

The wavenumber $\overline{v}$ of the radiation associated with the transition $V_0 \rightarrow V_1$ is loosely referred to as the vibrational frequency of the molecule. By treating the two atoms H and Cl as two masses connected by a spring and applying Hooke's law, the wavenumber is shown to be, as a first approximation

$$\overline{v} = \frac{1}{2\pi c} \sqrt{\left(\frac{k}{\mu} \right)}$$

where c = velocity of light; k = force constant; and μ = reduced mass of system.

$$\overline{v} = \frac{m_1 m_2}{m_1 + m_2} \times \frac{1}{N}$$

where m_1, m_2 = respective **atomic weights of the atoms**, and N = Avogadro number. $\overline{v}$ is obtained from the position of the Q branch, hence k may be determined. From the average separation of the peaks, say x cm^{-1}

$$x = \frac{h}{8\pi^2 I c} \times 2 \, (J + 1 = 1)$$

$$\therefore \quad I = \frac{h}{4\pi^2 x c}$$

Now

$$I = \mu r^2$$

where r = internuclear distance.

$$\therefore \quad r^2 = \frac{I}{\mu}$$

$$r = \sqrt{\left(\frac{I}{\mu}\right)}$$

$$= \sqrt{\left\{(I \times N)\frac{1.008 \times 35.46}{36.47}\right\}}$$

From Fig. 10.12 it may be seen that the separation between the peaks in the P and R branches is not strictly uniform and is therefore not in accord with the simplified discussion presented above. The discrepancy is due to coupling between the rotation and vibration of the molecule. Also the peaks vary in intensity due to the fact that the population of the various rotational energy levels varies in the lower state ($V = 0$). For details of these and other factors which complicate the spectra, see Banwell (1972) or Barrow (1962).

Application of the formula for the 'frequency' of vibration (in cm^{-1}) enables *approximate* figures to be obtained from some of the group frequencies, for example for C = O, CH, N—H, and so on. The value of k used in the calculation is 5, 10 and 15 newtons cm^{-1} for single, double and triple bonds respectively.

Experiment 2   *Determination of the absorption spectrum of a liquid – cyclohexanol*

Method Check the instrument for 100% and 0% transmission and select the slitwidth programme (minimum slitwidth). Place a drop of the liquid on a sodium chloride plate of a demountable cell and cover with another sodium chloride plate, thus forming a capillary film of liquid. Insert the cell into the sample beam. If the film is too thick, carefully tighten the screws holding the cell until a transmission of about 5% is obtained for the strongest absorption band. Do not overtighten or the plates may crack.

After use, dismantle the cell, rinse the plates with chloroform and allow to dry under a lamp to avoid condensation of water vapour as the chloroform evaporates.

The thickness of the film can be varied by means of lead spacers but the method adopted above proves satisfactory in most cases. The liquid chosen here illustrates, in conjunction with the results of Experiment 5, the effect of hydrogen bonding on the absorption spectrum. Phenols, alcohols and carboxylic acids show this effect, and for further discussion see Experiment 5.

Experiment 3   *Determination of the absorption spectrum of a solid using the mull technique*

Method Powder thoroughly about 15–20 mg of sample in a small agate mortar and add 2 drops of purified liquid paraffin. Continue the trituration until a smooth paste is obtained and transfer it to one of the sodium chloride plates of a demountable cell. Assemble the cell and continue as described for a liquid in Experiment 2.

It is essential to reduce the particle size of the sample to below $3 \, \mu m$ otherwise scattering of radiation will give rise to a poor absorption spectrum. With care, as little as $3 \, mg$ of material may be used with a small amount of liquid paraffin. In this method, unless the material dissolves in the paraffin, the spectrum obtained is that of the *solid* sample. Crystal forces and hydrogen bonding may therefore influence the trace obtained. Strong bands due to paraffin itself appear at $2920\text{–}2850$ and $1460\text{–}1380 \, cm^{-1}$.

Experiment 4   *To determine the pathlength of a sealed cell*

Method Record the spectrum of the *empty* cell versus air and note the wavelengths between which lie an integral number of fringes. Calculate the thickness (t) of the cell by means of the equations:

$$(a) \; t \text{ (in cm)} = \frac{n}{2}\left(\frac{1}{\bar{v}_1 - \bar{v}_2}\right)$$

where n = number of fringes, and $\bar{v}_1$, $\bar{v}_2$ = wavenumbers selected, or

$$(b) \; t \text{ (in } \mu m\text{)} = \frac{n}{2}\left(\frac{\lambda_1 \times \lambda_2}{\lambda_2 - \lambda_1}\right)$$

where λ_1, λ_2 = wavelengths selected (μm).

The 'spectrum' obtained in this experiment is that of a typical interference fringe pattern and may be obtained for cells up to about $0.5 \, mm$ in pathlength. To illustrate the calculation refer to the fringe pattern in Fig. 10.13.

$$t = \frac{6}{2}\left(\frac{5 \times 10.1}{10.1 - 5}\right)$$
$$= 29.7 \, \mu m$$
$$= 0.030 \, mm$$

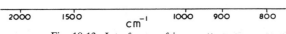

Fig. 10.13. Interference fringe pattern

The appearance of the pattern itself often indicates the quality of the cell: a poor fringe pattern is given by cells in which the plates are not exactly parallel or have become worn after use. Compare the patterns for *a* and *b* in Fig. 10.14.

For cells longer than about 0.5 mm the pathlength may be determined in several ways:

(a) by examination of a substance whose absorptivity is accurately known
(b) by use of a travelling microscope
(c) by comparison of a compound in the cell with the same compound in a previously calibrated variable pathlength cell.

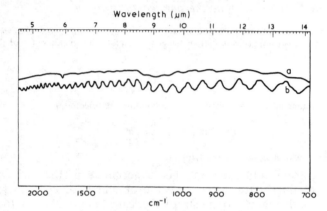

Fig. 10.14. Interference fringe patterns

Experiment 5 *Determination of the absorption spectra of solutions of cyclohexanol to show the effect of dilution on the O—H absorption*

Method Prepare two solutions (10% w/v and 1% w/v) of cyclohexanol in carbon tetrachloride. Obtain the absorption spectrum for each in a 0.1 mm cell (10% solution) and a 1 mm cell (1% solution) over the range 4000–2500 cm^{-1}.

Compare the curves with that obtained in Experiment 2. In the liquid state, a broad absorption band is recorded and this is retained in the strong (10%) solution. Note how this is reduced in the 1% solution; note also the increase in absorption at shorter wavelengths. This effect is due to a change from polymeric to dimeric and finally to monomeric hydroxyl absorption:

The formation of hydrogen bonds will tend to weaken the O—H bond; hence k, the force constant, will decrease, and the observed vibrational frequency will be less than that for the monomeric form. Carboxylic acids, however, remain dimeric even on dilution. If the hydrogen bond is intramolecular, no change in the O—H absorption will occur on dilution.

Solution spectra are useful in that difficulties inherent in solid-state spectra are absent. However, solvent effects may occur as a shift of the absorption frequency of a substance when one solvent is substituted for another. All solvents absorb in some part of the IR and several solvents are therefore necessary to cover the whole spectrum. Figures 10.15 to 10.17 show the absorption of three common solvents. In this experiment, carbon tetrachloride transmits most of the incident light and no reference cell is required.

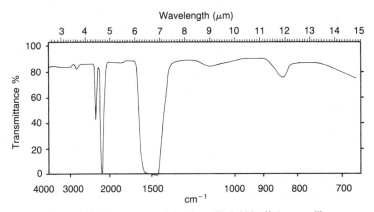

Fig. 10.15. IR spectrum of carbon disulphide (0.1 mm cell)

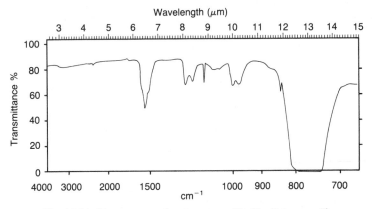

Fig. 10.16. IR spectrum of carbon tetrachloride (0.1 mm cell)

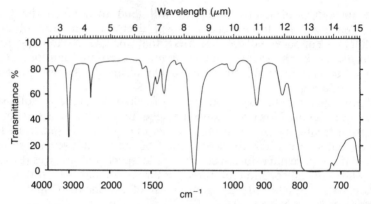

Fig. 10.17. IR spectrum of chloroform (ethanol-free) (0.1 mm cell)

The concentration needed for solutions will vary according to the substance and cell thickness, but a useful concentration for preliminary investigation is 10% w/v in a 0.1 mm cell.

Very thin films of aqueous solutions can be examined in cells of barium fluoride, Irtran-2 or silver chloride, but these solutions and, indeed, almost any kind of sample, can be handled by a technique called **attenuated total reflectance**, which is based upon total internal reflectance. The efficiency of the interface between two media as a mirror is dependent on the difference in refractive indices of the media. As the refractive index of a material changes rapidly in the region of an absorption band, these changes result in a decreased amount of energy reflected at the interface. The resulting spectra are, strictly, plots of reflected energy against wavelength and appear very similar to conventional absorption bands; they provide essentially the same information.

The technique requires an accessory for the IR spectrophotometer and the basic design for the unit handling micro-samples is shown in Fig. 10.18. Samples are held in contact with the flat face of the crystal

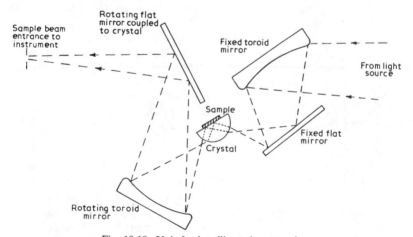

Fig. 10.18. Unit for handling micro-samples

hemicylinder and different crystals are required to provide a satisfactory range of refractive indices to cover all samples. Typical crystal materials are KRS-5, germanium and silver chloride, giving a high, medium and low refractive index respectively. The rotating mirrors are necessary to allow various angles of incidence on the crystal so that optimum attenuated total reflectance spectra are obtained.

COGNATE EXPERIMENTS
Benzoic acid
Catechol
o,o' and p,p'-Dihydroxydiphenyl

Experiment 6 *Recording the absorption spectrum of a solid using the potassium chloride disc technique*

Method Finely powder the sample (1 mg) and mix with dry finely powdered potassium chloride or potassium bromide (200 mg). Triturate the mixture thoroughly for about 2 min and transfer to a suitable die and press. Attach the die to a vacuum pump, allow to stand for 1 min under vacuum and compress the powder at 10–12 tons per square inch for 2 min. Release the pressure and vacuum, and extract the disc for insertion in the sample beam of the spectrophotometer.

For potassium chloride alone the disc should be completely transparent, but it may become translucent when the sample is present. The disc is about 13 mm diameter and the above quantities will therefore give a weight of sample between 10 and 15 $\mu g\,mm^{-2}$ and a concentration of 0.5%.

Attention has already been drawn to the care necessary when using this technique for comparison purposes. It has advantage, however, of giving spectra free from solvent peaks, and is therefore extremely useful as a routine method.

The disc method has not entirely superseded the mull technique for the spectra of solid samples, because differences have been observed in the results obtained by using two techniques; the differences are related to the treatment the sample received during the preparation of the disc. In finely ground material, for example, interaction may occur between vibrations of the sample and the potassium halide lattice. For a discussion of the IR analysis of solid substances see Duyckaerts (1959).

The method is also used in the examination of the very small quantities of material eluted from columns in gas chromatography. The fraction is trapped in about 300 mg of potassium chloride in the form of a short column placed immediately after the detector. The solid is powdered and pressed into a disc in the normal way, and the absorption spectrum of the trapped fraction is determined.

Experiment 7 *Determination of the amount of phenobarbitone in phenobarbitone tablets*

Method Prepare a series of solutions of phenobarbitone in chloroform containing 20, 40, 60, 80, and 100 mg in each 10 ml. Transfer each solution in turn to a 0.1 mm cell and measure the absorption spectrum over the region 4.4–6.5 μm. Calculate the absorbance using the base-line technique for each solution at the peak absorption (about $1730\,cm^{-1}$) and plot a graph of absorbance against concentration in the normal manner.

Weigh and powder 20 tablets. Extract with chloroform an accurately weighed quantity of powdered tablets equivalent to about 80 mg of phenobarbitone. Evaporate the extract to dryness and dissolve in sufficient chloroform to produce 10 ml of solution. Determine the absorption spectrum as described above and calculate the absorbance at the selected peak.

Read off the number of mg of phenobarbitone from the graph. Hence calculate the number of mg of phenobarbitone in a tablet of average weight.

Careful standardisation of instrumental conditions is essential in quantitative analysis. Of these the 100% and 0% transmittance and the slitwidth settings are very important. If the calibration curve obeys Beer's Law, a direct comparison of the absorbance of test and standard solutions at the peak absorption may be used. This direct comparison was adopted in the United States Pharmacopeia XVI for the assay of *Acetazolamide* and *Diethyltoluamide*.

Failure to obey Beer's Law means that reference must be made to a calibration curve for the determination of concentration.

Experiment 8   *To determine the percentage v/v of chloroform in Chloroform Spirit*

Method Prepare a series of solutions of ethanol-free chloroform in ethane-1,2-diol to contain 0.2, 0.5, 1.0, 1.5 and 2.0% w/v of chloroform. Treat each solution in turn in the following way. Transfer the solution (0.1 ml) to a flask approximately 1 ml in volume and attach to an evacuated gas cell (Fig. 10.19). Open tap A and carefully heat the flask for 2 min at 150° in an oil-bath. With the flask still in the bath close tap A. Insert the gas cell into the sample beam and record the absorption spectrum over the range $820{-}700\,cm^{-1}$. Using the base-line technique calculate the absorbance of each solution and plot the values against concentration of chloroform to construct a calibration curve.

Dilute the sample (5 ml) to 25 ml with ethane-1,2-diol and mix well. Treat 0.1 ml of solution as described above and determine the percentage w/v of chloroform in the sample by reference to the calibration curve. Convert the result to percentage v/v by dividing by the weight per ml of chloroform (1.48).

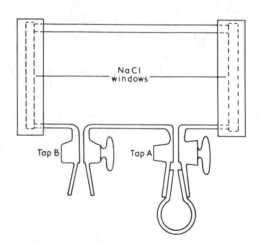

Fig. 10.19. Gas cell

The apparatus (Fig. 10.19) is a normal gas cell of about 150 ml volume and reproducible results are obtained by using the same cell, flask and pipette for each determination. An internal standard is not necessary providing extreme care is taken in using the pipette.

Tap B (Fig. 10.19) remains closed in this determination but when aqueous solutions are examined a small flask containing phosphorus pentoxide may be used to maintain a low level of water vapour during the determination.

COGNATE DETERMINATIONS
Compound Chloroform and Morphine Tincture
Chloroform Liniment

Experiment 9 *To determine the percentage of methanol in Orciprenaline Sulphate*

The determination of solvent residues in solid chemicals is a normal part of quality control and is included in such terms as loss on drying or volatile matter. Ordinary drying processes do not, however, guarantee complete elimination of water or organic solvent, and a number of examples exist where specific methods of determination are applied— water by titration with Karl–Fischer reagent, and acetone, dichloro-methane, propan-2-ol and methanol by GLC (Chapter 4).

Some of these are readily determined by vapour phase IR absorption with sample weights considerably less than those required for GLC. The IR spectrum also identifies the solvent.

Method Accurately weigh the sample (about 0.03 g) into the small bulb (Fig. 10.19), add ethane-1,2-diol (0.2 ml) and attach to an evacuated gas cell. Complete the determination as described for chloroform (above) but using aliquot portions of methanol in ethane-1,2-diol glycol for calibration purposes.

Note Methanol (1 mg) gives a suitable absorbance.

COGNATE DETERMINATIONS
Colchicine (0.1 g) for determination of ethyl acetate or chloroform.
Novobiocin Calcium (0.15 g) for determination of acetone.
Novobiocin Sodium (0.15 g) for determination of acetone.
Streptomycin Sulphate (0.15 g) for determination of methanol.
Warfarin Sodium (Clathrate) (0.1 g) for determination of propan-2-ol.

Experiment 10 *To determine the percentage of menthol in Peppermint Oil using the near IR region of the spectrum*

Method Prepare standard solutions of menthol in carbon tetrachloride to contain 0.5, 1.0, 1.5, 2.0 and 2.5% w/v of menthol. Measure the absorption spectrum of each solution in 4 cm glass or quartz cells over the region 8000–7000 cm^{-1}. Calculate the absorbance at the peak absorption (about 7200 cm^{-1}) using the base-line technique and construct a calibration curve.

Accurately dilute the peppermint oil with carbon tetrachloride to give a concentration of menthol of about 1.5% and determine the absorbance at about 7200 cm^{-1} as described above. By reference to the calibration curve, calculate the % w/v of menthol in the oil and finally convert to % w/w by using the weight per ml of the oil.

The near IR region of the spectrum is becoming of increasing interest with the advent of sensitive detectors. The spectra consist mainly of overtones of the fundamental bands and are less intense than those of the latter. Nevertheless, because solvents can be used in cells of long pathlength, the region has potential value for determination of those groups which show reasonable absorption. Fig. 10.20 illustrates the groups which may be determined and the regions of absorption.

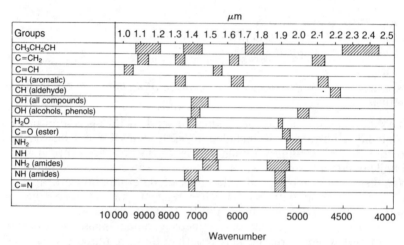

Fig. 10.20. Characteristic absorptions in near IR

The assay for menthol depends upon the determination of the hydroxyl group, and is therefore less specific than an assay which involves absorption due perhaps to a skeletal vibration characteristic of menthol. The result of the assay should be similar to that obtained by the Pharmacopoeial method.

Experiment 11   *To determine the E – Z isomer ratio in clomiphene citrate*

The shape of molecules often plays a significant role in the medicinal action of many compounds, and if isomers are formed in the synthesis it may be important to include in the specification a test for the relative proportions of one to the other. This control ensures uniformity of composition from batch to batch and is described for the *cis*-isomer of *Tranylcypromine* under Gas Chromatography (Chapter 4) and *Clomphene Citrate* under High Performance Liquid Chromatography (Chapter 4). It is well known that *cis*- and *trans*-substituted double bonds have slightly different IR absorption bands in the 13 μm region and this feature is the basis of the present determination.

Method Dissolve Z-clomiphene citrate (22.5 mg) and E-clomiphene citrate (52.5 mg) in water (10 ml) in a separator and add sodium hydroxide solution (5%, 1 ml). Extract the mixture of bases with ether (3 × 10 ml) and wash the combined ether extracts with water (2 × 10 ml). Dry the ether with anhydrous sodium sulphate, filter, evaporate to dryness and dissolve the residue in carbon disulphide (1 ml). Record the absorption spectrum in a 0.2 mm cell over the range 12.50 to 14.00 μm and calculate the absorbance for the peaks at about 13.16 and 13.51 μm using the base-line technique between the minima at about 12.66 and 13.89 μm.

Repeat the determination on a 1:1 mixture (75 mg) of Z and E-clomiphene citrates and on Clomiphene Citrate (75 mg). Calculate the ratio

$$\frac{\text{Absorbance at } 13.16 \, \mu\text{m}}{\text{Absorbance at } 13.51 \, \mu\text{m}}$$

for the sample and two standard mixtures and confirm that the ratio for the sample lies between the ratios for the standards, indicating that the sample contains 30–50% Z-clomiphene citrate.

Experiment 12 *To determine the percentage of thiotepa* $C_6H_{12}N_3PS$ *in Thiotepa Injection*

The preparation consists of a sterile powder of Thiotepa, Sodium Bicarbonate and Sodium Chloride in a sealed container to which Water for Injection is added before use.

Method Carry out the determination of Uniformity of Weight (Method B, Part 1, Chapter 14) and from the mixed contents weigh a quantity containing the equivalent of 75 mg of Thiotepa into a sinter-glass crucible. Extract with carbon disulphide (3 × 5 ml) (Note) and allow to filter through into a small flask. Evaporate the filtrates to low volume, transfer quantitatively to a 10 ml volumetric flask and dilute to volume with carbon disulphide. Measure the absorbance in a 0.1 mm cell over the region 9–12 μm and calculate the value at the maximum at about 10.75 μm using the base-line technique. Use Thiotepa (*BPCRS*) treated in the same way for comparison and calculate the content of $C_6H_{12}N_3PS$ in the average contents of one ampoule.

Note The utmost care should be exercised in the handling of carbon disulphide because of its toxicity and flammability.

Experiment 13 *Assay of meprobamate in Meprobamate Tablets*

The purpose of this experiment is to illustrate the use of potassium chloride discs in quantitative work by means of an internal standard. It is very difficult to obtain a reproducible thickness of disc (pathlength of sample) and, since the absorbance of the analyte depends on the thickness of the sample in addition to the concentration of the analyte, variation of the thickness will have a marked effect on the absorbance of the analyte. Addition of a known concentration of an internal standard to the analyte gives a mixture in which the ratio of absorbances at two wavelengths is independent of the thickness of the sample (p.286) and depends on the relative amounts of the analyte and internal standard.

The internal standard must necessarily have few absorption bands, and iron(III) thiocyanate meets this requirement. The method was described in the National Formulary XII (1965) for the assay of cyclophosphamide tablets.

Internal Standard Solution Extract a mixture of iron(III) ammonium sulphate (2.63 g) and ammonium thiocyanate (26.3 g) in water (250 ml) with ethyl acetate (6 × 100 ml). Dry the extracts (magnesium sulphate, dried) and filter. Make up to 1000 ml with ethyl acetate.

Standard Curve Prepare a standard solution of meprobamate (2 mg ml^{-1}) in methanol. Transfer an aliquot (2, 4 and 6 ml) to three 10 ml volumetric flasks containing 2 ml of the internal standard solution and dilute to volume with methanol. Pipette 0.5 ml of each solution into an agate mortar containing 500 ± 50 mg potassium chloride or potassium bromide and evaporate the solvent with the aid of a current of air and finally a vacuum (2 min). Thoroughly grind the dry powder to obtain a uniform distribution of the substances, indicated by a uniform pink colour of the internal standard. Prepare a disc of the powder (in replicate to improve precision) and record the IR spectrum. Calculate the absorbance of the internal standard at its λ_{max} around 4.8 μm and of the meprobamate band at its λ_{max} around 5.9 μm by the base-line technique. Calculate the ratio $A_{5.9}/A_{4.8}$ and construct a calibration graph.

Method Weigh and powder 20 tablets. Extract a quantity of powder equivalent to 200 mg of meprobamate with methanol (80 ml) by stirring for 15 min, dilute to 100 ml and filter through No. 1 filter paper. Transfer 5 ml to a 10 ml volumetric flask containing 2 ml of internal standard solution and continue as described for the standards. Calculate the ratio $A_{5.9}/A_{4.8}$ and read the concentration of meprobamate from the calibration graph. Calculate the content of meprobamate in a tablet of average weight.

Experiment 14   *The limit test for the β (inactive) polymorph of chloramphenicol palmitate in Chloramphenicol Palmitate Mixture*

Different polymorphs of a drug, particularly of a poorly soluble substance, which have different solubilities may show differences in pharmacological or therapeutic potency (Chapter 3).

Chloramphenicol palmitate exists in at least two polymorphic forms.

  (a) the α *polymorph* which is the biologically active polymorph
  (b) the β *polymorph* (called Polymorph A in the British and United States Pharmacopoeias) which has no antibiotic activity.

The α and β polymorphs of chloramphenicol palmitate can be distinguished by the small differences in their IR spectra, which are particularly apparent between 800 and 900 cm^{-1}. The different absorption bands in this region are due to differences in skeletal vibrations, i.e. those of the molecular framework of the crystal as a whole rather than of the characteristic chemical groups within the molecule.

The British and United States Pharmacopoeias specify a limit test which places an upper permitted limit of 10% on the level of the inactive β polymorph (Polymorph A) in *Chloramphenicol Palmitate Mixture (Chloramphenicol Palmitate Oral Suspension U.S.P.)*. The limit test is adapted from the work of Borka and Backe-Hansen (1968) who showed that the absorption bands at 858 cm^{-1} (11.65 μm) and 843 cm^{-1} (11.86 μm) are due to the α and β polymorphs respectively. If follows that the ratio of absorbance at 858 cm^{-1} to that at 843 cm^{-1} will be a high value for the α polymorph and a low value for the β polymorph, and that an increase in the proportion of the inactive β polymorph will result in a reduction of the ratio. The specification in the limit test is that the ratio of absorbance at 858 cm^{-1} to that at 843 cm^{-1} for a sample of chloramphenicol palmitate isolated by centrifugation from the suspension excipients is greater than the ratio of a standard mixture containing 90% α-polymorph and 10% β-polymorph.

Method

α-Polymorph Dissolve 2 g chloramphenicol palmitate in 20 ml dry ethanol by heating, and after the solution has cooled to room temperature pour it slowly into 200 ml of water, with vigorous stirring. Allow the recrystallisation to proceed overnight, filter the solid and dry under vacuum over phosphorus pentoxide. Prepare a thick mull in liquid paraffin (Note 1) (Experiment 3) and record the IR spectrum (Note 2).

β-Polymorph Dissolve 2 g chloramphenicol palmitate in 4 ml chloroform by heating and leave overnight. Filter the solid and dry as described for the α-polymorph. Record the IR spectrum of the β-polymorph in a mull.

10% β : 90% α mixture Weigh 5 mg of the β-polymorph and 45 mg of the α-polymorph onto an agate mortar, prepare a thick, evenly dispersed mull in 2 or 3 drops of liquid paraffin and record its IR spectrum.

Chloramphenicol Palmitate Mixture Transfer 10 ml of the suspension to a 50 ml stoppered centrifuge tube. Add 15 ml of water and mix well. Remove the stopper and centrifuge the suspension at full speed for 10 min. Pour off the supernatant liquid containing the suspending agents, add 20 ml of water, shake and centrifuge. Repeat the washing of the chloramphenicol palmitate until the supernatant liquid is reasonably free of suspending agents. Transfer the chloramphenicol palmitate to a desiccator containing phosphorus pentoxide and store for several days until dry. Record the IR spectrum of the sample in a mull.

Treatment of the results After calibration of the wavenumber scale (Note 2), identify the maximum absorption at $858\,cm^{-1}$ and the minimum absorption at $843\,cm^{-1}$ in the spectrum of the α-polymorph and the minimum absorption at $858\,cm^{-1}$ and maximum absorption at $843\,cm^{-1}$ in the spectrum of the β-polymorph. Confirm the presence of bands at 858 *and* $843\,cm^{-1}$ in the spectrum of the 10% standard mixture. In all four spectra, draw the base-line across the base of the absorption bands from the minimum at $885\,cm^{-1}$ to that at $790\,cm^{-1}$ and drop perpendiculars at 858 and $843\,cm^{-1}$. Calculate the absorbance at both wavenumbers by the base-line technique. Calculate the absorbance ratio A_{858}/A_{843} for each of the pure polymorphs, the 10% standard mixture and for the sample. A high value (> 4) should be obtained for the pure α-polymorph and a low value (< 0.05) for the β-polymorph. The sample contains less than the 10% inactive β-polymorph and therefore passes the limit test if the ratio A_{858}/A_{843} is *greater* than that of the 10% standard mixture.

Note 1 The preparation of the polymorphs in a liquid paraffin mull rather than a potassium bromide disc is adopted because grinding of the powder in the preparation of the latter has been shown to convert the β-polymorph to the α-form.

Note 2 The accurate identification of the absorption bands in the spectra of the polymorphs is facilitated by calibrating the wavenumber scale with the absorption bands of a superimposed spectrum of a polystyrene film at $1601\,cm^{-1}$ and $1028\,cm^{-1}$.

Note 3 In the B.P. (1980) the measured value is the ratio of **peak heights** of the absorption bands at $858\,cm^{-1}$ and $843\,cm^{-1}$ rather than ratio of **absorbances**.

Routine maintenance

Demountable cells

The cells for use in IR spectrophotometry require considerable care in storage and handling, as the sodium chloride or potassium bromide windows are fogged by traces of moisture and are easily scratched. A small cabinet fitted with a simple heater, e.g. a 60 watt lamp, and a dish of silica gel as drying agent should be used for storing all cells, windows or plates, and relevant apparatus such as the die for preparing potassium chloride discs. The cabinet and contents are thereby maintained at about 10–20° above room temperature and there is little risk of fogging from atmospheric moisture.

In spite of all precautions, however, the windows eventually become opaque to visible radiation, and it is necessary to repolish them if the transmission of IR radiation is much reduced by scatter. A polishing kit can be purchased which normally consists of a glass plate with a coarse and finely ground surface, a polishing cloth stretched on a firm base, polishing powder, e.g. jeweller's rouge, and an optical flat. They repay their cost in a short time as all types of defect, e.g. fogged or scratched surfaces in the windows, may be remedied.

The following system, however, has proved satisfactory in obtaining serviceable windows from sodium chloride 'blanks' within 6 to 7 min. Blanks offer a considerable saving in cost over the price of polished windows and they normally occur with surfaces similar to those obtained from the glass roughing and smoothing laps of a polishing kit.

Materials   Sodium chloride blanks.
Two polishing cloths.
Two thick glass or wooden bases on which the polishing cloths are firmly spread.
Polishing powder.
Ethanol containing a few crystals of sodium chloride.

Method   To one of the polishing cloths add a small amount of powder and saturate with ethanol. Place the blank over the powder and move it with firm pressure, rapidly, in a linear motion (Note). At short intervals rotate the blank and continue the polishing using more powder and ethanol. Examine the blank after 1 or 2 min by lightly drawing it over a clean portion of the cloth and continue the treatment if the surface is not smooth and transparent. Complete the polishing of the one side of the blank with rapid movements on the second clean and dry polishing cloth. Repeat for the other side of the disc.

Note   Some polishers prefer circular motion but this is a matter of personal choice.

This method will not remove deep scratches, for which ground glass surfaces or very fine sandpaper are necessary. Gloves are not essential for handling sodium chloride blanks.

Sealed cells

These consist of two sodium chloride or potassium bromide windows separated by a lead or teflon spacer, the whole being held together in a metal frame along with protective washers. Unless the transmission of the cell has deteriorated appreciably and the cell is in obvious need of renovation it is unwise to attempt to dismantle it. However, the use of teflon spacers instead of lead spacers simplifies considerably the dismantling and cleaning of cells.

Method Note the construction of the cell, isolate the window assembly and carefully separate the two windows and spacer. These are normally firmly stuck but the insertion of a razor blade into the join and *gentle* leverage at each corner will generally separate them. Peel off the spacer and polish the two windows in the normal way.

To reassemble the cell, place a new spacer of the required thickness on the polished window and add the second polished window. Press lightly together and incorporate into the metal frame using the washers (Note). Tighten the screws in the frame carefully to avoid cracking the windows. Test the cell for leakage, using chloroform, by filling and allowing to stand for an hour; the solvent should be retained.

Note Take care to line up the holes in the plate with those in the washer.

Potassium chloride

The method of Hales and Kynaston (1954) is satisfactory in yielding a uniform powder which compresses readily into discs.

Method Prepare a saturated solution of potassium chloride (AnalaR, 500 ml) and filter (sintered glass) from undissolved salt. Divide the filtrate into two (300 ml and remainder of filtrate) and add the 300 ml portion to ice-cold hydrochloric acid (300 ml) with stirring. Filter off the precipitate on a sintered-glass filter using a Buchner flask. Drain well and wash the precipitate with the reserved saturated solution in portions, draining well in between each wash. Transfer to filter paper and press gently to remove as much liquid as possible and dry in a large porcelain dish at 120° for 2 h. Break up the powder with a glass rod and complete the drying at 400° overnight in a muffle furnace (Note 1). Allow to cool and distribute the contents among weighing bottles each holding about 10 g. Store in a desiccator and use the contents of each bottle in turn (Note 2).

Note 1 Heating at 400° removes the last traces of water, and hydrochloric acid, which is difficult to remove completely, is reduced to negligible amounts.

Note 2 Distribution among several containers avoids constantly exposing all the material to the air during the period of its use.

References

Banwell, C.N. (1972) *Fundamentals of Molecular Spectroscopy,* McGraw-Hill, London.
Barrow, G.M. (1962) *Introduction to Molecular Spectroscopy,* McGraw-Hill, London.
Bellamy, L.J. (1958) *Infrared Spectra of Complex Molecules,* 2nd edn, Methuen, London.
Borka, L. and Backe-Hansen, K. (1968) *Acta Pharm. Succica* **5**, 271.
Conley, R.T. (1966) *Infrared Spectroscopy,* Allyn and Bacon, Boston.
Cross, A.D. (1964) *Practical Infrared Spectroscopy,* Butterworths, London.
Duyckaerts, G. (1959) *Analyst* **84**, 201.
Hales, J.L. and Kynaston, W. (1954) *Analyst* **79**, 702.

11
Nuclear magnetic resonance spectroscopy

R. T. PARFITT

Introduction

Nuclear magnetic resonance (NMR) spectroscopy is a technique that permits the exploration of a molecule at the level of the individual atom and affords information concerning the environment of that atom. Only about one half of known element isotopes, when placed in a magnetic field, absorb energy from the radiofrequency region of the electromagnetic spectrum. Of these isotopes ^{1}H and ^{13}C are the most important from the viewpoint of the organic and pharmaceutical chemist. The precise frequency from which energy is absorbed gives an indication of how an atom is bound to, or located spatially with respect to, other atoms. Thus NMR offers an excellent physical means of investigating molecular structure and molecular interactions. NMR spectra may also be used for compound identification, by a fingerprint technique similar to that employed in infrared spectroscopy, and sometimes as a specific method of assay for the individual components of a mixture.

Theory

Many atomic nuclei have an angular momentum, i.e. they spin. A charged spinning particle generates a magnetic dipole along its spin axis, and therefore an isolated nucleus stripped of its electrons may be thought of as a small bar magnet. From quantum mechanics the magnitude of the nuclear angular momentum (P) is given by the expression:

$$P = h/2\pi\sqrt{\{I(I + 1)\}}$$
$$= \hbar\sqrt{\{I(I + 1)\}} \qquad (1)$$

where h = Planck's constant; $\hbar = h/2\pi$ = modified Planck's constant; and I = spin quantum number.

The value of the spin quantum number (I) depends upon the particular nucleus or isotope and can be $0, \frac{1}{2}, \frac{2}{2}, \frac{3}{2}, \frac{4}{2}, \frac{5}{2}$ etc., i.e. integral or half integral multiples.

Furthermore, from quantum mechanical considerations there are $(2I + 1)$ possible orientations, and thus corresponding energy levels, for a nucleus with a magnetic moment under the influence of an external magnetic field. A comprehensive list of spin quantum values and other nuclear properties is provided by Pople *et al* (1959). To establish the value of I empirically, the following rules may be applied:

(a) $I = 0$ when both the mass and atomic number are even numbers. The nuclei of such isotopes have a spherically symmetrical charge distribution. They do not possess angular momentum, and do not give nuclear magnetic resonance spectra. They include ^{12}C, ^{16}O and ^{32}S.

(b) $I = \frac{1}{2}, \frac{3}{2}, \frac{5}{2}$ etc. (half integral multiples) when the isotope mass number is odd and atomic number is odd or even.

(c) $I = 1, 2, 3$ etc. when the mass number is even and the atomic number is odd.

By far the most important group of nuclei from the standpoint of organic chemistry and drug chemistry are those with $I = \frac{1}{2}$. Examples are ^{1}H, ^{3}H, ^{19}F, ^{31}P, ^{13}C, ^{15}N and ^{29}S (Harris and Mann, 1978). These nuclei possess spin and have a spherically symmetrical charge distribution. Examples where $I = 1$ are ^{2}H and ^{14}N and where $I > 1$ are ^{10}B, ^{11}B, ^{35}Cl, ^{17}O and ^{27}Al. Isotopes with a spin value equal to, or greater than, unity have an ellipsoidal charge distribution and posess spin. They have a **nuclear electric quadrupole moment (Q)**.

This chapter deals mainly with nuclei of spin $\frac{1}{2}$ and most examples and applications will be from proton magnetic resonance (PMR) spectroscopy.

In a homogeneous magnetic field an atomic nucleus ($I > 0$) will assume one of $(2I + 1)$ orientations. Thus nuclei of spin $\frac{1}{2}$ will have two modes of alignment with respect to the applied field. They can align themselves either 'with' the field or 'against' the field, thus corresponding to the spin quantum values of $I = -\frac{1}{2}$ and $I = +\frac{1}{2}$. The energetically favourable, or ground, state is that in which the nuclei are aligned 'with' the field, rather like a compass needle aligned naturally in the earth's magnetic field. The unfavourable $(I = +\frac{1}{2})$ orientation may be considered as the excited state.

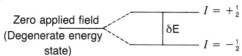

The energy difference between the ground and excited states may be expressed by the equation

$$\delta E = \gamma \hbar H_0 \tag{2}$$

where H_0 = applied magnetic field, and γ = magnetogyric ratio.

The **magnetogyric ratio**, a proportionality constant unique for each isotope, is given by the expression

$$\gamma = \frac{\mu}{I\hbar}$$

where μ is the maximum observable component of the magnetic moment.

A spinning nucleus in a magnetic field (H_0) experiences H_0 as a torque about an axis perpendicular to the axis of rotation. This causes the spin axis to precess about the direction of the field (Fig. 11.1). An increase in the strength of H_0 does not produce an energy transition of

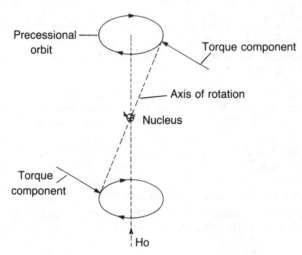

Fig. 11.1. Precession of spin axis of spinning nucleus

the nucleus, but simply increases the precessional frequency. For a given value of H_0 the precessional or **Larmor frequency** is given by the Larmor equation:

$$v_0 = \frac{\gamma}{2\pi} \cdot H_0 \qquad (3)$$

Introducing the Planck constant (h) to both sides, eq. (3) becomes:

$$hv_0 = \frac{\gamma h}{2\pi} \cdot H_0 \qquad (4)$$
$$= \gamma \hbar H_0$$

which from eq. (2) = δE.

Thus, if the nucleus is irradiated with radiofrequency energy where $hv_0 = \gamma \hbar H_0$ then, when it is in the ground state, it can absorb radiation energy. Radiofrequency energy equal to the precessional frequency of the nucleus will initiate resonance in the precessing spin axis and cause the nucleus to 'flip' into a higher energy state. The radiation energy required for such a transition falls within the radiofrequency region of the electromagnetic spectrum and, in an NMR experiment, is applied to the sample at right angles to the magnetic field H_0.

The NMR signal

When nuclei in a homogeneous magnetic field are scanned with a radiofrequency signal, then, as the frequency approaches the precessional frequency, becomes equal to it, and finally exceeds it, resonance develops, attains a maximum, and subsides (Fig. 11.2). Maximum resonance is observed when the radiofrequency (v) is equal to the precessional frequency of the nuclei. Likewise, a resonance signal results if v is kept constant and the magnetic field (H_0) is varied, since $v \propto H_0$. The magentic field sweep is usually employed in practice.

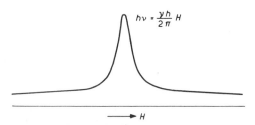

Fig. 11.2. Appearance of an NMR signal. Ordinate: absorption energy. Abscissa: strength H of the external magnetic field or frequency of the r/f generator.

All nuclei of a single isotope subjected to the above conditions do not automatically assume the most favourable alignment, but adopt a **Boltzmann distribution** with respect to the two adjacent energy levels. This distribution has a small excess (a few nuclei per million) of nuclei in the lower energy state. Only at absolute zero do all nuclei adopt the lower energy orientation; at higher temperatures thermal motion causes randomisation. It is the absorption of energy by this small excess of nuclei which is responsible for NMR spectra and this accounts for the low sensitivity of the technique. An equilibrium state is established by the applied radiofrequency working to reduce the lower spin state excess to zero, and relaxation processes acting to maintain the Boltzmann distribution. A state of **saturation**, when the number of nuclei in each energy state is equal, may be attained by irradiation of the sample with a powerful radiofrequency signal (see nuclear double resonance, p.000).

Relaxation processes

Mechanisms must exist whereby nuclei in an excited state can revert to the ground state, otherwise the energy levels would equalise rapidly and no further energy absorption would occur. The mechanisms responsible for the maintenance of an NMR signal may be classified under two general headings.

(a) Radiation emission

Radiofrequency energy emitted by those nuclei moving from an excited to the ground state under the influence of an external electromagnetic field is exceeded by the opposite transition, and this leads to the net absorption of energy. Spontaneous emission of electromagnetic energy in the radiofrequency region is negligible.

(b) Radiationless transitions

These are usually referred to as *relaxation processes*, and not only maintain a state of energy absorption but govern resonance line width. They are of prime importance in NMR spectroscopy. Relaxation processes are classified as either **spin-lattice relaxation or spin-spin relaxation**.

Spin-lattice or longitudinal relaxation

This involves the transfer of energy from a nucleus in an excited state to the molecular lattice. Here 'lattice' is used to describe the molecular structural environment in which a nucleus is situated irrespective of whether it is in a solid, a liquid or a gas. During this phenomenon, also known as spin-cooling, the nucleus returns to the ground state and the lattice gains thermal energy. The efficiency of spin-lattice relaxation (T_1) is expressed as the half-life required for the system to establish an equilibrium state. T_1 is a function of Brownian motion in fluids and lattice vibration in solids. Short relaxation times (T_1) imply efficient relaxation and lead to broad resonance bands. Organic liquids, for example, of T_1 approximately 1 s, give optimum line widths for practical purposes. Pure solids have large values of T_1, often many hours, and in the absence of other factors would result in very narrow absorption bands. Although for liquids T_1 is usually greater than 10^{-2} s and less than 10^2 s, the presence of paramagnetic impurities, such as oxygen, gives rise to very rapid spin-lattice relaxation and thus to broader resonance bands. Similarly nuclei with $I > \frac{1}{2}$ possess a **nuclear quadrupole moment** which causes local magnetic field fluctuations leading to short relaxation times and consequently line broadening.

Spin-spin or transverse relaxation

This occurs mainly in solids. It involves a mutual exchange of spins between an excited nucleus and a neighbour, without altering the overall spin state of the system. The time T_2 is a measure of the efficiency of spin-spin relaxation. The total magnetic field felt at any particular nucleus is H_0 plus or minus a factor for the small magnetic fields generated by the nuclei of its immediate environment. Thus, the greater the number of spin-state exchanges the greater are the local field fluctuations, leading to a greater range over which radiofrequency energy may be absorbed, and therefore to line broadening. Large values of T_2 result in narrow resonance lines, small values give broad lines. In liquids and gases local magnetic field variations average to zero, whereas in solids such interactions are finite and lead to very broad absorption bands.

Instrumentation

Several high resolution NMR spectrometers are available commercially; all may be considered as consisting of five major units:

(a) a magnet producing a strong homogeneous field;
(b) a means of varying the magnetic field or radiofrequency signal over a narrow sweep range;
(c) a radiofrequency oscillator;
(d) a radiofrequency receiver;
(e) recorder and integrator.

(a) The magnet

Either a permanent or an electromagnet may be used to supply a field of high homogeneity. Permanent magnets provide an ever-present field of good homogeneity and stability, but field variation is not possible. Electromagnets, on the other hand, require expensive field stabilisers but the field strength may be varied. Ideally, all points within the pole gap of the magnet should experience an identical value of H_0. Because the strength of a resonance signal depends upon the magnetic field strength, the latter should be as great as possible.

(b) Variation of the magnetic field or magnetic field sweep

This is now employed in only a minority of instruments to produce NMR spectra, and is accomplished by passing a direct current through coils wound around the magnet pole pieces or through a pair of Helmholz coils flanking the sample. The rate of sweep is important: too slow a sweep leads to saturation effects, whereas a fast sweep results in 'ringing' (Fig. 11.3). Provided ringing is not excessive and does not distort the resonance signal, it is indicative of good field homogeneity.

(c) A radiofrequency oscillator

This supplies the signal required to induce transitions in the nuclei of the sample from the ground to the excited state. The source is often a highly

Fig. 11.3. Ringing

stable crystal controlled oscillator, the output of which is multiplied to the desired frequency. The signal arises from a coil situated in the pole gap of the magnet and the sample rests within the coil.

(d) The receiver

The resonance signal is detected by one of the two methods. In **single coil** instruments a radiofrequency bridge, rather like a Wheatstone bridge, is employed. The applied signal is balanced against the received signal and the absorption or resonance signal is recorded as an out of balance emf, which may be amplified and recorded mechanically. In the **double coil** or **nuclear induction** method transmitter and receiver coils are set at right angles to each other about the sample. Figure 11.4 illustrates a double coil NMR spectrometer.

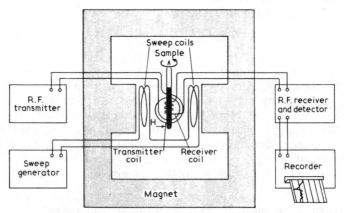

Fig. 11.4. Schematic diagram of an NMR spectrometer (Courtesy of Varian Associates Palo Alto, Calfornia)

(e) Recorder and integrator

Spectra from modern stable high resolution instruments are recorded mechanically either directly or *via* a dedicated computer, on pre-calibrated charts zeroed with respect to a reference compound (p.431). Resonance line intensity in an NMR spectrum is proportional to the number of nuclei responsible for the signal. The area under the signal is a direct measure of the intensity and is determined in a cumulative manner by an electronic integrator (Fig. 11.5).

Most instruments currently in use for proton studies operate at 60 MHz and 1.4092 Tesla (T) (1 Tesla = 10^4 Gauss), 90 MHz and 2.114 T or 100 MHz and 2.349 T. Higher resolution and sensitivity result from the use of fields produced by liquid helium cooled superconducting solenoids and operating at up to 600 MHz (Bothner-By and Dadok, 1982). The improvement in resolution between 60 MHz and 200 MHz is illustrated in Fig. 11.6.

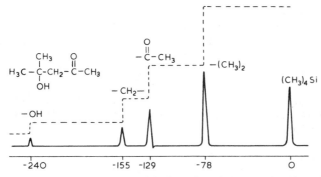

Fig. 11.5. Proton NMR spectrum of 2-hydroxy-2-methylpentan-4-one at 60MHz and 1.4092 Tesla with tetramethylsilane as internal standard (---): Integral absorption intensity. Abscissa: strength of the magnetic field referred to the position of the $(CH_3)_4Si$ resonance signal (Hz).

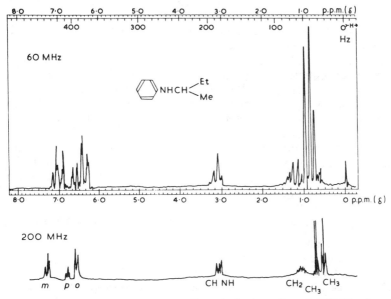

Fig. 11.6. NMR spectra of *N-sec*-butylaniline obtained at 60 and 200 MHz show that greater resolution is obtained at the higher frequency (lower spectrum). For example, the spectrum taken at 200 MHz shows well-separated resonance peaks for the *ortho*, *meta* and *para* aromatic protons

---

Practical considerations

Sample spinning

A simple method of increasing the effective homogeneity of the applied magnetic field is to spin the sample at approximately 30 rps about its longitudinal axis. Variation of the position of a nucleus in the pole gap effectively averages the field it experiences. Spinning at too high a rate may cause turbulences which seriously affect resolution. If the spin rate is too slow, field averaging is incomplete and resonance lines tend to broaden. Also, under certain circumstances strong reasonance bands are flanked by side-bands, one to each side of the parent. **Spinning side-bands** may be distinguished from true resonance signals by varying the spin rate, an increased rate causing the side-bands to move away from the parent signal.

Solvents

Samples normally encoutered in NMR studies are either liquids or solids; gases occur rarely and because of their low signal to noise ratio are examined under pressure. Liquids usually give excellent spectra 'neat', but samples of high viscosity benefit from dilution with a mobile solvent. Ideally, solids require a solvent in which they have 10% solubility, and, because a sample volume of 0.35–0.5 ml is needed, 35–50 mg of material must be available. Solutions of 20–25% afford excellent spectra with a high signal to noise ratio. Paramagnetic impurities in either sample or solvent cause resonance line broadening and efforts must be made to preclude all extraneous particles. Small amounts of oxgyen dissolved in a liquid sample or solution may cause considerable loss of spectrum resolution (line broadening). Oxygen may be conveniently removed by vigorously shaking the sample with about 0.3 ml of 10% sodium dithionite solution; a marked improvement in spectrum resolution results (Brophy *et al.*, 1968). Other methods of deoxygenation include vacuum distillation, repeated freezing, evacuation, or bubbling argon or nitrogen through the sample immediately before recording the spectrum.

The properties required of solvents for PMR spectroscopy are:

(a) chemical inertness
(b) magnetic isotropy
(c) volatility, to facilitate sample recovery
(d) absence of hydrogen atoms (this is not always possible).

Solvents commonly employed are CCl_4, $CDCl_3$, D_2O, CD_3OD, $(CD_3)_2SO$, CD_3COOD and CF_3COOH. (See solvent effects, p.426).

An excellent chart showing characteristic resonance bands of solvents used in NMR spectroscopy has been published (Henty and Vary, 1967).

Special techniques for small samples

Often sufficient sample for normal NMR investigation is lacking. Special techniques must then be used to obtain satisfactory spectra.

Microcells

Several types of microcell are available requiring as little as 1 mg of sample dissolved in 0.025 ml of solvent. The cell confines a small amount to the locality of the probe receiver coil. Microcells may be either spherical or cylindrical in design.

Time-averaging computer (Computer of Average Transients or CAT)

Signal to noise ratio enhancement can be effected by a computer, which stores the resonance signals from many passes of the spectrum and averages out the random noise. Sensitivity is increased by the square root of the number of scans.

An acceptable spectrum may be obtained from a fraction of a milligram sample. Either a whole spectrum or any desired portion of it may be studied in this way.

Fourier transform (FT) NMR (Shaw, 1976)

A disadvantage of the CAT technique is the length of time needed for the summation of numerous standard spectrum scans. This has been overcome by employing a strong r/f pulse of 50 μs duration over a broad bandwidth and recording the total spectral response in the memory of a computer. Each nucleus in the sample absorbs its characteristic frequency component and the resultant pattern recorded is called the **free induction decay**. From the total series of overlapping decay patterns (one from each absorbing nucleus) the spectrum is obtained via a computer by a mathematical operation known as **Fourier transformation**. Thus, by the accumulation of several hundred or even several thousand pulse responses spaced 0.1 to several seconds apart, the recording time for a spectrum may be reduced at least one hundredfold. A small storage computer is employed to add the spectra from a series of pulse irradiations. Enhancement of signals by a factor of about 10 has been achieved practically.

Fourier Transform NMR is of particular value in ^{13}C (p.461).

Where low solubility or a limited quantity of sample are restricting factors the following should be considered:

Insoluble sample	Small quantity of sample
(a) Use a more sensitive instrument, e.g. 100 MHz	(a) Use a more sensitive instrument
(b) Use a CAT or FT	(b) Use a CAT or FT
(c) Increase temperature	(c) Use a microcell
(d) Change solvent	

Temperature variation is often necessary in NMR spectroscopy, and many instruments have a variable temperature probe attachment able to operate between -100 and $+200°$. Devices for wider temperature ranges have been described by Pople *et al* (1959). Kinetic studies, conformational equilibria and hydrogen bonding investigations are facilitated by temperature variation.

Resolution

The resolution of a spectrum is its clarity of division into distinct bands corresponding to atomic or electronic transitions. Resonance line width is governed, in part, by magnetic field homogeneity, and therefore in assessing any spectrum an index of resolution is desirable. The resolution of a resonance band is conveniently expressed by its width in Hz at half height. The ethanol CHO quartet (Fig. 11.7), in the absence of atmospheric oxygen which causes paramagnetic line broadening, is often used as a standard for resolution in PMR studies.

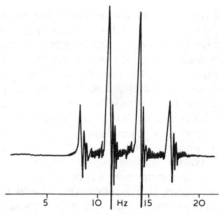

Fig. 11.7. NMR spectrum of the CHO group of acetaldehyde

Chemical shift

Dissimilar isotopes will require different amounts of energy to undergo transitions to a higher energy level and therefore they will absorb r/f energy at different frequencies. Nuclei of the same isotope, depending upon their molecular environments, also absorb energy at different frequencies. The foregoing discussion has been restricted to nuclei stripped of their electrons (unshielded nuclei). They must now be considered with regard to their environment. The electrons around each nucleus, when influenced by a magnetic field (H_0), circulate in such a way as to create a small local magnetic field of their own which opposes H_0. Thus, the resultant field experienced at the nucleus is H_0 less a factor for the electron cloud **diamagnetic shielding**. The nature of the electron cloud about the nucleus will therefore govern the region of the radiofrequency portion of the electromagnetic spectrum from which the nucleus absorbs energy. Protons in different parts of an organic molecule will usually give rise to separate resonance bands. The difference between the absorption position of a proton and the absorption position of the protons of a reference compound is known as the **chemical shift** of that particular proton. The most commonly employed reference compound is tetramethylsilane (p.431).

The field (H) experienced at a particular atomic nucleus in a molecule, where σ is the **shielding** or **screening** constant of the atom, subjected to an applied field H_0 is given by:

$$H = H_0(1 - \sigma)$$

and the resonance frequency for that particular atom, from eq. (3) becomes:

$$v = \frac{\gamma H_0(1 - \sigma)}{2\pi} \tag{5}$$

Thus the amount of shielding determines how much the applied field has to be increased in order to attain that field (H) which induces a transition in the nucleus. For protons, screening constants are smaller than for most other nuclei, since the electron densities about protons are usually smaller than about other nuclei. The relationship between the field independent scale for chemical shift (δ) and the screening constant is given by the expression:

$$\delta = \sigma - \sigma_{ref} = \frac{H_0 - H_{ref}}{H_{ref}} \qquad \text{(See p.430)}$$

Chemical shift values are proportional to the applied magnetic field (H_0). They cannot be calculated but must be determined empirically. Examples of chemical shift values for protons in different environments are found in Figs 11.8–11.11.

Factors influencing chemical shift

Often it is incorrectly assumed that only the election density about a proton is responsible for the shielding effect, although it is true that a proton on a carbon H—C—X resonates to lower field as X becomes more electronegative. In the series where X is Si, C, N and O respectively, then that is the order of decreasing magnetic field required to induce a nuclear transition. The proton becomes progressively more deshielded or has a diminished electron cloud density. Similarly, as the nature of carbon moves from sp^3 to sp^2, attached protons resonate at lower field. This tendency is not adhered to, however, since protons on sp hybridised carbon, alkynic protons, resonate upfield of alkenic protons; and cyclopropane protons which possess some sp^2 character are among the most shielded of protons.

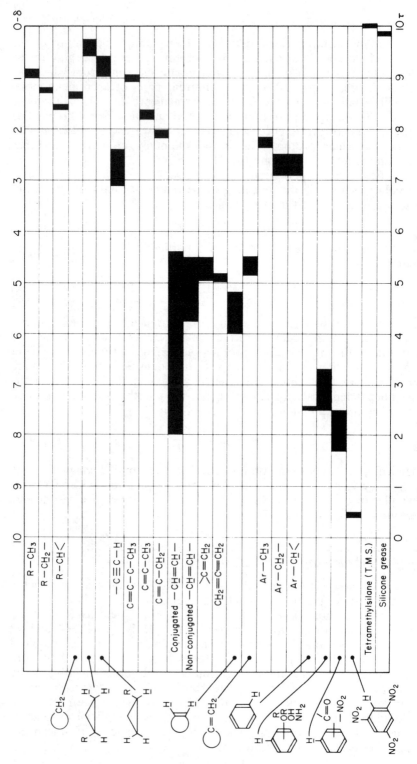

Fig. 11.8. Characteristic proton shift positions

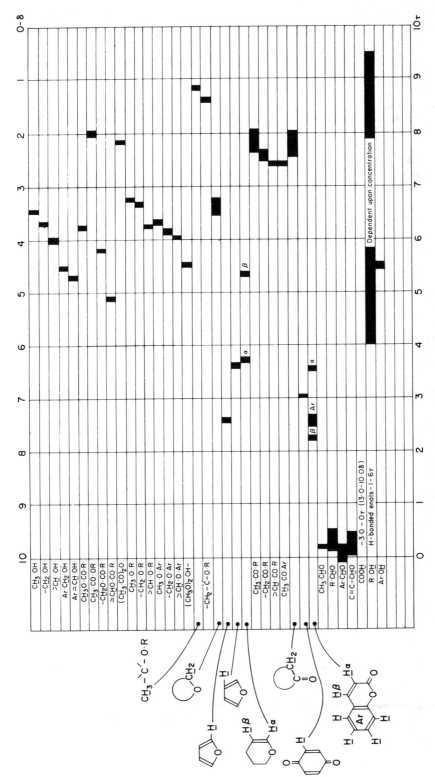

Fig. 11.9. Characteristic proton shift positions

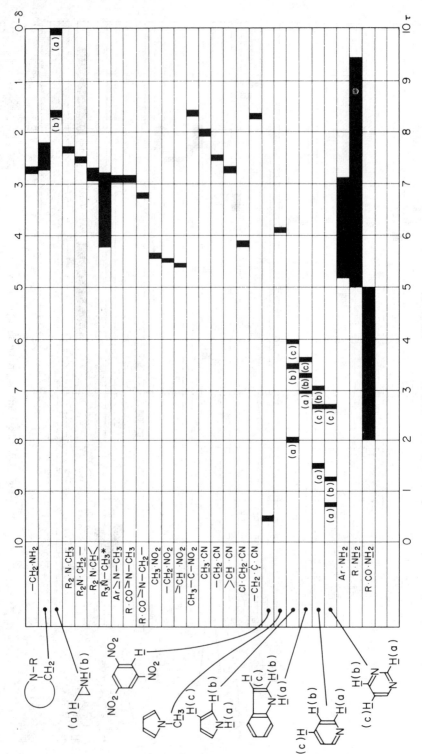

Fig. 11.10. Characteristic proton shift positions

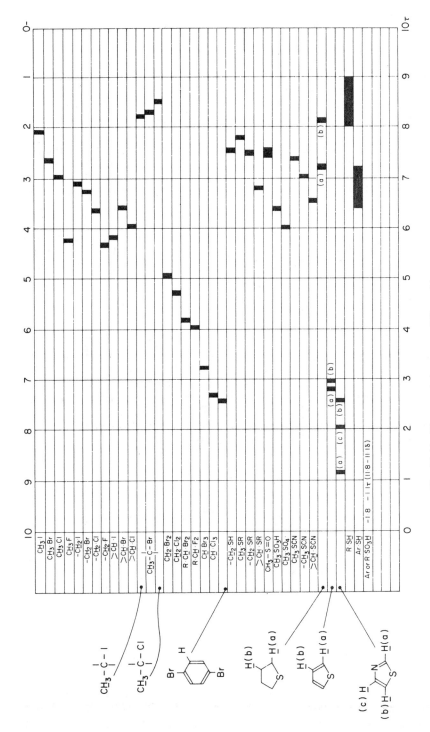

Fig. 11.11. Characteristic proton shift positions

In gases, liquids and solutions, a nucleus is only affected by localised diamagnetic fields from its own molecule; it is not usually influenced by those fields from neighbouring molecules, the effects of which are averaged by random motion. However, where association is possible, particularly in solutions where solute and solvent may form loose complexes, such effects may influence chemical shifts considerably (p.426).

Localised diamagnetic fields are induced about an individual nucleus by the **intraatomic** circulation of its electron cloud. In a molecule having groups of atoms with multiple bond character (π-electrons), **interatomic** electron circulation is possible. Here the circulation of electrons is only in certain preferred directions about the bond and, thus, such bonds exhibit **anisotropy of diamagnetic susceptibility** (diamagnetic anisotropy). The effect, either shielding or deshielding, on a nearby proton depends (*a*) upon its distance from the multiple bond, and (*b*) upon its orientation with respect to the bond. Ethene, ethyne and benzene show this phenomenon.

Ethene

When its double bond is orienated at right-angles to the applied field, ethene has a π-electron circulation about the bond (Fig. 11.12(a)). The induced magnetic field reinforces the applied field at the protons, which are consequently deshielded. Alkenic protons occur in the region of 7.6–4.5δ compared with 0.9δ for CH_3—C<. Figure 11.12(b) illustrates the shielding cones (+) and regions of deshielding (−); any proton held in these regions will be shielded or deshielded accordingly.
Carbonyl groups (>C=O) behave similarly.

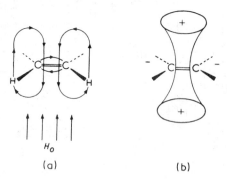

Fig. 11.12. Ethene

Ethyne

This is a linear molecule with an axis of symmetry passing through the triple bond. Orientation of the ethyne molecule with its longitudinal axis parallel to the magnetic field causes a diamagnetic circulation of π-electrons in such a way that its protons experience an induced magnetic field opposing the applied field (Fig. 11.13). The protons are therefore shielded by the anisotropy of the triple bond and resonate at 2.6δ.

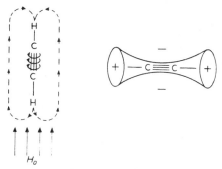

Fig. 11.13. Ethyne

Benzene and aromatic protons

At the aromatic protons the induced magnetic field reinforces the applied field and they therefore resonate at low field, about 7δ. Any proton lying above the plane of an aromatic ring will be shielded, and similarly a proton in the plane of the ring will be deshielded (Bovey and Johnson, 1958) (Fig. 11.14).

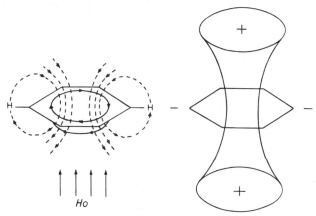

Fig. 11.14. Benzene

Ring current effects are well illustrated by the 60 MHz spectrum of C-18-annulene (I). The outer 12 protons experience a ring-current-induced reinforcing field and absorb at 530 Hz downfield from tetramethylsilane, whereas the inner six protons are subject to an H_0-opposing field and resonate at 115 Hz upfield from tetramethylsilane.

The deduction of the absolute configuration about C-9 of the analgesic 2-hydroxy-5,9-dimethyl-6,7-benzomorphans (II) (Fullerton *et al.*,

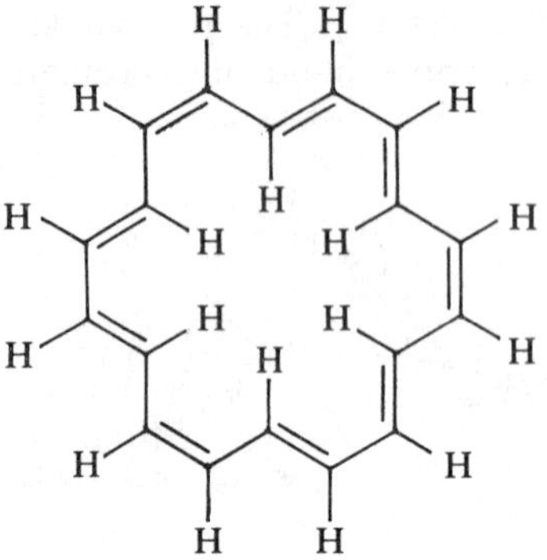

(I)

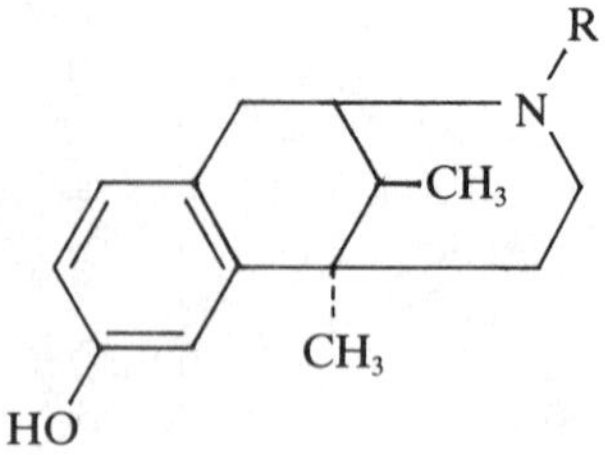

(II)

1962) was aided by a consideration of ring current effects. In the α-isomer where the 5-and 9-methyl groups are *cis*, the 9-methyl group overhangs the aromatic ring, and the ring diamagnetic anistropy produces a shift of 25 Hz (60 MHz) upfield from the 9-β-methyl signal. Resonance signals from the 5-methyl groups are little affected. These observations enabled absolute structures to be assigned to the α- and β-5,9-dialkyl-6,7-benzomorphans.

An aromatic compound has been defined as a compound that can support a ring current. This may not be so and the anisotropic nature of many cyclic systems could be due to other factors not yet understood (Jones, 1968).

It should be noted that significant diamagnetic anisotropy is exhibited by C—C single bonds.

Hydrogen bonds

Dilute solutions of compounds bearing protons capable of intermolecular hydrogen bonding have resonance lines for such protons which, on increasing solute concentration, shift downfield. That is, increased hydrogen bonding leads to greater proton deshielding. Similarly, the weakening of hydrogen bonds by raising the sample temperature causes an upfield shift of the proton resonance line. The chemical shifts of intramolecular hydrogen bonded protons are independent of temperature and solute concentration. NMR spectroscopy is an excellent means of studying hydrogen bonding (p.450).

Solvent effects　*(Bhacca and Williams, 1964; Ronayne and Williams, 1966; Becker, 1980).*

Solvent variation often results in dramatic chemical shift changes of certain groups of protons, a phenomenon which may be exploited in structural investigations. Magnetic non-equivalence may be induced where previously spectrum interpretation was severely impaired by resonance band overlap. In solutions of polar solutes, solvent molecules often adopt a specific orientation with respect to those of the solute.

Non-aromatic solvents, such as CS_2, CCl_4 and CH_3CN, in close proximity to a molecule of polar solute, are polarised; a local 'reaction field' is induced and the chemical shift values of solute protons are influenced to varying degrees. Although these effects are small they may be used to determine the proximity of a proton to a polar centre, such as carbonyl or ether oxygen.

Aromatic solvents often induce large 'solvent shift' effects, particularly in proton groups adjacent to a carbonyl or other polar function. The anisotropy of magnetic susceptibility of an aromatic solvent such as benzene will affect the proton groups of a solute differently, depending upon the mode of solute–solvent alignment. *N*-Methylformamide, represented by the resonance forms (III)⇌(IV), shows restricted rotation about the C—N bond and gives both *cis* (IIIa and IVa 92%) and *trans* (IIIb and IVb 8%) N—CH_3 PMR signals in a spectrum of the neat liquid. In benzene solution the *trans* N-C$\underline{H}_3$ signal is

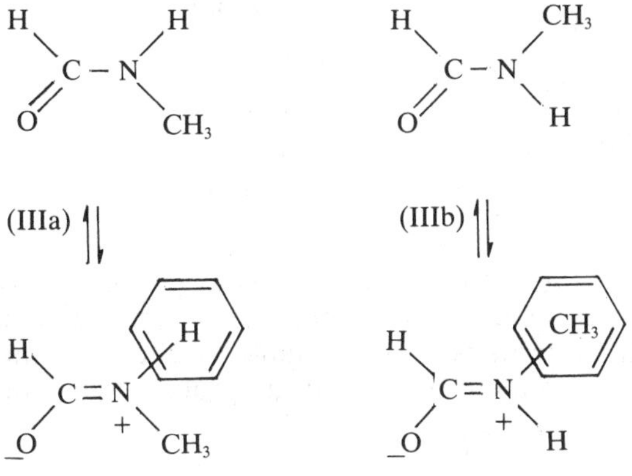

(IIIa)

(IIIb)

(IVa) *cis*

trans (IVb)

shifted upfield far more than that of the *cis* form, because of the alignment with the benzene solvent illustrated in (IVb). Here the partially positive nitrogen atom is close to the high π-electron density region of the benzene ring, which in turn is as far away as possible from the carbonyl oxygen bearing a fractional negative charge. In other words, orientation IVb is that preferred in the *trans* isomer and orientation IVa in the *cis* isomer. In the former orientation, the *N*-methyl group falls above the plane of the aromatic ring and is therefore shielded. In the latter no such shielding results. Polar compounds often exhibit good solvent shift effects with benzene; with pyridine somewhat larger spectral changes result. Solvent shift effects are best studied in weak (5%) solutions. Acidic solvents, such as CF_3COOH, CH_3COOH and CD_3COOD, also show useful solvent shifts, particularly with amines.

Lanthanide shift reagents   *(Sanders and Williams, 1972; Mayo, 1973; Sievers, 1973).*

Chemical shift changes induced by solvents are relatively small. A very much more effective method of selectively altering the magnetic environments of nuclei and thus changing their chemical shifts is by the addition of paramagnetic metal complexes to the sample.

The presence of paramagnetic ions usually causes rapid spin-lattice relaxation in nearby nuclei and this results in excessive line broadening of the NMR signal. However the europium ion (Eu^{3+}) has anomalously inefficient nuclear spin-lattice relaxation properties, and in the form of suitable complexes provides a valuable range of **shift reagents** affording well-resolved spectra.

The most useful complexes are β-diketone derivatives in which the lanthanide ion may increase its co-ordination number by interaction with lone pairs of electrons. An example of a commonly used shift reagent is tris(dipivalomethanato)europium, $Eu(DPM)_3$.

As a result of the formation of a new complex between the shift reagent and a sample molecule possessing suitable donor atoms, the chemical shift of protons of the sample is altered. The extent of the change is related to the proximity of the protons to the donor functional group (Fig. 11.15).

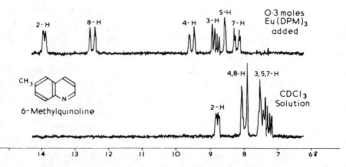

Fig. 11.15. 60 MHz spectra showing the effect of a lanthanide reagent on the chemical shifts of the aromatic protons of 6-methylquinoline (Courtesy Perkin-Elmer Ltd)

The paramagnetic metal species induces the shifts described by two merchanisms:

(a) **Contact shifts** arising from delocalisation of unpaired electron spin, *via* covalent bonds, to the affected nuclei.

(b) **Pseudocontact shifts** from secondary magnetic effects generated by the magnetic moment of the paramagnetic ion and transmitted through space.

Shift reagents of other lanthanides are also valuable in structural studies, e.g. the dipivalomethanato complex of praseodymium, $Pr(DPM)_3$. In contrast to shifts to low field induced by $Eu(DPM)_3$, the praseodymium reagent causes somewhat larger shifts, generally to high field. Ytterbium complexes, e.g. $Yb(DPM)_3$ are often useful reagents for N-heterocycles but they give rise to moderate line broadening.

Fluorinated β-diketone complexes of lanthanides often exhibit greater shifting power than non-fluorinated reagents. Examples of such complexes are:

$Eu(FOD)_3$ Tris(1,1,1,2,2,3,3-heptafluoro-7,7-dimethyloctan-4,6-dio-nato)europium.

$Eu(PFD)_3$ Tris(1,1,1,2,2-pentafluoro-6,6-dimethylheptane-3,5-dio-nato)europium.

$Eu(FHD)_3$ Tris(1,1,1-trifluoro-5,5-dimethylhexane-2,4-dionato)europium.

In practice known amounts of the lanthanide shift reagents are added successively to the sample dissolved, preferably in a non-polar solvent. The chemical shift of each proton or proton group of the sample molecule changes with each addition of reagent, and the extent of the induced shift is measured. A plot of the induced shift against the ratio of shift reagent to substrate will give a straight line at low values of the ratio. From plots for each proton group in a molecule valuable structural information may be deduced.

Clearly the application of lanthanide shift regents extends the utility of NMR spectroscopy. They may be used simply to resolve overlapping signals from different proton groups in a molecule or, by more quantitative studies, to provide information concerning molecular configuration. Chiral lanthanide shift reagents are of value in the quantitative determination of mixtures of enantiomers (Dewar *et al.*, 1982).

Shoolery's rules   *(Dailey and Shoolery, 1955)*

The approximate chemical shifts of protons in aliphatic methylene groups may be predicted by the application of Shoolery's additive constants. Additive constants for many other groups are also available.

Protons are described as **magnetically equivalent** when they have identical screening constants (σ). Such protons are almost always also chemically equivalent, as for example in a —CH$_3$ group. In propyne magnetically equivalent nuclei are not chemically equivalent; its spectrum consists only of a four-proton singlet at 1.8δ.

Much information concerning the chemical shifts of protons in organic molecules has been collected and compiled into correlation charts; examples are found in Figs 11.8–11.11. These charts should be used empirically like the corresponding compilations in infrared spectroscopy. Collections of spectra are also of value in structural studies. (Sadtler Standard NMR Spectra; Chamberlain, 1974; Brügel, 1979).

Scales of measurement

As protons in different environments in an organic compound resonate at different frequencies, a method of expressing the positions of resonance lines is required. Most proton absorption arises within a 600 Hz range on a frequency scale of 60×10^6 Hz (i.e. 60 MHz) and measurements are often required to 0.2 Hz. To measure frequency or magnetic field strength to this degree of accuracy on an absolute scale is not possible. A comparative scale is therefore employed. Resonance lines are measured with respect to the absorption from a reference compound, usually tetramethylsilane, $Si(CH_3)_4$, (TMS) (p.431). Chemical shifts are expressed in units downfield from TMS, usually at 60 MHz. In the PMR spectra of organic compounds, shift values upfield of TMS are rarely encountered.

There are three scales for expressing chemical shift values:

(1) v-Scale *(hertz)*

The hertz is the internationally accepted radiofrequency scale notation, where 1 Hz = 1 cycle per second. Chemical shifts quoted in Hz from a reference signal must be accompanied by a statement of the operating frequency of the instrument. The chemical shift is proportional to both the applied field (H_0) and the applied frequency (v_0).

Consider a chemical shift of 80 Hz downfield from TMS measured at 40 MHz; at 60 MHz this becomes $\frac{60}{40} \times 80 = 120$ MHz, and at 100 MHz it becomes $\frac{100}{40} \times 80$ or $\frac{100}{60} \times 120 = 200$ Hz. Thus, at 100 MHz, line separation and therefore resolving power is greater than at 60 MHz.

(2) δ-Values

It became necessary to provide a field-independent scale, thus δ-values are used where:

$$\delta = \frac{H_0 - H_{ref}}{H_{ref}} \times 10^6 \, \text{ppm}$$

H_0 = field strength for resonance of protons of the sample, and H_{ref} = field strength for resonance of protons of the reference compound. This provides a 0 to -10 scale with TMS as the zero marker (Fig. 11.16). δ is a dimensionless expression which is negative for most protons and is often referred to as parts per million (ppm).

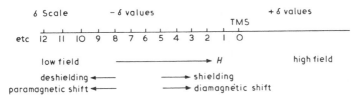

Fig. 11.16. δ-value scale

The International Union of Pure and Applied Chemistry define the δ-scale thus:

'Whenever possible the dimensionless scale should be tied to an internal reference which should normally be TMS. The proton resonance of TMS should be taken as zero; if some other internal reference is used that reference, and the conversion shift used to convert the measured shifts to TMS reference scale, should be explicitly stated. The dimensionless scale should be defined as positive in the high frequency (low field) direction. The scale in parts per million (ppm) based on zero for TMS should be termed the δ-scale.'

The relationship between the δ-scale and the v-scale is expressed by:

$$\delta = \frac{\text{Hz}}{\text{instrument frequency } v_0 \text{ (MHz)}}$$

(3) τ-Values

$$\tau = 10 - \delta$$

By defintion the TMS signal occurs at 10. Use of the τ-scale is now discouraged.

Spectrum calibration is accomplished by aligning the single sharp resonance line of TMS, in which all 12 protons are chemically and magnetically equivalent, with the 0δ mark on a precalibrated chart. TMS is the most commonly used internal standard, being dissolved to the extent of 0.5% in a solution of the sample. External references are necessary under certain circumstances, and then precision co-axial sample tubes are employed.

An internal reference should possess the following properties:

(a) chemical inertness
(b) magnetic isotropy
(c) it should give a single, sharp, easily recognised absorption signal
(d) miscibility with a wide range of solvents and organic liquids
(e) it should be volatile, facilitating its removal from valuable samples.

TMS has all these properties, and is an excellent reference for organic solvents and liquids. For solutions in water or deuterium oxide (D_2O), sodium 2,2-dimethyl-2-silapentane-5-sulphonate or DSS (V) is a reasonable internal reference, although no standard for aqueous solutions is really satisfactory.

$$CH_3$$
$$|$$
$$CH_3 - Si - CH_2 \cdot CH_2 \cdot CH_2 \cdot SO_3^- Na^+$$
$$|$$
$$CH_3 \qquad\qquad\qquad (V)$$

Spin-spin coupling

A high resolution spectrum is distinguished from one at low resolution by the presence, in the former, of fine structure. Many PMR spectra have resonance bands split into doublets, triplets, quartets etc. This phenomenon is known as spin coupling or spin-spin splitting.

Consider two protons on adjacent carbons:

$$\text{(A) (X)}$$
$$H \quad H$$
$$| \quad |$$
$$R''' - C - C - R$$
$$| \quad |$$
$$R'' \quad R'$$

The R groups are not protons and do not bear protons on nuclei adjacent to those under consideration. Then, provided that the R groups are not all the same, the signal for each proton will appear as a doublet (Fig. 11.17).

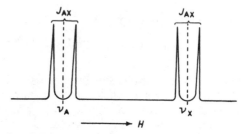

Fig. 11.17. Coupling between protons with widely different chemical shifts

The separation between the lines of each doublet, J_{AX}, is identical for each signal and is referred to as the coupling constant (J). The J value, which is usually measured in Hz, is independent of field strength. An increase in field strength causes v_A and v_X to shift further apart, but J_{AX} remains the same for any particular system.

By a consideration of the spin state of each proton with respect to that of its neighbour in a population of identical molecules, an explanation of spin-spin coupling is forthcoming. Proton A can have either a spin parallel or antiparallel with respect to that of X, and likewise proton X is aware of two spin forms of A. Each proton being aware of two spin states of its neighbour experiences two effective fields from it, and generates two corresponding resonance signals. Since the populations in each spin state are the same, resonance lines of equal intensity result. The 'awareness' of the spin state of one nucleus by an adjacent nucleus may be considered as being relayed by interceding bonding electrons. For further study, an excellent account of the theory of spin-spin interactions has been published by Roberts (1961).

Run under conditions of low resolution, the PMR spectrum of ethanol appears as three broad bands integrating for one, two and three protons respectively (Fig. 11.18(a)). In a high resolution spectrum, fine structure is apparent (Fig. 11.18(b)). Here the —CH_2— group is a quartet with an area-ratio of lines 1:3:3:1, and the —CH_3 signal a triplet of area-ratio 1:2:1. The spacing between the lines of the —CH_2— and CH_3— patterns is identical and is the **coupling constant** *J*. Figure 11.19 shows the spin arrangements possible for the three methyl protons and the two methylene protons. Four spin combinations of the CH_3 protons are possible with their statistical probabilities of occurrence being 1:3:3:1, and similarly three combinations of probability 1:2:1, are possible for

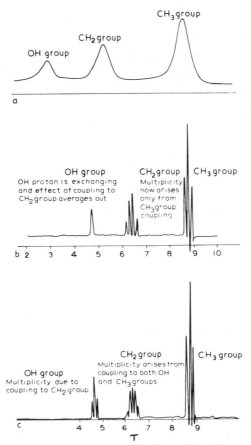

Fig. 11.18. (a) Low resolution NMR spectrum of ethanol; (b) High resolution spectrum of ethanol (trace of OH⁻ or H⁺); (c) High resolution spectrum of highly purified ethanol

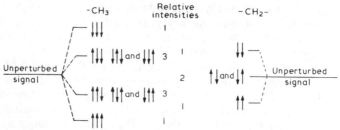

Fig. 11.19. Possible spin arrangements for three methyl protons and two methylene protons

the —CH_2— protons. Each proton group splits the adjacent group signal to its own tally of possible spin states.

The construction of multiplets arising from spin-spin coupling is governed by the following rules:

(a) Magnetically equivalent nuclei do not iteract, e.g. methane and ethane will each give only a single resonance line.

(b) Band multiplicity is determined by the neighbouring groups of equivalent nuclei, and is given by the expression:

$$\text{Number of lines} = (2nI + 1)$$

where I = the spin quantum number, and n = number of adjacent equivalent nuclei.

The integration of the whole multiplet is equal to the number of protons responsible for the signal, e.g., the 1:3:3:1 quartet in the ethanol spectrum integrates for two protons and the 1:2:1 triplet for three protons.

(c) If there are more than two interacting groups, the multiplicity of A as split by B and C is expressed by $(2n_B I_B + 1)(2n_C I_C + 1)$ etc. With protons of spin 1/2 this simplifies to $(n_B + 1)(n_C + 1)$ etc. For example, under first order coupling rules (p.435) consider

$$R—CH_2—CH_2—CH_2—R$$
$$\text{B} \qquad \text{A} \qquad \text{C}$$

the expression becomes $(2 + 1)(2 + 1) = 9$. Proton A resonance will therefore be seen as a pattern with a maximum of nine lines (three triplets), but fewer than nine lines may ensue from band overlap. If, however, R and R′ are identcial, then the B and C protons are chemically and magnetically equivalent and give rise to a five line pattern for the A protons, i.e. $n + 1 = 4 + 1 = 5$. This is not due to band overlap.

Note that the coupling constants J_{AB} and J_{AC} will most probably be different. Also, with nuclei other than protons the value of I may vary and, therefore, so will the band multiplicity.

(d) The intensities of lines in a first order multiplet are theoretically symmetrical about the mid-point of the absorption band, and follow the coefficients of a binomial expansion $(1 + x)^n$. The intensity ratios are best remembered by use of the Pascal triangle:

$$
\begin{array}{ccccccccc}
 & & & & 1 & & & & \\
 & & & 1 & & 1 & & & \dots\dots 1 \\
 & & 1 & & 2 & & 1 & & \dots\dots 2 \\
 & 1 & & 3 & & 3 & & 1 & \dots\dots 3 \\
1 & & 4 & & 6 & & 4 & & 1 \quad\dots\dots 4 \\
\end{array}
$$

```
                    1
                1       1        ............ 1
            1       2       1    ............ 2
        1       3       3       1   ............ 3
    1       4       6       4       1   ............ 4
  1     5      10      10      5      1   ............ 5
1     6     15      20      15      6      1 ............ 6
```

where *n* is the number of interacting nuclei.

Coupling is described as **first order** when:

    (a) the chemical shift separation of the nuclei ($\Delta v = v_A - v_X$), is at least six times the coupling constant, J_{AX}

    (b) each proton of a group is coupled equally to every proton of the other group.

Under these circumstances the simple multiplicity rules outlined above apply and the pattern is described as first order. The splitting between the CH_3— and —CH_2— groups of ethanol is an example of first order coupling.

In the two-proton picture referred to earlier (p.432), under ideal conditions of first order coupling all four line intensities will be the same. As the chemical shift difference between **A** and **X** is reduced, then deviation from the strict binomial intensities results, and the quartet collapses, ultimately, to a singlet (Fig. 11.20). This occurs when the chemical shifts of **A** and **X** are the same; in other words the protons are magnetically equivalent and resonate at the same frequency.

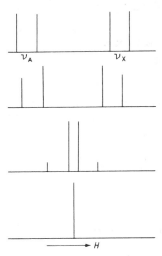

Fig. 11.20. Collapse of quartet to singlet with reduction of chemical shift difference between A and X

Exchange effects

The spectrum of ethanol run under normal circumstances exhibits a four-line pattern for the —CH$_2$— protons (Fig. 11.18(b)). This behaviour appears anomalous since further splitting by the hydroxyl proton would be expected. Such coupling becomes apparent in the spectrum of very pure ethanol where the —CH$_2$— signal is a double split quartet (8 lines) and the hydroxyl proton is the expected triplet (Fig. 11.18(c)). The addition of a trace of either acid or base to pure ethanol restores the simplified spectrum (Fig. 11.18(b)). Laboratory ethanol normally contains acid or base impurities which catalyse the exchange of hydroxyl protons between ethanol molecules. Thus, in a given time, any individual hydroxyl proton experiences a wide variety of spin arrangements from different —CH$_2$— group protons. The spin arrangements of the —CH$_2$— group protons as seen by the —OH protons, and *vice versa*, are effectively averaged, and no spin coupling is observed. The catalytic effect of H$^+$ and OH$^-$ is apparent in the spectrum of a mixture of ethanol and water which is acid and base free. Here separate resonance signals are seen for C$_2$H$_5$—O$\underline{H}$ and $\underline{H}_2$O, suggesting very slow exchange. The addition of acid causes coalescence of the two resonance lines to a sharp singlet, the signals being averaged by rapid exchange.

Chemical exchange may be demonstrated further by the addition of deuterium oxide (D$_2$O) to ethanol. Resonance from the ethanol hydroxyl proton disappears since the —OH proton is replaced by D; D$\underline{H}$O resonance occurs at about 4.88δ

$$C_2H_5OH + D_2O \rightleftharpoons C_2H_5OD + DHO$$

Protons bonded to O, N and S are exchangeable by deuterium, thus providing a good means of characterisation.

However, the rate of exchange of NH and SH protons in non-aqueous media is much slower than that of —OH protons.

Spin-spin coupling notation

An alphabetical notation for referring to types of nuclei engaged in spin-spin coupling has evolved. The following is a guide to this notation.

(1) Equivalent nuclei are referred to by the same letter of the alphabet, with sub-numbers indicating the number of equivalent nuclei.

$$\begin{array}{c} H \\ | \\ H-C-H \\ | \\ H \end{array} \qquad \text{is an } A_4 \text{ system}$$

and

$$\begin{array}{c} Cl \\ \diagdown \\ \diagup \\ Cl \end{array} C=C \begin{array}{c} H \\ \diagup \\ \diagdown \\ H \end{array} \qquad \text{an } A_2 \text{ system}$$

(2) Nuclei with resonance signals separated by only a small chemical shift difference are designated by letters from the same region of the alphabet.

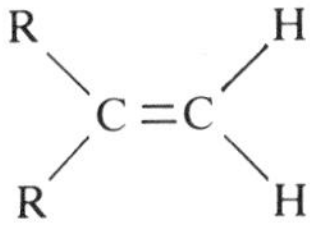

$$\begin{array}{c} R \\ \diagdown \\ R \diagup \end{array} C = C \begin{array}{c} H \\ \diagup \\ \diagdown H \end{array} \qquad \text{is AB}$$

and

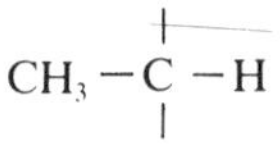

$$CH_3 - \overset{|}{\underset{|}{C}} - H \qquad \text{is A}_3B$$

(3) Where chemical shifts are widely separated, letters from quite different parts of the alphabet are employed.

$$H - \overset{\overset{F}{|}}{\underset{\underset{F}{|}}{C}} - H \qquad \text{is A}_2X_2$$

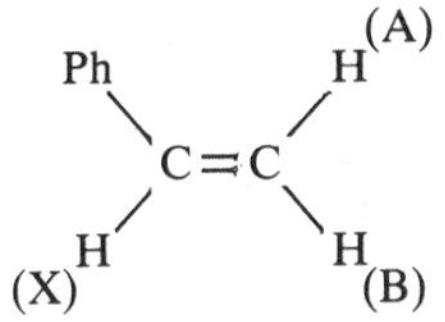

$$\begin{array}{c} Ph \\ \diagdown \\ {}_{(X)}H \diagup \end{array} C = C \begin{array}{c} H^{(A)} \\ \diagup \\ \diagdown H \\ {}_{(B)} \end{array} \qquad \text{is ABX}$$

$$\begin{array}{c} {}^{(A_3)} \quad H^{(M)} \quad H^{(N)} \\ CH_3 - \overset{|}{C} = C \diagup \\ \diagdown F(X) \end{array} \qquad \text{is A}_3MNX$$

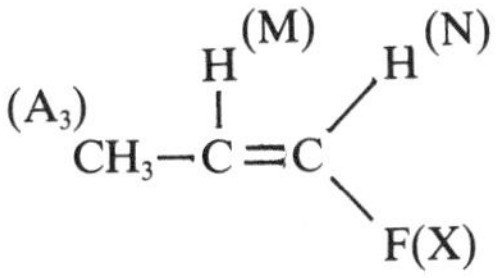

(4) Where more than one coupling constant between equivalent groups is possible, primed letters are resorted to:

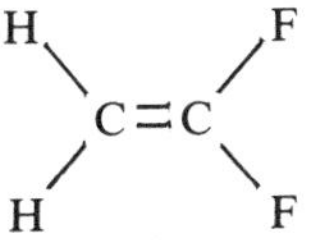

$$\begin{array}{c} H \\ \diagdown \\ H \diagup \end{array} C = C \begin{array}{c} F \\ \diagup \\ \diagdown F \end{array} \qquad \text{is an AA}'XX' \text{ system, not A}_2X_2$$

Each proton does not couple equally to each fluorine, and *vice-versa*, so the J_{cis} and J_{trans} values will be different. Similarly, in *para*-substituted aromatic compounds, e.g.

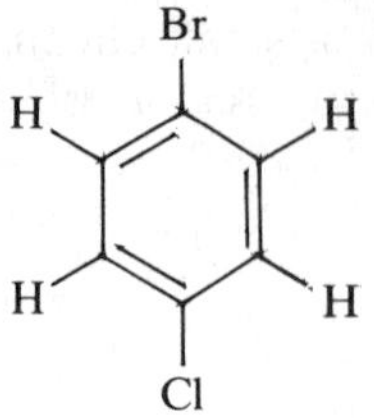

is AA'BB' and 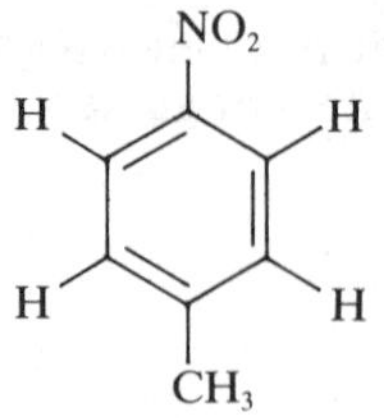 is AA'XX'

In both cases, *meta* and *para* couplings occur, as well as *ortho* coupling. Coupling to the CH_3 protons has been ignored.

Higher order spin-spin coupling

Values for chemical shifts and coupling constants are readily arrived at for spectra or portions of spectra arising from first order coupling; however, when first order coupling rules are not obeyed, these values are more difficult to deduce. Higher order spin systems, for example ABX, A_2B_2 etc., may be solved by recognising certain band spacings and substituting these values in established equations. In more complex cases the analysis of a spectrum may get very involved. For further reading on the analysis of non-first order spin systems reference should be made to specialist publications (Pople *et al.*, 1959; Roberts, 1961; Mathieson, 1967; Bible, 1965; Becker, 1965). Computer programs are available for the calculation of certain complex spectra and Wiberg and Nist (1962) have published a compilation of calculated spectra.

Factors influencing the value of coupling constants

It is not fully understood why coupling constants vary considerably with proximate structure. The following are some of the factors affecting the size of coupling constants.

(1) When only single bonds intercede between nuclei, coupling usually occurs only through three bonds.
In compounds such as

$$
\begin{array}{cc}
(A) & (B) \\
H & H \\
| & | \\
R - C - C - R''' \\
| & | \\
R' & R''
\end{array}
$$

J_{AB} is in the order of 5–8 Hz.
Occasionally, when molecule geometry is favourable, long range low order coupling can occur through four or even five bonds.
(2) Couplings through multiple bonds can occur over greater distances, presumably because of the mobility of π-electrons. Coupling values of alkenic protons decrease in the order:

trans > *cis* > *geminal*

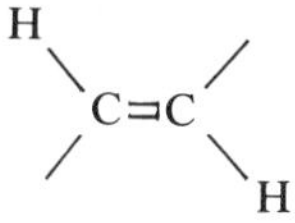

$J_{AB\ trans}$ 11–18Hz

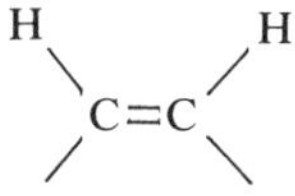

$J_{AB\ cis}$ 6–14Hz

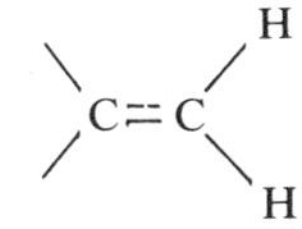

$J_{AB\ gem}$ 0–3.5Hz

In unsaturated compounds, coupling of the order of 1 Hz through four bonds is often observed.

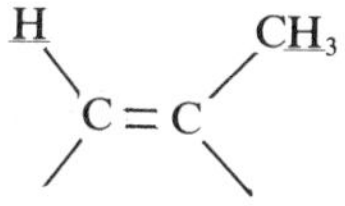

    $JA_{3x} \simeq 1Hz$

(3) In aromatic systems *ortho* coupling ($J \simeq$ 8 Hz) > *meta* ($J \simeq$ 2–3 Hz) > *para* ($J \simeq$ 0–1 Hz).

(4) coupling between geminal protons on saturated carbon is dependent, in part, upon the bond angle θ. J_{AB} decreases from about 20 Hz when θ is 105° to zero when θ is 125°.

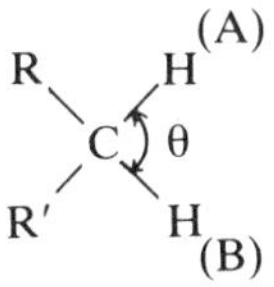

It must be emphasised that this is an approximate relationship since other factors influence J_{AB}. Cookson *et al.*, (1966) have published a comprehensive account of geminal constants and factors influencing them.

(5) The size of the vicinal proton coupling constant is dependent upon the dihedral angle θ. A plot of θ against J (Fig. 11.21) shows that J_{AB} is maximum ($\simeq$ 10 Hz) when the dihedral angle is 180°, and a minimum when it is 90° and about 8 Hz when θ is 0°. In cyclohexane axial-axial coupling has $J_{axax} \simeq$ 8–12 Hz, whereas $J_{eqax} \simeq$ 3–4 Hz. Karplus (1959), who has studied vicinal coupling extensively, warns against attempting to determine dihedral angles with any degree of accuracy from vicinal coupling constants.

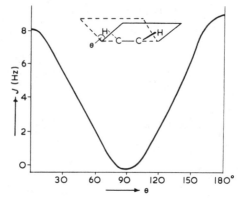

Fig. 11.21. Karplus relationship between the dihedral angle and coupling constant for vicinal protons

(6) The electronegativity of substituent X in a system:

$$-\overset{\displaystyle H}{\underset{\displaystyle |}{C}}-\overset{\displaystyle H}{\underset{\displaystyle |}{C}}-X$$

has little effect on the vicinal coupling constant (<2Hz). In the limited number of cases investigated increasing substituent electronegativity gives a decrease in the *J*-value.

(7) Coupling values for splitting between protons and other nuclei, e.g. ^{15}N, ^{31}P and ^{13}C, are often very large; an exception is proton-deuterium coupling, where the *J*-values are about one-seventh of those between protons. Advantage is taken of this in **spin-decoupling** by deuterium replacement of hydrogen.

Measurement of coupling constants

A major problem encountered in comparing NMR data from different published sources is the variability of conditions under which spectral parameters are measured. To minimise errors in the measurement of coupling constants, the use of scale expanded spectra is recommended.

In the spectrum of some molecules where certain H—H coupling constants are inaccessible from direct observation, ^{13}C satellite signals may be exploited. The 1.1% natural abundance of ^{13}C and the large ^{13}C—^{1}H coupling constant gives rise to weak '^{13}C satellites' almost symmetrically positioned about the proton signal being investigated. The signals are due to protons bonded to and therefore coupled to ^{13}C($I = \frac{1}{2}$) rather than ^{12}C. Although such satellite signals are weak, the sensitivity of modern NMR spectrometers renders them easily observable and H—H couplings of symmetrical molecules may often be extracted (see Becker (1969) and Batterham (1973)). For tables of coupling constants for protons see Mathieson (1967), Becker (1969), Batterham (1973), and Silverstein and Bassler (1968).

Spin-spin decoupling

A spectrum is often rendered very complex and difficult to interpret by spin-spin coupling. Simplification may be achieved in two ways.

(a) By using an instrument with a more powerful homogeneous magnetic field, e.g. a 100 MHz instrument in preference to one operating at 60 MHz. A greater line separation (higher resolution) results, but there is a practical limit to the magnet strength available (p.414).

(b) By a spin-spin decoupling technique.

Such technqiues not only reduce the amount of fine structure and clarify the spectrum, but enable interacting protons to be identified.

Isotope exchange

The hydroxyl protons of ethanol, as we have seen, are able to exchange with other protons or deuterium (^{2}H) causing decoupling and simplification of the —CH$_2$— proton resonance. Dimethylsulphoxide, as a solvent, facilitates the study of coupling through exchangeable protons of this sort by hydrogen bonding to them and slowing the exchange rate.

Isotopic exchange may be used to advantage in the deuteration of compounds with active (acidic) hydrogens bonded to carbon, as for example in protons on carbon *alpha* to a ketone or nitrile.

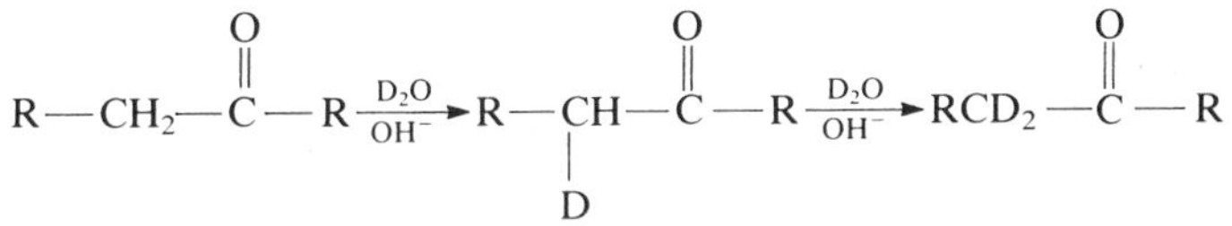

$$R-CH_2-\overset{\overset{\displaystyle O}{\|}}{C}-R \xrightarrow[OH^-]{D_2O} R-\underset{\underset{\displaystyle D}{|}}{C}H-\overset{\overset{\displaystyle O}{\|}}{C}-R \xrightarrow[OH^-]{D_2O} RCD_2-\overset{\overset{\displaystyle O}{\|}}{C}-R$$

and

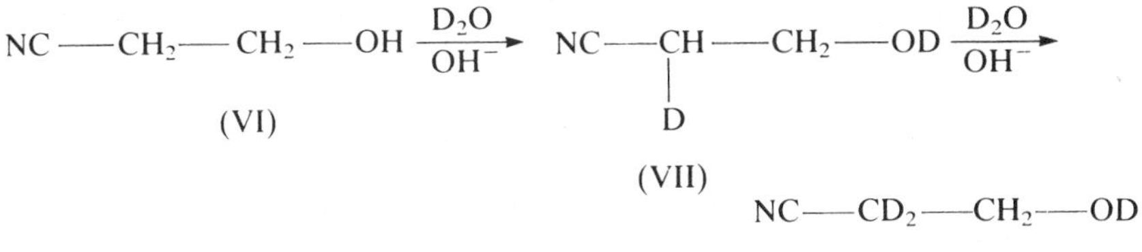

$$NC-CH_2-CH_2-OH \xrightarrow[OH^-]{D_2O} NC-\underset{\underset{\displaystyle D}{|}}{C}H-CH_2-OD \xrightarrow[OH^-]{D_2O}$$

$$(VI) \qquad\qquad\qquad (VII)$$

$$NC-CD_2-CH_2-OD$$

$$(VIII)$$

The scheme (VI–VIII) illustrates isotopic exchange in 2-hydroxy-propionitrile (Lapidot *et al.*, 1964). In D$_2$O its spectrum appears as an A$_2$X$_2$ pair of triplets, the C-1 and C-2 proton signals being at 2.68δ and 3.80δ respectively (Fig. 11.22). The addition of hydroxide ions to this solution causes the activated C-1 protons to exchange successively (VII and VIII) with deuterium and show the corresponding changes in the PMR spectrum. The rate of exchange is dependent upon the concentration of OH$^-$. Figure 11.22 illustrates the change in the spectrum of a solution of 2-hydroxypropionitrile (VI) in 1.44M NaOH in D$_2$O with time:

(a) At zero time only the A_2X_2 pattern of (VI) is present.
(b) and (c) A doublet due to the C-2 protons split by the single C-1 proton (VII) is superimposed upon the original low-field triplet giving a five-line pattern.
(d) Here there are contributions from structures (VI), (VII) and (VIII); a triplet, doublet and singlet respectively appear superimposed for the C-2 protons.
(e) contributions from only (VII) and (VIII) are apparent in the resonance of the C-2 protons.
(f) Finally, after 25 min almost complete exchange has occurred, the C-2 protons are almost entirely decoupled to yield a singlet and C-1 proton resonance is lost. Integration curves for C-2 and C-1 protons respectively are shown to the right of the spectra.

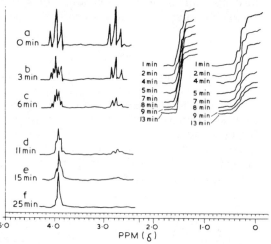

Fig. 11.22. Deuterium exchange of 2-hydroxypropionitrile with D_2O in the presence of sodium hydroxide. Left: Successive spectra. Centre: Integrated intensitites of β-proton triplets. Right: Integrated intensities of β-proton triplets (Courtesy *J. Chem. Ed.*)

The synthesis of deuterated compounds from appropriately deuterated starting materials also aids the elucidation of molecular fine structure. An example is the investigation of the conformational preference of the non-bonding electron pair in piperidines by the synthesis, and PMR examination of, the α-protons of 3,3,5,5-tetradeuteropiperidines (IX) (Lambert and Keske, 1966).

(IX)

Nuclear magnetic double resonance (NMDR) (von Philipsborn, 1971; Becker, 1980)

During a normal NMR experiment the sample is subjected to a single radiofrequency field. If a second (or sometimes more) radiofrequency field is applied in a specific manner, then a series of techniques collectively known as nuclear magnetic double resonance results. From the careful application of these techniques, particularly in complex spectra with overlapping resonance bands, much valuable information is forthcoming.

Spin-spin decoupling by NMDR

If during an NMR experiment on a simple AX system (Fig. 11.23(a)), the nucleus is irradiated with a second radiofrequency signal of frequency (v_x) equal to the precessional frequency of X, then the X-nuclei will undergo rapid changes of spin state. Nucleus A is aware of only one equivalent (averaged) spin state of X and therefore its resonance signal appears, not as a doublet, but as a sharp singlet. Irradiation of X effectively decouples the signal from A (Fig. 11.23(b)). Similarly if A is irradiated with a second radiofrequency signal equal to its own resonance frequency (v_A), then the signal from X will collapse to a singlet (Fig. 11.22(c)).

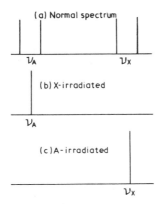

Fig. 11.23 Double irradiation

Double resonance experiments may be performed either by: (a) keeping both the applied frequency (v_1) and the irradiation frequency (v_2) constant, and varying the magnetic field (**field sweep**), or (b) by having a constant magnetic field (H_0) and irradiation frequency (v_2), and sweeping with the applied frequency (v_1) (**frequency sweep**). The irradiating frequency (v_2) may be changed after each scan, permitting an extensive investigation of spin-spin coupling. Nuclear triple resonance, where two irradiation frequencies are applied simultaneously and the signal due to a third nucleus is examined, is valuable in the investigation of multiple and long-range coupling.

The irradiation frequency (v_2), in order to decouple the nuclei completely, is a powerful radiofrequency signal; v_1 is weak in comparison.

Heteronuclear decoupling

This is the decoupling of different isotopes where the difference in resonance frequencies is large, e.g. ^1H and ^{19}F. The experiment is expressed as

$$^1H[^{19}F]$$

Nucleus being studied   Nucleus irradiated

Homonuclear decoupling

This involves nuclei of the same isotope, where the difference in resonance frequencies is small. It is expressed by $^1H[^1H]$. Similarly, the irradiation of the X nucleus of an ABX system is expressed as AB [X].

For effective double irradiation a good chemical shift difference between the interacting nuclei is required.

Spin decoupling techniques, as exemplified by ethanol spectra, are summarised in Fig. 11.24).

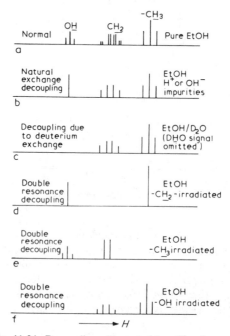

Fig. 11.24. Decoupling—illustrated by ethanol spectra

Decoupling experiments by NMDR are, in general, easy to conduct; however complete decoupling only occurs under rather restricted experimental conditions. True decoupling is only seen when the chemical shift separation is much greater than the coupling constant.

Spin tickling

When the second irradiating frequency (v_2) is relatively weak and spectrum collapse is incomplete extra resonance lines or 'residual splittings' are observed. These lines may be related to energy transitions of the spin system observed, and may aid the interpretation of complex spectra.

Nuclear Overhauser effect (NOE)   (Backers and Shaefer, 1971; Noggle and Schirmer, 1971)

Unlike spin decoupling and spin tickling, which are based upon changes in energy levels, the nuclear Overhauser effect results from population changes in the energy levels which cause corresponding intensity changes in that spectra. The NOE experiment consists of saturating one signal and observing the intensities of other signals. The resonance bands due to protons spatially close to the group being saturated will show an increase in intensity. Such intensity changes are best measured by integration rather than peak height. The theory of the NOE is dependent upon relaxation mechanisms, and the relaxation of any nucleus is affected by all surrounding magnetic nuclei. Since the contribution of any nucleus to the relaxation of a second nucleus is dependent upon the mean square value of the magnetic field produced by the former, then the magnitude of the nuclear Overhauser effect will relate to the internuclear distance.

In the spectrum of dimethyl formamide (X) (Fig. 11.26) it is not possible from shift or coupling parameters to assign the two methyl signals. Saturation of the low field methyl signal causes a 17% increase in the intensity of the CHO signal, whereas saturation of the high field methyl signal gives no change in the formyl proton band. Clearly the low field line is *cis* to the formyl hydrogen.

Interactions giving rise to nuclear Overhauser effects may be intermolecular or intramolecular. In dilute solutions in aprotic solvents only the latter will be observed. It is important from an experimental standpoint in all NOE experiments to de-gas samples. Oxygen, being paramagnetic, will affect and often dominate the relaxation process and its presence will invalidate results.

INDOR spectroscopy

Probably the most powerful double resonance technique for the observation of the multiplicity of signals hidden by band overlap is **IN**ternuclear **DO**uble Resonance (INDOR). Here signals are only observed from coupled protons.

In the INDOR experiment the magnetic field (H_0) and the observation frequency (v_1) are held fixed on one of the A signal lines in a spectrum where, for example, the X signal is obscured. The second weak perturbing frequency v_2 is now swept through the spectrum. As v_2 passes in turn through the hidden X lines a change in intensity of the A line may result, leading to a vertical movement of the recorder pen. Thus, a negative or positive signal will result at positions corresponding to the X resonance frequencies. The signals overlapping X will not register unless they too are coupled to proton A (Fig. 11.25).

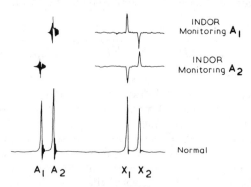

Fig. 11.25. 90 MHz spectra of $C_6H_5.CHBr.CHBr.NO_2$, 300 Hz scan, demonstrating the INDOR effect (Courtesy Perkin-Elmer Ltd)

Applications

Structure elucidation

The versatility of NMR spectroscopy as an analytical technique is illustrated here by a selection of its many applications. PMR is particularly useful in view of the abundance of ^1H in organic compounds, its strong resonance signal and convenient spin value of $\frac{1}{2}$. It has been applied extensively in structural elucidation.

The unequivocal proof of an organic structure by PMR spectroscopy is not often possible, but in conjunction with ultraviolet, infrared and mass spectrometry, it is an extremely valuable tool. Theoretically, all the resonance lines of a spectrum can be assigned to some molecular function, but in practice, for complex spectra, this proves difficult and often unnecessary.

An examination of a PMR spectrum must entail careful consideration of the following features.

(1) Resonance line positions, or **chemical shift** values, indicate the electronic environment of groups of equivalent protons. Correlation charts should be consulted to establish the probable nature of each

group. Resonance line overlap occurs frequently, particularly at high field, and spectrum simplification by spin-spin decoupling may be necessary. Examination of a spectrum in the presence of D_2O or CD_3OD will establish the identity of bands due to —O$\underline{H}$ or N$\underline{H}$; bases exhibit characteristic shifts in CF_3COOD solution.

(2) **Line intensity**, as determined by the electronic integrator, is proportional to the number of protons responsible for the signal.

(3) **Spin-spin coupling** patterns give the number of protons in interacting equivalent groups. Coupling patterns may be clarified by employing a lanthanide shift reagent or by INDOR spectroscopy.

A detailed account of structure elucidation by NMR is beyond the scope of this chapter and reference should be made to the excellent coverage of the subject by Bible (1965), Mathieson (1965, and 1967), Abraham and Loftus (1978), Becker (1980) and Harris (1983).

Investigation of dynamic properties of molecules

Conformational isomerism

At room temperature the cyclohexane molecule undergoes a rapid interconversion of chair forms, where the hydrogen atoms alternate between axial and equatorial conformations.

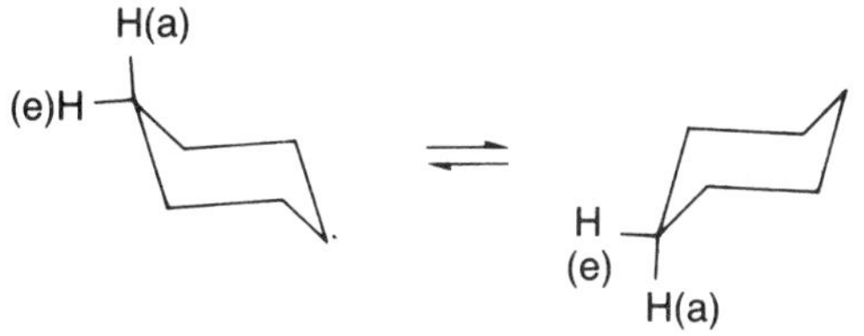

Above $-50°$ the PMR spectrum of cyclohexane in carbon disulphide exhibits a single resonance peak, a consequence of the average environment experienced by each proton. Below $-50°$ the signal broadens, as the conformational interconversion slows down, to a very broad band at $-65°$. At $-70°$ two ill-defined peaks are apparent and these develop until at $-100°$ they are seen as a pair of doublets ($J = 5$–$7\,Hz$) with a chemical shift difference of $27\,Hz$. As the temperature is lowered the protons spend increasingly longer times in axial and equatorial positions until ultimately the NMR spectrometer records signals due to each conformation. The low field doublet is assigned to the equatorial protons and that at high field to the more shielded axial protons. The inversion rate and the energy of inversion of such systems may be calculated from NMR data.

Further examples of the application of NMR in the investigation of conformational isomerism are given in Anderson (1965), Roberts (1966) and Jensen *et al.*, (1960).

Restricted rotation

Free rotation about the C—N bond of amides is impaired by the partial double bond character bestowed by resonance forms; dimethylformamide, for example, may be formulated thus:

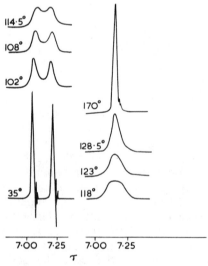

(X)   (XI)

In the absence of a contribution from structure (XI), free rotation about the C—N bond would render the N—CH$_3$ groups equivalent and their hydrogen resonances coincident. However, below 64° the N-groups are magnetically non-equivalent and two sharp resonance lines result (Fig. 11.26). Temperature increase enhances the rotation rate about the C—N bond until the methyl groups achieve magnetic equivalence. Both the rotation rate and the energy barrier to rotation may be calculated from PMR data. (Pople *et al.*, 1959). A more detailed discussion of the detection of hindered rotation and inversion by NMR spectroscopy may be found in Kessler (1970).

Fig. 11.26. The NMR spectrum of *N,N*-dimethylformamide (neat liquid) at room temperature shows a sharp peak for each methyl group. The peak at high field is *trans* to the formyl proton and is slightly broadened by coupling to it. When the temperature is raised, the peaks broaden and coalesce because the methyl groups exchange their environments by rotation about the C—N bond. This bond has partial double-bond character, as indicated by the 12-kcal activation energy required for rotation

Molecular asymmetry

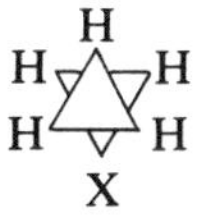

Under conditions of unrestricted rapid rotation about a C—C single bond in an open-chain system, the NMR spectrum reflects an average of the environments experienced by the nuclei concerned. In $CH_3.CH_2.X$, the CH_3 protons form one equivalent group and the $—CH_2—$ another, giving an A_3X_2 resonance pattern. A fixed configuration (XII) or slow rotation, on the other hand, would be observed in the PMR spectrum by the non-equivalence within the proton groups and thus a more complex splitting pattern would result. 1,2-Difluorotetrachloroethane, at room temperature, yields a single line ^{19}F spectrum; rotational isomers (XIII), (XIV) and (XV) appear as averaged.

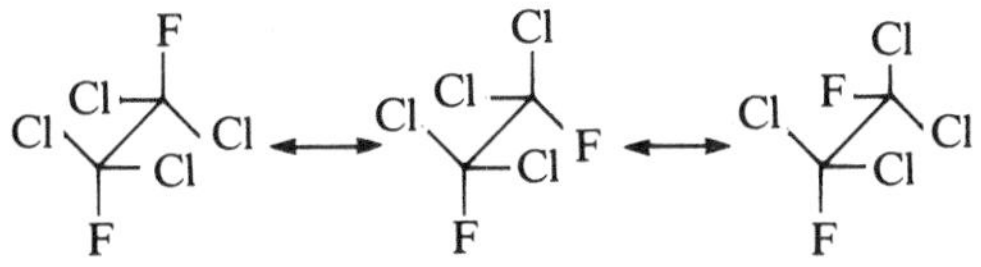

H
H⤺H
H⤹H
X
(XII)

(XIII)   (XIV)   (XV)

Rotation about the C—C bond decreases as the temperature is lowered, until at $-120°$ the fluorine nuclei exhibit two distinct signals separated by 51 Hz at 56.4 MHz.

Consider now the rotational isomers (XVI), (XVII) and (XVIII) of the compound

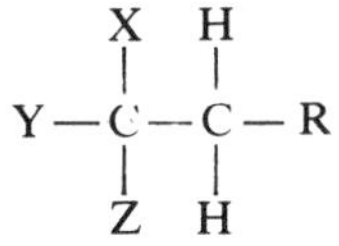

$$Y—\underset{\underset{Z}{|}}{\overset{\overset{X}{|}}{C}}—\underset{\underset{H}{|}}{\overset{\overset{H}{|}}{C}}—R$$

where the methylene group is attached to an asymmetrically substituted carbon. The protons H′ and H″ in the three rotational isomers do not experience equivalent averaged environments, even when rapid rotation about the C—C bond occurs, and they give rise to a geminal AB quartet. A similar splitting pattern would also result where the configuration is frozen by steric hindrance, even in the absence of an adjacent asymmetric carbon.

(XVI)   (XVII)   (XVIII)

Differentiation between *cis* and *trans* 2,6-dimethyl-1-benzylpiperidines (XIX and XX) exploits the principle of the non-equivalence of methylene protons through molecular asymmetry. The *trans* isomer, lacking a plane of symmetry, has magnetically non-equivalent benzylic methylene protons which appear as an AB quartet ($J = 13.2\,\text{Hz}$) centred at 3.41δ (Fig. 11.27(a)). On the other hand, the benzylic methylene protons of the symmetrical *cis* isomer appear (Fig. 11.27(b)) as a sharp singlet at 3.41δ (Hill and Chan, 1965).

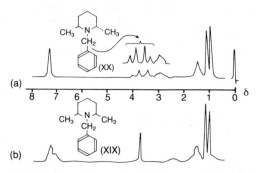

Fig. 11.27. NMR spectra of *cis*- and *trans*-2,6-dimethyl-1-benzylpiperidines

The term **intrinsic diastereotopism** (Mislow and Raban, 1967) is used to describe this phenomenon, which is seen not only in the methylene hydrogens of a benzylic group but also in sterically constrained methylene groups (Morris *et al.*, 1973).

A knowlege of the preferred conformation of a drug in aqueous solution may provide a useful insight into structure–biological activity relationships. In the case of acetylcholine (Culvenor and Ham, 1966), a consideration of the size of vicinal coupling constants from the A_2B_2 pattern of the —$CH_2.CH_2$— group favours a *gauche* conformation (XXI*a*), best represented by (XXII), rather than a *trans* conformation (XXI*b*).

(XXI*a*)   *gauche*   (XXI*b*) *trans*   (XXII)

Hydrogen bonding

Protons able to participate in hydrogen bonds can be identified and studied by means of PMR spectroscopy (Emsley *et al.*, 1965). The position of the hydroxyl proton resonance signal of ethanol is temperature dependent, temperature elevation causing a diamagnetic shift, whereas the chemical shifts of the —CH$_2$— and —CH$_3$ signals are unaffected. A consideration of the nature and origin of the hydroxyl proton signal provides an explanation of this phenomenon. Owing to rapid chemical exchange the O—H proton experiences many magnetic environments from both dissociated and hydrogen bonded molecules, and its resonance signal relects an average of these. As the temperature is raised, dissociation increases and the —OH proton's 'average' environment is changed. If hydrogen bonding is thought of as depleting the electron density around the bonded proton, then shielding, and a consequent upfield shift, results from dissociation. Similarly, a shift of the ethanol OH proton signal towards TMS (0.5δ–5δ) is observed when a sample is gradually diluted with a non-polar solvent, such as carbon tetrachloride. Dilution, like temperature elevation, favours the dissociated species, and intermolecular hydrogen bonded protons may be recognised by the dependence of their chemical shifts on these variables. Assignments of hydroxyl protons, however, should be verified by deuterium exchange. Protons engaged in intramolecular hydrogen bonds exhibit only small shifts on dilution or temperature elevation. Quantitative relationships between hydrogen bonding and the magnitudes of chemical shift changes have not been established.

A solution of chloroform in benzene provides an interesting example of the hydrogen bonding to the benzene π-cloud. The chloroform proton resonance signal occurs **upfield** of the normal position (7.3δ), because of shielding by the aromatic diamagnetic anisotropy.

High resolution PMR admirably complements infrared spectroscopy in the investigation of hydrogen bonding. However, unlike the latter, because of chemical exchange, it cannot detect monomers, dimers, polymers etc individually.

Keto–enol tautomerism

Nuclear magnetic resonance spectroscopy is probably the most powerful physical analytical method for qualitative and quantitative investigations of keto–enol equilibria (Smith, 1964). During such equilibria a specific proton experiences two distinct magnetic environments, which, if enough of each tautomer is present, will show up as separate resonance signals.

Acetylacetone (XXIII) at room temperature is a mixture of four parts enol and one part keto. Strong intramolecular hydrogen bonding stabil-

XXIII

ises the enolic form, and because chemical exchange is slow, both forms are seen in the PMR spectrum (Fig. 11.28). In the presence of triethylamine only the enol tautomer occurs; it is stabilised by strong hydrogen bonding between the amine nitrogen and the hydroxyl proton.

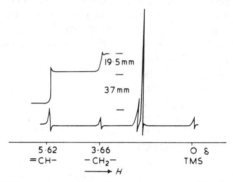

Fig. 11.28. NMR spectrum of liquid acetylacetone at 43° taken at 60 MHz

The concurrent upfield shift of the hydroxyl resonance reflects the strength of the intermolecular attraction. Tautomer ratios in keto–enol mixtures may be calculated from the integration curve. In the spectrum of acetylacetone taken at 43° (Fig. 11.28), e.g. the band at 5.62δ is assigned to the enol alkenyl hydrogen and that at 3.66δ to the keto —CH$_2$— group which correspond to integration values of 37 mm and 19.5 mm respectively. Thus

$$enol \equiv \; =CH— \; \equiv 1H \equiv 37 \; mm$$
$$keto \equiv —CH_2— \equiv 2H \equiv 19.5 \; mm$$
$$or \; 1H \equiv 19.5/2 \; mm$$

Therefore

$$\% \; enol = \frac{37}{37 + 19.5/2} \times 100$$
$$= 79.1\%$$

The enthalpy of keto–enol conversion may be established from variable temperature studies (Reeves, 1957).

Determination of optical purity

In pharmaceutical chemistry and particularly in drug-structure–activity studies, it is often desirable to determine the ratio of enantiomorphs in a mixture or to assess the optical purity of a compound without resorting to the physical separation of diastereoisomers. Since it is known that diastereoisomers differ in their NMR characteristics (Mateos and Cram, 1959), the technique has been investigated for practical applications (Raban and Mislow, 1965, 1966). If the mixture of enantiomorphs is quantitatively converted to the corresponding mixture of diastereo-isomers by reaction with an optically pure reagent, examination of the NMR spectrum of the mixture often permits assignment of bands to each isomer. Enantiomorph ratios can be calculated from integral values in the manner described for keto–enol tautomers. An instrument in good operating condition is able to detect less than 1% of optical impurity.

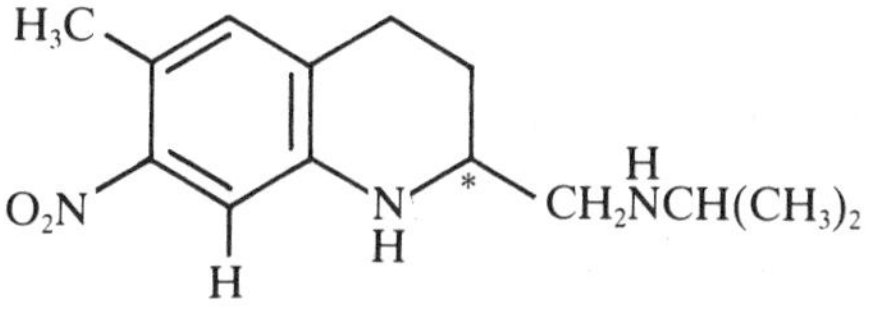

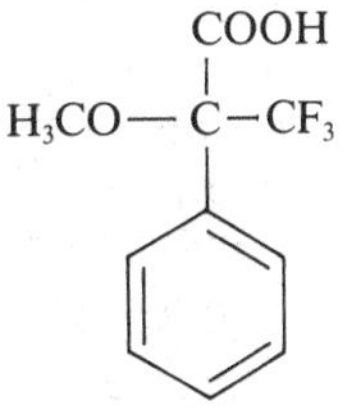

(XXV)

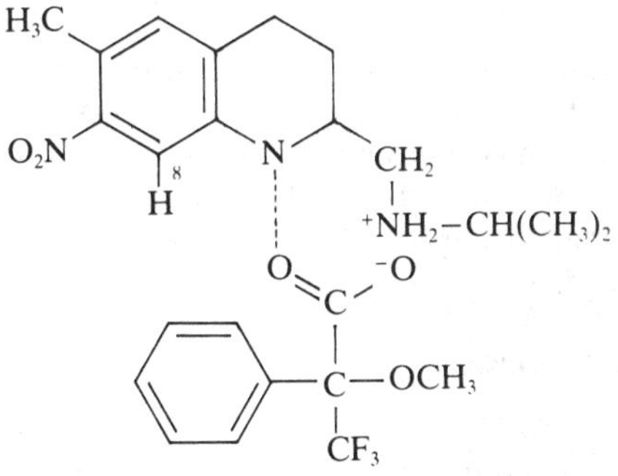

(XXVI)

Examination under normal solvent conditions fails to reveal differences in the spectra of enantiomers. However, the fact that strong solvent–solute interactions such as hydrogen bonding and dipolar forces often causes a specific preferred alignment of solute with solvent (p.426) may be exploited. In dilute solution in an optically pure chiral solvent such as (+)- or (−)- α-phenylethylamine, each of a pair of enantiomers may align itself differently with the solvent molecules, and chemical shift differences between corresponding groups in each isomer are then seen (Pirkle, 1966). Effectively, the solute molecules combine with solvent molecules to form 'loose' diastereomeric complexes. The use of a racemic solvent does not differentiate between optical isomers. Similarly, the addition of an optically pure chiral solute to a solution of enantiomers may result in a formation of diastereomeric complexes with observable shift differences (Anet *et al.*, 1968). The enzyme α-chymotrypsin, for example, interacts in solution with racemic N-trifluorophenylalanine, giving rise to two ^{19}F signals (Zeffren and Reavill, 1968). Chiral lanthanide shift reagents (p.429) often accentuate chemical shift differences in diastereomeric complexes (Whiteside and Lewis, 1970). This has been exploited in, for example, the determination of mixtures of *Dextromethorphan* and *Levomethorphan* (Wainer *et al.*, 1980) and non-steroidal anti-inflammatory agents such as *Ibuprofen* and *Naproxen* (Dewar *et al.*, 1982).

An interesting method for determining optical purity has been described for the schistosomicidal tetrahydroisoquinoline (XXIV) an oxamnaquine precursor, where the (+)- isomer is the more biologically active. A salt is made of the base isomeric mixture with an optically pure acid, (+)- α-methoxy-α-trifluoromethylphenylacetic acid (XXV). In the salts of the enantiomers there is a 0.05 ppm shift difference seen for the 8-position aromatic protons. The anisotropic effect of the acid aromatic ring in the complex (XXVI) appears to be responsible for this difference (Baxter and Richards, 1972).

Molecular interactions

Chemical shift differences arising from the interaction of solute and solvent, and solute and solute have been described, illustrating that NMR may provide an insight into the nature of intermolecular complexes. Thus, it should also be possible to explore drug-drug and drug-macromolecule complexation in a similar manner. Such phenomena need to be studied in dilute solution and, therefore, to overcome the limited sensitivity of NMR, a spectrum accumulation technique is often employed.

Micelle formation

Micelle formation by drugs in solution is a widespread phenomenon. With ionic compounds in aqueous solution, micellisation results in the removal of hydrophobic protons of the monomer from contact with water and concentration of the hydrophilic head groups at the surface of the micelle. The changes in the environment of these groups lead to changes in their PMR absorption characteristics. A series of phenothiazine drugs, which micellise at relatively high concentrations, e.g. 4×10^{-2} M (1.33%) for promethazine hydrochloride (XXVII) have been studied at concentrations above and below the critical micelle

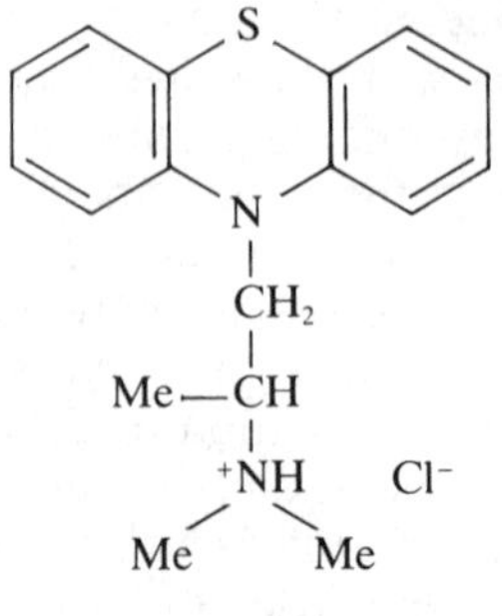

(XXVII)

concentration (CMC) (Florence and Parfitt, 1970, 1971). Chemical shifts were measured relative to appropriate signals of an external standard of a 20% promethazine solution (micellar). The shifts of the resonance line of the hydrophilic $\overset{+}{N}H(CH_3)_2$ group to higher fields on increasing concentration were explained by the increased dissociation of the group at the micelle surface. Diamagnetic shifts of the aromatic protons suggested a parallel stacking of the phenothiazine rings in the interior of the micelle (Figs 11.29 and 11.30).

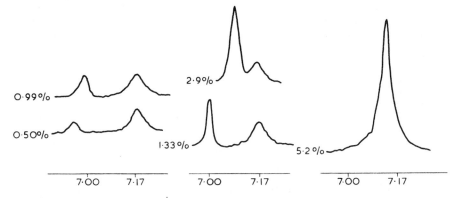

Fig. 11.29. PMR signal of the $\overset{+}{N}(CH_3)_2$ protons of promethazine hydrochloride in D_2O. The high field line of constant intensity is due to the external reference signal of 20% promethazine HCl in D_2O. The concentrations of the solution are marked (courtesy of *J. Pharm. Pharmacol.*)

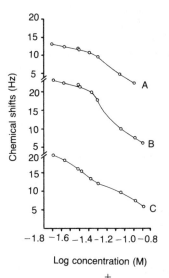

Fig. 11.30. Chemical shifts (Hz) of signal of A; $\overset{+}{N}(\underline{CH_3})_2$, B; CH—$\underline{CH_3}$ and C; aromatic ring protons, as a function of the concentration of solution, relative to the position of 20% promethazine hydrochloride signals from the respective groups (courtesy of *J. Pharm. Pharmacol.*)

Drug–macromolecule interactions   *(Jardetsky and Roberts, 1981; Dwek et al.*, 1977; Burgen and Metcalfe, 1970).

When considering interactions between molecules of low molecular weight and macromolecules, two NMR parameters, chemical shift and line width are usually observed. If a small drug (or substrate) molecule interacts and binds with some portion of a macromolecular surface then it effectively becomes a part of the macromolecule. Dependent upon the strength of the binding, the drug will only reside for a portion of any given time in the complex and thus during an NMR experiment an average picture is seen. Because the small molecule has become, for a time, part of a more rigid lattice, the spin-spin relaxation time (T_2) of proton groups will change, giving rise to changes in line width.

Drug or substrate	Enzyme or macromolecule	Complex
Large T_2	Small T_2	Bound substrate has smaller T_2 than free substrate
Narrow line	Broad line	Broader line than substrate alone

The magnetic environment of the bound substrate will also be different from that of the free molecule, so chemical shift changes will also occur. A consideration of the changes in chemical shift and relaxation times of specific proton groups in the drug when it moves from the free to the bound state will provide valuable information concerning the mode of binding.

A simple example is the broadening of the benzyl alcohol aromatic signal on moving from an aqueous solution to the same strength solution in a 1% suspension of erythrocyte membranes. The broadening is due to the partitioning of benzyl alcohol into the lipid membrane.

The storage of catecholamines, e.g. adrenaline in the adrenal medulla, is believed to be in the form of weak complexes with nucleotides. Relaxation time differences between solutions of adrenaline and of 1:1 adrenaline–adenosine triphosphate complex in D_2O indicated that the complex is formed by attachment of the catecholamine side chain to, probably, the phosphate of the nucleotide. Similar studies suggest that protein binding of benzylpenicillin occurs through the penicillin benzyl group interacting with some aromatic groups of protein.

A 360 MHz ^1H NMR study of free and complexed (K^+ and Rb^+) valinomycin within a perdeuterated phospholipid bilayer has demonstrated the conformational differences that occur in the cyclic peptide in its two forms, and the rapidity of their interchange (Feigenson and Meers, 1980).

Quantitative analysis

Attention has been focussed largely on NMR spectroscopy as an aid to structure elucidation and to the study of molecular dynamic processes.

Less heed has been paid to its applications to quantitative analysis (Kasler, 1973; Shoolery, 1977; Leyden and Cox, 1977). Automatic integration of resonance bands affords an easy and rapid quantitative means of determining the ratio of compounds in a mixture, provided that at least one resonance band from each constituent is free from extensive overlap by other absorption. The estimation of the keto–enol ratio in acetylacetone (p.451) is an example of the quantitative analysis of a mixture. An example (Smith, 1964) where a limited amount of

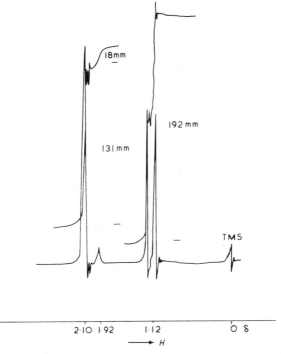

band overlap has been surmounted, is the analysis of a mixture of acetone (**XXVIII**), propan-2-ol (**XXIX**) and isopropenyl acetate (**XXX**), the high field spectrum of which is illustrated in Fig. 11.31.

Fig. 11.31. NMR spectrum of a mixture of isopropenyl acetate, propan-2-ol and acetone

The six-proton singlet of acetone at 2.10δ coincides with the acetate methyl three proton singlet of (XXX). Fortunately, the broad apparent singlet at 1.92δ is assignable to the methyl group of isopropenyl acetate attached to the double bond (low order coupling to the alkenyl protons produces line broadening). The band at 1.92δ integrates for 18 mm; therefore the acetone signal corresponds to an integral of $(131 - 18)$ mm. Thus for the mixture:

$$\text{Acetone (XXVIII)} \equiv 6\text{H}(2.10\delta) \equiv (131 - 18)\,\text{mm} \equiv 113\,\text{mm}$$
$$\text{Propan-2-ol (XXIX)} \equiv 6\text{H}(1.12\delta) \equiv 192\,\text{mm}$$
$$\text{Isopropenyl acetate (XXX)} \equiv 3\text{H}(1.92\delta) \equiv 18\,\text{mm}$$

therefore $\qquad\qquad\qquad\qquad\qquad\qquad\qquad 6\text{H} \equiv 36\,\text{mm}$

$$\%\,\text{Acetone} = \frac{113}{113 + 36 + 192} \times 100 = \frac{113}{341} = 33\%$$

$$\%\,\text{Propan-2-ol} \quad = \frac{192}{341} \times 100 \quad = 56\%$$

$$\%\,\text{Isopropenyl acetate} \quad = \frac{36}{341} \quad\quad\quad = 11\%$$

Similarly, mixtures of aspirin, phenacetin and caffeine and A.P.C. tablets have been assayed using the aspirin methyl ester protons (2.3δ), the protons of the phenacetin acetyl group (2.1δ) and for caffeine the 1- and 3-N—CH$_3$ protons (3.4δ and 3.6δ). The procedure takes 15–20 minutes and is reasonably accurate (Hollis, 1963).

The absolute concentration of a constituent in a pharmaceutical raw material or formulation may be obtained by adding a reference compound to the solution for NMR examination. Weighed amounts of both sample and reference are dissolved in an appropriate solvent and the spectrum and integrals are obtained.

Corticosteroids both in bulk and in formulations have been assayed in this manner (Avdovich *et al.*, 1970). Prednisone (XXXI), Prednisolone and Triamcinolone are 1,4-dien-3-ones and exhibit a characterisitc alkenyl-region spectrum (Fig. 11.32). The C-1 proton is seen as part of an AB quartet at 7.9δ, coupled to the C-2 proton at 6.4δ, which in turn is partly obscured by the C-4 (1H) resonance at 6.3δ. To the 1,4-dien-3-one sample in dimethylsulphoxide solution is added a known weight of a pure stable reference standard, fumaric acid, the alkene H resonance line of which is seen at 7.1δ. At 5.9δ a peak due to the 4-position proton of a 4-ene-3-one steroid is observed, but does not interfere with the assay.

Several integrals are run for each determination, and the integrals from the 7.9δ sample signal and the reference integral are measured. The amount of corticosteroid present is calculated from the simple expression:

Amount of Steroid (mg)

$$= \frac{\text{EW (steroid)}}{\text{EW (standard)}} \times \frac{\text{Integral (mm) steroid}}{\text{Integral (mm) standard}} \times \text{Weight of standard}$$

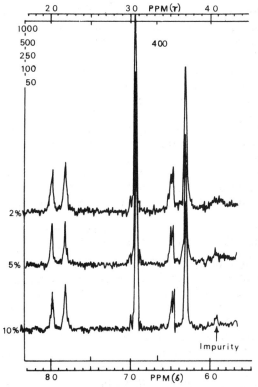

Fig. 11.32. Partial NMR spectrum of steroidal 1,4-dien-3-one in dimethyl sulphoxide containing fumaric acid (2, 5, or 10%) (Courtesy of *J. Pharm. Sciences*)

where

$$EW = \frac{\text{Molecule weight}}{\text{Number of hydrogens in signal chosen}}$$

Advantages of the NMR method over the ultraviolet and colorimetric methods in this determination are speed and specificity. Although the 4-en-3-one impurities may be detected, they are not measured.

(XXXI)

Many other examples of quantitative pharmaceutical analysis have now been published, including the determination of cephalexin in *Cephradine* (Warren *et al.*, 1978) and of *Penicillamine* in capsules (Nelson, 1981).

During quantitative studies the conditions under which spectra are run must be considered carefully. Saturation and relaxation effects are dependent upon the operating H_1 level and the speed of scan. The fastest practical speed of scan is limited not simply by the resolution required, but also by the necessity to allow sufficient time between scans to permit complete relaxation. Thus instrumental conditions must be carefully optimised for each type of assay.

Other spectrum factors which must be considered in quantitative studies are spinning side-bands and ^{13}C satellites. These are part of the main proton signal and must be treated consistently in each assay.

Surfactant chain-length determination

To establish the ethene ether chain length of a non-ionic surface-active agent by chemical methods is a lengthy procedure. Nuclear magnetic resonance provides a rapid, accurate and non-destructive method of analysis.

A spectrum and at least five repeat integrals are obtained for a 10% solution of the surfactant in carbon tetrachloride. The ratio of the ethene ether proton integral to the cetyl terminal chain of the surfactant is determined. From the knowledge that the cetyl end group integral is equivalent to 28H, the value of n for the $-O-(CH_2-CH_2-O)_n$ chain is obtained from a simple proportion calculation. Table 11.1 illustrates how the NMR results compare with the manufacturers specifications.

Table 11.1 Chain lengths of non-ionic surface-active agents

Detergent	Chain length (n)	
	Specification	NMR value
Texofor A10	10	10
Cetomacrogol 100	22–25	23
Texofor A45	~45	47
Texofor A60	~60	62

Iodine value

Triglycerides have in their PMR spectra four characteristic sets of signals from the resonance of alkenyl protons, the four C-1 and one C-2 glyceride methylene protons, methylene protons directly linked to a double bond, and the remaining protons on saturated carbons. The integration curve of the combined C-1 and C-2 glyceride methylene protons, occurring in isolation around 4δ, can be measured accurately, and, with these as an internal calibration, the alkenyl protons (the degree of unsaturation) and the total number of protons (a measure of the average molecular weight) may be determined. Iodine values, calculated from the alkenyl proton integration (Johnson and Shoolery, 1962) and the molecular weight, agree well, in general, with those estimated by Wijs method (Table 11.2). An exception is Tung oil which, because it possesses conjugated unsaturation, a function of a high oleostearic ester content, gives a low iodine value by Wijs method. The value calculated from PMR data corresponds well with that determined by vapour phase chromatography.

Table 11.2 Iodine values for a selection of natural oils

Oil	NMR method	Wijs method
Coconut	10.5±1.3	8.0–8.7
Olive	80.8±0.9	83.0–85.3
Peanut	94.5±0.6	95.0–97.2
Soybean	127.1±1.6	125.0–126.1
Sunflower seed	135.0±0.9	136.0–137.7
Safflower seed	142.2±1.0	140.0–143.5
Whale	150.2±1.0	149.0–151.6
Linseed	176.2±1.2	179.0–181.0
Tung	225.2±1.2	146.0–163.5

PMR spectroscopy may also be used to determine the percentage of hydrogen in an unknown sample, or the amount and nature of water adsorbed on biological materials (Pople *et al.*, 1959).

NMR imaging

Medical diagnosis is likely to be aided considerably by the development of NMR instruments with magnet bores of up to 100 cm, able to accommodate limbs or even the whole human body. Super-conducting magnets with a field homogeneity of 1 in 10^7 are sufficient to afford clear images of water (^{1}H) distributed in tissue or other biological material. Pulsed Fourier transform techniques have been used to overcome poor signal to noise ratios. A clear image of tissues and organs is given and displayed on a television screen, and the technique does not appear to present the dangers of X-ray diagnosis.

Carbon-13 NMR (CMR) spectroscopy

Carbon-13 NMR spectroscopy has developed to the stage where it rivals ^{1}H NMR in versatility and applications (Levy *et al.*, 1980; Wehrli and Wirthlin, 1976; Abraham and Loftus, 1978).

Almost all the foregoing discussion for ^{1}H NMR may be applied to CMR but there are important differences (Table 11.3)

Table 11.3 Nuclear properties of ^{1}H and ^{13}C

	^{1}H	^{13}C
Nuclear spin (*I*)	$\frac{1}{2}$	$\frac{1}{2}$
Resonance frequency at 235 Tesla	100 MHz	25.2 MHz
Natural abundance	99.9%	1.1%
Sensitivity for an equal number of nuclei	1.00%	0.016
Shift range	20 ppm	600 ppm

Sensitivity

A natural abundance of 1.1% for ^{13}C is clearly a factor that renders CMR less sensitive than PMR. However, because of its low abundance, the probability of a ^{13}C atom residing adjacent to another ^{13}C atom is low. Complications arising from ^{13}C—^{13}C coupling are therefore negligible.

A second factor affecting sensitivity is the low magnetogyric ratio (γ) of ^{13}C, about $\frac{1}{4}$ that of ^{1}H. Since sensitivity is proportional to γ^3, then ^{13}C affords about $\frac{1}{60}$ the sensitivity of ^{1}H in an NMR experiment. Effectively, ^{1}H is about 6000 times more sensitive than ^{13}C in NMR terms, i.e., ≈60 (γ factor) $\times$ 100 (abundance factor).

The low sensitivity of CMR is overcome by the use of large samples, up to 2 ml in 15 mm tubes, and by enhancement and decoupling techniques in conjunction with highly stable spectrometers operating at high fields.

Fourier Transformation techniques (p.417) have been described and are used to obtain ^{13}C spectra.

Coupling of ^{13}C to ^{1}H not only complicates spectra from the standpoint of interpretation but it gives, instead of a single sharp resonance signal, a multiplicity of bands often of low intensity. Heteronuclear decoupling of each ^{1}H individually from each ^{13}C signal would be a lengthy and tedious process. The problem is overcome by **proton noise decouping** or *off resonance decoupling*. During a noise decoupling experiment, all the protons in the sample are decoupled simultaneously and each ^{13}C resonance band is seen as a single line. Proton noise decoupling also disturbs the ^{13}C energy level populations and results in a nuclear Overhauser effect, affording an enhancement of each ^{13}C signal intensity by up to a factor of three.

By employing a combination of the above techniques, a ^{13}C natural abundance spectrum may be obtained from as little as 10 mg of sample. Much smaller amounts of compound are needed if the sample is ^{13}C-enriched.

Chemical shift

Chemical shift is usually the most important spectral parameter in CMR spectroscopy. Whereas ^{1}H resonances occur over a relatively narrow range (10–20 ppm), ^{13}C shifts cover a range of about 600 ppm, with most resonances falling within a 200 ppm range.

The signal spread and the application of proton noise decoupling results in spectra with each carbon atom in a molecule being displayed as a single discrete sharp band. As in PMR, ^{13}C shifts are measured relative to a TMS internal standard. In general, ^{13}C shifts fall into well-defined ranges according to the electronic and magnetic environment of the carbon (Fig. 11.33).

Lanthanide shift reagents are occasionally employed in CMR spectroscopy, but because of the low incidence of resonance band overlap, they are much less useful than in ^1H investigations.

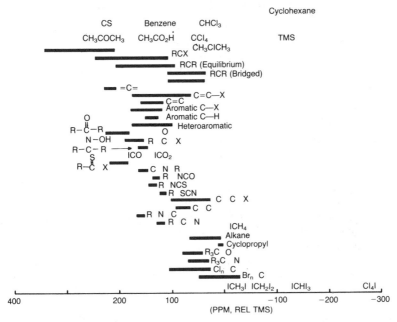

Fig. 11.33. General ^{13}C chemical shift chart (Courtesy of Wiley Interscience)

Spin-spin coupling

For the reasons outlined above, ^{13}C—^{13}C and ^{13}C—^1H couplings are not usually seen in CMR spectra. The former may be examined in ^{13}C-enriched samples and the latter can sometimes be extracted from ^{13}C satellites in ^1H spectra ($J_{^{13}C-^1H} \simeq 150$ Hz). In the absence of proton noise decoupling, ^{13}C spectra are very complex. However, selective decoupling techniques may help to determine specific ^{13}C-^1H coupling constants.

Integration

At present in normal CMR spectra, there is no correlation between integrated peak areas and the number of ^{13}C nuclei in each signal. Integrals are therefore not generally measured. A Fourier Transform derived spectrum is based upon the spin-lattice relaxation of each ^{13}C atom and the intensity of each band is not dependent solely upon the number of ^{13}C atoms responsible for the signal. Similarly, because of the variation in relaxation times, proton noise decoupling results in variable nuclear Overhauser enhancement.

Modification of the Fourier Transform technique is being explored as a means of quantifying ^{13}C signals.

Two-dimensional Fourier transform NMR (Ernst, 1982)

If data are collected as a function of two independent time domains and subjected to a double FT transformation, a two-dimensional spectrum results. For example, one spectrum may correspond to a frequency axis for ^{13}C chemical shifts and the second for ^1H shifts where only signals from coupled nuclei are displayed.

Applications

Structure elucidation is the most common application of ^{13}C spectroscopy (Gray, 1973; Wehrli and Wirthlin, 1976). PMR and CMR are often employed as complementary techniques. CMR is particularly useful in the investigation of the structures of complex natural products and biopolymers. Other applications in common with PMR include the study of the dynamic properties of molecules and molecular geometry.

The investigation of biosynthetic pathways, particularly in microorganisms, is being assisted by the increasing availability of ^{13}C-enriched substrates. Detection and identification of labelled sites may be accomplished by CMR directly or by examination of ^{13}C—^1H satellite signals, e.g. Cephalosporin C (Neuss, 1971) and the rifamycins (Martinelli *et al.*, 1973).

Quantitative analysis of mixtures of pharmaceuticals by ^{13}C NMR presents a number of difficulties. In contrast to ^1H NMR spectra, the intensities of ^{13}C resonances usually correlate poorly with spin populations. This is due largely to variations that occur in the relaxation times of carbon atoms of the molecule and differential nuclear Overhauser enhancement factors for the different carbon resonances. Relaxation time variations may be overcome by introducing a delay between pulses, long enough to permit complete relaxation of the analytical spin system. The differential NOE effects may be avoided by a procedure based upon the relationship between the peak height ratios of the analyte to a standard. By employing these techniques, commercial samples of gentamicin sulphate have been examined (Kountourellis *et al.*, 1983).

Although the order of accuracy achieved by ^{13}C NMR was low relative to an high performance liquid chromatography procedure, it was concluded that quantitative analysis by ^{13}C NMR had significant potential in pharmaceutical analysis.

References

Abraham, R.J. and Loftus, P. (1978) *Proton and Carbon-13 NMR spectroscopy*, Heyden, London.
Anderson, J.E, (1965) *Q. Rev.* **19**, 426.
Anet, F.A.L., Sweeting, L.M., Whitney, T.A. and Cram, D.J. (1968) *Tetrahedron Letters* 2617.
Avdovich, H.W., Hanbury, P. and Lodge, B.A. (1970) *J. Pharm. Sci.*) **59**, 1164.
Backers, G.E. and Shaefer, T. (1971) *Chem. Rev.* **71**, 617.
Batterham, T.J. (1973) *NMR Spectra of Simple Heterocycles*, Wiley, New York.
Baxter, C.A.R. and Richards, H.C. (1972) *Tetrahedron Letters* 3357.
Becker, E.D. (1965) *J. Chem. Educ.* **42**, 591–596.
Becker, E.D. (1980) *High Resolution NMR*, 2nd edn., Academic Press, New York.
Bhacca, N.S. and Williams, D.H. (1964) *Applications of NMR Spectroscopy in Organic Chemistry*, Holden-Day, San Franciso.
Bible, R.H. (1965) *Interpretation of NMR Spectra*, Plenum Press, New York.
Bothner-By, A.A. and Dadok, J. (1982) In Levy, G.C. (ed.) *NMR Spectroscopy. New Methods and Applications*, p.31 ACS Symposium Series 191, Washington D.C.
Bovey, F.A. and Johnson, C.E. (1958) *J. Chem. Phys.* **29**, 1012.
Brophy, G.C. Laing, O.N. and Sternhill, S. (1968) *Chem. and Ind.* 22.
Brügel, W. (1970) *Handbook of NMR Spectral Parameters*, Heyden, London.
Burgen, A.S.V. and Metcalfe, J.C. (1970) *J. Pharm. Pharmac.* **22**, 153.
Casy, A.F. (1971) *PMR Spectroscopy in Medicinal and Biological Chemistry*, Academic Press, London.
Chamberlain, N.F. (1974) *The Practice of NMR Spectroscopy*, Plenum Press, New York.
Chapman, D. and Magnus, P.D. (1966) *Introduction to Practical High Resolution Nuclear Magnetic Resonance Spectroscopy*, Academic Press, London.
Cookson, R.C., Crabb, T.A., Frankel, J.J. and Hudec, J. (1966) *Tetrahedron* Supplement No. 7, 355–390.
Culvenor, C.C.J. and Ham, N.S. (1966) *Chem. Comm.* 537.
Dailey, B.P. and Shoolery, J.N. (1955) **77**, 3877.
Dewar, G.H., Kwakye, J.K., Parfitt, R.T. and Sibson, R. (1982) *J. Pharm. Sci.* **71**, 802.
Dwek, R.A., Campbell, I.D., Richards, R.E. and Williams, R.J.P. (eds) (1977) *NMR in Biology*, Academic Press, New York.
Emsley, J.W., Feeney, J. and Sutcliffe, L.H. (1965) *High Resolution Nuclear Magnetic Resonance Spectroscopy*, Vol. 1, Pergamon Press, Oxford.
Ernst, R.R. (1982) In Levy, G.C. (ed.) *NMR Spectroscopy. New Methods and Applications*, p.47, ACS Symposium series 191, Washington D.C.
Farrar, T.C. and Becker, E.D. (1971) *Pulse and Fourier Transform NMR* Academic Press, New York.
Feigenson, G.W. and Meers, P.R. (1980) *Nature, Lond.* **283**, 313.
Florence, A.T. and Parfitt, R.T. (1970) *J. Pharm. Pharmac.* **20** (supplement), 121.
Florence, A.T. and Parfitt, R.T. (1971) *J. Phys. Chem.* **75**, 3554.
Fullerton, S.E., May, E.L. and Becker, E.D. (1962) *J. org. Chem.* **27**, 2144.
Gray, G.A. (1973) *Application of ^{13}C Nuclear Magnetic Resonance in Biochemistry— Critical Reviews in Biochemistry*, Chemical Rubber Company, Spring.
Harris, R.K. (1983) *Nuclear Magnetic Resonance Spectroscopy*, Pitman, London.
Harris, R.K. and Mann, B.E. (1978) *NMR and the Periodic Table*, Academic Press, New York.

Henty, D.N. and Vary, S. (1967) *Chem. and Ind.* 1782.

Hill, R.K. and Chan, T.H. (1965) *Tetrahedron* **21**, 2015.

Hollis, D.P. (1963) *Analyt. Chem.* **25**, 1682.

Jardetzky, O. and Roberts, G.C.K. (1981) *NMR in Molecular Biology,* Academic Press, New York.

Jensen, F.R., Noyce, D.S., Sederholm, C.H. and Berlin, A.J. (1960) *J. Am. chem. Soc.* **82**, 1256.

Johnson, L.F. and Shoolery, J.N. (1962) *Analyt. Chem.* **34**, 1136.

Jones, A.J. (1968) *Rev. Pure appl. Chem.* **18**, 253.

Karplus, M. (1959) *J. Chem. Phys.* **30**, 11.

Karplus, M. and Anderson, D.H. (1959) *J. chem. Phys.* **30**, 6.

Kasler, F. (1973) *Quantitative Analysis by NMR Spectroscopy,* Academic Press, New York.

Kessler, H. (1970) *Angew. Chem. (Int. Ed.)* **9**, 219.

Kountourellis, J., Parfitt, R.T. and Casy, A.F. (1983) *J. Pharm. Pharmacol.* **35**, 279.

Lambert, J.B. and Keske, R.G. (1966) *J. Am. chem. Soc.* **88**, 620.

Lapidot, A., Reuben, J. and Samuel, D. (1964) *J.Chem. Educ.* **41**, 570.

Levy, G.C., Lichter, R.L. and Nelson, G.L. (1980) *Carbon-13 Nuclear Magnetic Resonance,* 2nd edn, Wiley-Interscience, New York.

Leyden, D.E. and Cox, R.H(1977) *Analytical Applications of NMR,* Wiley, New York.

Martinelli, E,, White, R.J., Gallo, G.G. and Beynon, P.J. (1973) *Tetrahedron* **29**, 3441.

Mateos, J.L. and Cram, D.J. (1959) *J. Am. chem. Soc.* **81**, 2756.

Mathieson, D.W. (ed.) (1965) *Interpretation of Organic Spectra* Academic Press, London.

Mathieson, D.W. (ed.) (1967) *Nuclear Magnetic Resonance for Organic Chemists,* Academic Press, London.

Mayo, B.C. (1973) *Chem. Soc. Revs.* **1** 49.

Mislow, K. and Raban, M. (1967) *Topics in Stereochemistry* **1**, 1.

Morris, D.G., Murray, A.M., Mullock, E.B., Plews, R.M. and Thorpe, J.E. (1973) *Tetrahedron Letters* 3179.

Nelson, J. (1981) *J. Assoc. Offic. Analyt. Chem.* **64**, 1174.

Neuss, N. *et al.* (1971) *J. Am. chem. Soc.* **93**, 2337.

Noggle, J.H. and Schirmer, R.E. (1971) *The Nuclear Overhauser Effect,* Academic Press, New York.

Pirkle, W.H. (1966) *J. Am. chem. Soc.* **88**, 1837.

Pope, J.A., Schneider, W.G. and Bernstein, H.J. (1959) *High-Resolution Nuclear Magnetic Resonance,* McGraw-Hill, London.

Raban, M. and Mislow, K. (1965) *Tetrahedron Letters* 4249.

Raban, M. and Mislow, K. (1966) *Tetrahedron Letters* 3961.

Reeves, L.W. (1957) *Canad. J. Chem.* **35**, 1351.

Roberts, J.D. (1961) *An Introduction to Spin-Spin Splitting in High Resolution NMR Spectra,* Benjamin, New York.

Roberts, J.D. (1966) *Chemistry in Britain* **2**, 529.

Ronayne, J. and Williams, D.H. (1966) *Chem. Comm.* 712.

Sadlter Standard NMR Spectra, Sadlter Research Laboratories, Philadelphia, a continuing compilation.

Sanders, J.K.M. and Williams, D.H. (1972) *Nature, Lond.* **240**, 385.

Shaw, D. (1984) *Fourier Transform NMR Spectroscopy,* 2nd edn, Elsevier, Amsterdam.

Shoolery, J.N. (1977) *Prog. NMR Spectrosc.* **11**, 79.

Sievers, R.E. (ed.) (1973) *Nuclear Magnetic Resonance Shift Reagents,* Academic Press, New York.

Silverstein, R.M. and Basler, G.C. (1968) *spectrometric Identification of Organic Compounds,* Wiley, New York.

Smith, W.B. (1964) *J. chem. Educ.* **41**, 97.

von Philipsborn, W. (1971) *Angew. Chem. (Int. Ed.)* **10**, 472.

Wainer, I.W., Tischler, M.A. and Sheinin, E.B. (1980) *J. Pharm. Sci.* **69**, 459.

Warren, R.J., Zarembo, J.E., Staiger, D.B. and Post, A. (1978) *J. Pharm. Sci.* **67**, 1481.

Wehrli, F.W. and Wirthlin, T. (1976) *Interpretation of Carbon-13 NMR Spectra,* Heyden, London.

Whiteside, G.M. and Lewis, D.W. (1970) *J. Amer. chem. Soc.* **92**, 698.
Wibreg, K.B. and Nist, B.J. (1962) *The Interpretation of NMR Spectra,* Benjamin, New York.
Zeffren, E. and Reavill, R.A. (1968) *Biochem. Biophys. Res. Commun.* **32**, 73.

12
Mass spectrometry

R. T. PARFITT

Introduction

Many excellent reviews (Beynon, 1960; Biemann, 1962; Budzikiewicz *et al.*, 1967; Hill, 1966; McLafferty, 1963; Milne, 1971; Rose and Johnstone, 1982; Spiteller and Spiteller-Friedmann, 1965; Waller, 1972, Watson, 1976.) have been written on the theory of mass spectrometry and its applications in organic chemistry, and reference should be made to these for detailed treatment of the subject. The objective of this chapter is to make the student aware of its potential in qualitative and quantitative pharmaceutical analysis.

The mass spectrometer produces positive ion spectra which, unlike the overlapping band spectra from most other spectrometric methods, are line spectra. When an organic molecule is bombarded with electrons of sufficient energy (> 10 eV), it may lose an electron and so yield a positive ion:

$$M + e \rightarrow M^+ + 2e$$

M^+ is often unstable, having an energy excess, and in the mass spectrometer it fragments in a specific manner. During electron bombardment one of the molecules' valence electrons can be removed and the imparted energy excess surges through the molecule. Whenever sufficient energy accumulates in a particular bond, then that bond will cleave. Since different bonds require different amounts of energy to break them, each molecule will give rise to a unique fragmentation pattern. Among the fragments produced are further positive ions which are separated and recorded by the mass spectrometer according to their mass-to-charge ratios (m/z). Since the charge is usually unity, $m/z = m$, the mass of the fragment; occasionally, however, fragments of higher charge are encountered.

Wien, in 1898, produced the first crude mass spectra when he demonstrated that positive ions could be deflected according to their masses in electric or magnetic fields. This observation was developed by Thomson (1910) who used a combined electrostatic and magnetic field to observe the mass spectrum of a mixture of rare gases and thereby identified two neon isotopes. Instrumentation was developed further by Dempster (1918) and Aston (1919), the latter confirming the nature of Thomson's neon isotopes. By 1924, Aston had determined the isotopic constitution of about 50 elements.

The potential of mass spectrometry in the study of organic compounds was quickly realised, but it was not until 1940 that a commercial modification of the Dempster instrument was employed in the quantitative estimation of a complex mixture of hydrocarbons. Computer-assisted quantitative analysis of hydrocarbon fractions is now of considerable importance in the petroleum industry with the advent of high resolution mass spectrometry, the technique has been employed extensively for the elucidation of complex organic structures, by the rationalisation of their fragmentation patterns.

Theory

Figure 12.1 is a diagram of a single focussing mass spectrometer. The sample is introduced into the instrument in such a way that its vapour is bombarded by electrons having an energy of about 70 eV. Positive ions formed in the ion-source are accelerated between two plates by a potential difference of a few thousand volts (V). The ions pass through the source slit and are deflected by a magnetic field (H) according to their mass/charge ratios. They then pass through the exit or collector slit, and impinge upon the collector; the signal received is amplified and recorded. The height or intensity of the resulting peak is proportional to the ion abundance, i.e. the number of ions of identical mass received by the collector.

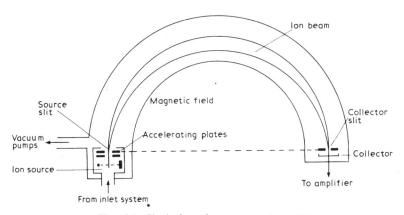

Fig. 12.1. Single focussing mass spectrometer

For an ion of unit charge z, mass m and velocity after acceleration v, the potential energy of the particle zV, is equal to its kinetic energy, i.e.

$$zV = \tfrac{1}{2}mv^2 \qquad (1)$$

where V = acceleration voltage.

In a magnetic field the ion experiences a force Hev at right angles both to the direction of the field and its direction of motion. It therefore moves in the arc of a circle, where the radius is given by the expression:

$$Hzv = \frac{mv^2}{r}$$

where r = radius of the ion path, and H = strength of the magnetic field. Therefore

$$r = \frac{mv}{zH} \tag{2}$$

Eliminating v between eqs (1) and (2) gives

$$m/z = \frac{H^2 r^2}{2V} \tag{3}$$

Thus the radius of the ion path may be changed by varying either the magnetic field (H) or the accelerating voltage (V). By either method, ions of different mass-to-charge ratios (m/z) can be made to impinge upon the collector in turn, thus giving rise to a spectrum (Fig. 12.2). Magnetic field variation enables a wide range to be covered in a single sweep, but, for very rapid scanning, a voltage sweep must be employed.

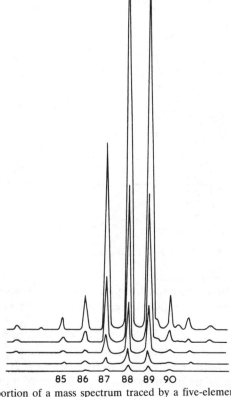

85 86 87 88 89 90

Fig. 12.2. A portion of a mass spectrum traced by a five-element galvanometer

The ion-source, ion path and collector of the mass spectrometer must be under high vacuum (10^7 mm Hg) for optimum operation. Any extraneous materials in the source, including atmospheric gases, will themselves ionise in the electron beam and their spectra will be recorded.

Instrumentation

The above description is that of a typical single focussing mass spectrometer, where a maximum resolution of 3000 is attainable(p.449). The principle of magnetic focussing falsely assumes that all ions are formed with zero kinetic energy, and in the magnetic field are focussed only according to their mass. Velocity or kinetic energy focussing is achieved by the introduction of an electrostatic analyser between the ion accelerating chamber and the magnetic analyser. The electrostatic analyser selects a beam of ions of very narrow energy range for final deflection in the magnetic analyser. The resolving power of this **double focussing mass spectrometer** (Fig. 12.3) is of the order of 30 000. Such a resolving capability enables high molecular weight fragments, which differ by only one mass unit, to be distinguished.

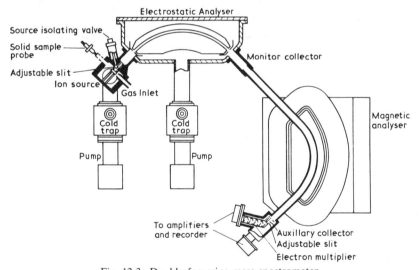

Fig. 12.3. Double focussing mass spectrometer

Double beam instruments, where two ion beams from independent sources pass side by side through a common mass analyser and are detected by separate collectors, are available. Such instruments may be used to compare samples directly, to investigate a single sample under different ionising conditions, or to compare a sample with a standard (e.g. perfluorokerosene) as a mass marker.

The **quadrupole mass spectrometer**, initially devised to separate uranium isotopes, has been adopted in organic mass spectrometry, particularly when in combination with a gas chromatograph. Focussing of ions, after acceleration from the ion source, is effected by a quadrupole mass filter where they are separated according to mass (Fig. 12.4), and detected by an electron multiplier. The mass filter consists of a quadrant of four parallel circular (or ideally hyperbolic) tungsten rods which focus ions by means of an oscillating and variable radiofrequency field.

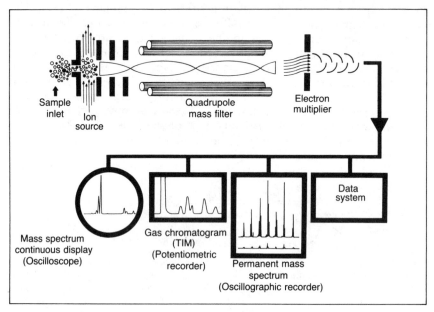

Fig. 12.4. Functional diagram of quadrupole mass spectrometer. (Courtesy of Finnegan Instruments Ltd.)

Advantages of the quadrupole mass spectrometer are:

(1) The mass of ions focussed is directly proportional to the voltage applied to the quadrupole filter. Since the voltage is a linear function, the mass spectrum display is also linear. Conventional sector instruments usually yield logarithmic outputs which may be rendered linear by the instrument's computer.

(2) Accelerating potentials in quadrupole instruments are in the order of 5–30 V, whereas a sector mass spectrometer generally requires 3000–6000 V. A high accelerating voltage requires a very low ion source pressure ($\simeq 10^{-7}$ mm Hg) to prevent arcing, whereas a low accelerating potential permits the use of a relatively high source pressure (10^{-4}–10^{-5} mm Hg). The latter is of particular advantage in gas chromatograph-mass spectrometer (GC/MS) combinations.

Quadrupole instruments are able to scan up to about 1000 atomic mass units (amu), but they do not give rise to metastable ions (p.484).

Practical considerations

Sample introduction

Physical and chemical properties govern the way in which a sample is introduced into the mass spectrometer. Comprehensive accounts of sample handling techniques are given by Beynon (1960), Biemann (1962), Hill (1966) and McLafferty (1980). The sample must be in the vapour state when bombarded by electrons. This requires it to have a minimum vapour pressure of about 10^{-6} mm Hg, since the ion-source is kept at an operating pressure of 10^{-7} mm Hg. Even compounds of low volatility, such as amino acids and peptides, may be investigated with the aid of modern inlet systems. Gases and volatile liquids are admitted to the source through a small leak from a **gas reservoir**. Solid samples and liquids of low volatility are introduced to within a short distance of the electron beam on the bakelite tip of a **direct-insert-probe**, which is inserted into the ion-source through a system of vacuum locks. The direct-insert-probe may be heated by platinum wire heating elements to aid sample vaporisation.

The **sample size** required to record a mass spectrum is dependent upon the method of sample introduction. As little as $0.1 \, \mu g$ is required for the direct-insert-probe, but as much as 1 mg may be necessary for introduction via the gas reservoir. The size of samples from vapour phase chromatographs is of the order of $0.01-1 \, \mu g$ or less.

Sample purity

Small amounts of impurities in the sample need not seriously affect the interpretation of a mass spectrum, although this depends largely upon the nature of the problem. The appearance of additional or more intense peaks facilitates the recognition of impurities. If the spectrum of a pure compound is available, then inspection of the mass spectra of other samples of that compound will show the presence of impurities by the additional peaks. Subtraction of the known spectrum from the spectrum of the impure sample furnishes a mass spectrum of the impurity (or impurities), and may lead to its identification. High molecular weight contaminants are readily detectable (provided they are sufficiently volatile) by peaks at higher mass than the known M^+.

The mass spectrometer does not distinguish readily an impurity which is stereoisomeric with the main component, since stereoisomers have very similar fragmentation patterns.

Temperature effects

Chemical changes induced in the sample by the elevated temperature of the inlet system may lead to variations in mass spectra. Alcohols, for example, may be dehydrated to alkenes, resulting in the absence of a mass peak for the alcohol, and a spectrum characteristic of the corresponding alkene. To overcome this either the spectrum is run at lower temperature or the sample is converted to a less sensitive derivative. Alcohols are often converted to their methyl ethers.

Spectra for comparison should be taken at, or as near as possibe to, the same operating temperature and voltage.

Spectrum resolution

The resolving power of a mass spectrometer is its ability to separate ions of different mass-to-change ratio (m/z). For two peaks corresponding to ions of mass m_1 and m_2 respectively, separated by Δm then:

$$\text{resolving power} = m_1/\Delta m$$

An illustration of definitions of resolution is given in Fig. 12.5.

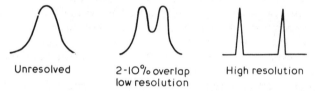

| Unresolved | 2-10% overlap low resolution | High resolution |

Fig. 12.5. Spectrum resolution

For a single focussing instrument of resolving power 1000, ions of mass of 999 and 1000 are just discernable as separate peaks. A high resolution instrument has a resolving power of over 20 000, but this does not imply that molecular weights of the order of 20 000 can be obtained, since such compounds are generally non-volatile. A high resolution instrument does, however, give a wider spread of peaks from ions of lower mass, and the fragment ions $C_2H_6^+$ (30.0469), CH_2O^+ (30.0105) and CH_4N^+ (30.0344), for example, give rise to three distinct lines.

Combined gas chromatography–mass spectrometry (GC/MS)

Gas–liquid chromatography (GC) is a very efficient method for separating a complex mixture into its components. The high sensitivity of mass spectrometry provides the necessary information for either identification of compounds by comparison with available spectra, or structure elucidation of a small quantity of compound. A combination of these techniques, with the introduction of GC effluents, after removal of most of the carrier gas, into a mass spectrometer is finding increasing use in analytical and structural organic chemistry and biochemistry (McFadden, 1973; Budd and Eichelberger, 1979). Efforts to combine the two techniques extended over many years. The most obvious method of combination is to condense the fraction emerging from the GC column into a capillary or onto a small metal surface; the fractions collected in this way are introduced, in the normal manner into the mass spectrometer source. Many operations are required for a multicomponent mixture, and losses are likely to occur during collection of the fraction; however, the mass spectrometer may be operated at high resolution and no GC carrier gas is admitted to the instrument.

A second, much more convenient, method is to feed the column effluent directly into the ion source after passage through an **interface** or **separator** which enhances the concentration of the sample by removing some of the carrier gas (Simpson, 1972).

The interface between the GC and MS has an important role to play in the overall efficiency of the instrument. It must be capable of providing an inert pathway from the column to the ion source without loss of chromatographic resolution, while at the same time removing the carrier gas and reducing the pressure from about one atmosphere at the column outlet to $10^{-5} - 10^{-6}$ mm Hg in the ion source. It is hardly surprising that the separator has been a major cause of technical problems encountered in GC/MS.

The **Watson-Biemann effusion separator** consists of a sintered glass tube, the surrounds of which are evacuated. The carrier gas, usually helium, passes preferentially through the sintered glass and the effluent is concentrated by a factor of up to 100. Two stage separators may enrich the effluent by a factor of 400 and be capable of dealing with gas flow rates in the order of 20–60 ml min^{-1}.

The **Ryhage jet separator** is based upon the differing rates of diffusion of different gases in an expanding supersonic jet stream. The heavier (sample) component concentrates in the centre of the gas jet. The gases pass at high speed through an orifice aligned with a second orifice a short distance away. The concentrate passes through the second orifice, and then on to the ion source, while the carrier gas is pumped away. Usually a Ryhage separator is two stage, although all glass single stage units (Fig. 12.6) are used, particularly in combination with quadrupole mass spectrometers.

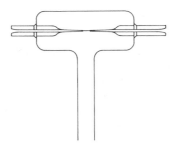

Fig. 12.6. The Ryhage jet separator

In the **Llwellyn-Littlejohn separator** (Fig. 12.7), separation of organic molecules from carrier gas molecules is achieved by means of the selective permeability of an elastomer membrane. Permeability is a function of both the solubility of the gas molecules in the membrane and their ability to diffuse through it. Gases such as hydrogen, helium, argon and nitrogen having a very low solubility and high diffusion rate, pass through the membrane much more slowly than organic vapours where the reverse is true. The vapour is therefore concentrated and the remaining carrier gas expelled into the atmosphere. The concentrate is further enriched by passage through a second semi-permeable membrane. Enrichment of the organic phase by a factor of greater than 10^5 has been achieved.

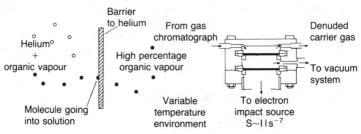

Fig. 12.7. The Llewellyn separator

No single separator is likely to give satisfactory results over a wide molecular weight spread.

After passage through an interface each component of the mixture is fed directly into the mass spectrometer in turn, and its mass spectrum is recorded. Overlapping spectra do occur, but by a process of spectrum subtraction the individual spectra of a partially resolved GC peak may often be deduced. In GC/MS analysis, metal chromatograph columns and metal parts and connections between the chromatograph and the ion source are to be avoided, since they may catalyse thermal breakdown of certain compounds.

Selected ion monitoring–mass fragmentography

In a GC/MS experiment, the gas chromatography may be obtained by simply splitting the GC effluent, part of which goes to a GC detector and the remainder of which passes into the MS. It is, however, more convenient to use the mass spectrometer as a GC detector, the total ion current from the MS collector being plotted against time to yield a chromatogram. Even more valuable is the use of **multiple ion detection** or specific ion detection. Here, a number of ions from the MS are monitored simultaneously and the resultant chromatograms displayed simultaneously. This technique is known as **selected ion monitoring** or **mass fragmentography** (Waller, 1972; McFadden, 1973).

Multiple ion detection is facilitated by an accelerating voltage alternator which permits a 'jump' scan where only ions of specific (selected) m/z are recorded. Several ions from a single compound may be monitored, affording a high confidence in assigning identity or structure, or, alternatively, ions from several components of a mixture may be monitored simultaneously.

Mass fragmentography is a technique endowed with a high level of specificity. It is most unlikely that two compounds would have not only the same GC retention time but also the same MS fragmentation with identical peak ratios. Perhaps the greatest advantage, however, is the sensitivity of the technique, particularly in the detection of trace compounds in biological fluids. By employing an internal standard, either a known amount of isotopically labelled isomer or even an unrelated compound added to the sample, quantification down to nanomole or picomole levels may be achieved.

Mass fragmentography is of particular value in drug metabolism investigations (Maume *et al.*, 1973; Ebbighausen *et al.*, 1973; Frigerio and Ghisalberti, 1976).

Data acquisition and processing

Mass spectrometers and GC/MS combinations produce a wealth of data rapidly. To process and interpret all of this data manually would be excessively time consuming. Clearly, in order to realise the maximum potential of mass spectrometry, extensive use must be made of computers. An account of mass spectrometer data acquisition and processing systems is beyond the scope of this chapter, but receives comprehensive treatment elsewhere (Waller, 1972; Henneberg *et al.*, 1972; Chapman, 1978).

High performance liquid chromatography–mass spectrometry (HPLC/MS)

Early in the development of combined GC/MS the major technical difficulty was to find an efficient gas separator or interface. A similar, though rather more difficult, problem arises in the removal of liquid carrier from an HPLC eluent before samples are passed into the MS source. Normal eluent flow rates of 0.5–2.0 ml min^{-1} cannot be handled by the MS pumping system. Moving belt inlet systems, jet separators and vacuum nebulisers are all techniques that are used to remove solvent and pass analytes into the source. Invariably there is a loss of detection sensitivity, and microbore HPLC/MS may offer a superior method for the introduction of liquid samples directly into the source (Games and Westwood, 1983; Games, 1980; Henion and Maylin, 1980).

HPLC/MS has been applied to peptide analysis (Yu *et al.*, 1981), the determination of endogenous leucine enkephalin and other endogenous neuropeptides with FD detection (Desiderio and Yamada, 1982; Desiderio *et al.*, 1981) and for the separation of mixtures of pharmaceuticals (Games, 1980).

Chemical ionisation mass spectrometry (CIMS)

The discussion so far has concerned spectra generated by electron impact (EI). Such spectra may suffer from the disadvantages of excessive fragmentaton and the lack of a molecular ion; problems which can sometimes be surmounted by lowering the electron impact energy.

Chemical ionisation (Waller, 1972; McFadden, 1973; Begg and Yergey, 1973; Watson,1976) has emerged as a valuable technique that generates simple positive ion spectra from a sample by low energy ion-molecule reactions. Chemical ionisation mass spectrometry evolved from fundamental physico-chemical studies of the very rapid reactions that occur when ions collide with neutral molecules.

In a mass spectrometer chemical ionisation source, a reactant (or ionising) gas is admitted in several thousandfold excess over the gaseous sample. The pressure within the ion chamber is about 1 mg Hg (1 Torr). The mixture is subjected to electron bombardment (100 eV) in the usual manner, whereon, because of the relatively low sample concentration, almost all primary ionisation occurs to the reactant gas. Reagent ions so formed undergo rapid reactions with their own neutral species to form a steady-state ion plasma, which in turn interacts in a specific manner with molecules of the sample.

If we consider methane as a typical reactant gas, electron impact first removes an electron from the molecule to give $CH_4^{+\cdot}$, which is then involved in ion–molecule reactions to yield the reagent plasma:

$$CH_4 + e \rightarrow CH_4^{+\cdot} + 2e$$
$$CH_4^{+\cdot} \rightarrow CH_3^+ + H^\cdot$$
$$CH_4^{+\cdot} + CH_4 \rightarrow CH_5^+ + CH_3^\cdot$$
$$CH_3^+ + CH_4 \rightarrow C_2H_5^+ + H_2$$
$$CH_3^+ + 2CH_4 \rightarrow C_3H_7^+ + 2H_2$$
$$CH_2^{+\cdot} + 2CH_4 \rightarrow C_3H_5^+ + 2H_2 + H^\cdot$$

and so on

Most sample molecules, particularly in biological studies, possess oxygen or nitrogen in the form of nucleophilic sites ideal for the acceptance of a proton. Reaction between reagent ions and sample may be summarised thus:

$$CH_5^+ + BH \rightarrow BH_2^+ + CH_4$$
$$C_2H_5^+ + BH \rightarrow BH_2^+ + C_2H_4$$
$$C_2H_5^+ + BH \rightarrow B^+ + C_2H_6$$

In CI spectra the molecular ion (M^+) is often weak. However, the spectrum base peak is usually $(M + 1)^+$, and is known as the **quasimolecular ion**. A comparison of the EI and CI spectra of ephedrine is seen in Fig. 12.8 (Fales *et al.*, 1970). From an interpretative standpoint ions occurring in methane CI spectra at $(M + 29)^+$ $(+ C_2H_5)$ and $(M + 41)^+$ $(+ C_3H_5)$ are often useful.

When the sample molecule is not a good proton acceptor, hydride ion

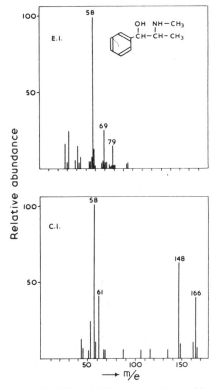

Fig. 12.8. A comparison of the EI and CI spectra of ephedrine (Courtesy of *J. Am. chem. Soc.*)

abstraction occurs affording a $(M - 1)^+$ quasi-molecular ion.

The most commonly used reagent gases in CIMS are methane, isobutane and ammonia, the ions from each of which have a specific proton affinity. Ion plasma with a high proton affinity, e.g. that from 2-methylpropane, causes little or no fragmentation and gives an uncomplicated spectrum of the sample with an intense quasi-molecular ion. Consequently, CI spectra of mixtures are relatively easy to interpret (Fig. 12.9).

Combined gas chromatography and mass spectrometry may be adapted to CI studies (GS/CIMS), thus providing a valuable tool for the investigation of complex mixtures of compounds, particularly from biological fluids. Quadrupole mass spectrometers are well-suited to this combination. The GC carrier gas and the CI reactant gas may be the same compound, e.g. methane. Once the GC effluent has left the column, it passes directly into the CI source without the intervention of a separator. The CI source is able to handle a high gas input, and the pressure gradient between the column outlet and the source is relatively shallow. If methane has an adverse effect on gas chromatographic resolution, when compared with, say helium, the latter may be employed as carrier, with the reagent gas being admixed with the

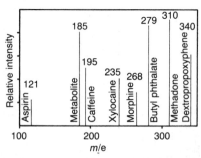

Fig. 12.9. Chemical ionisation spectrum of compounds detected in a chloroform extract of a urine sample from drug overdose patient. Reagent gas: 2-methylpropane (Adapted from *Industrial Research*)

column effluent before passage into the CI source. Selected ion monitoring is facilitated by the simplicity of spectra produced by GC/CIMS.

Field ionisation (FI) mass spectrometry

Field ionisation (Waller, 1972; McFadden, 1973; Beckey, 1977) is another low energy method of generating positive ions for mass spectrometric analysis, providing information complementary to EI ionisation. 'Soft' ionisation of molecules is induced by a very high positive electric field ($10^7 - 10^8$ V cm^{-1}) produced at a fine metal point, sharp metal edge or thin wire. The high electric field gradient between a sample molecule and the metal, usually platinum or tungsten, results in the loss of an electron by the molecule to the anode. The ions formed are repelled by the anode and accelerated into the MS analyser.

In field ionisation only sufficient energy (12–13 eV) is available to just ionise the molecule. Thus an intense 'parent ion' with little or no fragmentation usually results. Spectrum simplicity may be of value when operating in a GC/MS mode or in mass fragmentography. Disadvantages of FI are the relatively low (about $\frac{1}{10} - \frac{1}{100}$th EI) sensitivity of the source, its fragility and its susceptibility to being affected by previous samples.

Field ionisation is particularly useful in the determination of the structures of large molecules or molecules lacking a parent ion by EI, e.g. carbohydrates. Often a quasi-molecular ion $(M + 1)^+$ is observed due to interaction on the metal source surface between the sample and adsorbed water.

A modification of FI is **field desorption** (FD), a technique useful for the MS analysis of non-volatile or thermolabile compounds. The sample is applied in solution to the field ion emitter and the solvent allowed to evaporate. Ionisation ensues by application of the high electric field in the manner described above. Nucleotides and quaternary ammonium salts are among the compounds reported to exhibit a molecular ion by this technique (Brent *et al.*, 1973).

Figure 12.10 offers a comparison of the EI, FI and FD mass spectra of glutamic acid (Beckey *et al.*, 1970) and Fig. 12.11 illustrates the quasi-molecular ions of the three major gentamicin components in the FD spectrum of a commercial sample (Parfitt *et al.*, 1976).

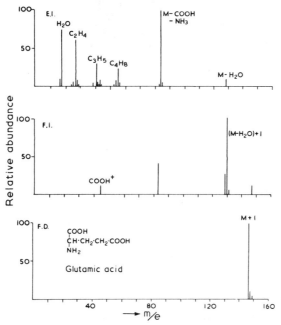

Fig. 12.10. Electron impact, field ionisation and field desorption mass spectra of glutamic acid. (Courtesy of Wiley-Interscience.)

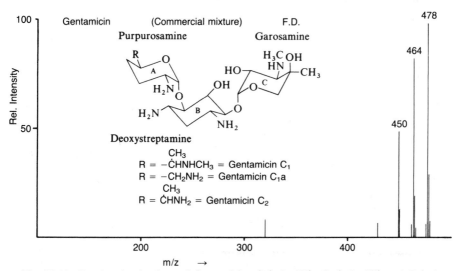

Fig. 12.11. Quasi-molecular ions of Gentamicins $C_1(m/z$ 478), $C_{1a}(m/z$ 450) and $C_2(m/z$ 464) in the FD mass spectrum of a commercial sample of Gentamicin (Courtesy Biomedical Mass Spectrometry.)

Fast atom bombardment (FAB) mass spectrometry

A more recent development (Barber *et al.*, 1981) facilitating the examination of thermolabile compounds and compounds of low volatility is fast atom bombardment mass spectrometry. In this technique, ionisation of sample from the solid/solution state is achieved by a beam of fast neutral atoms. The sample is applied to a copper stage within the source by evaporation of a solution or in solution in a non-volatile solvent such as glycerol. Argon gas is first ionised by a hot filament, and the ions generated are accelerated to 3–4 kV. The resultant Ar^+ beam is then passed into a chamber containing argon gas at about 10^{-3}–10^{-4} mm Hg.

$$Ar \xrightarrow{-e} Ar^+$$
$$\underset{\rightarrow}{Ar^+} + Ar \rightarrow \underset{\rightarrow}{Ar} + Ar^+$$

Charge exchange occurs and a beam of fast moving argon atoms is generated. When the beam impinges upon the target (sample), a complex series of ion–molecule reactions occurs and sample ions are sputtered from the target into the MS analyser. FAB has been exploited successfully in the examination of large peptides (e.g. insulin), aminoglycosides, nucleotides, phospholipids, vitamin B_{12} etc. Quasi-molecular ions $(M + H)^+$ and other addition ions such as $(M + Na)^+$ are frequently encountered in FAB/MS.

Applications

Structure elucidation   (Biemann, 1962; Hill, 1966; Budzikiewicz *et al.*, 1964; McLafferty, 1980; Waller, 1972; Spiteller and Spiteller-Friedmann, 1965)

The mass spectrum is a line spectrum, each line corresponding to a positive ion of specific mass. In the case of a compound which has simply lost one electron, the parent (P^+) or mass (M^+) peak correspond to the exact molecular weight of the compound. The 'nearest-whole-number' value for a molecular weight is arrived at rapidly by counting the number of lines (Fig. 12.2) from a reference line. Traces of air introduced into the instrument give rise to peaks corresponding to H_2O, m/z 18; N_2, m/z 28; O_2, m/z 32; Ar, m/z 40; and CO_2, m/z 44, which may act as references. The final ion recorded on the spectrum C^+, m/z 12, may also be used. Spectrum calibration is now usually effected by computer.

By a process known as **peak matching**, the molecular weight of an ion may be determined to six decimal places on a double-focussing mass spectrometer. The peak, corresponding to an unknown mass, is matched with a known fluorocarbon peak in that mass region, on an oscilloscope. Accuracy of this order permits the absolute identification of ions of identical 'rough' mass. For example, the ions $[CO]^+$, $[N_2]^+$, $[CH_2N]^+$ and $[C_2H_4]^+$ each correspond to a mass of 28; mass measurement to six decimal places establishes the identity and elemental composition of each ion:

CO	27.994 914
N_2	28.006 154
CH_2N	28.018 723
C_2H_4	28.031 299

Another method of determining the element make-up of an ion is by a consideration of isotope peaks. Isotopes of elements, which differ only by the number of neutrons in their nuclei, give rise to additional peaks in the mass spectrum of a compound. In a compound possessing only one carbon atom both ^{12}C and ^{13}C occur in the ratio 98.982 to 1.108. Compounds with more than one carbon atom have a correspondingly greater chance of possessing a ^{13}C nucleus. Thus, a molecular ion having a single carbon atom affords a mass peak (M) and mass plus 1 peak $(M + 1)$ in the ratio of approximately 99:1, i.e. corresponding to the natural isotopic abundance ratio of carbon. Important contributions to an $M + 1$ (or fragment ion + 1) peak are made by ^{13}C, ^2H, ^{15}N and ^{33}S, and to an $M + 2$ (or fragment ion + 2) peak by ^{18}O, ^{34}S, ^{37}Cl and ^{81}Br (Fig. 12.12). Because the number of each type of nucleus present in a molecule governs its contribution to the $M + 1$ and $M + 2$ peaks, the ratios of the intensities of $M + 1$ and $M + 2$ to the parent peak correspond to a specific elemental composition of the molecular ion.

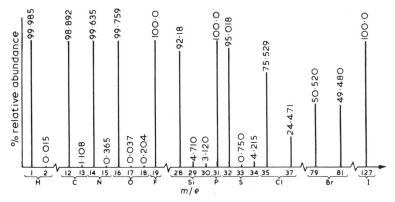

Fig. 12.12. Relative natural abundances of isotopes

The intensities of the $M + 1$ and $M + 2$ (or fragment ion + 1 and fragment ion + 2) peaks as a ratio of the parent or fragment ion should be established and the corresponding elemental composition of the molecule or fragment ascertained by reference to a table of **isotope abundance ratios** (Beynon, 1960; Beynon and Williams, 1963; Silverstein and Bassler, 1968). It should be remembered that fluorine and iodine consist of single atomic species of masses 19 and 127 respectively.

Nitrogen rule. If a molecular ion has an even molecular weight it must possess either no nitrogen or an even number of nitrogen atoms. An odd molecular weight compound requires an odd number of nitrogen atoms.

The **base peak** is the most intense line of the mass spectrum.

Metastable ions. Fragmentation ($m_1^+ \rightarrow m_2^+ +$ a neutral fragment), in general, occurs in the ion source of the mass spectrometer before the positive ions are accelerated and therefore a distinct peak results for each fragment ion. If an ion breaks up slowly during acceleration and flight then m_2^+, which has been generated in the ion source, will appear as a normal peak, whereas the same ion formed during flight occurs at a lower mass than m_2^+ and is designated m^*, a mestastable ion. The fragment ion m_2^+ formed after acceleration has less kinetic energy than one formed in the source because some of the kinetic energy received by m_1^+ during acceleration is carried off by the neutral fragment. The peak (m^*) due to such fragmentation therefore occurs at lower mass than m_2^+ and is generally broad and of low intensity (Fig. 12.13). The relationship of the position in the spectrum of m^*, with respect to the ions m_1^+ and m_2^+, is given by the expression:

$$m^* = (m_2)^2/m_1 \qquad (4)$$

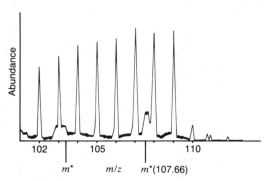

Fig. 12.13. Metastable peaks. m^* = metastable ion

Thus if a fragment m_2^+ is suspected as having arisen from an ion m_1^+, the observation of a metastable ion (m^*) at the mass value calculated from eq. (4) lends support to the fragmentation pathway. The absence of a metastable ion does not preclude the pathway.

The utility of mass spectrometry is not restricted to molecular and fragment weight determinations. The manner in which molecular ions fragment in the mass spectrometer is of the utmost importance and yields much information concerning the structure of the parent compound.

Some of the most important aspects of molecular fragmentation on electron impact are summarised below. A one electron shift is denoted by a 'fish-hook' arrow ⌒ and a two electron shift in the usual manner by ⌒ following the convention proposed by Budzikiewicz *et al.*, (1964).

(1) The parent peak (M^+) is the most intense in straight-chain compounds; the intensity diminishes with increased chain branching.

(2) In a homologous series the intensity of M^+ decreases with increase in molecule weight.

(3) In branched-chain hydrocarbons cleavage is favoured at the bond adjacent to the branch (Fig. 12.14), thus giving rise to tertiary ions rather than secondary, and secondary ions in preference to primary.

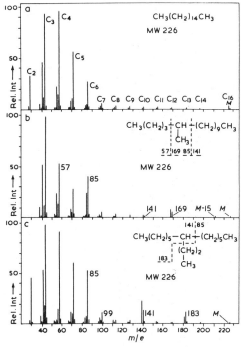

Fig. 12.14. Mass spectra of isomeric C_{16} hydrocarbons

This follows the order of stability of carbocations, tertiary > secondary > primary. For example, in the following methylpentanes:

$$\left[\begin{array}{c} CH_3 \quad CH_3 \\ | \qquad | \\ CH_3-C-CH-CH_2-CH_3 \\ | \\ CH_3 \end{array} \right]^{+\cdot} \longrightarrow \begin{array}{c} CH_3 \quad CH_3 \\ | \qquad | \\ CH_3-C^+ + \cdot CH-CH_2-CH_3 \\ | \\ CH_3 \end{array}$$

and m/z 57

$$\left[\begin{array}{c} CH_3 \\ | \\ CH_3-CH-CH_2-CH_2-CH_3 \end{array} \right]^{+\cdot} \longrightarrow \begin{array}{c} CH_3 \\ | \\ CH_3-C^+ + \cdot CH_2-CH_2-CH_3 \\ | \\ H \end{array}$$

m/z 43

(4) The probability of the existense of a strong M^+ peak is high when unsaturated or cyclic systems are present in the molecule. As in carbocation chemistry, the ions formed are stabilised.

(5) Alkylbenzenes cleave at the C—C bond β to the aromatic ring, resulting in a highly stabilised carbocation. Most alkylbenzenes have been shown to give the more stable tropylium ion (I) rather than the benzyl cation (II).

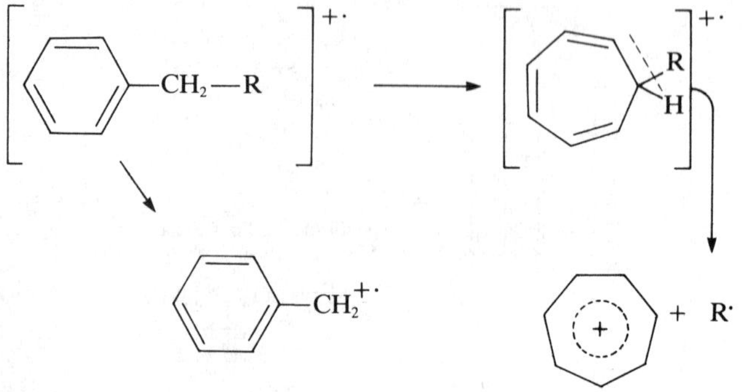

(II)　　　　　　　　　　(I)

Similarly, 3-alkylpyridines give an azatropylium ion (III):

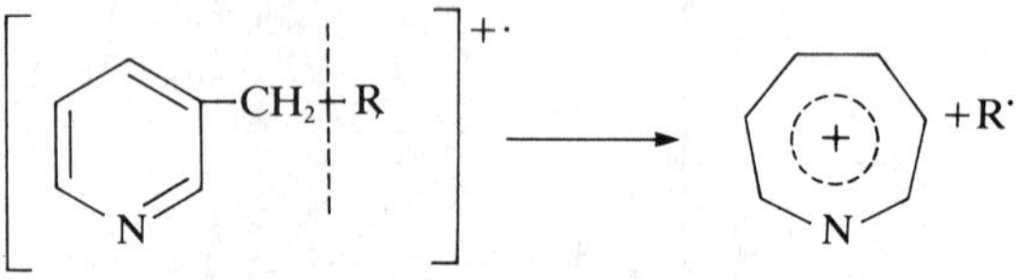

(III)

(6) The preferred cleavage in compounds possessing a carbon-carbon double bond is β to that bond, to yield the resonance stabilised allyl cation (IV).

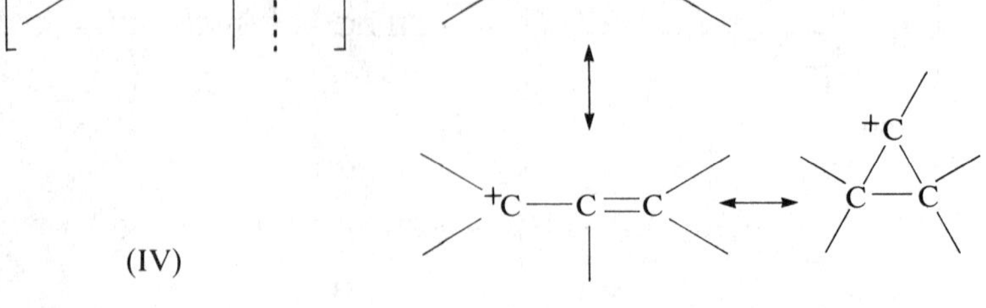

(IV)

Fragmentation of this sort should lead to considerable differences in the mass spectra of branched-chain alkenes; however, in the molecular ion the double bond appears to migrate readily. Branched-chain alkenes therefore give very similar mass spectra.

Cyclic alkenes, in which the double bond is fixed, often undergo a characterisic retro-Diels-Alder fission:

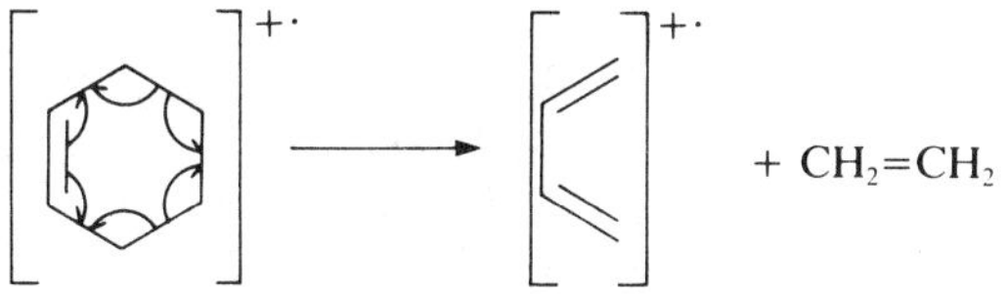

This fragmentation pathway is of importance in the rationalisation of the spectra of terpenes and unsaturated steroid molecules.

(7) Compounds such as alcohols, mercaptans, amines and esters cleave at the carbon-carbon bond β to the hetero-atom. Again, resonance stabilisation of the positive charge is possible by virtue of the hetero-atom lone pair.

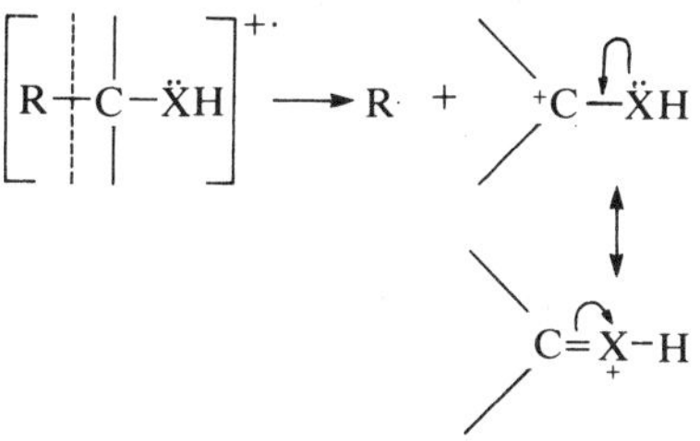

Where two hetero-atoms are present in the same molecule, the charge is retained predominantly by the fragment bearing the more electron donating hetero-atom (Fig. 12.15). Although cleavage is very much favoured in one direction, the reverse electron flow does give rise to some of the less stable ion.

m/z 31
30% abundance

m/z 30
57% abundance

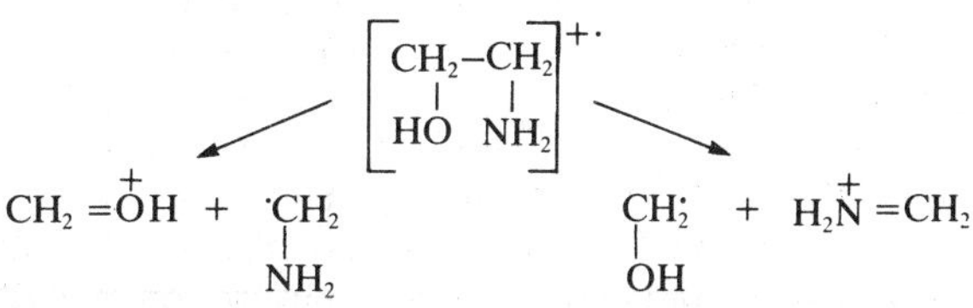

In ethers, halogenoalkanes and related compounds cleavage of the carbon-hetero-atom bond may occur leading to the more stable alkyl carbocation rather than the X^+ ion:

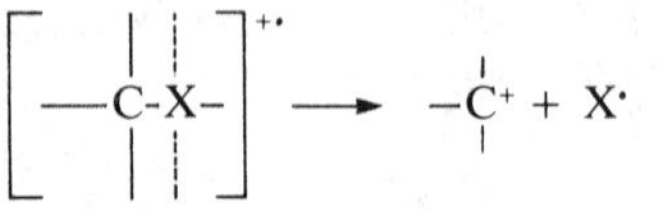

$$\left[-\overset{|}{\underset{|}{C}}-\overset{|}{\underset{|}{X}}- \right]^{+\bullet} \longrightarrow -\overset{|}{\underset{|}{C}}{}^+ + X^\bullet \quad (V)$$

where X = halogen, O—R, S—R, N—R_2. R can only be hydrogen when a tertiary carbocation is possible as a stable positive ion (V).

Carbonyl compounds cleave at the carbon-carbon bond α to the carbonyl group. In an asymmetrical ketone two such modes of fission are possible, the predominant positive ion produced being that which favours the elimination of the largest alkyl chain as the neutral fragment:

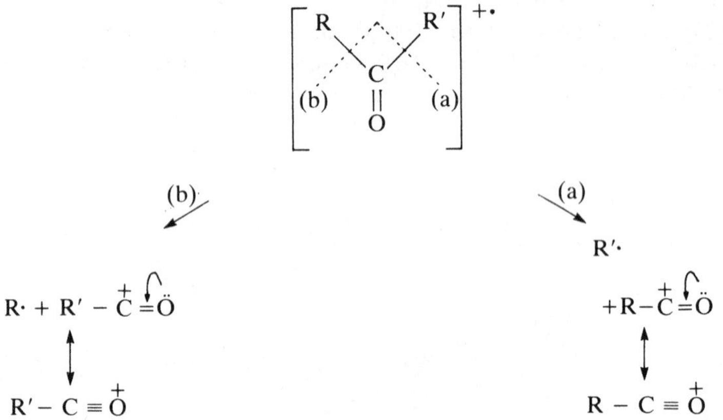

$$\left[\begin{array}{c} R \qquad\qquad R' \\ \searrow\quad\nearrow \\ C \\ (b)\quad \| \quad(a) \\ O \end{array} \right]^{+\bullet}$$

(b) ↙   (a) ↘

$$R'\bullet$$

$$R\bullet + R' - \overset{+}{C}\!=\!\ddot{O} \qquad\qquad +R - \overset{+}{C}\!=\!\ddot{O}$$

$$\updownarrow \qquad\qquad\qquad\qquad \updownarrow$$

$$R' - C \equiv \overset{+}{O} \qquad\qquad R - C \equiv \overset{+}{O}$$

If R = CH_3 and R' = C_3H_7 then CH_3—C≡O^+ (*m/z* 43) will occur in greater abundance than C_3H_7—C≡O^+ (*m/z* 71). In the case of ethanal, H—C≡O^+ (*m/z* 29) is formed in preference to CH_3—C≡O^+ (*m/z* 43), and the former appears as the base peak of the spectrum.

(8) Molecular or fragment ions may also rearrange, often with the elimination of a neutral fragment. Such rearrangements involve the transfer of hydrogen from one part of the molecular ion to another via, preferably, a six-membered cyclic transition state. The process is favoured energetically because as many bonds are formed as are broken.

Molecular ions possessing electronegative hetero-atoms often eliminate stable neutral molecules such as H_2O, CO, NO, HCl, H_2S, NH_3 or alkenes. The neutral fragments are not detected by the mass spectrometer but their loss is indicated by the nature of the positive ions formed:

$$\left[\begin{array}{c} \diagup\!\!\!\!\diagdown \\ C \qquad X \\ (C)_n \\ \diagdown\!\!\!\!\diagup \\ C \qquad H \end{array} \right]^{+\cdot} \longrightarrow \left[\begin{array}{c} \diagdown\,\diagup \\ C \\ (C)_n \\ \diagup\,\diagdown \\ C \end{array} \right]^{+\cdot} + HX$$

where X = —OH, —SH, ester, halogen and sometimes —C≡N.

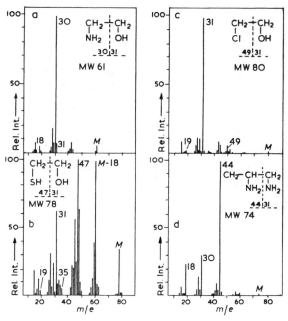

Fig. 12.15. Mass spectra of various simple, bifunctional molecules. (a) 2-aminoethanol; (b) 2-thioethanol; (c) 2-chloroethanol; (d) 1,2-diaminopropane

A rearrangement involving the transfer of a hydrogen atom from one part of an ion to another via a six-membered ring transition state is the McLafferty rearrangement. An example of this process is the elimination of ethene from pentan-2-one:

$$\left[\begin{array}{c} H \\ H_2C \quad O \\ \quad \| \\ H_2C \quad C—CH_2 \\ CH_2 \end{array} \right]^{+\cdot} \longrightarrow \begin{array}{c} H \\ \diagdown O^+ \\ C \\ \diagup\!\!\diagdown \\ CH_2 \quad CH_3 \end{array} + CH_2=CH_2$$

Many aldehydes, ketones, unsaturated hydrocarbons, amides, nitriles, and esters, will exhibit rearrangement peaks in their mass spectra when the stereochemistry of the molecule favours hydrogen transfer.

For a detailed treatment of fragmentation pathways the reader is referred to the works if Biemann (1962), Hill (1966), and McLafferty (1980).

Detection of impurities

The mass spectrometer is able to detect as little as a few parts per million of an impurity in a compound, particularly if the structure of the impurity is quite different from that of the main component. A good example is the detection of trace amounts of xylene in acetylacetone (Beynon, 1960; Nicholson, 1957) which had been purified for heat of combustion investigations. In the mass spectrum of the acetylacetone sample ($M^+ = 100$) extra peaks were observed at m/z 106, 105 and 91 suggesting the presence of a higher molecular weight impurity ($M^+ = 106$). Fragmentation of the acetylacetone molecular ion cannot give rise to the 91 peak since this would involve the loss of nine hydrogen atoms. The anomalous peaks may be explained by the presence of xylene as a contaminant, the parent ion (VI) of which, through loss of a proton, may rearrange to a tropylium ion (VII), or by loss of a methyl fragment yields the 91 peak (VIII) ($M^+ - 15$). The

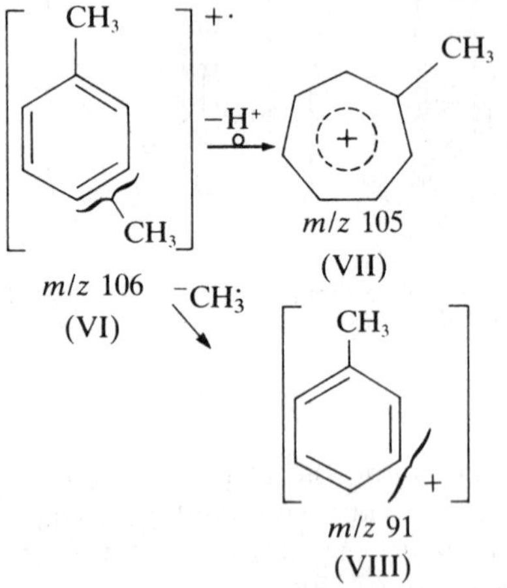

presence of xylene (the isomers could not be distinguished) was traced to the use of sodium, which had been stored over xylene, during the preparation of the sample. Since acetylacetone and xylene have approximately the same boiling point, distillation did not effect separation. The amount of xylene present in the sample was measured quantitatively, enabling the true heat of combustion to be calculated by the application of a correction factor. Contaminants of molecular weight higher than the major component are those most readily detected. The presence of impurities in a sample may also be established by lowering the intensity of the ionising electron beam. If all peak heights do not decrease proportionally then a contaminant (or contaminants) is present.

The advantage of such a sensitive means of impurity detection has obvious implications in quality control and in forensic science. From a consideration of the nature of the impurities present in a compound, it may be possible to rationalise its mode of manufacture.

Quantitative analysis (Millard, 1978)

The main stimulus for the development of early mass spectrometers came from the value of the technique in the quantitative analysis of multicomponent mixtures. The petroleum industry, in particular, used mass spectrometry for the analysis of complex hydrocarbon mixtures. A prerequisite of the technique is that spectra of pure samples of each component of the mixture must be available. The intensities of major peaks from each spectrum are calculated as a percentage of its base peak, and an intense peak from each component is chosen for the analysis. The partial pressure of each component in the gas reservoir is proportional to its amount in the mixture (Dalton's law of partial pressures) and therefore peak height is a quantitative measure of the constituents of the mixture. In a mass spectrometric assay of a mixture, account is taken of contributions of fragments from other components to the chosen peak. Fortunately, peaks are quantitatively additive and corrections are therefore easily applied. Today, spectra from multi-component mixtures are often programmed for feeding into a computer, permitting rapid and accurate analysis.

In the spectrum of a simple mixture of butane, 2-methylpropane, propane, ethane and methane, the first two possess peaks at m/z 58 and 57 (due to $C_4H_{10}^+$ and $C_4H_9^+$). Propane has a mass at m/z 44 which has to be corrected for contributions from butane and 2-methylpropane. At m/z 30 and 16 are the mass peaks of ethane and methane respectively, and these, in turn, must be corrected for contributions from the higher homologues. Problems such as this may be solved by the application of linear simultaneous equations (Kiser, 1965).

A similar method may be employed for compounds other than hydrocarbons. Amino acids (Biemann, 1960), for example, may be analysed successfully by their prior conversion to ethyl esters by treatment with boiling ethanol/hydrochloric acid. Less than 3 mg of the amino acid mixture is required, but care must be taken to avoid the formation of diketopiperazines. Amino acid ethyl esters exhibit only a few intense peaks in their mass spectra (Fig. 12.16) and for different amino acids these fall at quite different mass values, with a minimum of interference from contributions of other amino acid fragments. The base peak, and often the most valuable peak, of an amino acid ethyl ester spectrum is usually due to the fragment $R-CH=\overset{+}{N}H_2$, which corresponds in glycine to m/z 30, alanine, m/z 44 and valine, m/z 72

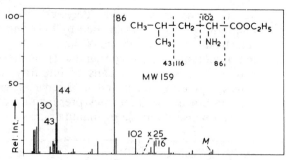

Fig. 12.16. Mass spectrum of leucine ethyl ester

etc. Thus, provided that the mass spectra of pure samples of all the amino acid ethyl esters which constitute the mixture are available, the quantitative composition of the mixture can be calculated. The instrument time involved is about 30–40 min, which permits rapid and accurate analysis of many mixtures during metabolism studies and protein structure investigation.

A more elegant method of determining quantitatively the components in a mixture of similar compounds is known as the **isotope dilution technique**. For this the compound or compounds to be assayed must be available in an isotopically labelled form, e.g. with ^{13}C, ^{15}N or ^{131}I. A known amount of labelled compound of known purity is added to the mixture to be estimated and a small quantity of compound(s) isolated; the isotope ratio of compound isolated is determined by MS. Isolation may be effected by standard chemical means, but it is far better if the complete exercise is performed by GC/MS. From the ratio of labelled compound to non-labelled, taking into account natural abundance and the purity of the labelled standard, the amount of compound(s) in the unknown mixture may be calculated. The technique has been applied widely in amino acid studies.

An interesting study has been reported by Horning (1973) where diphenylhydantoin, caffeine, pethidine and several barbiturates and barbiturate metabolites have been measured in the picogram and nanogram range in human body fluids, with the aid of stable isotope labelled drugs. Her objective was to establish analytical procedures that afforded rapid drug–body fluid profiles of patients under multiple drug therapy. Gas chromatography, GC/MS with CI and EI, selective or multiple ion detection and isotope dilution were used separately or in combination to examine samples of urine, plasma, breast milk and amniotic fluid in mother–infant pairs shortly after birth. Diphenylhydantoin-2,4,5-^{13}C and pentobarbitone-2,4(6),5-^{13}C were the internal standards. Body fluid samples as low as 50–200 μl were employed successfully to monitor profiles. Additionally, three metabolites of *Phenobarbitone*, 5-ethyl-5-(3,4-dihydroxy-1,5-cyclohexadien-1-yl) barbituric acid, 5-ethyl-5-(4-hydroxyphenyl)barbituric acid and 5-ethyl-5-(3,4-dihydroxyphenyl)-barbituric acid, and two metabolites of *quinalbarbitone*, 5-allyl-5-(3-hydroxy-1-methylbutyl)barbituric acid and 5-(2,3-dihydroxypropyl)-5-(1-methylbutyl)barbituric acid were identified.

Amino acid sequence analysis in peptides   (Waller, 1972; Biemann, 1960; Senn *et al.*, 1966; Biemann *et al.*, 1966; Jones, 1968; Morris *et al.*, 1975)

The determination of the structures of peptides and proteins is of major importance, and is often performed by a process of stepwise hydrolysis and much column chromatography. A method for overcoming the lack of appreciable volatility of peptides, a disadvantage in mass spectrometry, is to reduce them with lithium aluminium hydride to the corresponding polyamino alcohols.

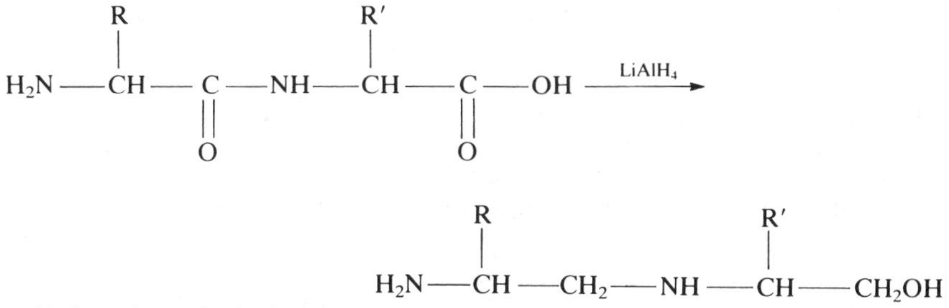

$$H_2N-CH(R)-C(=O)-NH-CH(R')-C(=O)-OH \xrightarrow{LiAlH_4}$$

$$H_2N-CH(R)-CH_2-NH-CH(R')-CH_2OH$$

Polyamino alcohols are volatile and upon electron impact fragment in a characteristic manner. The problem of determining an amino acid sequence may be simplified further by the use of gas chromatography. If a polyamino alcohol mixture derived from a protein-hydrolysate peptide mixture is subjected to gas chromatography separation, then the effluent fractions may be introduced into the mass spectrometer individually. The development of GC/MS has obviated the effluent isolation step and speeded up the process considerably. Derivatisation of amino alcohols also aids sequencing. Hydroxyl groups may be converted to trimethylsilyl ethers by treatment with *N*-(trimethylsilyl)diethylamine, a reagent that does not react with amino groups. Specific *O*-silyl derivatives exhibit a considerable enhancement of fragment ions from the C-terminal amino acid.

Mass spectrometric examination of peptides may be performed without prior reduction of the amide linkages. Compounds of sufficient volatility and stability can be prepared by esterificaton of the free carboxylic acid end-group and the conversion of the amine end-group to a characteristic derivative, e.g. acetyl, trideuteroacetyl, trifluoroacetyl and carbobenzoxy.

Amino acid sequencing by MS is being employed increasingly. The interpretation of spectra, however, although based upon simple arithmetic summations, is tedious. Fortunately, MS spectral data are amenable to computer analysis, thus speeding the sequencing process considerably (McFadden, 1973). FAB MS is being exploited increasingly for the sequencing of underivatized peptides (Barber *et al.*, 1981).

Drug metabolism   (Waller, 1972; Fenselau, 1972; Frigerio and Ghisalberti, 1976)

Drug metabolites usually arise from relatively minor structural modifications of the parent molecule. Oxidation to hydroxyl derivatives or *O*-demethylation, for example, often renders a foreign molecule more readily eliminated from the body. Mass spectrometry is a technique affording considerable structural information from a small sample and is consequently ideal for metabolism studies. The scope of the technique is extended further by its ability to detect drugs and metabolites that have been cold labelled, for example with ^{2}H, ^{13}C or ^{18}O.

The most obvious approach to identifying metabolites is to compare directly spectra of the pure drug with those of its biotransformation products. Direct insertion of metabolites, after isolation by column chromatography, preparative TLC, preparative GC or HPLC into the ion source of a high resolution instrument, followed by accurate mass measurement of each molecular ion, affords their molecular formulae. Differences in molecular formula indicate the transformation involved, and the fragmentation pattern of a metabolite will often yield information concerning the position of gain or loss of a molecular unit. Identification of the major metabolites of *Diazepam* (IX) was established in this manner (Schwarts *et al.*, 1967). Initial oxidation occurs at the positions indicated (*) in (IX) to give (X) and (XI). Nuclear magnetic resonance spectroscopy was used to assign unambiguously the positions of substitution. Secondary metabolism occurs by *N*-demethylation of (XI) to give (XIII) and by inserting a second oxygen (XII).

Combined GC/MS is now a common technique for the identification of organic compounds in admixture and in high dilution. It is often the method of choice in most metabolism studies.

For many years, 1-(2-acetylhydrazino)phthalazine (XV) was presumed to be a major metabolite of *Hydralazine* (XIV). However, attempts to synthesise it resulted in the isolation of the triazolophthalazine (XVI) only. Paper chromatogram of a metabolite, previously identified as the acetyl derivative (XV), was confirmed by a GC/MS study of the urine from three patients to be (XVI) (Zimmer *et al.*, 1973).

Hydralazine (XIV)   (XV)   (XVI)

A total ion current gas chromatogram is illustrated in Fig. 12.17, and Fig. 12.18 demonstrates the mass spectral characterisation of compound (XVI).

$C_{16}H_{13}N_2OCl$
$(+O)$

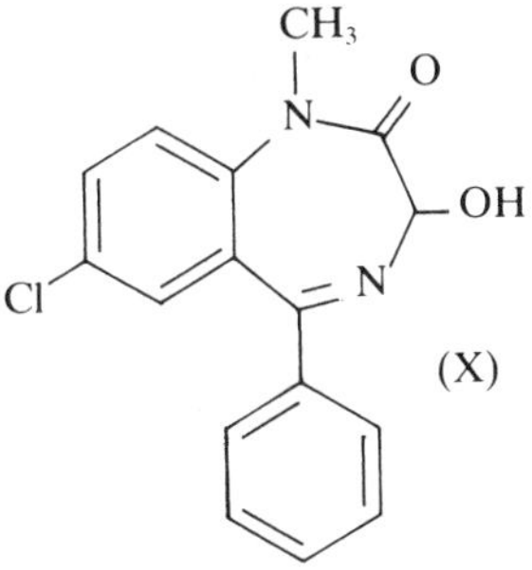

(X)

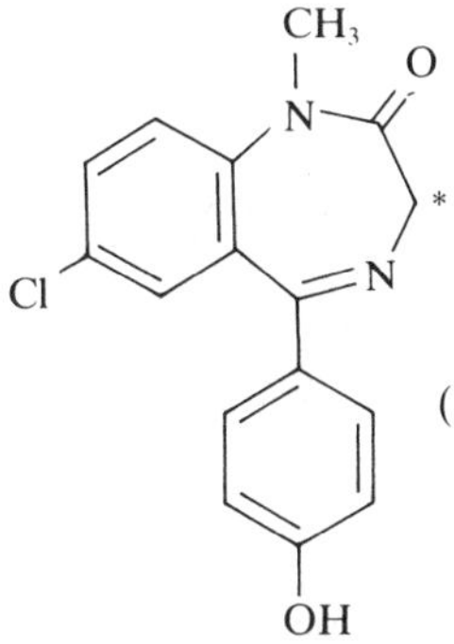

$C_{16}H_{13}N_2O_2Cl$
$(+O)$

(XI)

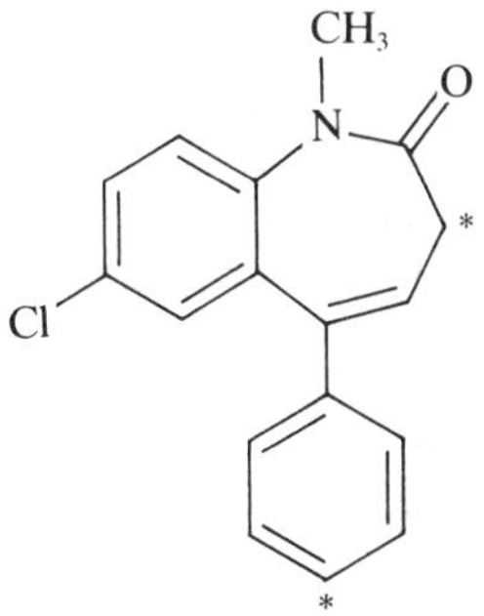

Diazepam(IX)

$C_{16}H_{13}N_2OCl$

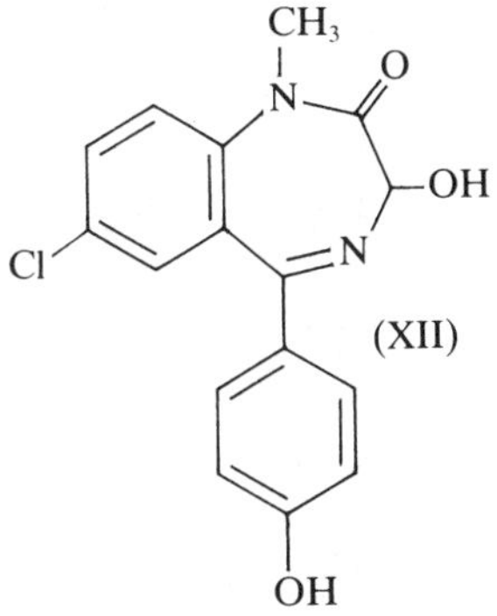

$C_{16}H_{13}N_2O_3Cl$
$(+20)$

(XII)

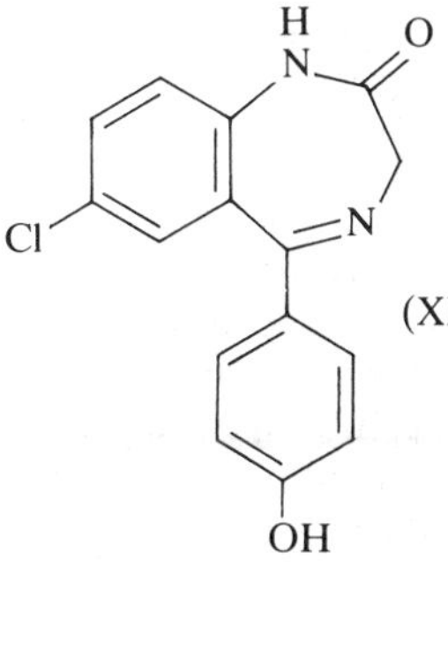

$C_{16}H_{11}N_2O_2Cl$
$(+O: -CH_2)$

(XIII)

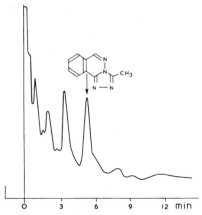

Fig. 12.17. Total ion current recording after injection at a $3\mu l$ aliquot of a urine extract from a patient treated with hydralazine. (Courtesy of *Arzneim. Forsch.*)

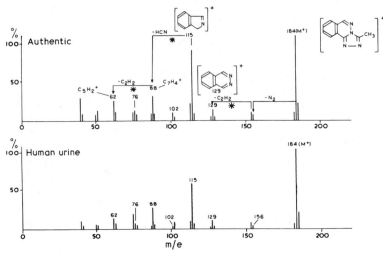

Fig. 12.18. Upper panel: Mass spectrum of authentic 3-methyl-s-triazolo[3,4-a]phthalazine. Lower panel: Mass spectrum of a compound with retention time of 5.5 min in a urine extract from a patient treated with hydralazine. (Courtesy of *Arzneim. Forsch.*)

A fully computerised method for the quantitative determination of picogram quantities of drugs by GC/MS has been described by Baczynskyj *et al.*, (1973).

The first application of mass fragmentography to the determination of drug metabolites was described for *Chlorpromazine* (XVII) (Hammer *et al.*, 1968).

A mass fragmentogram where the intensity of ions m/z 232 (A), m/z 234 (A^1) and m/z 246 (B) have been monitored with time is shown in Fig. 12.19. Fragmentation at bond x gives rise to ions A, A′ and fragmentation at y the ion B. The intensity of these ions in the mass spectrum of a compound related to chlorpromazine will vary according to the nature of the basic side chain. Preparation of trifluoroacetate (TFA) derivatives of the mixture of metabolites increases the specificity of the method. By a consideration of their GC retention times and the relative intensities of the three ions monitored as seen in the mass fragmentogram, desmethylchlorpromazine (XVIII) and didesmethyl-chlorpromazine (XIX) were identified as chlorpromazine metabolites.

Fig. 12.19. Mass fragmentogram, chlorpromazine metabolism study. Fragments monitored m/z 232, 234 and 246. TFA = trifluoroacetate derivative. (Adapted from *Biochemical Applications of Mass Spectrometry*, G. R. Waller, courtesy of Wiley-Interscience.)

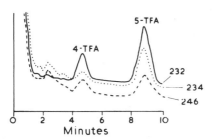

R = R′ = CH$_3$ = Chlorpromazine (XVII)
R = H; R′ = CH$_3$ (XVIII)
R = R′ = H (XIX)

A = ^{35}Cl
A′ = ^{37}Cl

B = ^{35}Cl

More recent applications include the determination of the human urinary excretion of *Pethidine* and metabolites (Lindberg *et al.*, 1980), the identification of *Fentanyl* metabolites in rat urine (Goromaru *et al.*, 1982), the simultaneous determination of *Imipramine* and related compounds in body fluids (Alkalay *et al.*, 1979) and the quantification of *Amitriptyline* and *Nortryptyline* by GC/CIMS (Garland, 1977).

Clinical and forensic applications   (Waller, 1972; McFadden, 1973; Fenselau, 1972; Watson, 1976)

A delicate balance of biochemical reactions is necessary to ensure the health of the human organism. Specific disease states cause changes in body chemistry and excretion products, and the latter may be detected in body fluids, expired air or sweat. Mass spectrometry and the GC/MS combination may be employed to monitor these changes and aid diagnosis.

The health of an individual may also be impaired by accidental or delibrate abuse or overdosage of drugs or toxic chemicals. Forensic science and toxicology are therefore areas where MS and GC/MS are finding increasing application. Finkle (1973), for example, has reported the case of a male with a history of drug abuse found dead with narcotic overdosage being suspected as the cause. Analysis by GC/MS showed that his blood and urine contained codeine, and that bile, and veins at the injection site, contained morphine, supporting the probable cause of death.

Chemical ionisation (p.478) often gives peaks of much higher intensity in the molecular ion region than do corresponding electron impact spectra. In the identification of mixtures of drugs in body fluids, this enhanced sensitivity coupled with spectrum simplicity is of special value.

Barbiturates have been prescribed so widely as sedatives that accidental or deliberate overdosage is relatively common. Because of the multiplicity of barbiturate analogues, identification by standard procedures is difficult, if not impossible. Fales *et al.*, (1970) have reported a method of barbiturate identification by GC/MS. The detection of *Quinalbarbitone* and *Pentobarbitone* in an extract of gastric contents is illustrated in Fig. 12.20.

GC/CIMS has been employed for the detection of *Morphine*, related alkaloids and their metabolites in the urine of opium eaters (Cone *et al.*, 1982), and the very high toxicity of 2,3,7,8-tetrachlorobenzodioxin has led to MS detection techniques sensitive down to about 5 femtograms (Taylor and Gooch, 1980).

Mass spectrometric techniques are advancing rapidly and are exploited in an increasing number of medicial and related disciplines. Identification and quantification of drugs, metabolites and toxins will become more sophisticated, rapid medical diagnosis by MS will spread, and the monitoring of the environment will develop.

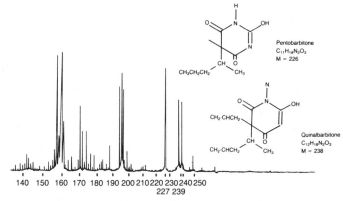

Fig. 12.20. Chemical ionisation mass spectrum of extract of gastric contents. (Courtesy of *Analytical Chemistry*.)

References

Alkalay, D., Volk, J. and Carlsen, C. (1979) *Biomed. Mass Spec.* **6**, 200.

Baczynskyj, L., Duchamp, D.J., Zieserl, J.F. and Axen, U. (1973) *Anal. Chem.* **45**, 479.

Barber, M., Bordoli, R.S., Sedgwick, R.D. and Tyler, A.N. (1981) *Nature (Lond.)* **293**, 270.

Beckey, H.D. (1977) *Principles of Field Ionization and Field Desorption Mass Spectrometry*, Pergamon Press.

Beckey, H.D. Heindricks, A. and Winkler, H.W. (1970) *Int. J. Mass Spectrom. Ion. Phys.* **3**, 9.

Begg, D. and Yergey, A. (1973) *Industrial Research* 46.

Beynon, J.H. (1980) *Mass Spectrometry and its Applications to Organic Chemistry*, Elsevier, Amsterdam.

Beynon, J.H. and Williams, A.E. (1963) *Mass Abundance Tables for use in Mass Spectroscopy*, Elsevier, Amsterdam.

Biemann, K. (1960) *Chimia (Switz.)* **14**, 393.

Biemann, K. (1962) *Mass Spectrometry*, McGraw-Hill, New York.

Biemann, K., Cone, C., Webster, B.R. and Arsenault, G.P. (1966) *J. Am. chem. Soc.* **88**, 5598.

Brent, D.A., Rouse, D.J., Sammons, M.C. and Bursey, M.M. (1973) *Tetrahedron Letters* 4127.

Budde, W.L. and Eichelberger, J.W. (1979) *Organic Analysis Using GC/MS*, Ann Arbor Science.

Budzikiewicz, H., Djerassi, C. and Williams, D.H. (1964) *Structure Elucidation of Natural Products by Mass Spectrometry*, 2 vols, Holden-Day, San Francisco

Budzikiewicz, H., Djerassi, C. and Williams, D.H. (1967) *Mass Spectrometry of Organic Compounds*, Holden-Day, San Francisco.

Chapman, J.R. (1978) *Computers in Mass Spectrometry*, Academic Press, London.

Cone, E.J., Gorodetzky, C.W., Yeh, S.Y., Darwin, W.D. and Buchwald, W.F. (1982) *J. Chromatog.* **230**, 57.

Desiderio, D.M. and Yamada, S. (1982) *J. Chromatog.* **239**, 87.

Desiderio, D.M., Yamada, S., Tanzer, F.S., Horton, J. and Trimble, J. (1981) *J. Chromatog.* **217**, 437.

Ebbighausen, W.O.R., Mowat, J. and Vestergarrd, Per. (1973) *J. Pharm. Sci.* **62**, 146.

Fales, H.M., Lloyd, H.A. and Milne, G.W.A. (1970) *J. Am. chem. Soc.* **92**, 1590.

Fales, H.M., Milne, G.W.A. and Axenrod, R. (1970) *Anal. Chem.* **42**, 1432.

Fenselau, C. (1972) In Chignell, C.F. (ed) *Methods in Pharmacology*, vol.2, *Physical Methods*, Appleton-Century Crofts, New York.

Finkle, B.S. (1973) *Techniques of Combined Gas Chromatography/Mass Spectrometry*, p.385, Wiley, New York.

Frigerio, A. and Ghisalberti, E.L. (1976) (eds) *Mass Spectrometry in Drug Metabolism*, Plenum Press, New York.

Games, D.E. (1980) *Anal. Proc.* **17**, 322.

Games, D.E. and Westwood, S.A. (1983) *Europ. Spec. News* **48**, 14.

Garland, W.A. (1977) *J. Pharm. Sci.* **66**, 77.

Goromaru, T., Matsuura, H., Furuta, T., Baba, S., Yoshimura, N. and Miyawaki, T. (1982) *Drug Metab. Dispos.* **10**, 542.

Hammer, C.G., Holmstedt, B. and Ryhage, R. (1968) *Anal. Biochem.* **25**, 532.

Henion, J.D. and Maylin, G.A. (1980) *Biomed. Mass Spec.* **7**, 115.

Henneberg, D., Casper, K., Ziegler, E. and Wiemann, B. (1972) *Angew. Chem. (Int. Ed.)* **11**, 357.

Hill, H.C. (1966) *Introduction to Mass Spectrometry*, Heyden, London.

Horning, M.G., Nowlin, J., Lertratanangkoon, K., Stillwell, W.G. and Hill, R.M. (1973) *Clini. Chem.* **19**, 845.

Jones, J.H. (1968) *Q. Revs.* **22**, 302.

Kiser, R.W. (1965) *Introduction to Mass Spectrometry and its Applications* Prentice-Hall, Englewood Cliffs, N.J.

Lindberg, C., Bondesson, H. and Hartrig, P. (1980) *Biomed. Mass Spec.* **7**, 88.

McFadden, W. (1973) *Techniques of Combined Gas Chromatography/Mass Spectrometry*, Wiley, New York.

McLafferty, F.W. (1963) *Mass Spectrometry of Organic Ions*, Academic Press, New York.

McLafferty, F.W. (1966) *Chem. Comm.* 78.

McLafferty, F.W. (1980) *Interpretation of Mass Spectra*, 3rd edn, University Science Books, Calfornia.

Maume, B.F., Burnot, P., Lhuguenot, J.C., Baron, C., Barber, F., Maume, G., Prost, H. and Padieu, P. (1973) *Anal. Chem.* **45**, 1073.

Millard, B.J. (1978) *Quantitative Mass Spectrometry,* Heydon, London.

Milne, G.W.A. (1971) *Mass Spectrometry*, Wiley, New York.

Morris, H.R., Dell, A. and Batley, K.E. (1975) *ASTM 23rd Conf. on Mass Spectrometry and Allied Topics*, Houston.

Nicholson, G.R. (1957) *J. Chem. Soc.* 2431.

Parfitt, R.T., Games, D.E., Rossiter, M., Rogers, M.S. and Weston, A. (1976) *Biomed. Mass Spec.* **3**, 232.

Rose, M.E. and Johnstone, R.A.W. (1982) *Mass Spectrometry for Chemists and Biochemists* Cambridge University Press.

Schwarts, M.A., Bommer, P. and Vane, F.M. (1967) *Arch. Biochem. Biophys.* **121**, 508.

Senn, M., Venkataraghavan, R. and McLafferty, F.W. (1966) *J. Am. chem. Soc.* **88**, 5593.

Silverstein, R.M. and Bassler, G.C. (1968) *Spectrometric Identification of Organic Compounds*, Wiley, New York.

Simpson, C.F. (1972, Sept.) *Critical Reviews in Analytical Chem.* 1.

Spiteller, G. and Spiteller-Friedmann, N. (1965) *Angew. Chem. (Int. Ed.)* **4**, 383.

Taylor, K.T. and Gooch, G. (1980) *Kratos: Mass Spec.* data sheet No. 121.

Waller, G.R. (1972) *Biomedical Applications of Mass Spectrometry*, Wiley, New York.

Watson, J.T. (1976) *Introduction to Mass Spectrometry,* Raven Press, New York.

Yu, T.J., Schwartz, H., Giese, R.W., Karger, B.L. and Vouros, P. (1981) *J. Chromatog.* **218**, 519.

Zimmer, H., Kokosa, J. and Gorteiz, D.A. (1973) *Arzneim. Forsch.* **23**, 1028.

13
Radiochemistry and radiopharmaceuticals

T. L. WHATELEY

Introduction

Radionuclides, mainly ^{3}H and ^{14}C, are widely used as tracers in analysis, and in distribution and metabolism studies of drugs in animals. Such isotopes with long half-lives are not suitable for use in human medicine, but a number of radioisotopes with comparitively short half-lives, measured in hours or days, are now widely used in radiopharmaceutical preparations as diagnostic agents and in the treatment of neoplastic diseases. These include ^{32}P, ^{51}Cr, ^{57}Co, ^{59}Fe, ^{75}Se, ^{99m}Tc, ^{125}I, ^{131}I, and ^{133}Xe.

The main subjects of this chapter are

(a) radiochemical methods of analysis
(b) the quality control of radiopharmaceuticals
(c) the use of radionuclides as tracers in distribution and metabolism

illustrated by schemes of experiments.

These subjects are approached by a general introduction to radiochemistry which covers the following topics:

(a) Fundamental properties of radionuclides and of radiation
(b) Safety aspects—radiation hazards and protection
(c) Measurement of radioactivity
(d) Radiopharmaceuticals and radionuclide generators.

Fundamentals of radioactivity

A **radionuclide** is a nuclide which is radioactive, i.e. radiation is emitted by the spontaneous transformation of the nucleus. The transformation may involve emission of charged particles (α and β particles) which may be accompanied by the emission of γ-radiation. Gamma radiation may also arise from the processes of isomeric transition, electron capture and positron (β^+) emission (β^+ particles being annihilated on contact with electrons to produce two 511 kev gamma photons).

Isotopes are nuclides of the same element, i.e. with the same atomic number (A) but different numbers of neutrons and hence different mass numbers (Z).

Radioactive decay is a random first order process independent of temperature and the chemical state of the radionuclide. The following laws of radioactive decay apply to all forms of radiation emission.

The fundamental radioactive decay law is

$$A = A_0 e^{-\lambda t}$$

where A is activity at time t; A_0 is initial activity; and λ is the decay constant.

The form of such a decay curve is shown in Figs 13.1 and 13.2. The **half-life** of a radionuclide is defined as the time for the activity to decay to one-half of its initial value: this is illustrated in Fig. 13.1. Thus, in two half-lives the activity has fallen to $\frac{1}{4}$ its initial activity.

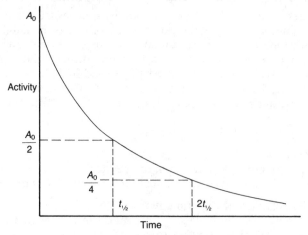

Fig. 13.1. Radioactive decay

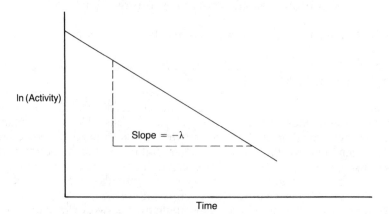

Fig. 13.2. Radiocative decay with log scale

As a very good approximation it can be shown that in 10 half-lives activity falls to 10^{-3} of its initial value (i.e. to 0.1% of initial activity). This is a useful guide when considering the disposal of radioactive waste. After 20 half-lives, activity will have fallen to 10^{-6} of initial value.

It is more usual to tabulate the decay properties of radionuclides in terms of their half-lives $(t_{1/2})$ rather than their decay constants: the relationship is required in order to use the fundamental decay law equation in dose calculations (dosages of radiopharmaceuticals must always be specified **at the time of administration**, rather than at the time of measurement; for short-lived radionuclides these two quantities are very different).

The relationship between $t_{1/2}$ and λ is arrived at by substituting the values $A = A_0/2$ when $t = t_{1/2}$ into the fundamental decay equation:

$$A_0/2 = A_0 e^{\lambda t_{1/2}}$$
$$\therefore \quad \tfrac{1}{2} = e^{-t_{1/2}}$$
$$\therefore \quad \ln 0.5 = -\lambda t_{1/2} \text{ or } \ln 2 = \lambda t_{1/2}$$
$$\therefore \quad \lambda = \frac{\ln 2}{t_{1/2}} = \frac{0.693}{t_{1/2}}$$

Some definitions

Specific activity: the radioactivity per unit mass of material, e.g. $Bq(\mu mol)^{-1}$, $\mu Ci\, mg^{-1}$.

Radioactive concentration: the radioactivity per unit volume, e.g. $\mu Ci\, cm^{-3}$ (i.e. applicable only to liquids and solutions and should be carefully distinguished from specific activity).

Bequerel (Bq): defined as 1 disintegration per second (dps).

Curie (Ci): defined as 3.7×10^{10} disintegrations per second. Therefore, $1\, Ci = 3.7 \times 10^{10}\, Bq$ and $1\, mCi = 37\, MBq$.

Calculation to determine the mass of sodium pertechnetate, $Na^{99m}TcO_4$, associated with 37 MBq (1 mCi) of radioactivity

There are a number of reasons for following through this calculation:

(1) 37 MBq is the order of activity injected routinely into patients: it is interesting to see what mass of technetium-99 m is involved. Toxicological and pharmacological aspects may then be considered.

(2) The sensitivity of radiochemical methods in both analysis and research is clearly illustrated.

(3) The problems involved in the handling and transferring of such masses can be considered.

Use the fundamental equation:

$$\frac{dN}{dt} = \lambda N$$

$$\therefore \quad N = \frac{dN}{dt} \lambda^{-1}$$

Take $\dfrac{dN}{dt}$ to be 37 MBq or 37×10^6 dps, and

$$\lambda = \frac{0.693}{t_{1/2}} = 0.693/6.0 \times 60 \times 60 \text{ s}^{-1} = 3.21 \times 10^5 \text{ s}^{-1}$$

$$N = \frac{37 \times 10^6}{3.21 \times 19^{-5}}$$

$$= 1.15 \times 10^{12}$$

This is the number of atoms of $^{99\,m}$Tc present.
Thus:

$$\text{Number of moles} = \frac{1.15 \times 10^{12}}{6.023 \times 10^{23}}$$

$$\text{Mass of Na}^{99\,m}\text{TcO}_4 = \frac{1.15 \times 10^{12} \times 186}{6.023 \times 10^{23}} \text{ g}$$

$$= 3.5 \times 10^{-10} \text{ g}$$

$$= 0.35 \times 10^{-9} \text{ g}$$

$$= 0.35 \text{ ng}$$

Thus the mass associated with 37 MBq (1 mCi) of $\text{Na}^{99\,m}\text{TcO}_4$ is 0.35 ng. Such a mass will have no toxicological or pharmacological effect on the patient: one is only concerned, in imaging, with the final location of this radioactivity in the body, and that is determined by the nature of the chemical entity to which the $^{99\,m}$Tc is attached.

Thus we have seen that 0.35 ng is associated with 37×10^6 dps. It is very easy to detect and measure 37 dps on modern counters, and thus one is measuring 0.35×10^{-6} ng or 0.35×10^{-15} g or 0.35 femtogram (fg), illustrating clearly the sensitivity of radiochemical methods.

The handling of such extremely small masses can be a problem in that such amounts may adsorb onto the walls of containing vessels and hence may not be transferred as expected. Fortunately, there is no real problem with $\text{Na}^{99\,m}\text{TcO}_4$ solutions in glass or plastic containers. If adsorption is a problem (e.g. ^{113}Sn in glass containers) the answer is to add a **carrier**: this is the non-radioactive form of the same chemical species. The carrier will, in effect, saturate the adsorption sites, allowing the trace amount of radioactive form to remain in solution.

Table 13.1 Some radionuclides used in radiopharmaceuticals and radiochemical methods

Radionuclide	Half-life	Principal γ-energy(keV)	Maximum β-energy(keV)
Technetium-99m	6.0 hours	140	–
Technetium-99	2.1×10^5 years	–	259
Iodine-125	60 days	28,35	–
Iodine-131	8.05 days	364	606
Indium-111	2.8 days	173,247	–
Indium-113m	99.8 minutes	392	–
Gallium-67	78 hours	92,182,300	–
Gallium-68	1.14 hours	510	1890
Chromium-51	27.8 days	323	–
Phosphorous-32	14.3 days	–	1710
Hydrogen-3(tritium)	12.26 years	*n*	18.6
Carbon-14	5730 years	–	156
Cobalt-57	270 days	122	–
Cobalt-60	5.26 years	1170,1330	313
Caesium-137	30.0 years	660	514,1170
Yterbium-169	32 days	63,177,198	–
Thallium-201	74 hours	167	–

Properties of radiation

Alpha particles

These are helium nuclei: $^4He^{2+}$, mass 4, charge $+2$. They have low penetrating power, being stopped by 1–2 cm air, and by a sheet of paper. They cause intense ionisation, 10^4–10^5 ion pairs per cm in air.

Beta-particles

These are fast-moving electrons, (β^-). They penetrate up to 50 cm in air, and up to 0.5 cm in aluminium. Ionisation in air is 10^2–10^3 ion pairs per cm.

From a collection of identical nuclei (e.g. ^{14}C) β particles with a range of energies are emitted. This leads to the concept of a β-spectrum as shown in Fig. 13.3 for ^{14}C and ^3H. This phenomenon led to the hypothesis that the neutrino carries away the excess energy in those cases when the emitted β-particle has energy less than the maximum value. It is E_{max} values that are given in Table 13.1, so that shielding arrangements can be determined based on the presumption that if the fastest moving electrons can be stopped, all other lower energy β-radiation will also be stopped.

The fact that different radionuclides have different (albeit overlapping) β-spectra enables mixtures of radionuclides (e.g. ^{14}C and ^3H, both of which are of widespread use in tracer and drug metabolism studies) to be measured simultaneously. This is illustrated in Experiment 2.

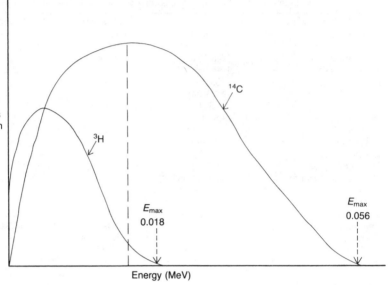

Fig. 13.3. Typical β-spectra

Gamma-radiation

This is electromagnetic radiation similar to light and x-rays but of higher energy (up to 3 MeV) and shorter wavelength.

γ-radiation is very penetrating: for example, 2 inches of steel is required to reduce the flux of 1.17 MeV γ-rays from ^{60}Co to one half the incident value. Because they penetrate through air they have low ionising power, of the order 1–10 ion pairs per cm. Unlike β-radiation, all γ-rays from a collection of identical nuclei have the *same* energy. Thus, γ-spectra appear as in Fig. 13.4.

Gamma-spectra can therefore be used in a similar manner to infrared spectra, i.e. to identify impurities in a sample and to quantify the amounts present (based on peak heights and the use of standards). This is illustrated in Experiment 4.

Radiation protection

The unit of absorbed dose is the **gray** (Gy) defined as the generation of 1 J kg^{-1} (the previously used unit for absorbed dose was the rad, equivalent to 0.01 J kg^{-1}, i.e. 100 rad = 1 gray).

To take account of the greater biological damage caused by α-rays (due to their being absorbed in a very small volume when ingested and causing much ionisation and therefore damage in this small volume) the

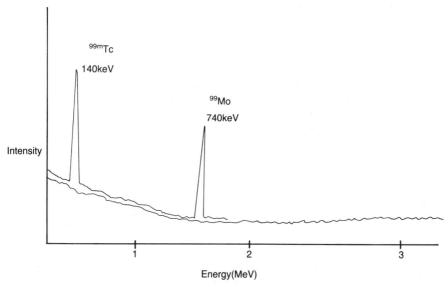

Fig. 13.4. Typical γ-spectra

sievert (Sv) is defined as grays × RBE, where RBE is **Relative Biological Effectiveness** (also known as Quality Factor, QF). RBE has the value of 10 for α-particles, and the value of 1 for β and γ-radiation.

The Ionising Radiations Regulations 1985 (and the associated Approved Code of Practice) which came into force in 1986 require workers to be designated as 'classified workers' if they are likely to receive a dose of ionising radiation which exceeds $\frac{3}{10}$ of the dose limit. This is set at 50 mSv per annum. For members of the general public the dose limit is set at 5 mSv per annum; this is approximately twice the average background radiation dose, but certain locations, e.g. Aberdeen, have background radiations at about the 5 mSv per annum level. Areas have to be designated as 'controlled area' if radiation doses exceeding the $\frac{3}{10}$ dose limit are likely to be received by workers, i.e. 15 mSv per annum (thus for a 40 h week and 50 week working year the dose rate would be 7.5 μSv h^{-1}). Areas where the annual dose is likely to exceed 5 mSv (but not exceed 15 mSv) need to be designated as 'supervised areas'.

Detailed figures are given for hundreds of radionuclides for permitted levels of air and surface contamination as well as levels for notification of occurrences etc. Both the Regulations and the Code of Practice should be consulted. It should be remembered that all uses of radioactive materials must be based on three general principles:

(1) every activity resulting in exposure to ionising radiation shall be justified by the advantages it produces
(2) all exposures shall be kept as low as is reasonably achievable
(3) doses received shall not exceed certain, defined limits.

Practical radiation protection

With α-emitters and low energy β-emitters (e.g. ^{3}H and ^{14}C), **contamination** (rather than radiation) is the major hazard to be guarded against. Thus, regulations within the radiochemical/radiopharmacy area must be such that ingestion, inhalation or contamination cannot occur, i.e. smoking, drinking, eating, pipetting by mouth etc. must be forbidden; gloves must be worn while handling radioactive material; operations must be performed in fume hoods where possible; personnel must wash hands and monitor themselves on leaving the area.

With γ-emitters, in addition to avoiding any form of contamination as outlined above, there is also the radiation **hazard**. The three key factors in minimising the radiation dose received are **time, distance** and **shielding**.

(a) The **time** in performing the operation should be **minimised**; practice with non-radioactive material is invaluable here.

(b) The **distance** between the radioactive source and the operator should be **maximized**, if necessary by the use of forceps and handling tongs. As radiation flux is inversely proportional to the **square** of the distance, large reductions in received dose can be achieved, e.g. if the distance is doubled, the dose received is reduced to $\frac{1}{4}$.

(c) **Shielding**, normally using lead bricks, may also be necessary in order to reduce the dose rate at the operator's body to less than $7.5\ \mu\text{Sv h}^{-1}$.

Basic laboratory organisation

(a) All floor coverings should consist of a single sheet of non-absorbent material; gaps and joins should be avoided; if unavoidable, joins should be sealed with a suitable sealant. An additional coating of a non-absorbent, non-permeable coating, e.g. polyurethane, is an added precaution.

(b) Walls and ceiling should be covered with a non-absorbent, gloss coating, preferably strippable for ease of removal in the event of contamination.

(c) Bench tops should be of non-absorbent material (i.e. Formica and not wood). Joints to walls should be sealed. All edges, e.g. around sinks, should also be sealed.

(d) In a radiochemical laboratory (as opposed to the radiopharmaceutical preparation and dispensing area) fume hoods should be utilised wherever possible. The air flow and venting of such fume hoods (and the possible filtering of the exhaust) requires specialist advice.

(e) If possible, the counting room should be separate from (but adjoining) the radiochemical laboratory; only samples and sources appropriately prepared and contained for counting should be taken into the counting room.

(f) The requirements for the radiopharmacy include all of the above, but, in addition 'good manufacturing practice' and aseptic handling facilities are required, and this needs to be considered with the safety aspects involved in handling relatively large amounts of ^{99m}Tc, a γ-emitter.

Measurement of radioactivity

General points

(a) All measurements of activity must have associated with them the time and date (the reference date/time) at which the measurement is relevant. Radionuclides with longer half-lives need only have the date specified.

(b) A background measurement must always be taken under identical conditions and subtracted from the sample measurement. This is further considered under the heading 'Statistics of counting' later in this section.

(c) The **efficiency** of a counter is defined as:

$$\frac{\text{counts per min}}{\text{disintegrations per min}} \times 100 = \% \text{ efficiency}$$

(d) A **proportional** counter is defined as one where the **output signal** is **proportional** in intensity to the **energy** of the radiation causing that signal.

In general, methods of detecting and measuring radiation may be divided into three categories, depending on the particular property of radiation which is being utilised:

(1) the property of radiation to cause **ionisation** (in a gas)
(2) the property of radiation to cause **scintillation** (fluorescence)
(3) the property of radiation to **effect chemical changes**.

Only methods for the detection and measurement of β and γ-radiation will be considered, as the handling of α-emitting radionuclides requires specialised containment facilities. Methods for the counting of β-radiation will be considered first. ('Counting' is the term generally used to describe the combination of detection and display on a ratemeter or scaler.)

Geiger–Müller counting

This method is based on the ability of radiation to cause ionisation in a gas; an easily ionisable gas such as argon is used, and the schematic construction of a Geiger–Müller tube is shown in Fig. 13.5.

A β-particle enters the Geiger tube through the thin mica end window and causes some argon atoms to become ionised; the argon ions are attracted to the outer negative elecrode by the substantial potential gradient present. Further ions are formed by the movement of these heavy charged ions, and these are in turn attracted to the anode, causing further ionisation to occur. The net result of this 'avalanche'

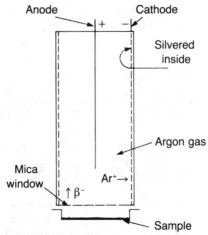

Fig. 13.5. Geiger–Müller counter

effect is to ionise the whole volume of the detector tube; this is equivalent to the flow of a pulse of current which can be amplified readily and presented as an analog or digital output and/or used to give an audio output. Detailed aspects of operation such as recovery and quenching will not be considered here.

The Geiger–Müller tube has the advantage of providing a high output. However, all pulses produced are of the same value irrespective of the energy of the β-particle initiating the avalanche, i.e. it is a **non-proportional** counter. Efficiencies are not high for low energy β-emitters, and it may not be possible to detect ^{3}H. Although the efficiency for γ-radiation is very low ($< 1\%$), it is sufficient, for the use of Geiger–Müller detectors to be practical in, for example, contamination monitoring.

Liquid scintillation counting

The measurement of low energy β-emitters such as ^{3}H and ^{14}C, which have wide application in tracer work, e.g. in drug metabolism, was greatly improved by the introduction of liquid scintillation counting.

In this method the sample is dissolved in a solvent (e.g. toluene or dioxane, but not water) together with some scintillant material; this will be a compound such as diphenyloxazole (PPO). Such a molecule has the property that it can be raised to an excited electronic state from which it can fall to the ground state, emitting in the process visible (or near ultraviolet) light, i.e. fluorescence can occur. The term 'scintillation' is used to cover the whole sequence of events whereby the energy of the fast-moving electron (i.e. the emitted β-particle) is transferred via the solvent to the scintillant molecules, which are thereby raised to the excited electronic level and which then fluoresce with the emission of light which is detected by photomultipliers (very sensitive light detectors). The experimental arrangement is shown in Fig. 13.6.

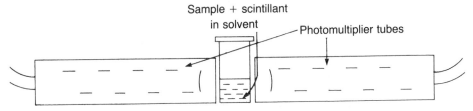

Fig. 13.6. Liquid scintillation counting

Because of the very low levels of light intensity involved, the photomultiliers need to be operated at high sensitivity, with consequent increase in electronic noise generated within the photomultiplier tubes. To reduce this background electronic noise, two photomultipliers are used in the 'coincidence counting' mode, i.e. only output pulses coincident in time are counted. Scintillation events will give coincident output pulses in both photomultipliers, but the random noise pulses developed independently in each photomultiplier will not, in general, be coincident.

The main problems associated with liquid scintillation counting are:

(1) sample preparation
(2) variable quenching.

Sample preparation can be a problem due to the requirement of having a clear (or at least translucent) solution or dispersion in a non-aqueous solvent such as toulene and dioxane. Many methods are used to overcome this problem (e.g. use of detergents); the best source of advice on this matter is probably the booklet produced by Amersham International plc on 'Preparation of Samples for Liquid Scintillation Counting'. Quenching and quench correction are discussed in Experiment 3.

Measurement of gamma-radiation

A solid scintillation process is used here: a single crystal of sodium iodide with a trace impurity level of thallium (about 0.1%) is used, NaI(Tl). Fig. 13.7 shows a cross-sectional diagram of a typical γ-detector. The hygroscopic sodium iodide needs to be coated with a thin film of aluminium for protection against atmospheric moisture.

The sample, in a counting vial, sits in a 'well', a hole drilled into the sodium iodide which will typically be about 2 inches diameter by 2 inches in height. This 'well' minimises the fraction of radiation from the sample which does not pass through the detecting crystal. Interaction of a γ-photon with the high density sodium iodide crystal results in the

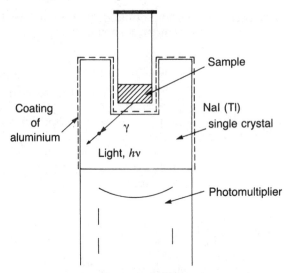

Fig. 13.7. Gamma scintillation detector

production of a 'centre' in an excited electronic level (the thallium plays a role in this process but will not be considered in detail here). Decay back to the ground state results in the emission of visible light which travels through the transparent or translucent single crystal and is detected by the photomultiplier as in liquid scintillation counting. The process is a proportional counting method, and hence gamma energies and gamma spectra can be determined by the process of pulse height analysis (PHA).

Statistics of counting

Radioactive decay is a random process, and so there is a variation in the number of nuclei disintegrating in a unit time interval. If a large number of replicated counts are made on a given sample (with a sufficiently long half-life so that there is no change in the activity over the time scale of the measurements), there will be a variation about the mean. This distribution of counts is strictly a Poisson distribution, which is asymmetric when the mean count, $\bar{x}$, is small (e.g. $\bar{x} < 20$) but which rapidly tends to a normal distribution at high values of $\bar{x}$. For large values of $\bar{x}$ the standard deviation, σ, is given by $\sqrt{\bar{x}}$, i.e.

$$\sigma = \sqrt{\bar{x}}$$

Thus, 68.4% of observations should lie between $\bar{x} \pm \sqrt{\bar{x}}$, 95.5% should lie between $\bar{x} \pm 2\sqrt{\bar{x}}$, and 99.7% should lie between $\bar{x} \pm 3\sqrt{\bar{x}}$. In general, a mean of 10 000 counts should be collected, as this results in the coefficient of variation, $\sigma/\bar{x}$ or $\sqrt{\bar{x}}/\bar{x}$ or 100/10 000 or 0.01, being not greater than 1%.

The count rate for the sample will include a contribution from the background, and this, or course, must be subtracted. The standard deviation of the resultant count will be given by

$$\sigma = (\sigma_s^2 + \sigma_b^2)^{1/2}$$
$$\sigma = (\overline{x}_s + \overline{x}_b)^{1/2}$$

where the subscripts s and b refer to sample and background respectively. It is only when the sample count rate is of the same order of magnitude as the background count rate that σ is affected by the background count. When this is the case, the most efficient distribution of counting time (t) in order to reduce the overall error is given by

$$\frac{t_s}{t_b} = \left(\frac{\overline{x}_s}{\overline{x}_b}\right)^{1/2}$$

This gives the best precision on the estimate of the required $(\overline{x}_s - \overline{x}_b)$.

Paralysis time

When measurements are made at high count rates, there may be a reduction in the indicated count rate as compared to the true rate due to loss by coincidence of pulses due to the finite resolving time of the detector and its associated electronic equipment. With Geiger–Müller detectors the paralysis time (the time during which the detector is effectively not operative following a detection event) may be of the order of 400 μs. Corrections can be made using the following equation:

$$N = \frac{N_0}{1 - N_0 T}$$

where N is true count rate per second; and N_0 is observed count rate per second; and T is paralysis time (in seconds)

Such a correction equation (based on evenly spaced pulses) is only approximate and should only be used when the product $N_0 T$ is small; count rates greater than 10 000 cpm should not be used with Geiger–Müller detectors.

Radiopharmaceuticals and radionuclide generators

Radiopharmaceuticals

The requirements for a radiopharmaceutical can be described under three headings,

1. Properties of the radionuclide

(a) Suitable half-life , i.e. a few hours, such that a reasonable dose can be administered to the patient, counting and/or imaging can be performed within an hour or so with good counting statistics, and the radioactivity will then decay away within a day so that the radiation dose to the patient is minimised. The total radiation dose to the patient is of course determined by a combination of both the physical (nuclear) half-life of the radionuclide and of the biological half-life of the imaging agent. In general very little is known of the metabolism and excretion of technetium scanning agents, and calculations of patient dose are based on the physical half-life.

(b)   Pure gamma emitter of suitable energy . There must be no α and β emissions and the γ-energy must be sufficiently high for the radiation to penetrate out from internal organs, e.g. liver, but not too high to prevent efficient detection. The range 100–200 KeV is optimal.

(c)   The radionuclide must be capable of being produced daily, on a routine basis, in the hospital radiopharmacy (the requirement of a short half-life precludes supply from any distance).

(d)   The radionuclide must be capable of being converted to a range of chemical entities for imaging a range of organs.

The radionuclide ^{99m}Tc fulfils all of the above criteria and is the radionuclide of choice for most radiopharmaceuticals. The half-life of 6 h, and γ-energy of 140 KeV, are optimum; the ^{99}Mo/^{99m}Tc generator allows production in the radiopharmacy, and technetium chemistry is sufficiently versatile (if not fully understood) to allow for a range of imaging agents to be produced.

2. Pharmaceutical properties

Normal requirements for an injection must be applied, e.g. sterility, apyrogeneity, pH, and absence of particulates (or particle size of a particulate preparation).

3. Chemical properties

The correct chemical (or physico-chemical) form is needed to target to the required organ. Pharmacological and toxicological aspects are important, but because of the extremely small masses injected are generally not a problem.

Radionuclide generators

The system which is widely used in hospital radiopharmacies will be used to illustrate the principle radionuclide generators. An understanding of the operation of the ^{99m}Tc generator is essential to the understanding of the quality control and quality assurance of radiopharmaceuticals.

Generators are based on the concept of radioactive equilibrium . If one radionuclide decays to give another radioactive material, then a radioactive equilibrium can be established when the ratio of the two half-lives is appropriate. Consider the following scheme:

$$^{99}\text{Mo} \xrightarrow[66\,\text{h}]{t_{1/2}} {}^{99\text{m}}\text{Tc} \xrightarrow[6\,\text{h}]{t_{1/2}} {}^{99}\text{Tc}$$

Table 13.2 Some radiopharmaceutical preparations in use

Preparation	Application
Na$^{99\text{m}}$TcO$_4$	Thyroid uptake studies
	Brain scanning (block thyroid with perchlorate)
$^{99\text{m}}$Tc-Colloid (50–1000 nm) e.g. tin colloid	Liver and spleen scanning
$^{99\text{m}}$Tc-DTPA (diethylenetetramine pentacetic acid)	Kidney function
	Brain scanning
$^{99\text{m}}$Tc-HIDA	Biliary tract imaging
$^{99\text{m}}$Tc-Macroaggregated Albumin (20–50 μm)	Lung imaging (capillary bed entrappment)
$^{99\text{m}}$Tc-phosphate complexes e.g. —MDP (methyl diphosphonate) —HMDP (hydroxymethylene diphosphonate) —PYP (pyrophosphate)	Bone imaging $^{99\text{m}}$Tc-PYP also used for myocardial infraction imaging, 1–4 days after occurrence
^{201}Thallous Chloride	Myocardial infarction imaging
^{67}Gallium Citrate	Localises in abscesses in kidney, liver
^{169}Yb-DTPA	Cisternography
^{111}In-DTPA	Cisternography

A radioactive equilibrium will be established whereby the rate of formation of $^{99\text{m}}$Tc (from the decay of ^{99}Mo) will be equal to the rate of decay of $^{99\text{m}}$Tc (by isomeric transition to ^{99}Tc).

Initially (with pure ^{99}Mo) there is no $^{99\text{m}}$Tc, but this is formed from the decay of ^{99}Mo and consequently builds up; at the same time, $^{99\text{m}}$Tc is decaying to ^{99}Tc via γ-emission. After about 24 h an equilibrium is established and the activities of both ^{99}Mo (the 'parent') and $^{99\text{m}}$Tc (the 'daughter') appear to decay with the decay constant of the parent ^{99}Mo.

The fact that the parent ^{99}Mo and daughter $^{99\text{m}}$Tc have different chemistry can be utilised to produce the 'generator' which allows the short-lived $^{99\text{m}}$Tc ($t_{1/2}$ 6 h) to be produced daily in the hospital radiopharmacy, a procedure which has revolutionised the practice of nuclear medicine in the last 15 years and has led to the close involvement of pharmacists in the preparation and quality assurance of radiopharmaceuticals.

The ^{99}Mo parent can be irreversibly adsorbed onto an alumina (Al$_2$O$_3$) column as the ^{99}MoO$_4^=$, molybdate ion. The daughter product, in the form of the $^{99\text{m}}$TcO$_4^-$, pertechnetate ion, is not adsorbed strongly to the alumina column. Thus, when radioactive equilibrium has been established, elution of the column with a suitable eluent, such as sterile, isotonic saline (i.e. suitable for intravenous injection) causes the $^{99\text{m}}$Tc (as pertechnetate) to be eluted from the column, while the parent ^{99}Mo remains on the column, and is able to 'generate' more ^{99m}Tc until radioactive equilibrium is again achieved after about 24 h. Thus the generator may be eluted (or 'milked') every day; after a week the parent ^{99}Mo activity will have decayed to approximately $\frac{1}{4}$ of its initial activity and may not be adequate to provide a sufficiently high radioactive concentration of ^{99m}Tc. Hence, hospitals in general will buy a new generator every week. The process of repeated elution and the 'growing in' of more ^{99m}Tc is illustrated in Fig. 13.8.

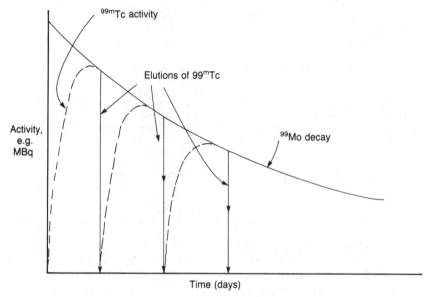

Fig. 13.8. Principle of ^{99m}Tc generator

It is not necessary to wait for radioactive equilibrium to be established before elution is carried out: $\frac{1}{2}$ the equilibrium activity of ^{99m}Tc appears in about 6 h. Elutions more frequent than once per day result in a greater net total output of ^{99m}Tc from a generator, but the radioactive concentrations of such eluents will be lower. The construction of a typical generator is shown in Fig. 13.9 although more sophisticated designs are now available.

Quality control of radiopharmaceuticals

Radionuclidic purity

This is defined as the percentage of the total radioactivity due to the specified radionuclide. Terms which have been used in the past to describe this concept include 'radioisotopic purity' and 'radioactive purity', but these terms should not be used.

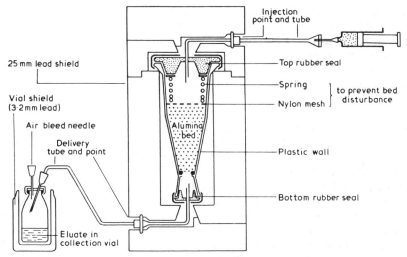

Fig. 13.9. Construction of a typical ^{99m}Tc sterile generator

This concept is readily understood by two examples. Sodium Iodide (^{125}I) should not have ^{126}I (a β-γ emitter with $t_{1/2} = 13$ days) present to an amount greater than 1%. Sodium Pertechnetate (^{99m}Tc) Injections should not have present greater than 0.1% ^{99}Mo (all ratios calculated on the basis of radioactivities at the time of administration).

The standard method of determining radionuclidic purity is by γ-spectrometry, whereby the characteristic γ-ray emissions of radionuclides can be used to (a) detect and (b) quantify impurity levels.

It should be remembered that the radionuclidic purity value will, in general, vary with time because of the differing half-lives of impurity and main constituents.

Experiment 4 illustrates the principles of γ-spectrometry, its use to identify unknown γ-emitting radionuclides, and its use to determine the radionuclidic purity of a sample of Sodium Iodide (^{125}I). If ^{99m}Tc is available, the **rapid test** for ^{99}Mo in ^{99m}Tc is a simple experiment based on the fact that the 140 keV γ-photon of ^{99m}Tc is essentially entirely absorbed by a 6 mm thickness of lead, whereas the more energetic 740 keV γ-photon of ^{99}Mo is essentially not attenuated. If a calibration standard of ^{99}Mo is available, then the method can be made precise; an alternative and more readily available (on account of its long half-life) standard is ^{137}Cs with a γ-photon of 660 keV.

Radiochemical purity

This is defined as the amount of radioactivity in the specified chemical form expressed as a percentage of the total radioactivity. This concept must be clearly distinguished from radionuclidic purity.

For example, a radiochemical impurity in Sodium Iodide (^{125}I) would be iodate (^{125}I) ion (^{125}IO$_3^-$). Here, the radioactivity is associated with a chemical form (iodate) different from the specified chemical form (iodide). Iodate (^{126}IO$_3^-$) would be both a radiochemical and a radionuclidic impurity. Iodate (^{127}IO$_3^-$, the non-radioactive iodine isotope) would be simply a chemical impurity, while iodide (^{127}I) would be termed a 'carrier', i.e. the non-radioactive isotope in the same chemical form.

The radiochemical purity of a radiopharmaceutical (and of any radioactive material) can change (usually decrease) with time, due to radiation-induced decomposition (radiolysis) where H· and ·OH free radicals produced by the interaction of emitted radiation with water attack the radioactive material, producing other chemical species (radiochemical impurities). The very small masses of radioactive materials involved make this problem of significant proportions: care should be taken regarding storage conditions (i.e. solvent, temperature, light; addition of, for example, reducing agent in the case of Sodium Iodide (^{131}I and ^{125}I) solutions to maintain the iodide ion in the reduced state).

Radiochemical purity may be determined by any chemical separation method, and measurement of radioactivity in separated fractions. Chromatographic methods are used most commonly, i.e., thin-layer chromatography, (TLC), paper chromatography, column chromatography and, increasingly in recent years, high performance liquid chromatography (HPLC) with a radiochemical detector. TLC plates and paper chromatograms require scanning with a radiochromatogram scanner to determine the location (and amount) of the radioactive areas. A number of commercial radiochromatogram scanners are available (e.g. Berthold, Panax) and involve moving a gas-flow Geiger–Müller detector tube (with a narrow-slit collimator) over the chromatogram and coupling the output to a recording device which can be correlated to the chromatogram. A typical output trace is shown in Fig. 13.10.

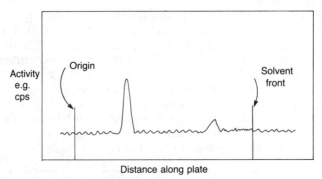

Fig. 13.10. Radiochromatogram scan

Alternatively, the chromatogram strip may be cut (i.e. for paper) into small sections (e.g. 5 mm) and each section counted separately using an appropriate counter and the count rate plotted *vs* position on the strip.

Experiments 7, 8 and 9 illustrate the determination of radiochemical purity using paper chromatography (a cheap and simple technique in the teaching situation).

Radiochemical methods in analysis

Isotope dilution analysis (IDA)

Isotope dilution analysis (IDA) is the homogeneous dilution of a labelled compound of known specific activity with the same unlabelled compound. The application of IDA results from the fact that the compound need not be isolated quantitatively. It is necessary only to isolate a **pure** sample of the compound.

Direct IDA

A known weight of labelled compound of known specific activity is added to a mixture which contains an unknown amount of unlabelled compound.

Let A_0 be the original activity of added labelled compound, M_0 be the original amount of added labelled compound, and S_0 be the original specific activity of added labelled compound. Then

$$S_0 = \frac{A_0}{M_0}$$

Let M_u be the amount of unlabelled compound present (unknown), and S_1 be the specific activity of the final isolated sample. Then

$$S_1 = \frac{A_0}{M_0 + M_u} \quad \text{(assume \textit{whole} sample isolated)}$$

Substituting for A_0 ($A_0 = S_0 M_0$ from above):

$$S_1 = \frac{S_0 M_0}{M_0 + M_u}$$

Solving for M_u:

$$M_u = \left(\frac{S_0}{S_1} - 1\right) M_0$$

Inverse or reverse IDA

A known quantity of non-labelled compound is added to a mixture containing an unknown amount of labelled compound.

In contrast to the direct method, the inverse method allows the addition of a relatively large quantity of unlabelled carrier compound; this facilitates isolation and purification. This is especially useful in **metabolism studies** of labelled drugs.

Experiment 14 illustrates the use of isotope dilution analysis to determine the benzoic acid content of *Camphorated Tincture of Opium* B.P.

Radioimmunoassay (RIA)

The essential principle of radioimmunoassay is that of competition between the labelled and unlabelled compound for binding to a specific antiserum prepared against the compound. The proportion of labelled compound bound to the antiserum is related to the concentration of unlabelled compound (in the sample). The compound–antiserum complex is separated and its radioactivity determined. A calibration curve gives the concentration in the sample.

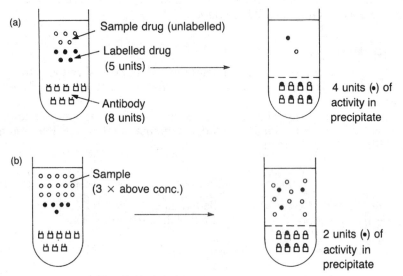

Fig. 13.11. Principle of radioimmunoassay

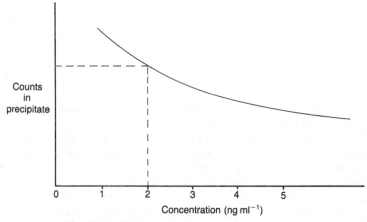

Fig. 13.12. Calibration curve for radioimmunoassay

The various methods of radioimmunoassay differ mainly in the manner in which hormone bound to antiserum is separated from hormone free in solution. Some examples are:

(1) Double Antibody technique, which involves the use of a second antiserum (e.g. guinea pig antiserum prepared in rabbits) to precipitate the hormone–hormone/antiserum complex.

(2) Adsorption(e.g. on charcoal or ion exchange resin) of hormone–hormone/antiserum complex of hormone-free insulin, e.g. digoxin radioimmunoassay.

Figs 13.11 and 13.12 illustrate the principle for insulin RIA.

Neutron activation analysis (NAA)

This uses neutrons to activate nuclides which become radioactive and which can be analysed by γ-spectrometry, e.g.

$$^{75}\text{As} \ (n, \gamma) \quad ^{76}\text{As}$$

normal nuclide of   radioactive nuclide of arsenic
arsenic $t_{1/2} = 1.1$ day

Sources of neutrons can be nuclear reactors, particle accelerators or isotopic neutron sources

The method is **very sensitive**. For example, some detection limits are as follows: As, $0.003 \ \mu g$; Sb, $0.001 \ \mu g$; Ba, $0.01 \ \mu g$.

The method is also **specific**, but there can be interferences, e.g. $^{76}\text{Se}(n,p)^{76}\text{As}$, and $^{79}\text{Br}(n,d)^{76}\text{As}$. Thus, determination of arsenic is uncertain in the presence of larger quantities of bromine and selenium. The method depends on nuclear reactions, which are independent of the atom's chemical situation. This technique is useful for matching of hair, paint samples etc. in forensic science.

Practical experiments

Experiment 1 *Geiger–Müller counting*

Introduction Two considerations are particularly important with G–M counting.

(a) *Geometry*. The closer the sample is to the G–M tube the better, since the probability that a particle will enter the G–M tube will be higher, i.e. the counting efficiency will be higher. At best the counting geometry can be only 2π (as opposed to 4π in liquid scintillation counting, i.e. radiation emitted in all directions is detected) so that the counting efficiency cannot in general be greater than 50%.

(b) *Self absorption*. This occurs when the upper part of the sample absorbs the β-particles before they reach the G–M tube. The ideal sample should be either 'infinitely thin' when no self-absorption occurs, or 'infinitely thick', when β-particles from the lower part of the sample are completely absorbed and the count-rate becomes independent of sample thickness.

Operation of G–M counter (the detailed instructions given here apply to the Panax Low Background G–M Counter). For planchet counting, use the upper slot position. Handle planchets only with tweezers and keep planchets in the plastic boxes (closed) when not being counted to avoid contamination of the counting room and yourself. Remember that β-emitting nuclides are **extremely dangerous if ingested**; there is little hazard from the **radiation** from ^{14}C and ^{3}H.

Determination of the efficiency of the G–M counter for β-radiation

Method

(a) The ^{90}Sr source used must be a 'sealed' source for safety reasons. Use a pre-set time sufficient to give about 10 000 counts. Count this source five or six times and note the statistical variation. As with all counting procedures, determine the background count under identical conditions.

(b) Prepare 2 planchets with 100 μl of a benzoic acid (^{14}C-labelled) solution. Have the infrared lamp (in the fume-hood) about 18 inches above the planchet; when dry, take the planchets (in a closed container) to count on the G–M counter. Also determine the radioactive concentration of this solution by counting 100 μl aliquots (in duplicate) in the liquid scintillation counter. You know the efficiency for ^{14}C counting from the liquid scintillation experiment, and hence the radioactive concentration (expressed in μCi per ml) can be found. This result can be used to determine the efficiency of the G–M counter for ^{14}C β-radiation.

Similar experiments can be carrierd out using solutions of ^{32}P-or ^{3}H-labelled compounds (choose those which are non-volatile).

Experiment 2   *Liquid scintillation counting*

Introduction The prime use of this technique is the counting of low energy β-particles (e.g. from ^{3}H, ^{14}C, ^{35}S), and its development was an important step forward in this field. Low energy β-particles rapidly lose energy in passing through air and other material, and this results in low efficiency when counting with G–M counters. Use of a liquid scintillant, with which the sample can be intimately mixed, minimises losses due to absorption and escape of the β-particles from the detector.

However, due to the small pulses of light produced, high photomultiplier gain is required and this results in high background noise. The use of two photomultipliers counting in coincidence eliminates much of this noise.

Figure 13.3 shows the β-ray spectra of ^{3}H and ^{14}C, and the settings of the level controls at the various points need to be given. The channel width to count a given isotope should be wide enough to include all β-rays of that isotope, but no wider as this would merely increase the background.

There are two parts to the experiment:

(1) Counting of single isotopes, e.g. ^{3}H and ^{14}C and determination of the efficiency of the process.

(2) Determination of ^{3}H and ^{14}C in a mixture of these two isotopes.

Operation of the liquid scintillation counter (detailed instructions required; the following example applies to the Nuclear Enterprises Model 6500).

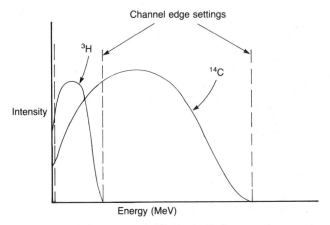

Fig. 13.13. β-spectra: liquid scintillation counting

About 10 min should be allowed for the instrument to warm up after putting on the mains and high voltage switches. After the warm-up period, the H.V. (high voltage) should be adjusted to 1120 V. Readjust to this value at intervals if necessary. Check that the Coincidence switch is at 'IN' position; that the P.M. switch at 'OFF' position (central position of 3-position switch); and that the Delay switch is 'OUT'.

If the liquid scintillation counter is in the 'COUNT' situation then, in order to load the instrument, the handle on the lid must be turned fully clockwise and the 'LOAD' button pushed. This drives the photomultiplier tubes out of the sample compartment into a dark chamber so that they are not exposed to the relatively high light level of the room. (This would damage them as an excessive current would flow.)

When the green 'LOAD' button lights up, the sample compartment lid may be opened (turn handle on lid anticlockwise) and the sample vial inserted (the lead shielding makes the lid heavy). Please ensure that sample vials are *clean* and *dry* before putting them in.

After closing the lid and turning the lid handle fully clockwise, pressing the 'COUNT' button brings the photomultiplier tubes back to the counting position. Counting can start when the yellow 'COUNT' button lights up. Safety interlocks prevent the sample compartment lid being opened.

With the 'Threshold' control set to maximum and the Upper and Lower gate levels set to appropriate values (see later) you can select to count in either Channel 1 (Output switch to '1') or Channel 2 (Output switch to '2').

(1) *Determination of efficiency for* ^3H *and* ^{14}C
The efficiency is equal to counts per min divided by disintegrations per min in the sample:

$$\% \text{ Efficiency} = \frac{\text{cpm}}{\text{dpm}} \times 100$$

Method Prepare, in duplicate, a ^{14}C standard by accurately weighing in a scintillation vial, 2 drops of a *n*-hexadecane-^{14}C standard (available from Amersham International plc) and adding 5 ml of scintillant from the automatic dispenser.

With the channel lower and upper levels set to values so that all ^{14}C β-radiation is included, determine the efficiency of the liquid scintillation counter for ^{14}C.

The background level of radiation, due to cosmic rays, radioactivity in the vial glass or scintillant solution, must be counted under identical conditions and allowed for. A vial containing scintillant only is used (a standard volume of scintillant is used in all cases).

Similarly, determine the efficiency for counting tritium, ^{3}H. A ^{3}H standard must be provided and it must be noted that, as the half-life of ^{3}H is 12.26 years a decay calculation may be necessary. Channel settings need to be chosen for optimum efficiency.

(2) *Determination of the amounts of ^{3}H and ^{14}C in a mixture of these two radionuclides*

It is often advantageous to use doubly labelled compounds in metabolic fate investigations (i.e. one group of a drug molecule can be labelled with ^{3}H and a separate part of the molecule labelled with ^{14}C). In such a case it is necessary to be able to measure quantitatively both ^{3}H and ^{14}C in the same sample. The fact that ^{3}H and ^{14}C have different maximum β-energies (0.018 MeV for ^{3}H and 0.156 MeV for ^{14}C) allows this to be done. This method can also be used to determine mixtures of other β-emitting radionuclides, e.g. ^{99}Tc (E_{max} of 0.285 MeV) with ^{3}H or with ^{14}C.

The calculation will not be discussed in detail; there are various methods, all of which will be acceptable. All are based on the fact that there is an energy region available (above 0.018 MeV and below 0.156 MeV) where only ^{14}C β-particles are present. Thus, by measuring in this region (and comparing with a ^{14}C standard), the ^{14}C content of the mixture may be determined.

The energy region from zero to 0.018 MeV contains contributions from both ^{3}H and ^{14}C; in order to subtract the contribution of ^{14}C to this region, a standard (pure) ^{14}C sample is measured in both regions (i.e. 0–0.018 and 0.018–0.156 MeV) to determine the ratio of counts in these two regions. Hence, as the ^{14}C count in the mixture in the 0.018–0.156 MeV region is known, the contribution which ^{14}C makes to the 0–0.018 MeV region can be calculated and subtracted from the total count in that region, the resultant count being due, of course, to only ^{3}H. The efficiency of ^{3}H in this region will be known from the first part of this experiment, and hence the dpm figure for ^{3}H calculated.

Experiment 3   *Quenching*

Quenching is the name given to the effect whereby the output of light to the photomultipliers is reduced either by interference with the transfer of energy to the scintillant (chemical quenching) or by the absorption of light due to a coloured solution (colour quenching).

Quenching can be corrected for by counting in two channels—the so-called 'Channels Ratio' method (Fig. 13.14). The net effect of quenching is to shift the β-ray spectrum to lower energy values as shown by the curve in the Fig. 13.14. It is clear that the ratio of counts in Channel A to counts in Channel B is increased in the quenched sample.

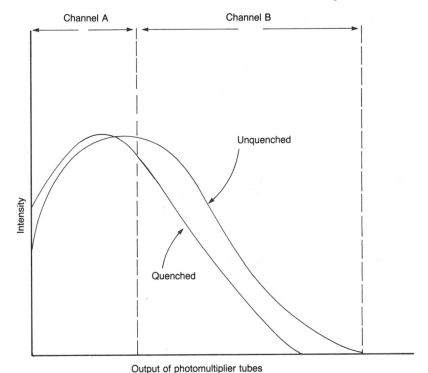

Fig. 13.14. Quench correction by channels ratio method

The efficiency is also decreased (total counts has decreased but dpm figure is unchanged). Thus, by determining the ratio and plotting this *vs* the efficiency, when a sample with unknown quenching is to be counted, the 'channels ratio' is determined and then the efficiency read off from the above plot.

Method Prepare four ^{14}C samples with variable amounts of a quenching agent (chloroform) added (e.g. 0.1, 0.2, 0.3, 0.5 ml). Use these to draw up a 'Channels ratio' curve.

Do colour quenched samples fall on the same curve as the above chemically quenched samples? A convenient approach here is to prepare a ^{14}C standard, add a very small amount of iodine and again measure the channels ratio as above. Further amounts of iodine can be then added progressively to the same vial, the channels ratio being determined at each stage.

A similar pair of experiments can be performed with ^{3}H samples.

Experiment 4 *Determination of radionuclidic purity by gamma-spectroscopy*

Introduction γ-Spectrometry is used to determine the radionuclidic purity of radiopharmaceuticals, e.g. to show that ^{125}I labelled compounds have less than 1% of their activity due to ^{126}I, and that ^{57}Co- and ^{58}Co-labelled compounds have less than 1% of their activity due to ^{60}Co.

Unlike β-radiation, γ-radiation is emitted in narrow energy peaks, and the γ-spectrum of a radionuclide can be used to detect and determine that nuclide, as illustrated in Fig. 13.4. The most common γ-detection method involves the use of a sodium iodide crystal as a scintillation medium. A photomultiplier tube is optically coupled to the sodium iodide crystal and the output of the photomultiplier tube fed to counting and spectrometry equipment.

Each γ-ray produces an output pulse from the photomultiplier tube proportional to its energy, and by measuring only those pulses which are of a certain, narrowly defined height, a γ-spectrum can be built up.

As the precise experimental procedure will depend on the type of γ-spectrometer available, only an outline of the experiment will be given.

(1) *Identification of an unknown radionuclide*
For calibration purposes (if necessary) the following radionuclides are useful:

	γ-peak(MeV)
^{113}Sn	0.39
^{22}Na	0.51
^{54}Mn	0.84

A suitable long-lived ($t_{1/2}$ = 30 year) unknown would be ^{137}Cs (γ peak of 0.66 MeV emitted via the metastable daughter product ^{137}Ba, with a half-life of 2.5 min).

(2) *Qualitative search for radionuclidic impurities in a sample of NaI (^{125}I)*
A sample of pure NaI(^{125}I) standard should be provided in order to show that there are no γ peaks other than the low energy x-rays at 28 and 35 KeV (these may not be resolved and/or may simply appear as a large peak near to the origin or zero energy level).

The presence of a radionuclidic impurity, ^{126}I, which has a relatively short half-life and γ-energy of 0.39 and 0.65 MeV can be simulated for long term use by a mixture of ^{113}Sn ($t_{1/2}$ 118 days and γ-energy of 0.39 MeV) and ^{137}Cs($t_{1/2}$ 30 years and γ-energy of 0.66 MeV). The two peaks of ^{126}I are of approximately the same intensity, and a mixture of ^{113}Sn and ^{137}Cs should be prepared such that the two peaks appear to be of similar intensity on the particular γ-spectrometer being used for the experiment.

Using the following information for the γ-energies in MeV and relative intensities (as given by the % figures in brackets) of the various iodine radioisotopes, it should be possible to identify the impurity in the above 'sample':

^{132}I	: 0.52(22%)	; 0.65(26%)	; 0.67(100%)	; 0.78(84%)
^{131}I	: 0.28(63%)	; 0.36(79%)	; 0.64(9.3%)	
^{130}I	: 0.41(35%)	; 0.53(100%)	; 0.66(100%)	; 0.74(88%)
^{126}I	: 0.39(34%)	; 0.48(5%)	; 0.65(33%)	; 0.75(4%)

(3) *Quantitative determination of the radionuclidic purity of the NaI (^{125}I) sample*

The specification for the radionuclidic purity of NaI(^{125}I) solutions and Injections is that not more than 1% of the total activity is due to ^{126}I.

The basis of this quantitative determination is that the γ- and x-ray spectrum of the sample is measured in comparison with the spectra of standardised solutions of ^{125}I and ^{137}Cs. The relative amounts of ^{125}I and ^{126}I are determined on the assumption that the 0.65 MeV photon of ^{126}I is emitted in 33% of disintegrations, and that the 0.66 MeV γ-photon of ^{137}Cs is emitted in 86% of disintegrations.

Caesium-137 is chosen as the standard for the ^{126}I determination because ^{126}I standard solutions are not available (it also has a short $t_{1/2}$ of 13 days) and ^{137}Cs has a similar γ-photon energy and a long half-life (30 years). Standardised solutions of both ^{125}I and ^{137}Cs need to be provided (a decay calculation may be necessary to determine the activity at the time of the experiment). The important point is that both sample and standard are measured under identical instrumental conditions.

Experiment 5 *Simulated plasma volume determinations*

If a known volume of radioactive solution is added to an unknown volume, and if, after complete mixing, the radioactivity of a similar volume is determined, the decrease in radioactivity is directly proportional to the dilution that has occurred and hence the total volume can be calculated. This method is useful where the volume of liquid cannot be readily determined by other means, e.g. plasma volume.

The tracer radionuclide for measuring plasma volume is attached to human serum albumin, e.g. ^{125}I. The quantity of labelled albumin required is 1–10 μCi, depending on the sensitivity of the equipment and on the volume to be measured. The dose needs to be kept to a minimum for the safety of the patient, consistent with obtaining the required precision for the measured volume.

This simple dilution technique is related to Isotope Dilution Analysis (IDA), where a known weight of radioactive tracer compound of known specific activity is mixed with the sample containing an unknown amount of the inactive compound. If a pure sample of the compound is isolated (the yield is not important), the new, lower specific activity, caused by dilution with the inactive material, enables the amount of inactive material originally present to be calculated.

Notes Radioactive disintegration is a random process. If a large number of observations of the counts in time, t, are made, it is found that the distribution of values obtained corresponds to a Poisson distribution. This is very similar to a normal distribution, provided that the observed counts are not very small in numbers. Such a normal distribution is characterised by its standard deviation: 95% of the observations will be within two standard deviations of the mean.

There is a simple relationship between the number of counts recorded and the standard deviation: the standard deviation, d, the square root of the number of counts, $\bar{n}$

$$d = \sqrt{\bar{n}}$$

where d is the standard deviation, and $\bar{n}$ is the square root of the number of counts Thus, if the counts in time, t, are observed many times, 95% of the values will lie in the range $\bar{n} \pm 2\sqrt{\bar{n}}$. For example, the standard deviation on an observed count of 100 is 10, and the percentage standard deviation is 10%. With repeated counts, 5% of the observations (or 1 in 20) will be below 80 or above 120.

If the number of counts recorded is increased to 10 000, the standard deviation becomes 100 or 1%. This is a useful figure to remember, since there is usually no point in exceeding this number of counts, as other errors such as those of pipetting are then likely to set a limit to accuracy.

It is this factor which is to be applied in this experiment to determine the volume of radioactive solution ($Na^{125}I$) to be used. The volume required is the **minimum** that will give a result with a standard deviation not greater than 1% (assuming that counting statistics are the limiting factor, i.e. that pipetting accuracy is at least as good).

Method Accumulate 10 000 counts from 1 ml of diluted sample in a reasonable counting time, say 10 min. Given that the unknown volume is 5 ± 0.5 l, calculate the minimum volume of the iodide-125 solution required.

It is essential that the counting of the standard solution and the final diluted sample should be carried out under identical conditions, i.e., with identical volumes of liquid in the counting tubes. To directly count 1 ml of the standard solution would result in an extremely high count rate with possible loss in accuracy due to the paralysis time of the gamma counter; therefore, take a smaller volume ($100\,\mu l$) but ensure that the final total volume is the same as the volume you will use for the final diluted sample, say 1 ml. Your background count should be with 1 ml of water. Count your background sample for at least 1000 s. Mixing must, of course, be uniform and complete.

After calculating the unknown volume measure the volume using a measuring cylinder and note how this value compares with that from the radioactive dilution procedure.

Volumes between 2 and 5 l are suitable for this experiment.

Experiment 6   *Preparation of indium ($^{113\,m}In$) chloride from a simple laboratory generator*

The ideal arrangement for the teaching situation is to make available a $^{99\,m}Tc$ generator so that students can have 'hands on' experience of the operation of such a generator and the quality control of the eluate and derived preparations. Unfortunately, the purchase of such generators every few weeks would be prohibitively expensive, and the levels of γ-radiation and radioactivity would be unnecessarily high for a teaching environment. The ideal alternative arrangement is to have a local hospital radiopharmacy provide its spent generators when the activity has decayed to the 100 MBq level or less (i.e. a few mCi).

In the absence of such an arrangement, the ^{113}Sn–^{113m}In generator constructed in the laboratory provides a satisfactory alternative source of the short-lived ^{113m}In($t_{1/2}$ = 99.8 min). The parent ^{113}Sn has a half-life of 118 days so that the generator only requires re-charging every 6–12 months.

Method Set up a shielded glass chromatography column (about 20 × 2 cm) charged with about 20 g chromatography grade silica dispersed in 0.05M hydrochloric acid. Tin-113 (10–20 MBq) in 0.05M hydrochloric acid is adsorbed irreversibly onto the silica column and elution is carried out with 0.05M hydrochloric acid.

The eluate of ^{113m}In (as ^{113m}InCl$_3$) can be used in experiments to determine its half-life and its radiochemical purity and to prepare an In–DTPA complex.

Experiment 7 *Preparation and quality control of indium (^{113m}In) chloride*

All these preparations use 'open' systems, and all final preparations have to be autoclaved.

Method The ^{113m}In generator is a simple chromatographic column with ^{113}Sn adsorbed onto silica. Elution with 0.05M hydrochloric acid yields a carrier-free solution of indium (^{113m}In) chloride. Collect about 10 ml.

Radiochemical purity This can be determined at the same time (on the same plate) as the ^{113m}In–DTPA complex. TLC using 0.2 mm silica gel plates and 0.9% sodium chloride solution as eluent is suitable. (Spot 5 μl in 1 μl portions.) Use an appropriate radiochromatogram scanner to determine/location and hence R_f value of radioactivity.

Radionuclidic purity The determination of ^{113}Sn (the parent radionuclide and most likely radionuclidic impurity) is difficult, as ^{113}Sn emits only a low intensity γ-ray at 0.255 MeV.

Preparation and quality control of ^{113m}In-DTPA complex for brain scanning
Method Use a 20 ml liquid scintillation vial for this preparation.
  (a) Add 1 ml of DTPA (1.6 mg ml^{-1}) in acetic acid (1.02 mg ml^{-1}) into vial.
  (b) Add 5 ml of ^{113m}In eluent into vial.
  (c) Add 1 ml of phosphate buffer into vial.

Radiochemical purity Spot 5 μl (1 μl portions) and run at the same time (on same plate as the ^{113m}In). It is free ^{113m}In that you are looking for in this preparation.

Experiment 8 *Technetium-99 m generator and quality control of the eluate—sodium pertechnetate (^{99m}Tc) injection*

If a low level generator (100 MBq or less) is available, the manufacturer's detailed instructions for elution should be followed.

Radiochemical purity All the radioactivity (^{99m}Tc) should be in the form TcO$_4{}^-$. Use TLC on 0.25 mm silica gel plates and 0.9% sodium chloride solution as eluent. 5 μl can be spotted in 1 μl portions. Scan the plate for activity using, for example, an automatic radiochromatogram scanner which moves the plate under a Geiger tube with a narrow entrance slit.

Radionuclidic purity Record the γ-spectrum using standard conditions depending on the type of γ-spectrometer available. A more sensitive search for long-lived radionuclidic impurities (i.e. the dangerous ones) can be made after the ^{99m}Tc activity has decayed away—say, after 3 days.

Check for ^{99}Mo (the parent radionuclide and the most likely radionuclidic impurity) using a 6 mm lead shield between the sample and the sodium iodide crystal. This thickness of lead absorbs essentially all the 0.14 MeV γ-rays of ^{99m}Tc but does not absorb the higher energy 0.74 MeV γ-rays of ^{99}Mo to an appreciable extent. Compare the count from 1 μCi of ^{99}Mo with the count from 1 mCi of ^{99m}Tc.

Experiment 9 *Radiochemical purity of sodium pertechnetate (^{99}Tc)*

Rather than using the ^{99m}Tc isotope, the 6 h half-life of which is inconveniently short for setting up laboratory experiments, the ^{99}Tc isotope, with a half-life of 2.1 $\times$ 10^5 years can be used to apply the method for radiochemical purity. Although ^{99}Tc is a β-emitter (E_{max} 0.259 MeV), this is readily detected using methods for similar low energy β-emitters such as ^{14}C.

Method Dilute with water to a suitable radioactive concentration and apply 5 μl to a strip of chromatographic paper. Develop using methanol : water (4 : 1) in the bottom of the tank as mobile phase. After drying, detect the areas of radioactivity by a suitable method. Not less than 95% of the total radioactivity is present in the principal spot, which has an R_f value of about 0.6.

Extension With technetium compounds there can be no non-radioactive reference material or carrier, as there are no non-radioactive isotopes of technetium. As well as a radiochemically pure sample of Na^{99}TcO$_4$ (obtainable from Amersham International plc) a sample with some radiochemical impurity can also be prepared by the addition of a small amount of SnCl$_2$ to a sample (so-called 'reduced hydrolysed' technetium species, are produced which, in general, will remain at the origin in the system described above).

The radiochemical purity may be quantified either by measuring the area under the peaks on the radiochromatogram or by cutting out the areas of the radiochromatogram where the radioactivity is found and counting these strips on a Geiger–Müller counter.

Experiment 10 *Radiochemical purity determination of NaI (^{125}I)*

Method Dilute the preparation being examined with water until its activity is appropriate for the radiochromatogram scanner being used. Add an equal volume of a solution containing 0.1% (w/v) potassium iodide, 0.2% (w/v) potassium iodate and 1% (w/v) sodium bicarbonate, mix, place 5 μl of the mixture on a strip of chromatographic paper (Notes 1 and 2) and allow to dry (Note 3). On the same paper, place separate 2 μl aliquots of a 1% (w/v) solution of potassium iodide and a 2% (w/v) solution of potassium iodate and allow to dry. Develop the chromatogram by ascending chromatography. Allow to dry (Note 3) and determine the positions of the inactive potassium iodide and potassium iodate by the application of filter paper impregnated with acetic acid and potassium iodate, and acetic acid and potassium iodide respectively (Note 4). Determine the position of the radioactive spot by a suitable radiochemical method (Notes). *The radioactivity should appear in one spot only, corresponding in R_f value to the potassium iodide.*

Note 1 Draw a very light pencil line at one end of each strip of chromatography paper, about 2 cm from the end, and parallel to the end of the paper. Place all samples on these lines. Apply 5 μl of the active solution plus carriers to one strip, using a Hamilton type syringe. Apply only 1 μl at a time, allowing this to dry under the infrared lamp before applying a further 1 μl aliquot. This procedure is to keep the area of the spot small and hence improve the resolution.

Note 2 Use Whatman 3MM paper with a development solvent of acetone, 65%, and water, 35%.

Note 3 Under the infrared lamp, see Note (a).

Note 4 While the sample strip is on the radiochromatogram scanner, the R_f values of potassium iodide and potassium iodate standards may be determined also by spraying with acidified potassium iodate and acidified potassium iodide respectively. Mark the liberated iodine spot (as it will fade).

Note 5 Preferably using a radiochromatogram scanner; alternatively, the paper may be cut into 5 mm stripes, each of which can be counted separately and the activity plotted *vs* position on the chromatogram.

Experiment 11 *Determination of the half-life of a short-lived radionuclide*

An ideal radionculide for this experiment is $^{113\,m}$In derived from the ^{113}Sn/$^{113\,m}$In generator as described in Experiment 7.

Method Count the sample at known times over a period of hours. A plot of $\ln_e$ (count rate) *vs* time should be a straight line of slope $-\lambda$, from which the half-life can be calculated from the relationship:

$$t_{1/2} = \frac{0.693}{\lambda}$$

Indium-113 m has a half-life of 99.8 min.

Experiment 12 *Determination of a very long half-life*

The natural abundance of ^{40}K is 0.0199%. Eighty-nine per cent of the disintegrations of ^{40}K yield a β-particle (energy 1.32 MeV).

The efficiencies of the G–M counter for these β-particles has to be known.

Using this data (and Avogadro's Number, 6.023×10^{23}) the half-life of ^{40}K can be calculated from the fundamental radioactive decay equation $dN/dt = \lambda N$

Suggested experimental procedure (using a low background anti-coincidence G–M counter). Measure the background with a 1 inch flat, aluminium planchet in position (the longer this is measured for, the more accurate the result—count for at least 400 s).

Re-count (again for at least 400 s) in duplicate with between 100 and 200 mg of AnalaR potassium chloride spread evenly over the planchet.

Experiment 13 *Absorption of β-radiation*

The energies of β-rays emitted by a radioactive decay are distributed from zero to a maximum which is characteristic of the decay. For the identification of an unknown β-emitting isotope it is often necessary to determine by experiment the maximum energy of the β-ray spectrum. The absorption of β-rays is a complicated phenomenon and there is no exact mathematical expression analogous to the exponential law describing the absorption of mono-energetic γ-rays. It is found experimentally, however, that the proportional of β-particles (produced by radioactive decay) which emerges through a given thickness of absorber decreases nearly exponentially with increasing thickness. The absorber thickness is conveniently expressed in mg cm^{-2}, and the results are then almost independent of the nature of the absorber.

Beta absorption curve The determination of the β absorption curve and hence the calculation of the coefficient of mass absorption (μ) is discussed under 'Radiopharmaceutical Preparations' in the British Pharmacopoeia. When aluminium screens of increasing thickness are placed between a β-emitting source and the detector (e.g. G–M tube) the count rate is progressively decreased. The resulting change in count rate depends on the energy of the β-radiation and therefore allows β-emitters to be identified. Fig. 13.15 illustrates the method.

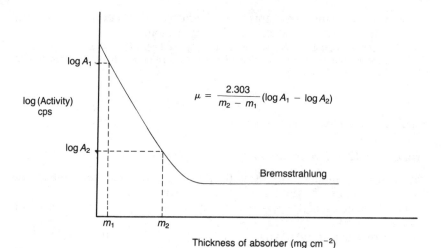

$$\mu = \frac{2.303}{m_2 - m_1}(\log A_1 - \log A_2)$$

Fig. 13.15. β-absorption curve

Method Determine the variation in counting rate with absorber thickness for all the absorbers provided and make one measurement without any absorber in place. Plot the results on semi-log graph paper and make the following observations from the graph.

(a) Verify the roughly exponential nature of the absorption.

(b) Determine the half-thickness. This is defined as the thickness of absorber necessary to reduce the count-rate to half its original value.

(c) Determine the maximum energy of the β-rays in mg cm^{-2}.

(d) Identify the 'tail' due to secondary bremsstrahlung radiation (Note).

Note Bremsstrahlung radiation is the name given to the x-rays produced by collisions of electrons with atoms of the absorber. This x-radiation is more penetrating than the original electrons.

Experiment 14   *Radioactive isotope dilution analysis (IDA)*

This method is very useful for the quantitative determination of a compound in a mixture, when it is not possible to isolate **quantitatively** the pure compound. Using isotope dilution analysis, it is only necessary to isolate a **sample** of the compound in **high purity**. That is, the yield of the isolation or separation procedure is not important, but it is important that the sample of the compound isolated is pure.

The basis of isotope dilution analysis is that, if a known weight of radioactive compound (of known specific activity) is mixed with the inactive compound (in the material to be analysed), the specific activity of the compound is reduced. By determining this new, reduced specific activity, it is possible to calculate the amount of inactive material causing this reduction in specific activity. The theory and equations have been given earlier in this chapter (p.519).

Method Place 50 ml of Camphorated Opium Tincture in a 100 ml separating funnel, acidify with 1 ml concentrated hydrochloric acid and add the ^{14}C labelled benzoic acid solution (5 ml) (pipette and pipette filler). The ^{14}C label is in the carboxylic acid group. Extract with chloroform (2 × 25 ml). Extract the combined chloroform extracts with 0.1M sodium hydroxide (1 × 25 ml). Acidify the alkali extract with 1 ml concentrated hydrochloric acid and extract with chloroform (2 × 20 ml). Evaporate the combined chloroform extract to dryness in a 100 ml beaker on the hot-plate (set below the 60° mark) with a stream of nitrogen directed at the liquid surface. Recrystallise the residue from water—the volume you need will depend on your yield. (Do not add more than 15 ml initially.) Add a small amount of activated charcoal when you are dissolving the residue (to remove final traces of coloured impurities) and filter the hot solution using a warm filter funnel and receiving flask. Wash the beaker and then the filter paper with a small volume of hot water. Crystals of benzoic acid will appear on cooling the filtrate. Filter this crystalline benzoic acid using a Buchner funnel and wash the crystals with *cold* water.

To determine the specific activity of the benzoic acid it is necessary to determine the activity of a given weight.

Take the benzoic acid crystals and dry them in an oven (about 90°) for at least 10 min. *Take care with this dry radioactive material—cover the container* when transferring to the oven from the fume-hood and vice versa. Accurately weigh 10–20 mg of the dried benzoic acid into a scintillation vial (in duplicate), adding the radioactive benzoic acid to the scintillation vial *in the fume-hood.* Replace the cap and then re-weigh. Add 5 ml of scintillation fluid.

Count by liquid scintillation counting. Also take 100 μl aliquots of the initial ^{14}C-benzoic acid solution in order to determine its specific activity: the mass concentration of bezoic acid of this solution must be known.

Experiment 15 *Protein binding of warfarin and its displacement by phenylbutazone*

This experiment uses an ultrafiltration technique to separate albumin (plus bound drugs) from the lower molecular mass drug molecules in solution. 'Amincon' ultrafiltration cones and membranes are suitable, together with a bench top centrifuge.

Solutions ^{14}C-labelled warfarin is used at a concentration of $2.5 \times 10^{-4} \, \mathrm{mol \, dm^{-3}}$. All solutions are prepared in phosphate buffer, pH 7.4 and ionic strength 0.1.

Prepare the following solutions:
warfarin: $2.5 \times 10^{-4} \, \mathrm{mol \, dm^{-3}}$
bovine serum albumin (BSA): $2.5 \times 10^{-4} \, \mathrm{mol \, dm^{-3}}$
phenylbuazone: $1.65 \times 10^{-4} \, \mathrm{mol \, dm^{-3}}$
Prepare the following dilutions of the 'cold' warfarin solution:
$1.0 \times 10^{-5} \, \mathrm{mol \, dm^{-3}}$
$5.00 \times 10^{-6} \, \mathrm{mol \, dm^{-3}}$
$2.5 \times 10^{-6} \, \mathrm{mol \, dm^{-3}}$
$1.0 \times 10^{-6} \, \mathrm{mol \, dm^{-3}}$
$5.0 \times 10^{-7} \, \mathrm{mol \, dm^{-3}}$
To each of 5 ultrafiltration membranes add:

0.5 ml BSA solution
0.1 ml ^{14}C-Warfarin solution
3.4 ml buffer
1.0 ml of each of the 'cold' warfarin dilutions

When determining the displacement of warfarin by phenylbutazone, substitute 1 ml of the phenylbutazone solution for 1 ml of buffer in the above formulations.

Method Take 0.1 ml aliquots of all mixtures for liquid sciltillation counting. Centrifuge the cones and membranes for a time adequate to achieve at least 2–3 ml filtrate. Take 0.1 ml aliquots for counting under the same conditions as the initial 0.1 ml aliquots.

Hence, calculate the *extent* of warfarin binding and its *displacement*. The reciprocal of free warfarin concentration should be plotted against the reciprocal of bound warfarin per mole of albumin (further details can be found in Solomon and Schrogie, *Biochem. Pharmacol.* **16**, 1219 and **17**, 143).

14
The application of spectroscopic techniques to structural elucidation

G. A. SMAIL

Introduction

Two approaches to illustrating the use of four principal spectroscopic techniques (ultraviolet and infrared spectrophotometry, mass spectrometry and proton magnetic resonance spectrometry) are adopted in this chapter. For a number of compounds the molecular formula and relevant spectroscopic data are furnished and in these instances the objective is to deduce the probable structure of each compound. This may be termed the classical application of spectroscopic techniques in structural elucidation, and is the one found in many texts. The second approach is to assign structural features to the spectroscopic data of a compound of known structure. In these examples, since the structure is established, the objective is to fully interpret the spectroscopic data.

The first approach is the one applied in a research setting where a reaction gives rise to an unexpected product, or when a natural product is isolated from a plant or other source, or in the identification of metabolites of drugs. Once the product has been purified, ascertained to be a single entity and not a mixture, and subjected to elemental analysis, then application of a combination of these spectroscopic techniques should permit deduction of at least a partial structure. The second approach would be applied in many settings where the expected structure of a compound is established by the appearance of spectral characteristics totally compatible with the structural features of the molecule.

The British Pharmacopoeia in its monographs makes considerable use of ultraviolet data in identification tests (e.g. *Diphenoxylate Hydrochloride*). Similarly, in many monographs the infrared (e.g. *Amiloride Hydrochloride*) and nuclear magnetic resonance (e.g. *Dexamethasone Sodium Phosphate*) spectral data of Pharmacopoeial substances are required to be concordant with those of the pure reference compound.

Aids to spectral interpretation

Wherever possible, the assignment of structural features to spectroscopic data has been cross-referenced to appropriate chapters and pages of this textbook. Fuller explanations of individual points must, however, where necessary, be sought by reference to the specialized sources cited in the individual chapters relating to each principal spectroscopic technique (Chapters 6 to 12).

The following generalizations apply to the presentation and interpretation of the spectral data in this chapter.

Ultraviolet (UV) spectra

Wavelengths of maximal absorption obtained in a particular solvent are quoted in nanometres ($\lambda_{max}^{solvent}$ nm). The absorptivity at maxima is quoted either as $A_{1cm}^{1\%}$ or ε, and conversion of one method of expression to the other should be practised (Chapter 7, p.276). Bathochromic or hypsochromic shifts with associated hyperchromic or hypochromic effects which occur with changes of pH are usually indicative of auxochromes (e.g. NH_2, OH, COOH groups) attached to the chromophoric system (p.320).

Infrared (IR) spectra

The infrared spectra of compounds liquid at ambient temperature were prepared from capillary films of the sample on sodium chloride plates (Chapter 10, p.394). Solid compounds were examined using the potassium chloride disc technique (p.399). The principal absorption bands (v_{max}) are given in wavenumbers (cm^{-1}). Only those bands which are diagnostic of a particular structural feature or which can be assigned with certainty are quoted. Reproduction of full infrared spectra has deliberately been omitted to prevent the common temptation to overinterpret spectra and to assign molecular features to all bands.

The interpretation of the infrared data should rely, in the first instance, on the charts in Chapter 10 which correlate band frequency with molecular structure; s = strong intensity, m = medium intensity and w = weak intensity.

Mass spectra (MS)

The quoted mass spectra data were derived from a double-focussing high resolution mass spectrometer. The spectra were generated by electron impact (70 eV). The particular instrument used gives a computer print of the molecular weight of each ion by peak matching (Chapter 12, p.482) and, by providing the computer with the elemental composition of the parent compound, the possible elemental composition of each ion (Table 14.1). Operation of the instrument in this mode does not give metastable ions (p.484), and hence when fragmentation pathways are given in the solution to the structure of a compound the mechanisms are plausible (largely by literature analogy) rather than proven.

Relative abundances of ions are given as percentages of the base

Table 14.1 Calculation of high resolution mass spectrum

Found	Theory	Spectrum Reference No 9748 Run No 1							Error	Area
		C	H	13C	N	O	35Cl	37Cl		
186.0280	186.0232	6	7	2	1	2	1	0	25.5	0.0
	186.0329	7	8	1	2	1	0	1	-26.6	0.0
	186.0284	6	7	2	2	1	0	1	-2.5	0.0
	186.0248	7	7	0	3	1	0	1	17.0	0.0
	186.0203	6	6	1	3	1	0	1	41.0	0.0
	186.0262	9	9	0	0	2	0	1	9.8	0.0
	186.0217	8	8	1	0	2	0	1	33.8	0.0
	186.0334	4	11	1	3	0	1	1	-29.2	0.0
	186.0289	3	10	2	3	0	1	1	-5.1	0.0
	186.0347	6	13	1	0	1	1	1	-36.4	0.0
	186.0303	5	12	2	0	1	1	1	-12.4	0.0
	186.0266	6	12	0	1	1	1	1	7.2	0.0
	186.0222	5	11	1	1	1	1	1	31.2	0.0
186.0117	186.0092	12	0	0	3	0	0	0	13.3	18.9
	186.0106	14	2	0	0	1	0	0	6.1	18.9
	186.0111	11	5	0	1	0	1	0	3.5	18.9
	186.0196	7	7	0	2	2	1	0	-42.5	18.9
	186.0070	6	5	0	3	2	1	0	25.1	18.9
	186.0050	12	5	0	0	0	1	1	35.9	18.9
	186.0136	8	7	0	1	2	0	1	-10.1	18.9
	186.0141	5	10	0	2	1	1	1	-12.7	18.9

peak, which is given the value 100%. The relative abundances of molecular and fragment ions containing the elements chlorine or sulphur have, however, been idealized to permit ready interpretation of the data.

Proton magnetic resonance (PMR) spectra

The proton magnetic resonance spectra, unless otherwise stated, were obtained on a 90 MHz instrument using tetramethylsilane as internal standard (Chapter 11, p.430). The instrument was operated in the field-locked mode and thus the resonance at $\delta 0$ given by the internal ·standard (Fig. 14.A) is replaced by a beat pattern. The chemical shift (δ) of a proton or group of protons is readily measured on the spectra, since the x-axis of the chart paper is calibrated in δ units. Protons with a chemical shift greater than 10 are displayed in an offset spectrum of the region $\delta 10$–15. The area under each peak (or group of peaks for a multiplet) is proportional to the number of protons giving rise to the resonance signal. These areas are measured by the integration unit of the instrument, the height of the integral step (Chapter 11, Fig.414) being proportional to the number of protons responsible for each peak (or group of peaks).

An attempt should be made to assign each resonance signal or group of signals in the spectrum to possible functional groups, using the correlation charts (Chapter 11, Figs. 11.8–11.11) and using the value of the coupling constant (J; p.432) in spin multiplets (doublets, triplets, quartets etc.) to gain information on the relative positions of the coupled protons within the molecule. The 90 MHz PMR spectrum of ethanol containing a trace of mineral acid (Fig. 14.A) demonstrates the essential features of the measurement of integrals and coupling constants (a set of dividers is useful).

The following abbreviations are used for the PMR data: s = singlet; d = doublet; t = triplet; q = quartet; m = multiplet; dd = double doublet; br = broad. In the solutions the integration and the protons responsible for a particular resonance signal are shown respectively in brackets and underlined, e.g. (2H, —CH$_2$CH$_2$OH). Coupling constants (J) are quoted in hertz (Hz).

Exercises

Compounds 1a–d

Figures 14.1(a)–(d) give the PMR spectra of the four isomeric butanols (C_4H_9OH). The spectra were derived from 50% v/v solutions of each alcohol in CDCl$_3$, and the solutions contain a trace of acid to prevent coupling of the hydroxyl proton (Chapter 11, Figs. 11.18(a) and (b)).

Objective: assign each spectrum to one of the isomers. (One way to approach this problem is to predict the spectrum of each isomer and match it with one of the authentic spectra.)

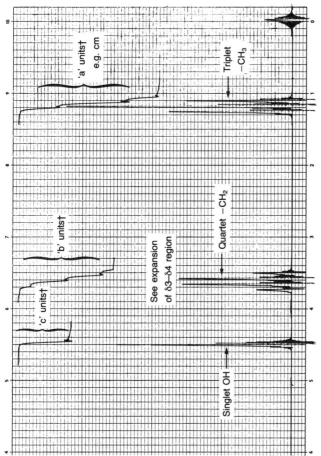

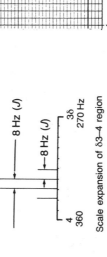

Fig. 14.A The 90 MHz spectrum of ethanol showing measurement of integrations and a coupling constant (J) in hertz (Hz). *At 90 MHz each δ unit equals 90 Hz; at 250 MHz each δ equals 250 Hz. †The 'height' of an integral can be measured with a rule in any convenient unit (e.g. cm); a : b : c = 3 : 2 : 1.

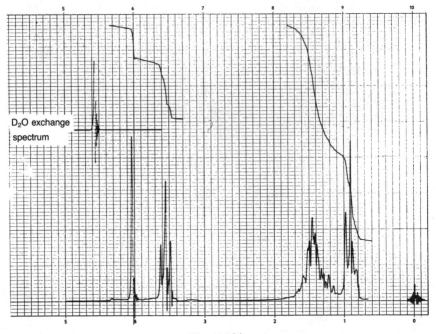

Fig. 14.1(a).

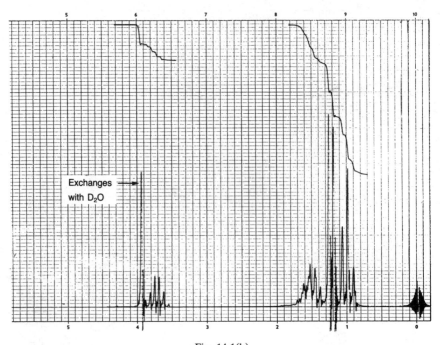

Fig. 14.1(b).

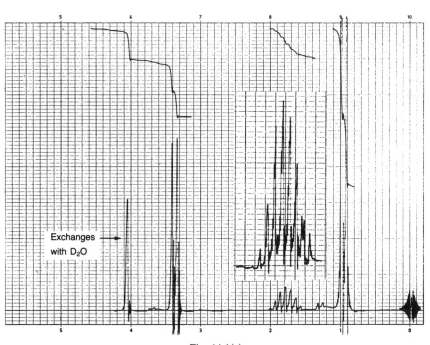

Fig. 14.1(c).

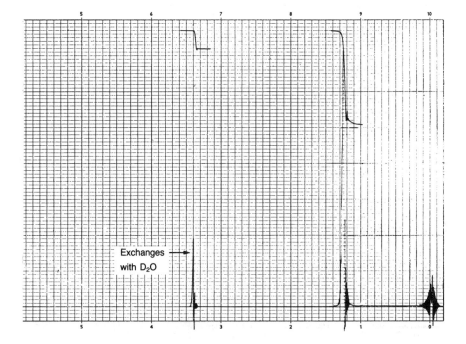

Fig. 14.1(d).

Compound 2

Description: colourless liquid, b.p. 80°
Molecular formula: C_4H_8O
IR, v_{max}^{film} (cm^{-1}): 1715 (s)
PMR (neat liquid): Fig. 14.2
Objective: deduce the structure of compound 2 and predict the principal ions in its electron impact mass spectrum

Compound 3

Description: colourless liquid, fruity odour, b.p. 77°
Molecular formula: $C_4H_8O_2$
IR, v_{max}^{film} (cm^{-1}): 1740 (s), 1240 (m)
PMR (neat liquid): Fig. 14.3
Objective: deduce the structure of compound 3

Compound 4

Description: colourless liquid, ammoniacal odour, b.p. 68–69°
Molecular formula: $C_4H_{11}N$
IR, v_{max}^{film} (cm^{-1}): 3330 (m), 3250 (m), 2960 (s), 1390 (m), 1365 (m)
MS, m/z: 73 (M$^+$), 30 (base peak)
PMR (neat liquid): Fig. 14.4
Objective: deduce the structure of compound 4

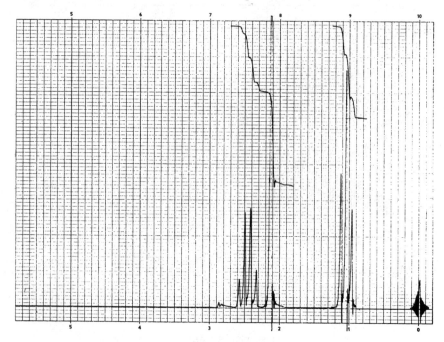

Fig. 14.2.

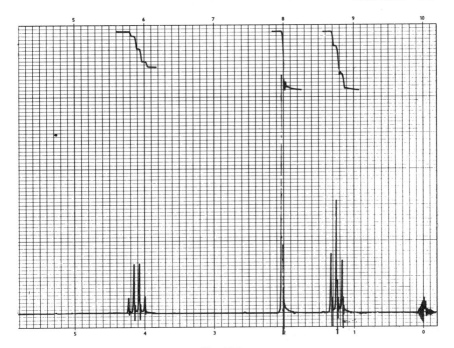

Fig. 14.3.

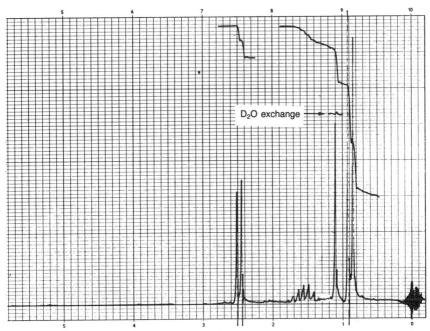

Fig. 14.4.

Compound 5

Description: colourless liquid, b.p. 211–213°
Molecular formula: $C_9H_{10}O_2$
IR, v_{max}^{film} (cm^{-1}): 1720 (s), 1602 (m), 1581 (m), 1270 (s), 1105 (s)
PMR (neat liquid): Fig. 14.5
Objective: deduce the structure of compound 5

Compound 6

Description: oily liquid, b.p. 248°
IR, v_{max}^{film} (cm^{-1}): 1695 (s), 1600 (m), 1580 (m), 1505 (m)
MS, m/z: 136 (M$^+$, 71%), 135 (100%)
PMR: Fig. 14.6
Objective: deduce the structure of compound 6

Compound 7

Description: white crystalline solid: m.p. 88–90°
Molecular formula: $C_9H_{11}NO_2$
UV, λ_{max} (nm): in 0.1M NaOH 286 (A$_{1\,cm}^{1\%}$ 1300); in H$_2$O 286 (A$_{1\,cm}^{1\%}$ 1300); in 2M HCl 225 (A$_{1\,cm}^{1\%}$ 800), 271 (A$_{1\,cm}^{1\%}$ 100)
IR, v_{max}^{KCl} (cm^{-1}): 3410 (m), 3340 (m), 3200 (m), 1675 (s), 1590 (s), 1570 (m), 1505 (m), 1280 (s)
PMR (CDCl$_3$): Fig. 14.7
Objective: deduce the structure of compound 7

Compound 8

Description: yellowish oily liquid, b.p. 246° (some decomposition)
Molecular formula: C_9H_8O
UV, $\lambda_{max}^{ethanol}$ (nm): 220 (ε 12,500), 285 (ε 25,000)
IR, v_{max}^{film} (cm^{-1}): 1675 (s), 1612 (s), 1601 (m), 1575 (m), 1492 (w), 755 (s), 695 (s)
MS, m/z: 132 (M$^+$, 66%), 131 (100%), 103 (46%), 77 (65%), 51 (45%)
PMR (CDCl$_3$): Fig. 14.8
Objective: deduce the structure and stereochemistry of compound 8

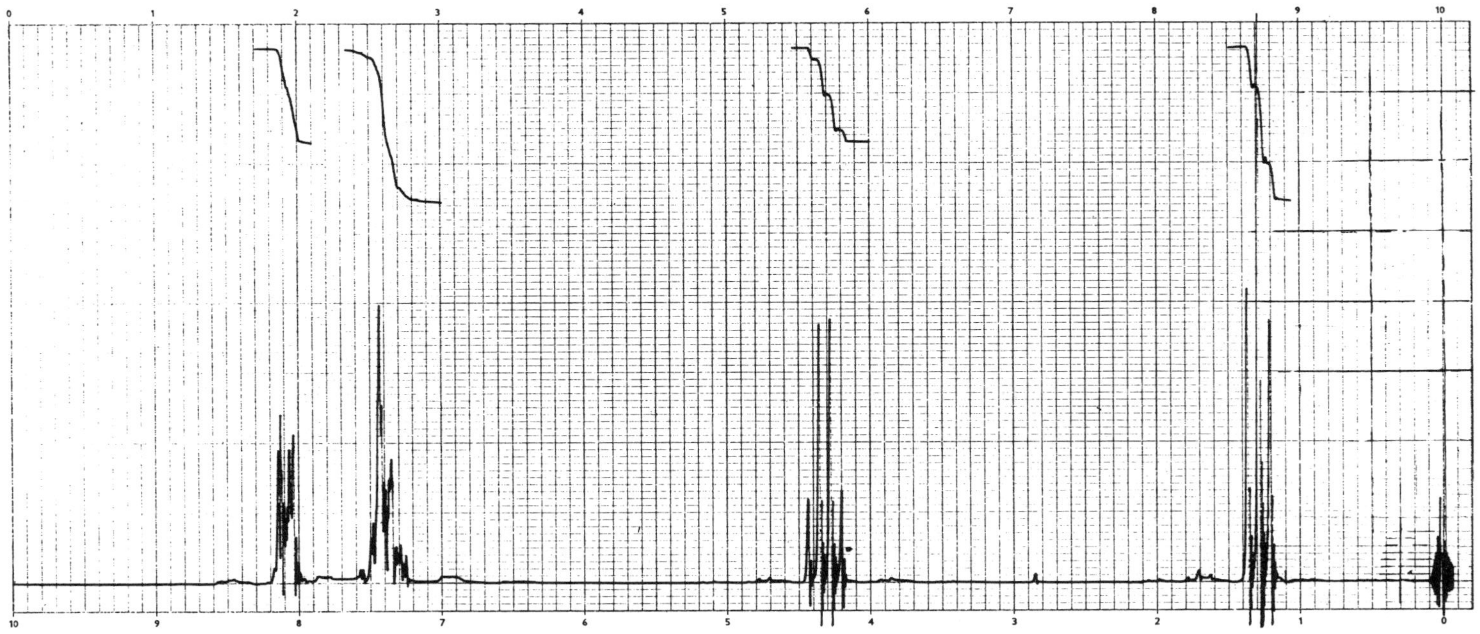

Fig. 14.5.

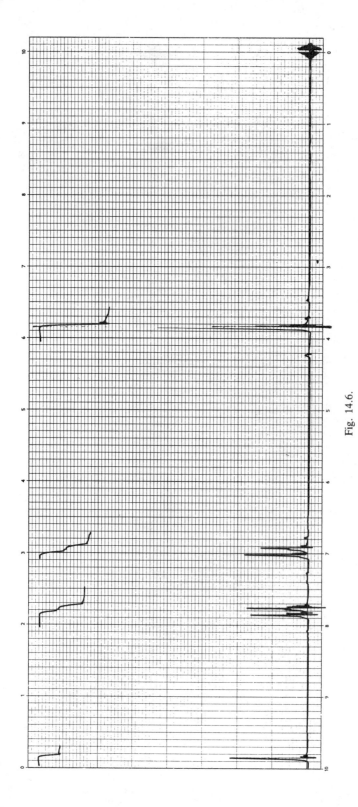

Fig. 14.6.

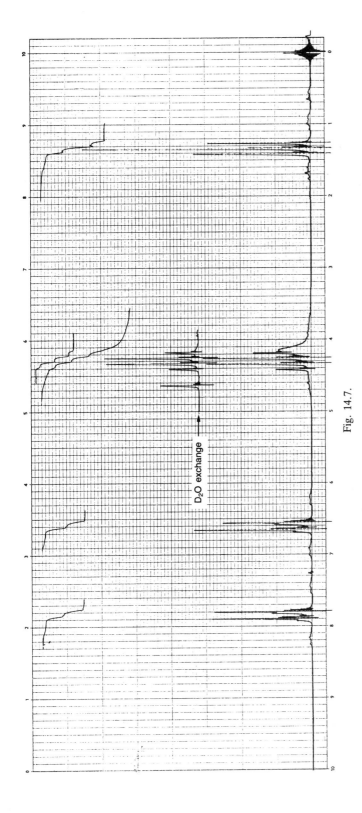

Fig. 14.7.

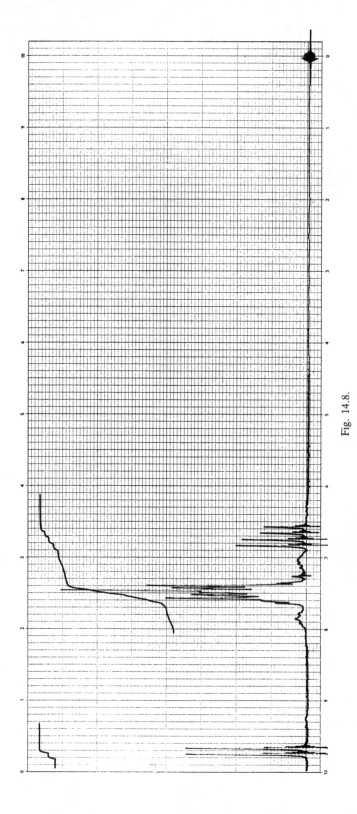

Fig. 14.8.

Compound 9

Description: colourless crystalline solid, m.p. 138–140°
Molecular formula: $C_9H_8O_4$
UV, $\lambda_{max}^{0.05M\ H_2SO_4}$ (nm): 229 ($A_{1\,cm}^{1\%}$ 484), 276 ($A_{1\,cm}^{1\%}$ 66)
IR, v_{max}^{KCl} (cm^{-1}): 3000–2500 (m; series of small ragged bands), 1750 (s),
    1690 (s), 1600 (s), 1575 (w), 1480 (s)
MS, *m/z*: 180 (M$^+$, 7%), 138 (65% $C_7H_6O_3$), 120 (100%, $C_7H_4O_2$), 92
    (16%, C_6H_4O)
PMR (CDCl$_3$): Figs. 14.9 and 14.9(a)
Objective: deduce the structure of compound 9

Compound 10   **Propyl-*p*-hydroxybenzoate**

UV, λ_{max} (nm): in water 253 ($A_{1\,cm}^{1\%}$ 380); in 0.1M NaOH 294 ($A_{1\,cm}^{1\%}$
    1204)
IR, v_{max}^{KCl} (cm^{-1}): 3250 (s, broad), 1670 (s), 1601 (s), 1585 (s), 1505 (m),
    1280 (s)
MS, *m/z*: 180 (39%, $C_{10}H_{12}O_3$), 138 (86%, $C_7H_6O_3$), 121 (100%,
    $C_7H_5O_2$), 93 (49%, C_6H_5O)
PMR (CDCl$_3$): Fig. 14.10
Objective: assign structural features to the spectroscopic data

Compound 11   **Ibuprofen**

IR, v_{max}^{KCl} (cm^{-1}): 3000–2600 (s, broad series of ragged bands), 1709 (s)
MS, *m/z*: 206 (63%, $C_{13}H_{18}O_2$), 163 (90%, $C_{10}H_{11}O_2$), 161 (100%,
    $C_{12}H_{17}$), 119 (73%, C_9H_{11}), 91 (82%, C_7H_7)
PMR (CDCl$_3$): Fig. 14.11
Objective: assign structural features to the spectroscopic data

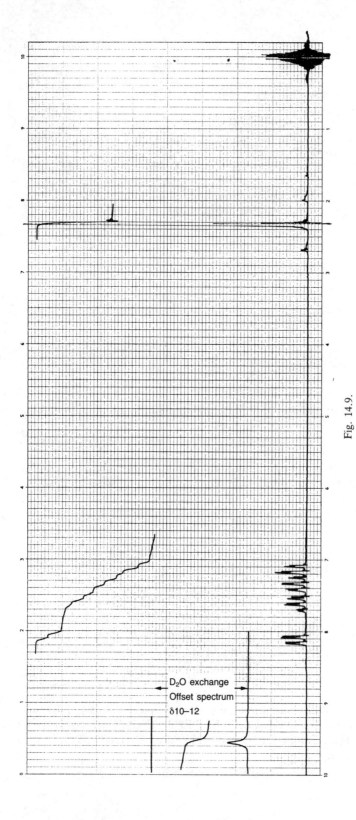

D₂O exchange

Offset spectrum

δ10–12

Fig. 14.9.

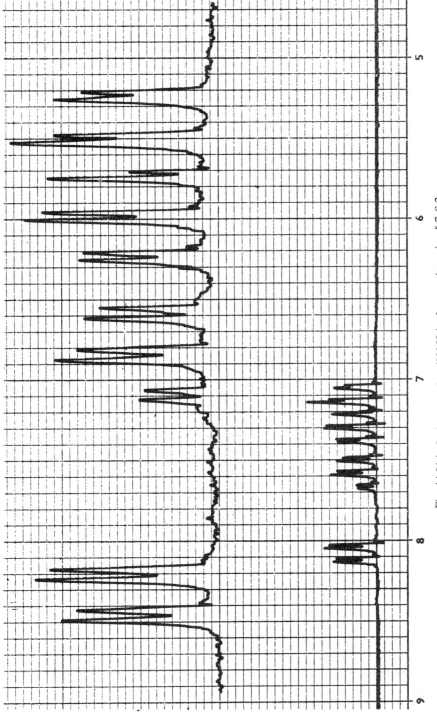

Fig. 14.9(a). Scale expansion (300 Hz) of aromatic region δ 7–8.2.

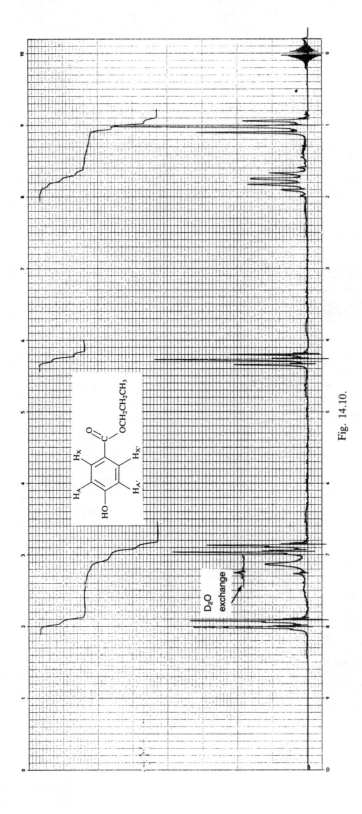

Fig. 14.10.

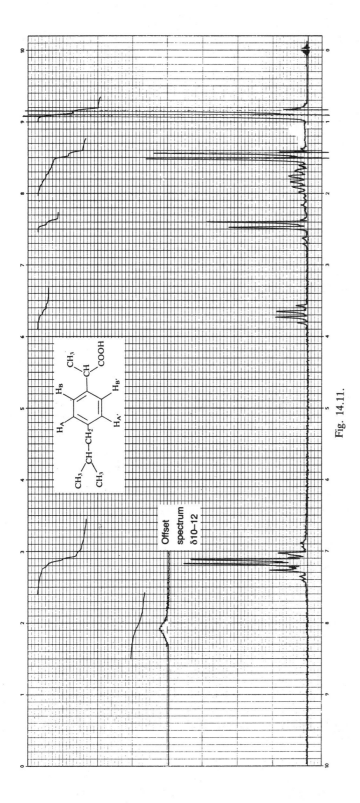

Fig. 14.11.

Compound 12   Clofibrate

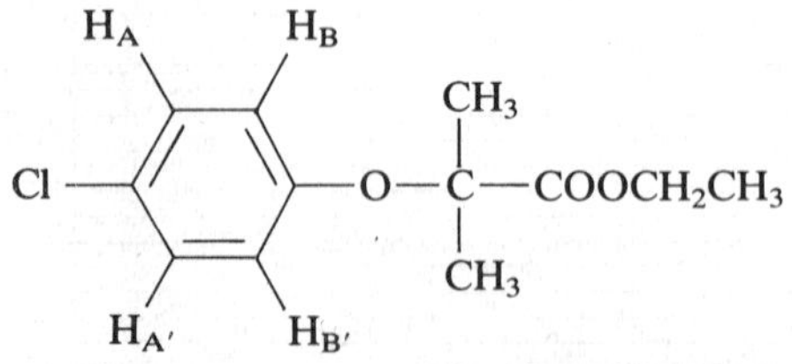

UV, λ_{max}^{EtOH} (nm): 226 (ε 11,700), 275 (ε 1700)

IR, ν_{max}^{KCl} (cm^{-1}): 2990 (s), 2960 (m), 1740 (s), 1595 (m), 1580 (w), 1485 (s), 1290 (s), 1245 (s), 1185 (s), 1140 (s)

MS, m/z: 244 (33%), 242 (91%), 171 (36%), 169 (100%)

PMR (CDCl$_3$): Fig. 14.12

Objective: fully interpret the spectroscopic data

Compound 13   Ethacrynic acid

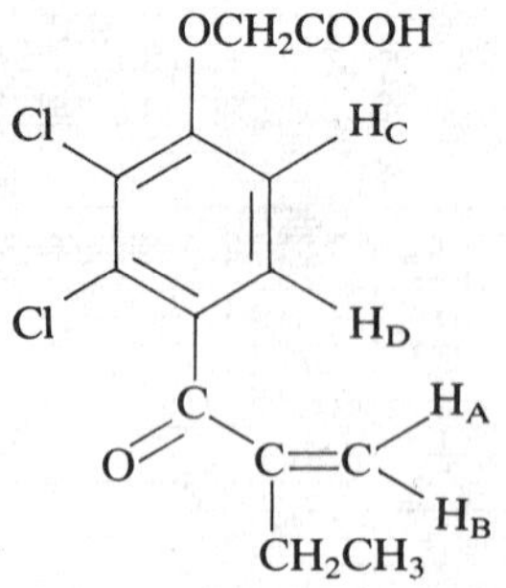

UV, $\lambda_{max}^{0.1M\ NaOH}$ (nm): 227 (A$_{1\,cm}^{1\%}$ 470), 280 (A$_{1\,cm}^{1\%}$ 150)

IR, ν_{max}^{KCl} cm^{-1}: 3000–2500 (m, broad ragged band), 1710 (s), 1670 (s), 1658 (s), 1250 (s), 1080 (s)

MS, m/z: 306 (3%), 304 (11%), 302 (19%), 250 (9%), 248 (44%), 246 (64%)

PMR (CDCl$_3$): Fig. 14.13

Objective: assign structural features to the spectroscopic data

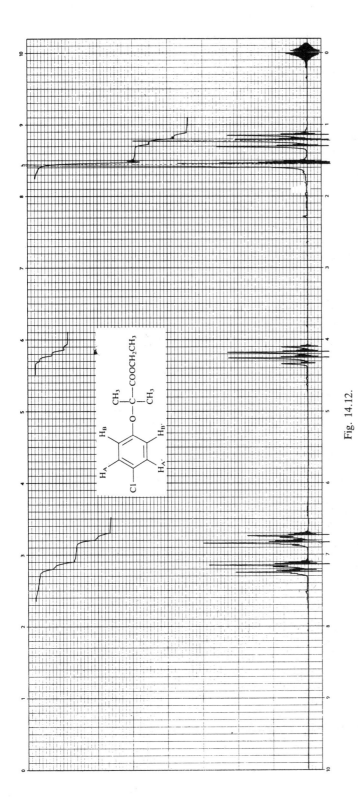

Fig. 14.12.

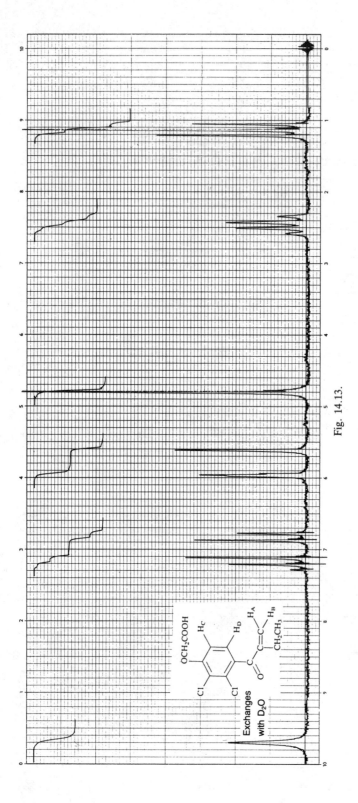

Fig. 14.13.

Compound 14   Indomethacin

UV, λ_{max}^{EtOH} (nm): 208, 235, 325
IR, ν_{max}^{KCl} (cm^{-1}): 3000–2500 (m, series of ragged bands), 1715 (s),
                   1685 (s), remainder of spectrum very complex
MS, *m/z*: 359 (33%), 357 (100%), 314 (5.6%), 312 (17.3%)
PMR (CDCl$_3$): Fig. 14.14
Objective: fully interpret the spectroscopic data

Compound 15   Carbimazole

UV, $\lambda_{max}^{0.1M\ HCl}$ (nm): 227 (ε 7600), 291 (ε 10,360)
IR, ν_{max}^{KCl} (cm^{-1}): 3190 (w), 3150 (w), 3120 (w), 2960 (w), 2940 (w),
                   1750 (s), 1705 (m)
MS, *m/z*: 188 (4.4%), 186 (100%), 116 (4%), 114 (96%), 115 (2.7%),
          113 (62.5%)
PMR (CDCl$_3$): Fig. 14.15
Objective: fully interpret the given spectral data

Compounds 16 and 17

These two compounds have been chosen to illustrate the application of high resolution nuclear magnetic resonance spectrometry to structural elucidation. In both examples the 90 MHz spectra are largely interpretable, but the spectra obtained from a higher frequency instrument (250 MHz) permit finer points of interpretation and accurate measurement of coupling constants.

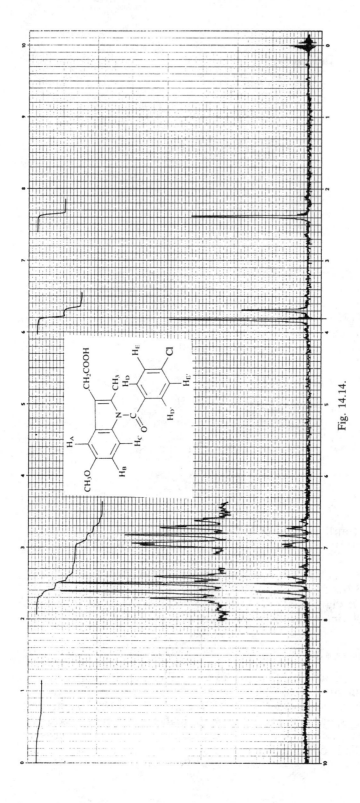

Fig. 14.14.

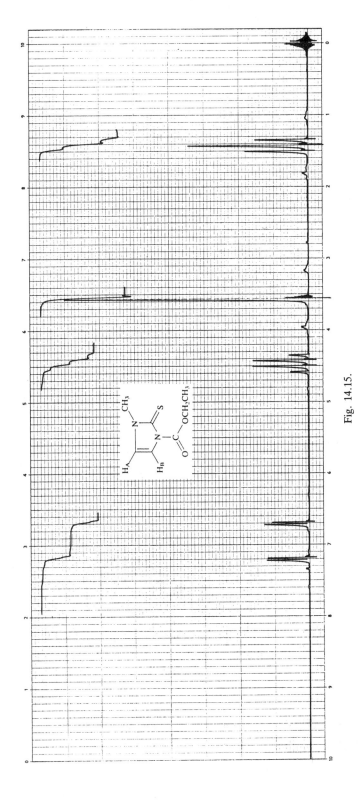

Fig. 14.15.

Compound 16   Crotamiton

PMR(CDCl$_3$): (a) 90 MHz, Fig. 14.16
       (b) 250 MHz,   Fig.   14.16(a);   scale   expansion
       (20 Hz cm^{-1}) 1360–1840 Hz (δ 5.44–7.36), Figs.
       14.16(b) and (c)
Objective: assign structural features to the spectra

Compound 17   Mexenone

PMR (CDCl$_3$): (a) 90 MHz, Fig. 14.17
       (b) 250 MHz, scale expansion (20 Hz cm^{-1}) 1590–
       1900 Hz (δ 6.36–7.6), Figs. 14.17(a) and (b)
Objective: assign structural features to the spectral data

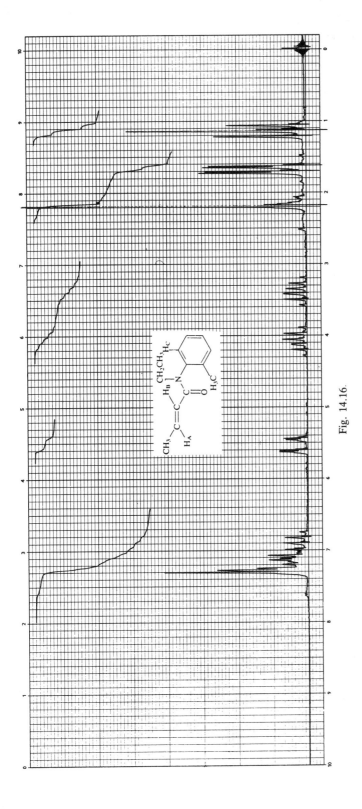

Fig. 14.16.

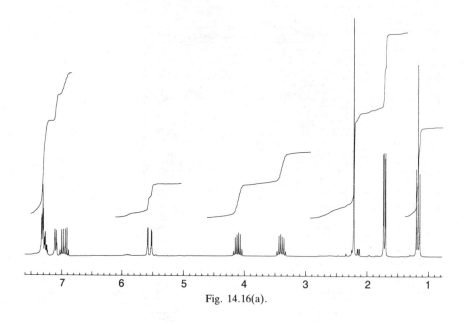

Fig. 14.16(a).

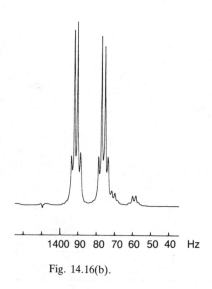

Fig. 14.16(b).

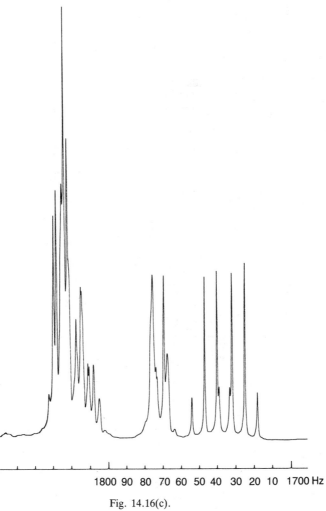

1800 90 80 70 60 50 40 30 20 10 1700 Hz

Fig. 14.16(c).

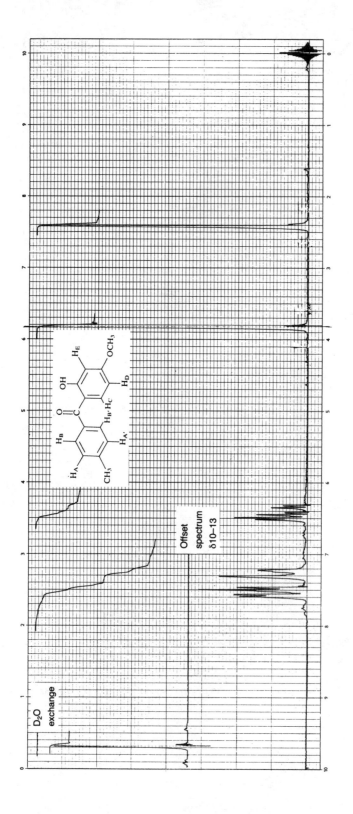

D₂O
exchange

Offset
spectrum
δ10–13

Fig. 14.17.

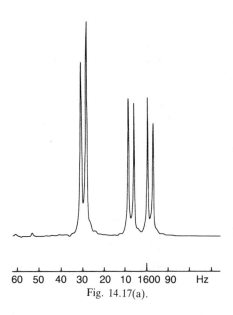

60 50 40 30 20 10 1600 90   Hz

Fig. 14.17(a).

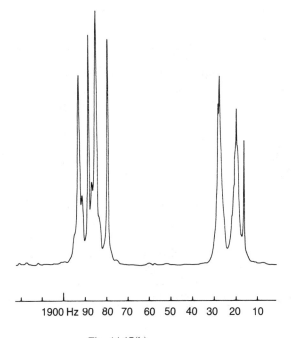

1900 Hz 90 80 70 60 50 40 30 20 10

Fig. 14.17(b).

Solutions

Compound 1a   *(Fig. 14.1(a))*

The signal at $\delta 4.03$ exchanges with D_2O, and in the superimposed exchange spectrum the water (HOD, HOH; Chapter 11, p.436 and 441) arising from the exchange appears at $\delta 4.56$. The peak at $\delta 4.03$ is therefore assigned to the O<u>H</u> group of the alcohol and the integral step at $\delta 4.03$ thus represents one proton. From the remaining integral steps it follows that the triplet centred at $\delta 3.55$ arises from two protons, the multiplet at $\delta 1.45$ from four protons and the signal at $\delta 0.9$ from three protons.

The high field resonances ($\delta 0.9$ and 1.45) are characteristic of saturated hydrocarbon entities, alkyl groups in this case. The lower field signal ($\delta 3.55$) must therefore arise from two protons (**C**) on a carbon atom which is adjacent to an electronegative atom (p.421), a —C<u>H$_2$</u>OH moiety in this instance. Since these methylene protons appear as a triplet, and are not coupled to the hydroxyl proton, there must be two protons ($n + 1 = 3$, therefore $n = 2$, p.434) on the adjacent carbon atom which indicates the presence of the entity —CH_2CH_2OH. It therefore follows that Fig. 14.1(a) is the PMR spectrum of *n*-butanol (1a).

$$\overset{\textbf{A}}{CH_3} - \overset{\textbf{B}}{CH_2} - \overset{\textbf{C}}{CH_2} - CH_2 - OH$$
$$(1a)$$

Comparison of structure (1a) with that on p.434 predicts that the **B** protons will appear as a pattern with a maximum of nine lines and the **A** protons with a maximum of twelve lines $[(3 + 1) (2 + 1)]$. Band overlap ($J_{AB} \approx J_{BC}$) produces the four-proton multiplet at $\delta 1.45$. The terminal methyl group would be expected as a $1 : 2 : 1$ triplet ($n + 1 = 3$) but the pattern observed is frequently shown by the terminal methyl of higher straight-chain alkyl groups.

Compound 1b   *(Fig. 14.1(b))*

The sharp singlet at $\delta 3.94$ which exchanges with D_2O is the proton of the O<u>H</u> group of the alcohol. The sextet centred at $\delta 3.72$ also integrates for one proton. The multiplet centred at $\delta 1.5$ therefore arises from two protons, the doublet at $\delta 1.22$ from three protons, and the triplet at $\delta 0.98$ from three protons. The chemical shifts of the doublet and triplet indicate that both sets of resonances arise from methyl groups, one of which is adjacent to a methine group (giving the doublet) and the other adjacent to a methylene group (giving the triplet). The only isomer compatible with these structural features is *s*-butanol (butan-2-ol) (1b).

$$CH_3 \overset{\text{A}}{—} CH_2 \overset{\text{B}}{—} \overset{\overset{\text{H}}{|}\,\text{C}}{C} — OH$$
$$\overset{|}{CH_3}$$
$$\text{D}$$

(1b)

Methyl group **D** is closer to the electronegative OH group than methyl group **A** and hence resonates at lower field. Proton H_C would be expected as a pattern with a maximum of 12 lines, but band overlap reduces the number. The splitting pattern and peak intensities are as expected from an equally coupled ($J_{BC} = J_{CD} = 8$ Hz) triplet and quartet as shown in the following theoretical coupling diagram:

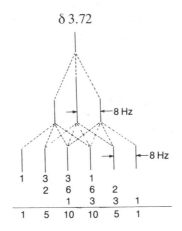

However, as a first approximation it is useful to consider proton H_C to have five equivalent protons on adjacent carbon atoms, and its resonance would therefore be expected as a sextet.

The resonance signal for the **B** protons would be expected as a pattern with a maximum of eight lines. Assuming that $J_{AB} = J_{BC}$, band overlap would produce five lines with peak intensities in the ratio $1:4:6:4:1$. However, J_{AB} and J_{BC} are not exactly equal, and, since $\Delta\delta$ (i.e. $\delta A - \delta B$) is less than six times J_{AB} (p.434), perturbation from first order treatment occurs to give the observed multiplet at $\delta 1.5$.

Compound 1c *(Fig. 14.1(c))*

From the two previous spectra (Figs. 14.1a and 14.1b) it can be concluded that the resonance at $\delta 4.05$ which exchanges with D_2O arises from the OH of the alcohol. The doublet at $\delta 3.37$ therefore integrates for two protons, the multiplet at $\delta\,1.78$ for one proton and the doublet at $\delta\,0.95$ for six protons. The doublet at $\delta\,0.95$ ($J = 7$) may be confidently assigned, in this series, to two equivalent methyl groups attached to a methine group. The doublet at $\delta\,3.37$ ($J = 7$) arises from a methylene group adjacent to a methine group and, from its chemical shift, also adjacent to oxygen. These observations indicate that Fig. 14.1(c) is the PMR spectrum of isobutanol (2-methylpropan-1-ol) (1c).

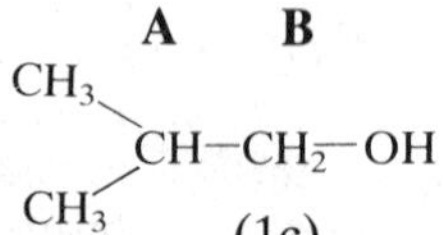

A   B

$$CH_3 \diagdown$$
$$CH-CH_2-OH$$
$$CH_3 \diagup$$

(1c)

Both methyl groups are magnetically equivalent and therefore resonate at the same frequency; both are equally coupled to the **A** proton, giving the observed six-proton doublet ($J = 7$). Similarly, the **B** protons give a doublet with the expected chemical shift. The **A** proton resonance appears as the predicted multiplet whose complexity is revealed in the inset produced at higher sensitivity.

Compound 1d   *(Fig. 14.1(d))*

This spectrum shows the three equivalent methyl groups of *t*-butanol (2-methylpropan-2-ol) (1(d)) as a sharp singlet at $\delta\,1.25$ and the O<u>H</u> group of the alcohol at $\delta\,3.40$.

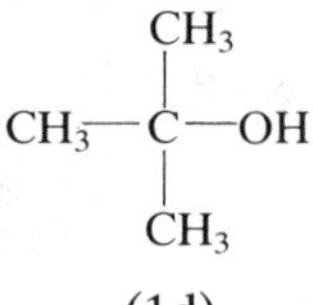

$$CH_3$$
$$|$$
$$CH_3-C-OH$$
$$|$$
$$CH_3$$

(1d)

The integrations are not, however, in the expected $9:1$ ratio. The sample is contaminated with water, and under the experimental conditions the protons of both the water and the hydroxy group of the alcohol resonate at the same frequency.

Compound 2

The IR band at $1715\,\text{cm}^{-1}$ indicates that compound 2 possesses a carbonyl group. This evidence, taken with the molecular formula (C_4H_8O), suggests that the compound is either an aliphatic aldehyde or ketone (Chapter 10, Fig. 10.4).

The PMR spectrum immediately excludes an aldehyde (no resonance at about $\delta\,9.5$; Chapter 11, Fig. 11.9) and confirms the structure as butan-2-one (methyl ethyl ketone). The integrations, from high field to low field, are in the ratio $3:3:2$ and therefore the resonance signals are assigned as follows:

	A O B
δ 1.06 (3H, t, $J = 8$)	$\underline{CH_3}CH_2 \overset{\overset{\displaystyle O}{\|\|}}{—C—} CH_3$
δ 2.14 (3H, s)	$CH_3CH_2 \overset{\overset{\displaystyle O}{\|\|}}{—C—} \underline{CH_3}$
δ 2.43 (2H, q, $J = 8$)	$CH_3\underline{CH_2} \overset{\overset{\displaystyle O}{\|\|}}{—C—} CH_3$

Methyl group **A** and the methylene group appear respectively as the expected triplet and quartet. Methyl group **B** and the methylene group are both deshielded (moved to lower field) by the adjacent carbonyl group and their proton chemical shifts equate well with the range for such protons given in Chapter 11 (Fig. 11.9).

The principal ions in the mass spectrum of butan-2-one are $CH_3—C \equiv O^+$ (m/z 43) and $CH_3CH_2C \equiv O^+$ (m/z 57) the former appearing in greater abundance (Chapter 12, p.483).

Compound 3

The IR band at 1740 cm^{-1} points to a carbonyl group. and further consideration of the absorption at 1240 cm^{-1} and the molecular formula and description indicates that the compound is an aliphatic ester. The band at 1240 cm^{-1} strongly suggests an acetate (Chapter 10, Fig. 10.4).

The PMR spectrum confirms the structure as ethyl ethanoate (ethyl acetate) and the resonance signals are assigned as follows:

δ 1.25 (3H, t, $J = 8$)	$CH_3 \overset{\overset{\displaystyle O}{\|\|}}{—C—} O—CH_2\underline{CH_3}$
δ 2.03 (3H, s)	$\underline{CH_3} \overset{\overset{\displaystyle O}{\|\|}}{—C—} O—CH_2CH_3$
δ 4.12 (2H, q, $J = 8$)	$CH_3 \overset{\overset{\displaystyle O}{\|\|}}{—C—} O—\underline{CH_2}CH_3$

The PMR spectrum of ethyl acetate has the same overall appearance as that of butan-2-one (Fig. 14.2) in displaying a triplet, singlet and quartet. The methyl groups attached to the carbonyl groups of the two compounds have almost identical proton chemical shifts; the methylene groups have, however, distinctly different chemical shifts. In ethyl acetate the methylene protons are more deshielded (resonate at lower field) by being bonded to an oxygen atom which is more electronegative and which has a greater deshielding effect than the carbonyl group of butan-2-one (Chapter 11, p.424). Comparison of the proton chemical shift of the second methyl group (—CH_2CH_3) in the two compounds shows that these protons are more deshielded in ethyl acetate, and this too is a reflection of the inductive effect of the oxygen atom (—O—←CH_2—←CH_3).

Compound 4

The IR bands at 3330 and 3250 cm^{-1} are compatible with a primary amino group and the remaining bands with an alkyl group (Chapter 10, Fig. 10.4).

The mass spectrum confirms that compound 4 is a primary amine, the base peak (*m/z* 30) arising through carbon–carbon bond cleavage β to the hetero-atom (Chapter 12, p.487).

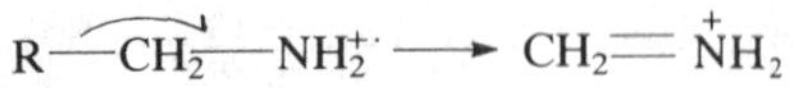

$$R—CH_2—NH_2^{+ \cdot} \longrightarrow CH_2 = \overset{+}{N}H_2$$

The PMR spectrum shows the integrations to be in the ratio 6:2:1:2. The molecular formula reveals that compound 4 contains 11 protons, and thus the integrations correspond to 6H, 2H, 1H and 2H respectively. The two–proton singlet at δ 1.10 exchanges with D_2O (Chapter 11, p.436) and is therefore assigned to the primary amino group (NH_2). The pattern of the remaining resonance signals is virtually identical to that displayed by the alkyl group of isobutanol (Fig. 14.1c) from which it follows that this compound is isobutylamine (1-amino-2-methylpropane); (4).

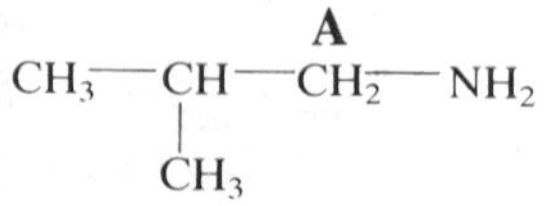

$$\overset{\textbf{A}}{CH_3—CH—CH_2—NH_2}$$
$$\underset{CH_3}{|}$$

(4)

The two–proton doublet (δ 2.48) of the methylene protons **A** is upfield of the corresponding methylene protons (δ 3.37) of isobutanol, since nitrogen is less electronegative than oxygen. There is obviously no coupling between the **A** protons and those of the primary amino group.

Compound 5

The carbonyl absorption at $1720\,cm^{-1}$ in the IR spectrum of compound 5 and the bands at 1270 and $1105\,cm^{-1}$ are compatible with an aromatic ester, the bands at 1602 and $1581\,cm^{-1}$ confirming the presence of an aromatic ring.

The integrations in the PMR spectrum are in the ratio $3:2:3:2$ ($\equiv 10H$). The triplet at $\delta\,1.29$ ($3H$, $J = 8$) and the quartet at $\delta\,4.35$ ($2H$, $J = 8$) clearly indicate (compare compound 3) the moiety $-O-CH_2-CH_3$, and the compound is therefore ethyl benzoate (5).

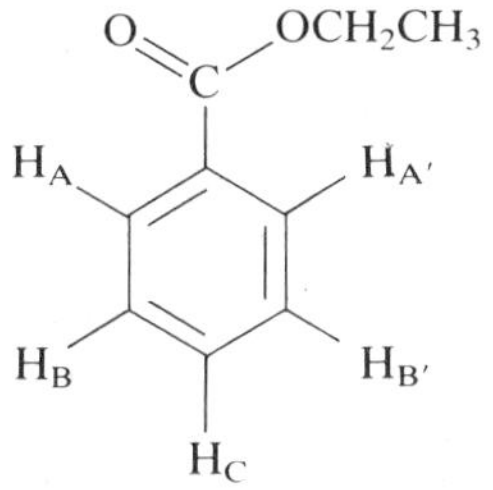

$$
\begin{array}{c}
\text{O}\diagdown\diagup\text{OCH}_2\text{CH}_3 \\
\text{C} \\
H_A \diagup\diagdown H_{A'} \\
\\
H_B H_{B'} \\
H_C \\
(5)
\end{array}
$$

The aromatic protons appear as two groups of signals centred at $\delta\,7.4$ and $\delta\,8.1$, which respectively integrate for three and two protons. The lower field group arises from the protons, H_A and $H_{A'}$, *ortho* to the carbonyl group which has the expected deshielding effect on these proximate protons (Chapter 11, Fig. 11.8). The pattern may be approximately interpreted as a double doublet by considering H_A and $H_{A'}$ to be magnetically equivalent and H_B and $H_{B'}$ to be magnetically equivalent but of different chemical shift from the **A** protons. Thus the **A** protons are *ortho* coupled (Chapter 11, p.432) to the **B** protons ($J = 8$) and *meta* coupled to H_C ($J = 2$), giving the following theoretical coupling diagram:

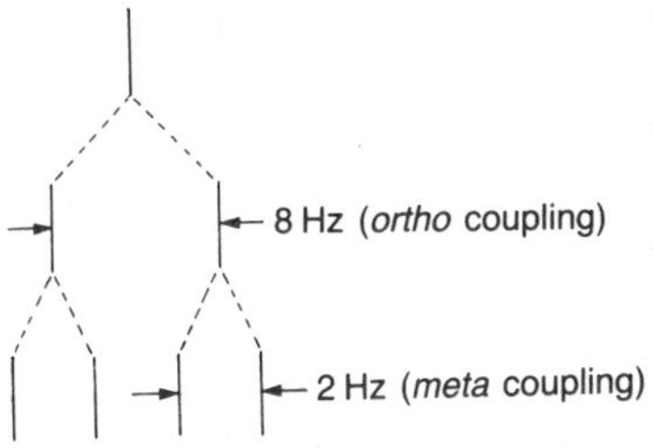

The **B** and **C** protons, however, appear as an unresolved multiplet in this 90 MHz spectrum.

Compound 6

The IR data suggests a carbonyl group (1695 cm^{-1}) attached to an aromatic ring (1600, 1580, 1505 cm^{-1}). The molecular weight of the compound is 136 (M$^+$). The base peak in the mass spectrum, *m/z* 135, taken with the infrared evidence, clearly indicates that the compound is an aryl aldehyde and gives (6a) as a partial structure. The functional group R therefore has a formula weight of 31 mass units.

(6a)   *m/z* 135

The resonance at δ 9.85 in the PMR spectrum of compound 6 confirms the aldehyde moiety, and the integral therefore corresponds to one proton. The singlet at δ 3.85 thus arises from three protons, and the chemical shift is characteristic of an aromatic methoxy group; a methoxy group also satisfies the required 31 mass units for the substituent R in (6a). The two doublets, each integrating for two protons, centred at δ 6.95 (*J* = 8) and δ 7.80 (*J* = 8) are characteristic of a *para*-disubstituted aromatic system (AA'XX', Chapter 11, pp.439 and 441) the lower field doublet being assigned to the aromatic protons *ortho* to the carbonyl group (compare compound 5). Inspection of the signals of the aromatic protons reveals that they do not present as a simple pattern *J ortho* = 8; there are a number of inner lines associated with long-range *meta* and *para* couplings (Chapter 11, p.439). The same feature is apparent in the double doublet (δ 8.1) of compound 5.

Compound 6 is therefore *para*-methoxybenzaldehyde.

Compound 7

The hypsochromic and hypochromic effects observed in the UV spectra on changing from neutral or alkaline conditions to acidic solution indicate the presence of an aromatic amino group (Chapter 7, pp.316 and 320).

The IR bands at 3410, 3340 and 3200 cm^{-1} may be due to either OH or NH stretching vibrations. The strong absorption at 1675 and 1590 cm^{-1} may be interpreted as an amide, but the UV requirement for an aromatic amino group excludes this possibility. The strong band at 1280 cm^{-1}, taken in conjunction with the carbonyl absorption (1675 cm^{-1}), indicates that compound 7 is more likely to be an ester. The bands at 1590, 1570 and 1505 cm^{-1} confirm the presence of a benzene ring.

Comparison of the PMR spectrum of compound 7 with those of compounds 5 and 6, and consideration of the above UV and IR data, clearly identifies this compound as ethyl *para*-aminobenzoate (benzocaine, (7)). The broad peak at δ 4.15, which overlaps the ester

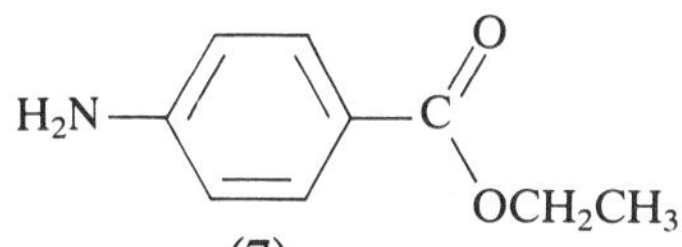

(7)

methylene quartet, integrates for two protons and these exchange with D$_2$O. This signal therefore arises from the amino protons (N$\underline{H}_2$). In the exchange spectrum the signal at δ 4.65 is a water peak (HOD or HOH) arising from the exchange.

Compound 8

The UV absorption bands at 220 and 285 nm and the ε values indicate an aromatic ring conjugated to group(s) which extend the chromophore (Chapter 7, p.316).

The IR band at 1675 cm^{-1} may be correlated with an aryl or α,β-unsaturated carbonyl compound (the molecular formula excludes amide absorption) and the band at 1615 cm^{-1} with an alkene moiety conjugated with an aryl or carbonyl group.

Consideration of the combined UV and IR interpretations suggests (8a), (8b) and (8c) as potential structures; isomer (8c), in displaying cross-conjugation rather than extended conjugation, is less likely to be compatible with the UV data.

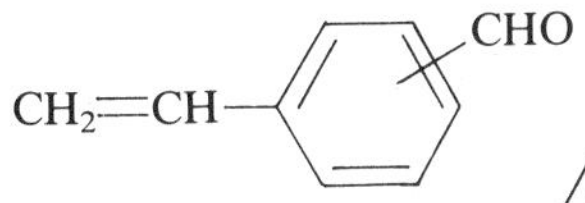

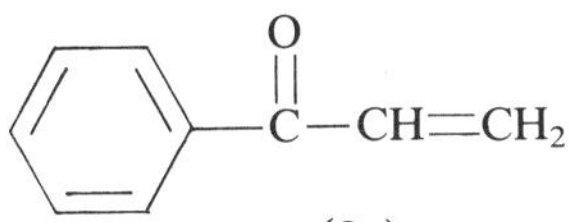

The IR bands at 1601, 1575 and 1492 cm^{-1} corroborate the aromatic ring. The two strong bands at 755 and 695 cm^{-1} are indicative of a monosubstituted benzene derivative, and the infrared evidence therefore favours (8b). The bands in the 900–700 cm^{-1} region are not, however, always reliable indicators of the substitution pattern in a benzene ring.

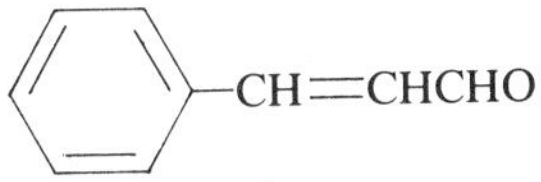

The resonance at δ 9.7 (1H, d) in the PMR spectrum clearly indicates an aldehyde group (C$\underline{H}$O). Since the aldehydic proton appears as a doublet ($J = 8$), there must be one proton on the adjacent carbon atom (CH—CHO) and this confirms (8b) as the correct structure. The PMR data further shows that the compound is *trans*-cinnamaldehyde (8d).

Proton H$_B$ is *trans*-coupled to H$_C$ ($J = 16$; Chapter 11, p.439) and vicinally coupled to H$_A$ ($J = 8$), giving the four–line pattern centred at δ 6.7. The doublet for H$_C$ ($J = 16$) straddles the signal from the aromatic protons, and the individual lines of the doublet are at δ 7.4

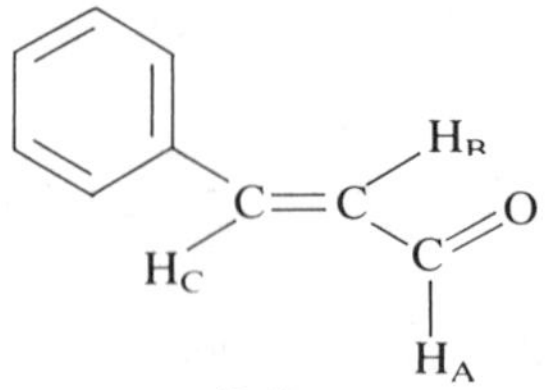

(8d)

and $\delta\,7.57$. The total resonance from $\delta\,7.3$–7.7 integrates for six protons (H_C + five aromatic) and thus the remaining lines of this multiplet arise from the protons of the aromatic ring. The broad low intensity signals ($\delta\,6.9$–7.2 and $\delta\,7.7$–8.0) are spinning side bands (Chapter 11, p.416) of the six-proton multiplet, and the small peak at $\delta\,7.27$ is due to chloroform ($C\underline{H}Cl_3$) in the solvent ($CDCl_3$).

The mass spectrum supports the assigned structure. The fragmentation pathway is as follows:

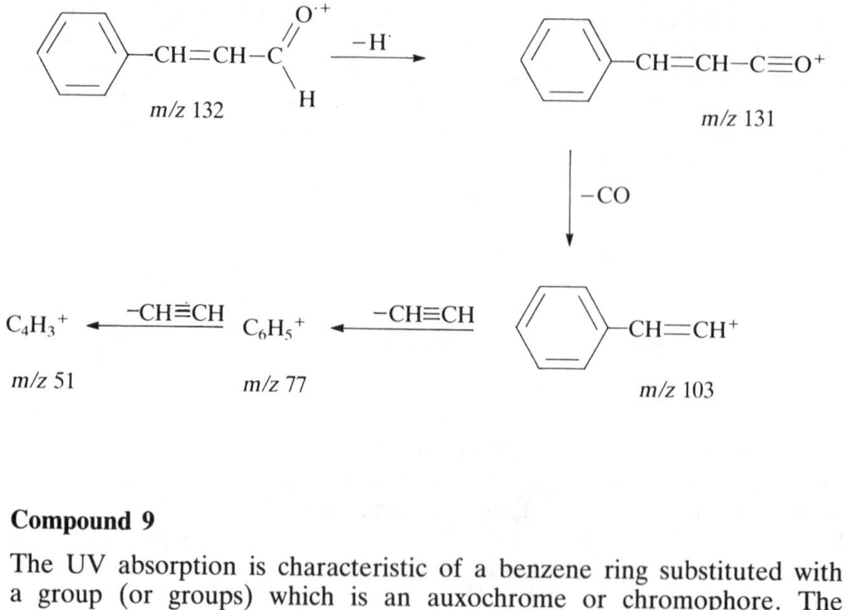

Compound 9

The UV absorption is characteristic of a benzene ring substituted with a group (or groups) which is an auxochrome or chromophore. The series of IR bands at 3000–2500 cm^{-1} is attributable to O—H stretching of a carboxyl group, and the strong band at 1690 cm^{-1} provides confirmatory evidence of an aromatic carboxylic acid. An aromatic acid is also compatible with the UV data. The IR bands at 1600–1480 cm^{-1} corroborate the aromatic nature of compound 9, and partial structure (9a) is indicated. The absorption at 1750 cm^{-1} shows that the compound possesses a second carbonyl group which can be assigned to the acetate (9b) rather than the methyl ester (9c), since the latter would be expected to show a band at about 1720 cm^{-1} (compare compound 5).

The infrared spectrum of the compound has too many bands in the

$900-650\ cm^{-1}$ region to permit unambiguous deduction of the substitution pattern.

(9a)   (9b)   (9c)

The loss of 42 mass units from the molecular ion to give the radical ion $C_7H_6O_3$ represents the loss of ketene (CH_2=C=O) and, in this instance, confirms that compound 9 is a phenolic acetate. The ions m/z 120 (base peak) and 92 represent further sequential loss of water and carbon monoxide.

The combined UV, IR and MS data have given (9b) as a partial structure, but the substitution pattern on the aromatic ring remains unknown. The PMR spectra show that the compound is acetylsalicylic acid (aspirin, 9d).

(9d)

The sharp singlet at $\delta\,2.31$ (3H) arises from the methyl group of the ester (compare compound 3), and the broad singlet at $\delta\,11.55$, which exchanges with D_2O, from the carboxylic proton. The aromatic region ($\delta\,7-8.2$), which is shown as a 300 Hz scale expansion in Figure 14.9(a), clearly indicates *ortho*-disubstitution. The double doublet ($J=8$ and 2) centred at $\delta\,8.12$ arises from H_A, which is the most deshielded of the aromatic protons, through its proximity to the carbonyl group of the carboxylic acid moiety; proton H_A is *ortho* coupled to H_B ($J=8$) and *meta* coupled to H_C ($J=2$) (compare compound 5). Proton H_D appears at highest field in the aromatic region by being *ortho* to an electron-donating oxygen atom (Chapter 11, Fig. 11.8). The resonance signal for H_D is centred at $\delta\,7.15$ and also appears as a double doublet ($J=8$ and 1.5) through *ortho* coupling to H_C ($J=8$) and *meta* coupling to H_B ($J=1.5$). Proton H_B presents as a double triplet ($J=8$ and 1.5) centred at $\delta\,7.33$, the origin of which is shown in the following coupling diagram.

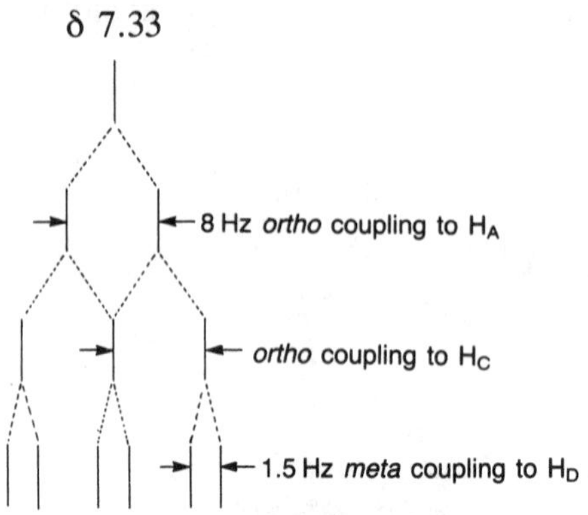

δ 7.33

Proton H_C is observed as a similar double triplet centred at δ 7.61 $(J_{BC} = J_{CD} = 8$ and $J_{AC} = 2)$. The slight difference in the values of the *meta* coupling constants $(J_{AC}$ and $J_{BD})$, which permits the unequivocal assignment of protons H_B and H_C, can be observed by simple visual inspection of Fig. 14.9(a).

Interpretation of spectral data

Compound 10

UV. The molar absorptivities can be calculated (Chapter 7, p.277) as being about 6800 at 253 nm in water and about 21750 at 294 nm in alkali. The bathochromic shift and hyperchromic effect observed on raising the pH is characteristic of phenolic compounds (Chapter 7, p.320).

IR. The broad band at 3250 cm^{-1} is assigned to the hydrogen bonded OH group. The ester carbonyl band at 1670 cm^{-1} is lower than that characteristic for aromatic esters (compare compound 5) but the phenolic hydroxyl group lowers the bond order of the carbonyl group by the mesomeric effect shown below. Similar consideration applies to the

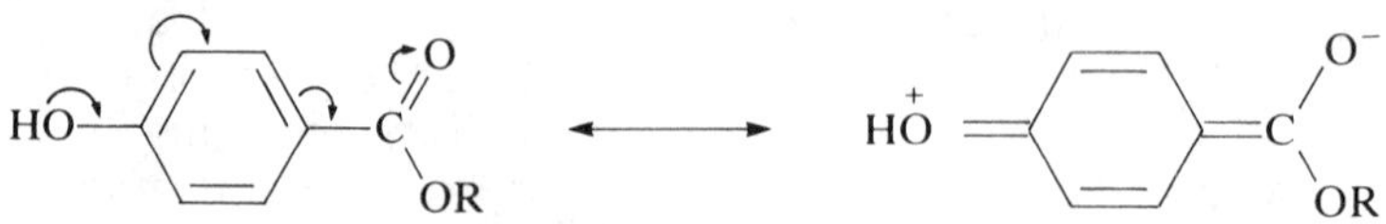

(10)   (10a)

carbonyl stretching frequency observed in compound 7. The bands at 1601, 1585 and 1505 cm^{-1} arise from skeletal vibrations of the aromatic ring. The absorption at 1280 cm^{-1} originates from asymmetric C—O stretching of the ester. Neither the symmetrical stretching band nor the bands associated with *para*–substitution can be identified with certainty.

MS. The following fragmentation pathway accounts for the principal ions observed in the mass spectrum:

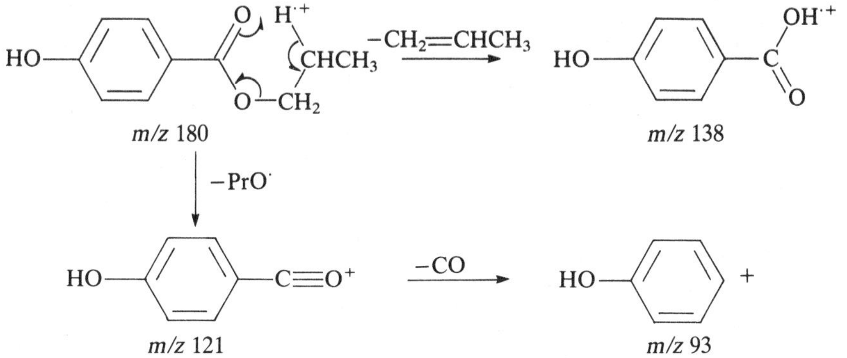

m/z 180   *m/z* 138

m/z 121   *m/z* 93

The fragmentation of the molecular ion to the ion *m/z* 138 is an example of a McLafferty rearrangement (Chapter 12, p.489).

PMR. The assignments (δ) are 1.01 (3H, t, —$CH_2CH_2C\underline{H}_3$); 1.78 (2H, m, —$CH_2C\underline{H}_2CH_3$), 4.28 (2H, t, $OC\underline{H}_2$); 6.92 (2H, d, H_A and $H_{A'}$); 7.13 (s, br, exchanges with D_2O, $O\underline{H}$); 7.95 (2H, d, H_X and $H_{X'}$).

Compound 11

IR. The series of bands at 3000–2600 cm^{-1} and the band at 1709 cm^{-1} are respectively associated with the OH and C=O of the carboxyl group.

MS. The fragmentation can be interpreted as follows:

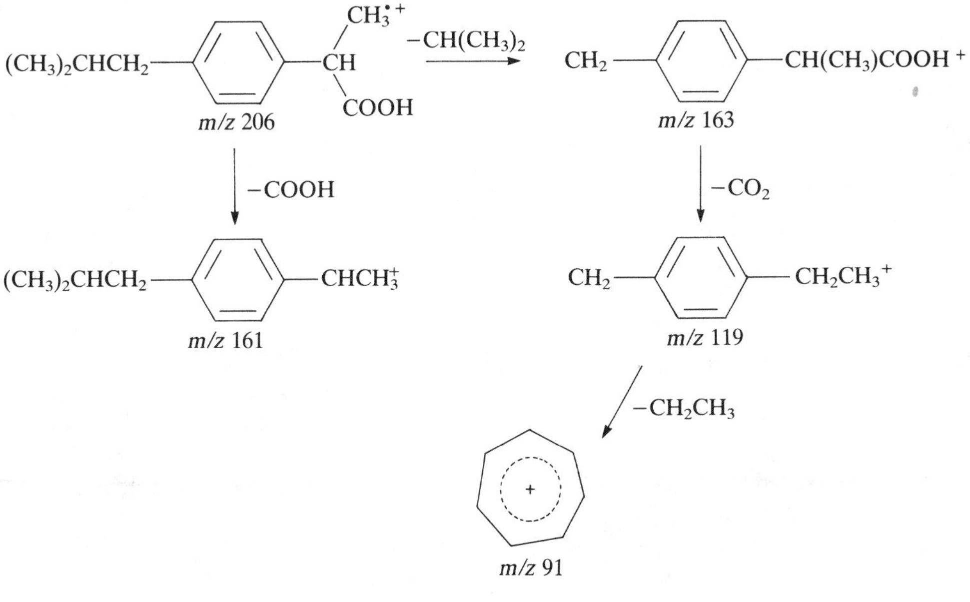

m/z 206   *m/z* 163

m/z 161   *m/z* 119

m/z 91

PMR. The assignments (δ) are 0.88 (6H, d, $(\underline{C}H_3)_2CH$); 1.48 (3H, d, $\underline{C}H_3CH$); 1.8 (1H, m, $(CH_3)_2\underline{C}H$); 2.45 (2H, d, $\underline{C}H_2$); 3.69 (1H, q, $\underline{C}H\!-\!COOH$); 7.15 (4H, q, aromatic protons); 11.1 (1H, s, br, $COO\underline{H}$).

Since both substituents on the aromatic ring have similar electronic effects, the aromatic protons present as an AA′ BB′ quartet (Chapter 11, p.438) since the chemical shift difference between the **A** protons (H_A and $H_{A'}$) and the **B** protons (H_B and $H_{B'}$) is reduced (Fig. 11.20; p.435). This situation is to be contrasted with that in compounds 6, 7 and 10.

Compound 12

UV. The example shows benzenoid absorption typical of this class of compound with the more intense band at shorter wavelength (Chapter 7, p.319).

IR. The bands at 2990 and 2960 cm^{-1} arise from C—H stretching in the geminal methyl groups. The carbonyl absorption at 1740 cm^{-1} is consistent with an aliphatic ester (compare compound 3). The aromatic ring vibrations occur at 1595, 1580 and 1485 cm^{-1}. The four bands in the region 1290–1140 cm^{-1} derive from asymmetrical and symmetrical C—O stretching of the ether and ester moieties.

MS. The two pairs of ions m/z 244 (M$^+$ + 2) and 242 (M$^+$) and m/z 171 and 169 arise because chlorine consists of the isotopes ^{37}Cl and ^{35}Cl in the approximate ratio of $1:3$ (Chapter 12, p.483). These principal ions have the structures (12a) and (12b).

(12a)
m/z 244 $C_{12}H_{15}{}^{37}ClO_3$
m/z 242 $C_{12}H_{15}{}^{35}ClO_3$

(12b)
m/z 171 $C_9H_{10}{}^{37}ClO$
m/z 169 $C_9H_{10}{}^{35}ClO$

PMR. The interpretation (δ) is 1.22 (3H, t, $-CH_2C\underline{H}_3$); 1.56 (6H, s, $(C\underline{H}_3)_2C\!-\!$); 4.22 (2H, q, $OC\underline{H}_2$); 6.78 (2H, d, H_B and $H_{B'}$); 7.20 (2H, d, H_A and $H_{A'}$).

Compound 13

UV. Comparison of the UV data of compound 13 with those of compound 12 again reveals typical benzenoid absorption. The hyperchromic effect observed at the longer wavelength ($\varepsilon\,1700$ in compound 12 and $\varepsilon\,4525$ in compound 13) may be due to the conjugated chromophore O=C—C=C but, because of the different solvents employed, the data are not directly comparable.

IR. The series of ragged bands at 3000–2500 cm^{-1} are associated with O—H stretching of the carboxyl group (compare compounds 9 and 11).

The carbonyl bands at 1710 cm^{-1} and 1670 cm^{-1} arise respectively from the carboxyl group and the 2-ethylacryloyl group. The C=C stretching vibration of the side chain is observed at 1658 cm^{-1}, and the bands at 1250 and 1080 cm^{-1} are assigned to C—O stretching in the ether moiety.

MS. The isotopic composition of chlorine (see compound 12) and the presence of two chlorine atoms in this compound gives a molecular ion at m/z 302, an M + 2 ion (m/z 304) and an M + 4 ion (m/z 306) in the approximate ratio 6 : 4 : 1. Similarly, the fragment ion arising from the side chain cleavage (13→13a) retains both chlorine atoms and hence presents a similar pattern.

(13)

m/z 306, 304, 302

(13a)

m/z 250, 248, 246

PMR. The assignments (δ) are 1.13 (3H, t, C$\underline{\text{H}}_3$); 2.46 (2H, q, C$\underline{\text{H}}_2$CH$_3$); 4.80 (2H, s, OC$\underline{\text{H}}_2$); 5.61 (1H, s, H$_B$); 5.96 (1H, s, br, H$_A$); 6.83 (1H, d, H$_C$); 7.16 (1H, d, H$_D$); 9.70 (1H, s, exchanges with D$_2$O, COO$\underline{\text{H}}$).

The lines for the quartet of the methylene protons (δ 2.46) are somewhat broadened, as is the signal for proton H$_A$ δ 5.96, the latter also displaying some fine structure. This is a result of the long-range coupling through four bonds (*J* about 1, Chapter 11, p.438) shown in

(13b)

(13b). An unequivocal assignment of the chemical shifts of H$_A$ and H$_B$ cannot be made, but the *trans*-proton (H$_A$) is more frequently observed to show allylic coupling.

Proof of the long-range coupling is provided by the double resonance (Chapter 11, p.443) experiments shown in Figs. 14.18(a) and 14.18(b). Irradiation of H$_A$ (δ 5.96) decouples the signal from the C$\underline{\text{H}}_2$CH$_3$ protons and narrows the lines of the quartet (Fig. 14.18(a)) and irradiation of the quartet (Fig. 14.18(b)) decouples the signal from H$_A$ which collapses to a sharp singlet.

Geminal coupling between the terminal methylene protons (Chapter 11, p.439) is absent.

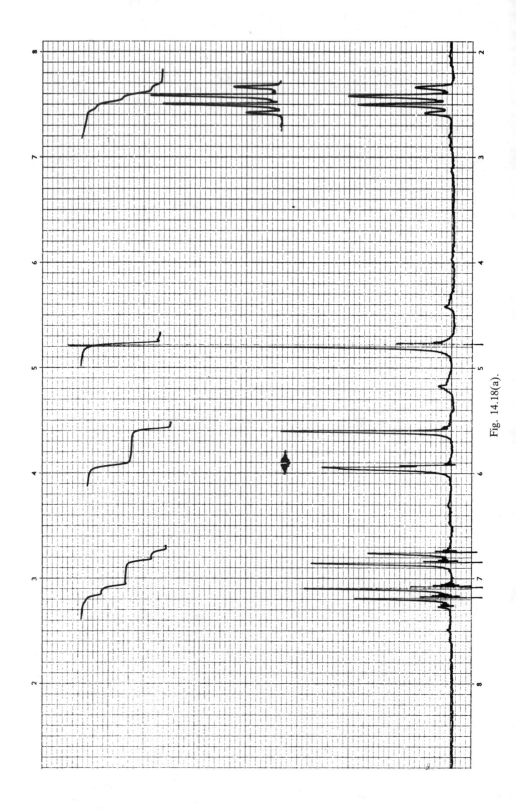

Fig. 14.18(a).

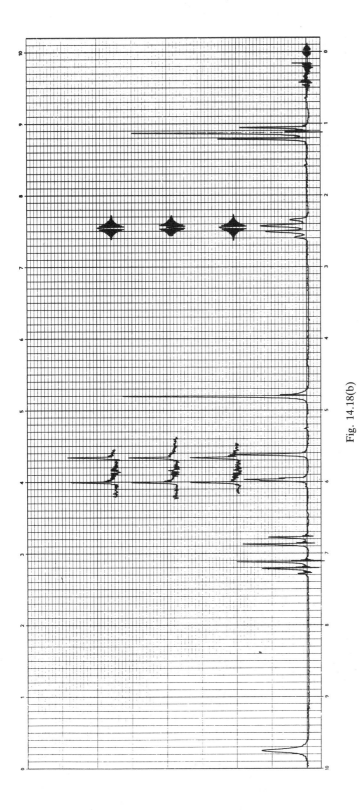

Fig. 14.18(b)

Compound 14

UV. The ultraviolet spectra of heteroaromatic compounds are less amenable than those of substituted benzenes to the simple treatment used in this text. The parent indole (14a) shows $\lambda_{max}^{cyclohexane}$ (nm) 220 (ϵ 6300), 280 (ϵ 5600) and 288 (ϵ 4200). Substituents, as in benzenoid compounds, have profound effects on the spectrum of the parent, and in interpretation recourse frequently has to be made to published catalogues of ultraviolet spectra to identify a model chromophoric system.

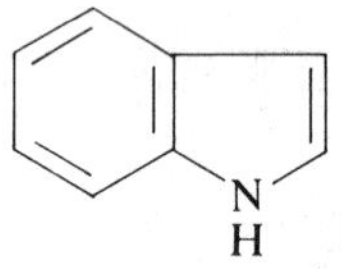

(14a)

IR. The bands at 3000–2500 cm^{-1} and the band at 1715 cm^{-1} are readily identified with the carboxyl group (compare compounds 11 and 13). The carbonyl stretching vibration of the tertiary amide is observed at 1685 cm^{-1}.

MS. The presence of one chlorine atom in indomethacin gives two molecular ions m/z 359 and 357 (compare compound 12). The ions m/z 314 and 312 similarly contain one chlorine atom. Accurate mass measurement (peak matching, Chapter 12, p.482) of these ions gives the structures shown in the following fragmentation pathway:

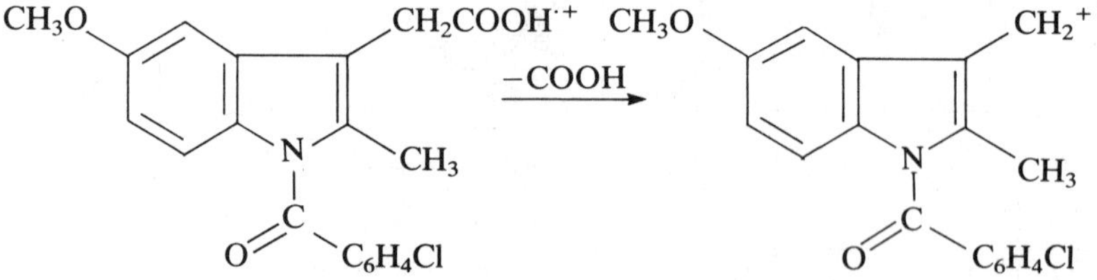

m/z 359.0741 = $C_{19}H_{14}{}^{37}ClNO_4$
357.0728 = $C_{19}H_{14}{}^{35}ClNO_4$

m/z 314.0776 = $C_{18}H_{15}{}^{37}ClNO_2$
312.0784 = $C_{18}H_{15}{}^{35}ClNO_2$

PMR. The interpretation (δ) is 2.38 (3H, s, indole C$\underline{H}_3$); 3.68 (2H, s, C$\underline{H}_2$); 3.81 (3H, s, OC$\underline{H}_3$); 6.67 (1H, dd, J = 8 and 2, H$_B$); 6.85* (1H, d, J = 8, H$_C$); 6.95 (1H, d, J = 2, H$_A$); 7.26 (C$\underline{H}$Cl$_3$ solvent); 7.45 (2H, d, H$_E$ and H$_{E'}$); 7.65 (2H, d, H$_D$ and H$_{D'}$); 9.1–10.0 (1H, br, COO$\underline{H}$).

---

*The second peak of this doublet overlaps the doublet of proton H$_A$ as is more readily seen in the inset spectrum obtained at higher sensitivity.

Compound 15

UV. As in the case of compound 14, identification of a model chromophoric system, through either published catalogues of ultraviolet spectra or a literature search, is necessary. For example, 1-methyl-2-thioxo-4-imidazoline (15a) shows maxima at about 210 nm and 250 nm, the longer wavelength band being the more intense. Addition of the chromophore —COOEt to this ring system would be expected to cause a bathochromic shift of the maxima, an observation which is compatible with the UV characteristics of carbimazole.

$$CH_3—N\underbrace{\qquad}_{S}NH$$

(15a)

IR. The series of bands at 3190–2490 cm^{-1} are associated with C—H stretching vibrations of the methyl groups and the ethenic CH groups; the higher frequency bands are more likely to arise from the ring CH groups. The carbonyl band is observed at 1750 cm^{-1} and C=C stretching at 1705 cm^{-1}.

MS. The series of ions separated by two mass units and appearing in the ratio 0.04 : 1 within each pair approximately reflects the relative isotopic abundance of ^{34}S and ^{32}S. Peak matching gave the molecular weight of each ion (Chapter 12, p.483) and a computer program (see introduction) its elemental composition. The following fragmentation pathway is proposed:

m/z 116.0212 $C_4H_6N_2{}^{34}S$
114.0234 $C_4H_6N_2{}^{32}S$

m/z 188.0430 $C_7H_{10}N_2O_2{}^{34}S$
186.0431 $C_7H_{10}N_2O_2{}^{32}S$

m/z 115.0134 $C_4H_5N_2{}^{34}S$
113.0169 $C_4H_5N_2{}^{32}S$

PMR. The interpretation (δ) is 1.42 (3H, t, CH$_2$C$\underline{H}_3$); 3.57 (3H, s, N—C$\underline{H}_3$); 4.46 (2H, q, OC$\underline{H}_2$); 6.7* (1H, d, $J = 2$, H$_A$); 7.2* (1H, d, $J = 2$, H$_B$).

*Unequivocal assignment of H$_A$ and H$_B$ is difficult, since both protons are attached to a carbon atom which is adjacent to nitrogen and many canonical forms contribute to the resonance hybrid.

The small vicinal coupling constant ($J = 2$) is also observed in many five-membered heteroaromatic and heteroethenic compounds as is illustrated in (15b), (15c) and (15d).

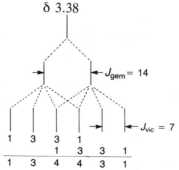

(15b)	(15c)	(15d)
$J_{AB} = 2.6$	$J_{AB} = 1.8$	$J_{AB} = 2$

Compound 16

The 90 MHz spectrum (Fig. 14.16) of crotamiton shows at $\delta 1.14$ an apparent triplet ($J \sim 7$) for the CH_3 of the N-ethyl group. The double doublet ($J \sim 7$ and 2) at $\delta 1.67$ arises from the propenyl CH_3, which is vicinally coupled to H_A ($J \sim 7$) and allylically coupled to H_B ($J \sim 2$). The three-proton singlet of the aromatic methyl group appears at $\delta 2.2$. The two sextets centred at $\delta 3.38$ and $\delta 4.1$, which each integrate for one proton, show that the two protons of the methylene group ($N{-}CH_2$) are not magnetically equivalent. Thus, each exhibits geminal coupling ($J \sim 14$) and further vicinal coupling ($J \sim 7$) to give overlapping quartets as shown in the following theoretical coupling diagram, which uses the signal at $\delta 3.38$ as illustration. The non-equivalence probably arises from hindered rotation in the tertiary amide moiety. The CH_3 protons of the N-ethyl group ($N{-}CH_2{-}CH_3$) are equally coupled ($J = 7$) to each of the non-equivalent methylene protons, and the apparent triplet ($\delta 1.14$) is in fact composed of two overlapped doublets.

$$\delta\ 3.38$$

$$J_{gem} = 14$$

$$J_{vic} = 7$$

1	3	3	1

	1	3	3	1

1	3	4	4	3	1

The resonances at $\delta 5.45$ and 5.62 together integrate for one proton and are thus assigned to H_B which is *trans*-coupled to H_A ($J \sim 15$) and allylically coupled to the propenyl CH_3 ($J \sim 2$); the multiplicity is not readily discernible, however, in the 90 MHz spectrum. The protons H_A and H_C and the aromatic protons appear in the region $\delta 6.75{-}7.4$, but are unresolved in Fig. 14.16.

The 250 MHz spectrum of crotamiton (Fig. 14.16(a)) shows the same features as the 90 MHz spectrum (Fig. 14.16) but the region δ 6.75–7.4 is now partly resolved to reveal at δ 6.95 the signal from proton H_A (a sextet) and a broad doublet at δ 7.1 from proton H_C. The 250 MHz spectrum also shows an impurity in the sample by the appearance of the small doublet at δ 2.15; the impurity has probably arisen through storage of the sample under inappropriate conditions after receipt (i.e., in a small, well-closed container protected from light).

Scale expansion of regions of interest presents a clearer picture and simultaneously permits accurate measurement of coupling constants. As an illustrative example, 20 Hz cm^{-1} scale expansions of the regions 1360–1420 Hz (δ 5.44–5.63) and 1700–1900 Hz (δ 6.80–7.60) are shown in Figs. 14.16(b) and (c). Proton H_B (1360–1410 Hz) is now readily observed as the expected doublet of quartets in which the *trans*-coupling constant is 15 Hz and the long-range (allylic) coupling constant is 1.6 Hz. High field instruments can print out the frequency of each line in a spectrum, which permits still better accuracy in coupling constant measurement, and in this instance J_{trans} is 14.99 Hz and $J_{allylic}$ is 1.64 Hz. Proton H_A (1710–1760 Hz) is clearly observed as two overlapped quartets in which J_{trans} is 15 Hz (14.99 Hz) as before and J_{vic} is 7 Hz (6.94 Hz accurately). The following coupling diagram accounts for the observed pattern:

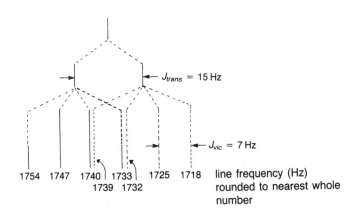

Proton H_C appears as a broadened doublet at δ 7.1 (J_{ortho} about 8 Hz, J_{meta} about 2 Hz). The remaining aromatic protons are left unresolved.

Similar treatment of the two sextets arising from the non-equivalent methylene protons reveals these also to be overlapped quartets in which J_{gem} is 13 Hz (13.34 accurate value) and J_{vic} is 7 Hz (6.94 accurate value). Construction of a coupling diagram, drawn to scale and using the values of 13 and 7 Hz, establishes the arrangement of the lines in these overlapping quartets.

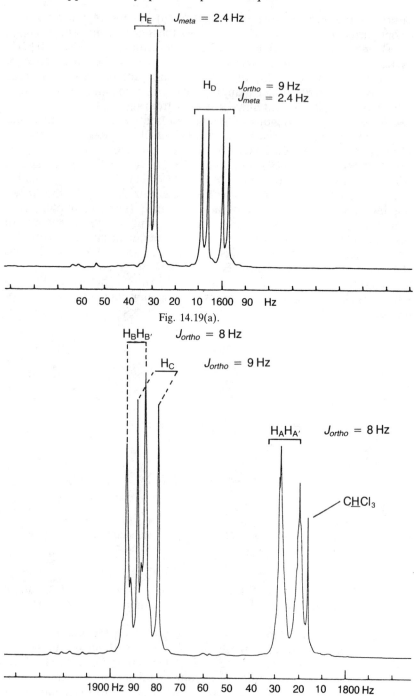

H_E $J_{meta} = 2.4\,Hz$

H_D $J_{ortho} = 9\,Hz$
        $J_{meta} = 2.4\,Hz$

60 50 40 30 20 10 1600 90   Hz

Fig. 14.19(a).

$H_B H_{B'}$ $J_{ortho} = 8\,Hz$

H_C $J_{ortho} = 9\,Hz$

$H_A H_{A'}$ $J_{ortho} = 8\,Hz$

CHCl$_3$

1900 Hz 90 80 70 60 50 40 30 20 10 1800 Hz

Fig. 14.19(b).

Compound 17

In the 90 MHz PMR spectrum (Fig. 14.17) of mexenone the protons of the aromatic methyl group (δ 2.41) and of the methoxy group (δ 3.83) are readily identifiable. The phenolic proton (O$\underline{H}$), which exchanges with D$_2$O, appears at δ 12.78, which is well down field of the phenolic proton of compound 10. This observation is explained by the intramolecular hydrogen bonding (cf hydrogen bonded enols (Chapter 11, Fig. 11.33)) shown in (17a).

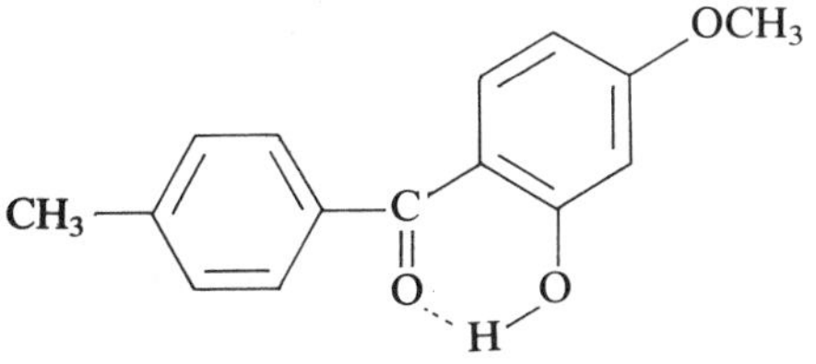

(17a)

The aromatic protons appear as three distinct groups of signals centred at δ 6.4, 7.25 and 7.55, which respectively integrate for two, two and three protons. The resonances at δ 6.4 arise from protons H$_D$ and H$_E$, those at δ 7.25 from H$_A$ and H$_{A'}$ and those at δ 7.55 from H$_B$, H$_{B'}$ and H$_C$. Interpretation, although comparatively simple from the 90 MHz spectrum, is facilitated by the 20 Hz cm^{-1} scale expansion of the aromatic region obtained from a 250 MHz instrument. The assignments and associated coupling constants for Figs. 14.17(a) and (b) are shown in Figs. 14.19(a) and (b).

Acknowledgments

The skilled technical assistance of Mrs Aileen Bosanquet in the preparation of ultraviolet, infrared, and nuclear magnetic resonance spectra is gratefully acknowledged. Gratitude is also expressed to The Boots Company (ibuprofen), Boehringer Ingelheim (mexenone), Ciba Geigy (crotamiton), Merck Sharp and Dohme (ethacrynic acid) and ICI Ltd (clofibrate) for generously supplying samples of the indicated compounds.

Index